Cat Version

Laboratory Manual for

Human Anatomy
& Physiology

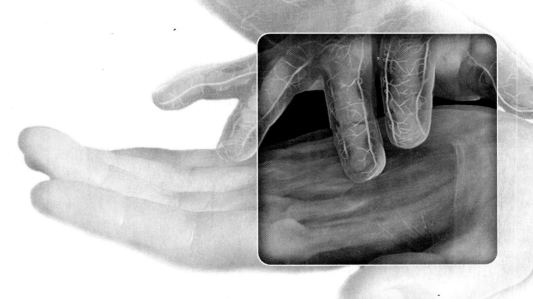

TERRY R. MARTIN

Kishwaukee College

 Higher Education

Boston   Burr Ridge, IL   Dubuque, IA   New York   San Francisco   St. Louis
Bangkok   Bogotá   Caracas   Kuala Lumpur   Lisbon   London   Madrid   Mexico City
Milan   Montreal   New Delhi   Santiago   Seoul   Singapore   Sydney   Taipei   Toronto

 **Higher Education**

LABORATORY MANUAL FOR HUMAN ANATOMY & PHYSIOLOGY: CAT VERSION

1 2 3 4 5 6 7 8 9 0 QPD/QPD 0 9

ISBN 978–0–07–340357–1
MHID 0–07–340357–1

Publisher: *Michelle Watnick*
Senior Sponsoring Editor: *James F. Connely*
Vice-President New Product Launches: *Michael Lange*
Developmental Editor: *Fran Schreiber*
Marketing Manager: *Lynn M. Breithaupt*
Lead Project Manager: *Peggy J. Selle*
Lead Production Supervisor: *Sandy Ludovissy*
Senior Media Project Manager: *Tammy Juran*
Designer: *Laurie B. Janssen*
Cover Designer: *Ron Bissell*
(USE) Cover Image: *Anatomical Travelogue / Photo Researchers, Inc.*
Senior Photo Research Coordinator: *John C. Leland*
Photo Research: *Danny Meldung/Photo Affairs, Inc*
Supplement Producer: *Mary Jane Lampe*
Compositor: *Precision Graphics*
Typeface: *10/12 Times LT Std*
Printer: *Quebecor World Dubuque, IA*

The credits section for this book begins on page 591 and is considered an extension of the copyright page.

Some of the laboratory experiments included in this text may be hazardous if materials are handled improperly or if procedures are conducted incorrectly. Safety precautions are necessary when you are working with chemicals, glass test tubes, hot water baths, sharp instruments, and the like, or for any procedures that generally require caution. Your school may have set regulations regarding safety procedures that your instructor will explain to you. Should you have any problems with materials or procedures, please ask your instructor for help.

www.mhhe.com

Contents

*These laboratory exercises are available online only at
www.mhhe.com/martinseries1 (click on this laboratory manual
cover).

Preface

In Touch
with Anatomy & Physiology Lab Courses

Author Terry Martin's thirty years of teaching anatomy and physiology courses, authorship of three laboratory manuals, and active involvement in the Human Anatomy and Physiology Society (HAPS) drove his determination to create a laboratory manual with an innovative approach that would benefit students. The *Laboratory Manual for Human Anatomy & Physiology* includes a cat version and a fetal pig version. Each of these versions includes sixty-one laboratory exercises, three supplemental labs found online, and six cat or fetal pig dissection labs. A main version with no dissection exercises is also available. All three versions are written to work well with any anatomy and physiology text.

Martin Lab Manual Series . . .
IN TOUCH with Anatomy & Physiology Lab Courses

▶ Available in **3 Versions:** main (no dissection), cat dissection, fetal pig dissection.

▶ Incorporates **learning outcomes and assessments** to help students master important material!

▶ **Pre-Lab** assignments available on corresponding website (www.mhhe.com/martinseries1) will help students be more prepared for lab and save instructors time during lab.

▶ **Clear, concise,** writing style facilitates more thorough understanding of lab exercises.

▶ **BIOPAC**© exercises use hardware and software for data acquisition, analysis, and recording.

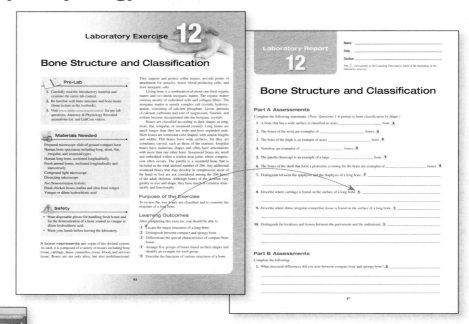

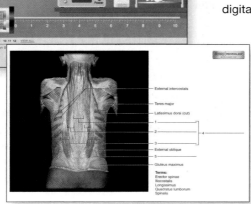

▶ **Ph.I.L.S. 3.0 CD-ROM** included and physiology lab simulations interspersed throughout make otherwise difficult & expensive experiments a breeze through digital simulations.

▶ Cadaver images from **Anatomy & Physiology Revealed**® **(APR)** are incorporated throughout the lab. Cadaver images help students make the connection from specimen to cadaver.

▶ **Micrographs** incorporated throughout the lab aid students' visual understanding of difficult topics.

▶ **Instructor's Guide** is annotated for quick and easy use by adjuncts and available online at www.mhhe.com/martinseries1.

In Touch
with Student Needs

▶ The procedures are clear, concise, and easy to follow. Relevant lists and summary tables present the contents efficiently. Histology micrographs and cadaver photos are incorporated in the appropriate locations within the associated labs.

▶ A pre-lab section directs the student to carefully read the introductory material and the entire lab to become familiar with its contents. If necessary, a textbook or lecture notes might be needed to supplement the concepts. A visit to **www.mhhe.com/martinseries1** will provide a list of animations from **Anatomy & Physiology Revealed®** **(APR)** and LabCam videos before answering five fundamental laboratory questions and the hypothesis question for that particular lab.

▶ **Terminologia Anatomica** is used as the source for universal terminology in this laboratory manual. Alternative names are included when a term is introduced for the first time.

▶ Laboratory reports immediately follow each laboratory exercise.

▶ Histology photos are placed within the appropriate laboratory exercise.

▶ A section called "Study Skills for Anatomy and Physiology" is located in the front material. This section was written by students enrolled in a Human Anatomy and Physiology course.

▶ A list of terms is provided to assist in the labeling of most figures; the first example is figure 2.2.

▶ "Critical Thinking Activities" are incorporated within most of the laboratory exercises to enhance valuable critical thinking skills that students need throughout their lives.

▶ Cadaver images are incorporated with dissection labs.

In Touch
with Instructor Needs

▶ The instructor will find digital assets for use in creating customized lectures, visually enhanced tests and quizzes, and other printed support material.

▶ A correlation guide for **Anatomy & Physiology Revealed®** **(APR)** and the entire lab manual is located on the lab manual's website at **www.mhhe.com/martinseries1**. Cadaver images from APR are included within many of the laboratory exercises.

▶ Some unique labs included are "Scientific Method and Measurements," "Chemistry of Life," "Fetal Skeleton," "Surface Anatomy," "Diabetic Physiology," and "Genetics."

▶ The annotated instructor's guide for *Laboratory Manual for Human Anatomy and Physiology* describes the purpose of the laboratory manual and its special features, provides suggestions for presenting the laboratory exercises to students, instructional approaches, a suggested time schedule, and annotated figures and assessments. It contains a "Student Safety Contract" and a "Student Informed Consent Form".

▶ Each laboratory exercise can be completed during a single laboratory session.

In Touch
with Educational Needs

▶ Learning outcomes with icons ⊙ have matching assessments with icons Ⓐ so students can be sure they have accomplished the laboratory exercise content. Outcomes and assessments include all levels of learning skills: knowledge, comprehension, application, analysis, synthesis, and evaluation.

▶ Assessment rubrics for entire laboratory reports are included in Appendix 2.

In Touch
with Technology

▶ Physiology Interactive Lab Simulations (Ph.I.L.S. 3.0 CD-ROM) is included with the lab manual. Eleven lab simulations are interspersed throughout the lab manual. The correlation guide for all of the simulations is included in Appendix 3.

▶ **BIOPAC** Systems, Inc. BIOPAC© exercises are included on four different body systems. BIOPAC© systems use hardware and software for data acquisition, analysis, and recording of information for an individual.

Guided Tour Through A Lab Exercise

The laboratory exercises include a variety of special features that are designed to stimulate interest in the subject matter, to involve students in the learning process, and to guide them through the planned activities. These particular features include the following:

Pre-Lab The pre-lab directs the student to carefully read introductory material and examine the entire laboratory contents after becoming familiar with the topics from a textbook or the lecture. Students will also be directed to visit **www.mhhe.com/martinseries1** to obtain a list of correlated **Anatomy and Physiology Revealed®** animations and LabCam videos and to answer pre-lab questions. After successfully answering the pre-lab questions, the student is prepared to become involved in the laboratory exercise.

Materials Needed This section lists the laboratory materials that are required to complete the exercise and to perform the demonstrations and learning extensions.

Safety A list of safety guidelines is included inside the front cover. Each lab session that requires special safety guidelines has a safety section following "Materials Needed." Your instructor might require some modifications of these guidelines.

Introduction The introduction describes the subject of the exercise or the ideas that will be investigated. It includes all of the information needed to perform the laboratory exercise.

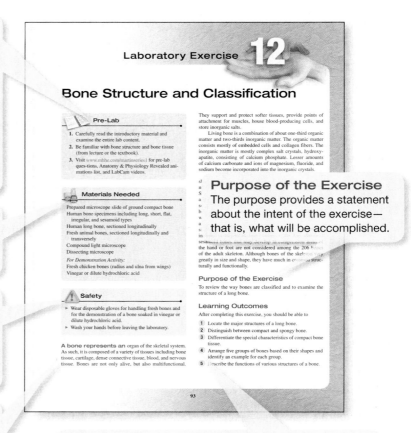

Purpose of the Exercise The purpose provides a statement about the intent of the exercise—that is, what will be accomplished.

Learning Outcomes The learning outcomes list what a student should be able to do after completing the exercise. Each learning outcome will have matching assessments indicated by the corresponding icon A in the laboratory exercise or the laboratory report.

Procedure The procedure provides a set of detailed instructions for accomplishing the planned laboratory activities. Usually these instructions are presented in outline form so that a student can proceed efficiently through the exercise in stepwise fashion.

The procedures include a wide variety of laboratory activities and, from time to time, direct the student to complete various tasks in the laboratory reports.

There are also separate procedures in 11 labs that utilize Ph.I.L.S. 3.0.

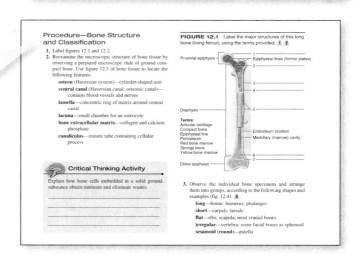

Demonstration Activities Demonstration activities appear in separate boxes. They describe specimens, specialized laboratory equipment, or other materials of interest that an instructor may want to display to enrich the student's laboratory experience.

Learning Extension Activities Learning extension activities also appear in separate boxes. They encourage students to extend their laboratory experiences. Some of these activities are open-ended in that they suggest the student plan an investigation or experiment and carry it out after receiving approval from the laboratory instructor. Some of the figures are illustrated as line art or in grayscale. This will allow colored pencils to be used as a visual learning activity to distinguish various structures.

Illustrations Diagrams similar to those in a textbook often are used as aids for reviewing subject matter. Other illustrations provide visual instructions for performing steps in procedures or are used to identify parts of instruments or specimens. Micrographs are included to help students identify microscopic structures or to evaluate student understanding of tissues.

In some exercises, the figures include line drawings suitable for students to color with colored pencils. This activity may motivate students to observe the illustrations more carefully and help them to locate the special features represented in the figures.

Laboratory Reports A laboratory report to be completed by the student immediately follows each exercise. These reports include various types of review activities, spaces for sketches of microscopic objects, tables for recording observations and experimental results, and questions dealing with the analysis of such data.

As a result of these activities, students will develop a better understanding of the structural and functional characteristics of their bodies and will increase their skills in gathering information by observation and experimentation. By completing all of the assessments in the laboratory reports, students will be able to determine if they were able to accomplish all of the learning outcomes.

Histology Histology photos placed within the appropriate exercise.

Demonstration Activity

Examine a fresh chicken bone and a chicken bone that has been soaked for several days in vinegar or overnight in dilute hydrochloric acid. Wear disposable gloves for handling these bones. This acid treatment removes the inorganic salts from the bone extracellular matrix. Rinse the bones in water and note the texture and flexibility of each (fig. 12.5a). Based on your observations, what quality of the fresh bone seems to be due to the inorganic salts removed by the acid treatment? **3**

Learning Extension Activity

Use colored pencils to differentiate the bones illustrated in figures 14.1 and 14.2. Select a different color for each bone in the series. This activity should help you locate various bones shown in different views in the figures.

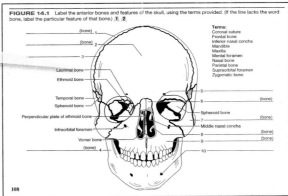

FIGURE 14.1 Label the anterior bones and features of the skull, using the terms provided. (If the line lacks the word bone, label the particular feature of that bone.) **1 2**

108

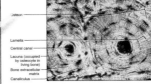

Name _____

Laboratory Report

12

Date _____

Section _____

The corresponds to the Learning Outcome(s) listed at the beginning of the laboratory exercise.

Bone Structure and Classification

Part A Assessments

Complete the following statements: (*Note:* Questions 1–6 pertain to bone classification by shape.)

1. A bone that has a wide surface is classified as a(an) _____ bone. **4**

2. The bones of the wrist are examples of _____ bones. **4**

3. The bone of the thigh is an example of a(an) _____ bone. **4**

4. Vertebrae are examples of _____ bones. **4**

5. The patella (kneecap) is an example of a large _____ bone. **4**

6. The bones of the skull that form a protective covering for the brain are examples of _____ bones. **4**

7. Distinguish between the epiphysis and the diaphysis of a long bone. **1** _____

8. Describe where cartilage is found on the surface of a long bone. **1** _____

9. Describe where dense irregular connective tissue is found on the surface of a long bone. **1** _____

10. Distinguish the locations and tissues between the periosteum and the endosteum. **1** _____

_____ between compact bone and spongy bone? **2** _____

Acknowledgments

I value all the support and encouragement by the staff at McGraw-Hill, including Michelle Watnick, Jim Connely, Fran Schreiber, Lynn Breithaupt, Peggy Selle, and Cathy Schmitt. Special recognition is granted to Colin Wheatley for his insight, confidence, wisdom, warmth, and friendship.

I am grateful for the professional talent of Lurana Bain, Terri Bidle, and Marian Langer for their significant involvement in the development of some of the laboratory exercises. The main contents of the surface anatomy lab were provided by Lurana Bain, Ph.I.L.S. lab simulations components by Terri Bidle, and BIOPAC® labs by Marian Langer.

I am particularly thankful to Dr. Norman Jenkins and Dr. David Louis, retired presidents of Kishwaukee College, and Dr. Thomas Choice, president of Kishwaukee College, for their support, suggestions, and confidence in my endeavors. I am appreciative for the expertise of Womack Photography for numerous contributions. The professional reviews of the nursing procedures were provided by Kathy Schnier.

I am also grateful to Laura Anderson, Rebecca Doty, Michele Dukes, Troy Hanke, Jenifer Holtzclaw, Stephen House, Shannon Johnson, Brian Jones, Marissa Kannheiser, Morgan Keen, Marcie Martin, Angele Myska, Sparkle Neal, Susan Rieger, Eric Serna, Robert Stockley, Shatina Thompson, Nancy Valdivia, Marla Van Vickle, Jana Voorhis, Joyce Woo, and DeKalb Clinic Chartered for their contributions. There have been valuable contributions from my students, who have supplied thoughtful suggestions and assisted in clarification of details.

To my son Ross, I owe gratitude for his keen eye and creative suggestions. Foremost, I am appreciative to Sherrie Martin, my spouse and best friend, for advice, understanding, and devotion throughout the writing and revising.

Terry R. Martin
Kishwaukee College
21193 Malta Road
Malta, IL 60150

Focus Group

The direction and development of this laboratory manual were enhanced by the discussions and input from the focus group members who met with the author and editors of the McGraw-Hill Companies. They include the following:

Raja Bhandari
Chattanooga State University

Terri Bidle
Hagerstown Community College

Mary Bonine
Northeast Iowa Community College

Mark Jaffe
Nova Southeastern University

Dean Kruse
Portland Community College

Leigh Levitt
Union County College

Gregory Reeder
Broward Community College–Central

Amy Fenech Sandy
Columbus Technical College

Reviewers

I would like to express my sincere gratitude to all reviewers of the laboratory manual who provided suggestions for its improvement. Their thoughtful comments and valuable suggestions are greatly appreciated. They include the following:

Michael Aaron
Shelton State Community College

M. Abdel-Sayed
CUNY/New York

John V. Aliff
Georgia Perimeter College

Emily Allen
Gloucester County College

Kathy Pace Ames
Illinois Central College

Merrilee G. Anderson
Mount Aloysius College

Penny P. Antley
University of Louisiana—Lafayette

Emmanuel Ayoade
Fayetteville Technical Community College

Isaac Barjis
New York City College of Technology

Janice G. Barney
Mount Wachusett Community College

Verona A. Barr
Heartland Community College

Marilynn Bartels
Black Hawk College

Anne Marie Basso
St. John's University

Andrew A. Beall
University of North Florida

Moges Bizuneh
Ivy Tech Community College

Lois Brewer Borek
George State University

Ronny K. Bridges
Pellissippi State Technical Community College

Pearl-Ann T. Brown
New York College of Podiatric Medicine

Winnifred Bryant
University of Wisconsin, Eau Claire

Nishi Sood Bryska
UNC Charlotte

Rebecca R. Burton
West Kentucky Community & Technical College

Claire Michelle Carpenter
Yakima Valley Community College

Michael O. Casey
Gaston College

Maurice M. Culver
Florida Community College-Jacksonville

Marc DalPonte
Lake Land College

Mary Elizabeth Dawson
Kingsborough Community College

Danielle Desroches
William Paterson University of New Jersey

Richard Doolin
Daytona Beach Community College

Joyce Festa
St. John's University

Purti Gadkari
Wharton County Junior College

Ewa Gorski
Community College of Baltimore County

Clare Hays
Metropolitan State College of Denver

Johanna Kruckeberg
Kirkwood Community College

Michael S. Kopenits
Amarillo College

Walter Johnson
Merritt College

Geri Mayer
Florida Atlantic University

John P. McNamara
Jefferson College of Health Sciences

Susan Caley Opsal
Illinois Valley Community College

Dee Ann S. Sato
Cypress College

Kathleen Sellers
Columbus State University

Judith Shardo
Middle Tennessee State University

K. Dale Smoak
Piedmont Technical College

Bonnie J. Tarricone
Ivy Tech State College

Delon Washo-Krupps
Arizona State University

Janice M. Webster
Ivy Tech Community College

Students

We also are grateful to students who participated in class tests:

Northeast Iowa Community College

Rachel Baumhover	Kelli Guidebeck	Kyla E. Reamon
Jessica L. Beck	Ryan Kieffer	Missy Richard
Amanda Brant	Jodi Knaeble	Maggie Schweitzer
Shaun Cavanaugh	Kristina Kurtz	Sarah Simon
Erin Cullen	Kim E. Lewis	Sharon Slaght
Jessica Fleckenstein	Sara Murphy	

Hagerstown Community College

Megan Crouse	Kristin E. Patrie
Karen Dillard	Aaron Vittini
Jennifer Foster	Mindy Waltz
Tami Hardesty	

About the Author

This laboratory manual series is by Terry R. Martin of Kishwaukee College. Terry's teaching experience of over thirty years, his interest in students and love for college instruction, and his innovative attitude and use of technology-based learning enhance the solid tradition of his other well-established laboratory manuals. Among Terry's awards are the 1972 Kishwaukee College Outstanding Educator, 1977 Phi Theta Kappa Outstanding Instructor Award, 1989 Kishwaukee College ICCTA Outstanding Educator Award, 1996 and 2004 Who's Who Among America's Teachers, 1996 Kishwaukee College Faculty Board of Trustees Award of Excellence, and 1998 Continued Excellence Award for Phi Theta Kappa Advisors. Terry's professional memberships include the National Association of Biology Teachers, Illinois Association of Community College Biologists, Human Anatomy and Physiology Society, former Chicago Area Anatomy and Physiology Society (founding member), Phi Theta Kappa (honorary member), and Nature Conservancy. In addition to writing many publications, he co-produced with Hassan Rastegar a videotape entitled *Introduction to the Human Cadaver and Prosection,* published by Wm. C. Brown Publishers in 1989. Terry revised the *Laboratory Manual to Accompany Hole's Human Anatomy and Physiology,* Twelfth Edition, revised the *Laboratory Manual to Accompany Hole's Essentials of Human Anatomy and Physiology,* Tenth Edition, and authored *Human Anatomy and Physiology Laboratory Manual, Fetal Pig Dissection*, Third Edition. A series of seven LabCam videos of anatomy and physiology laboratory processes were produced at Kishwaukee College and published by McGraw-Hill Higher Education in 2007. Cadaver dissection experiences have been provided for his students for over twenty years. Terry teaches portions of EMT and paramedic classes and serves as a Faculty Consultant for Advanced Placement Biology examination readings. During 1994, Terry was a faculty exchange member in Ireland. The author locally supports historical preservation, natural areas, scouting, and scholarship. We are pleased to have Terry continue the tradition of authoring laboratory manuals for McGraw-Hill Higher Education.

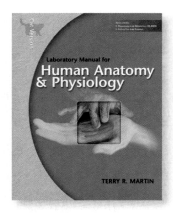
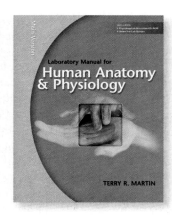
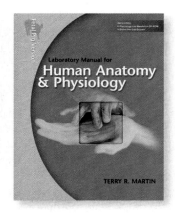

To the Student

The exercises in this laboratory manual will provide you with opportunities to observe various anatomical structures and to investigate certain physiological phenomena. Such experiences should help you relate specimens, models, microscope slides, and your body to what you have learned in the lecture and read about in the textbook.

Frequent variations exist in anatomical structures among humans. The illustrations in the laboratory manual represent normal (normal means the most common variation) anatomy. Variations from normal anatomy do not represent abnormal anatomy unless some function is impaired.

The following list of suggestions and study skills may make your laboratory activities more effective and profitable.

1. Prepare yourself before attending the laboratory session by reading the assigned exercise and reviewing the related sections of the textbook and lecture notes as indicated in the pre-lab section of the laboratory exercise. It is important to have some understanding of what will be done in the lab before you come to class.

2. Visit the website **www.mhhe.com/martinseries1** and view suggested *Anatomy & Physiology Revealed* animations, view LabCam videos, and answer the pre-lab questions.

3. Be on time. During the first few minutes of the laboratory meeting, the instructor often will provide verbal instructions. Make special note of any changes in materials to be used or procedures to be followed. Also listen carefully for information about special techniques to be used and precautions to be taken.

4. Keep your work area clean and your materials neatly arranged so that you can locate needed items. This will enable you to efficiently proceed and will reduce the chances of making mistakes.

5. Pay particular attention to the purpose of the exercise, which states what you are to accomplish in general terms, and to the learning outcomes, which list what you should be able to do as a result of the laboratory experience. Then, before you leave the class, review the outcomes and make sure that you can perform all of the assessments.

6. Precisely follow the directions in the procedure and proceed only when you understand them clearly. Do not improvise procedures unless you have the approval of the laboratory instructor. Ask questions if you do not understand exactly what you are supposed to do and why you are doing it.

7. Handle all laboratory materials with care. These materials often are fragile and expensive to replace. When-

ever you have questions about the proper treatment of equipment, ask the instructor.

8. Treat all living specimens humanely and try to minimize any discomfort they might experience.

9. Although at times you might work with a laboratory partner or a small group, try to remain independent when you are making observations, drawing conclusions, and completing the activities in the laboratory reports.

10. Record your observations immediately after making them. In most cases, such data can be entered in spaces provided in the laboratory reports.

11. Read the instructions for each section of the laboratory report before you begin to complete it. Think about the questions before you answer them. Your responses should be based on logical reasoning and phrased in clear and concise language.

12. At the end of each laboratory period, clean your work area and the instruments you have used. Return all materials to their proper places and dispose of wastes, including glassware or microscope slides that have become contaminated with human blood or body fluids, as directed by the laboratory instructor. Wash your hands thoroughly before leaving the laboratory.

Study Skills for Anatomy and Physiology

My students have found that certain study skills worked well for them while enrolled in Human Anatomy and Physiology. Although everyone has his or her learning style, there are techniques that work well for most students. Using some of the skills listed here could make your course more enjoyable and rewarding.

1. **Time management:** Prepare monthly, weekly, and daily schedules. Include dates of quizzes, exams, and projects on the calendar. On your daily schedule, budget several short study periods. Daily repetition alleviates cramming for exams. Prioritize your time so that you still have time for work and leisure activities. Find an appropriate study atmosphere with minimum distractions.

2. **Note taking:** Look for the main ideas and briefly express them in your own words. Organize, edit, and review your notes soon after the lecture. Add textbook information to your notes as you reorganize them. Underline or highlight with different colors the important points,

To the Student

major headings, and key terms. Study your notes daily, as they provide sequential building blocks of the course content.

3. **Chunking:** Organize information into logical groups or categories. Study and master one chunk of information at a time. For example, study the bones of the upper limb, lower limb, trunk, and head as separate study tasks.

4. **Mnemonic devices:** An *acrostic* is a combination of association and imagery to aid your memory. It is often in the form of a poem, rhyme, or jingle in which the first letter of each word corresponds to the first letters of the words you need to remember. **S**o **L**ong **T**op **P**art, **H**ere **C**omes **T**he **T**humb is an example of such a mnemonic device for remembering the eight carpals in the correct sequence. *Acronyms* are words formed by the first letters of the items to remember. *IPMAT* is an example of this type of mnemonic device to help you remember the phases of the cell cycle in the correct sequence. Try to create some of your own.

5. **Note cards/flash cards:** Make your own. Add labels and colors to enhance the material. Keep them with you in your pocket or purse. Study them often and for short periods. Concentrate on a small number of cards at one time. Shuffle your cards and have someone quiz you on their content. As you become familiar with the material, you can set aside cards that don't require additional mastery.

6. **Recording and recitation:** An auditory learner can benefit by recording lectures and review sessions with a cassette recorder. Many students listen to the taped sessions as they drive or just before going to bed. Reading your notes aloud can help also. Explain the material to anyone (even if there are no listeners). Talk about anatomy and physiology in everyday conversations.

7. **Study groups:** Small study groups that meet periodically to review course material and compare notes have helped and encouraged many students. However, keep the group on the task at hand. Work as a team and alternate leaders. This group often becomes a support group.

Practice sound study skill during your anatomy and physiology endeavor.

The Use of Animals in Biology Education*

The National Association of Biology Teachers (NABT) believes that the study of organisms, including nonhuman animals, is essential to the understanding of life on Earth. NABT recommends the prudent and responsible use of animals in the life science classroom. NABT believes that biology teachers should foster a respect for life. Biology teachers also should teach about the interrelationship and interdependency of all things.

Classroom experiences that involve nonhuman animals range from observation to dissection. NABT supports these experiences so long as they are conducted within the long-established guidelines of proper care and use of animals, as developed by the scientific and educational community.

As with any instructional activity, the use of nonhuman animals in the biology classroom must have sound educational objectives. Any use of animals, whether for observation or dissection, must convey substantive knowledge of biology. NABT believes that biology teachers are in the best position to make this determination for their students.

NABT acknowledges that no alternative can substitute for the actual experience of dissection or other use of animals and urges teachers to be aware of the limitations of alternatives. When the teacher determines that the most effective means to meet the objectives of the class do not require dissection, NABT accepts the use of alternatives to dissection, including models and the various forms of multimedia. The Association encourages teachers to be sensitive to substantive student objections to dissection and to consider providing appropriate lessons for those students where necessary.

To implement this policy, NABT endorses and adopts the "Principles and Guidelines for the Use of Animals in Precollege Education" of the Institute of Laboratory Animals Resources (National Research Council). Copies of the "Principles and Guidelines" may be obtained from the ILAR (2101 Constitution Avenue, NW, Washington, DC 20418; 202-334-2590).

*Adopted by the Board of Directors in October 1995. This policy supersedes and replaces all previous NABT statements regarding animals in biology education.

Scientific Method and Measurements

Pre-Lab

1. Carefully read the introductory material and examine the entire lab.
2. Be familiar with the scientific method (from lecture or the textbook).
3. Visit www.mhhe.com/martinseries1 for pre-lab questions.

Materials Needed

Meterstick
Calculator
Human skeleton

Scientific investigation involves a series of logical steps to arrive at explanations for various biological phenomena. This technique, called the *scientific method,* is used in all disciplines of science. It allows scientists to draw logical and reliable conclusions about phenomena.

The scientific method begins with *observations* related to the topic under investigation. This step commonly involves the accumulation of previously acquired information and/or your observations of the phenomenon. These observations are used to formulate a tentative explanation known as the *hypothesis.* An important attribute of a hypothesis is that it must be testable. The testing of the hypothesis involves performing a carefully controlled *experiment* to obtain data that can be used to support, reject, or modify the hypothesis. An *analysis of data* is conducted using sufficient information collected during the experiment. Data analysis may include organization and presentation of data as tables, graphs, and drawings. From the interpretation of the data analysis, *conclusions* are drawn. (If the data do not support the hypothesis, you must reexamine the experimental design and the data, and if needed develop a new hypothesis.) The final presentation of the information is made from the conclusions. Results and conclusions are presented to the scientific community for evaluation through peer reviews, presentations at professional meetings, and published articles. If many investigators working independently can validate the hypothesis by arriving at the same conclusions, the explanation becomes a **theory.** A theory verified continuously over time and accepted by the scientific community becomes known as a **scientific law** or **principle.** A scientific law serves as the standard explanation for an observation unless it is disproved by new information. The five components of the scientific method are summarized as

Observations

↓

Hypothesis

↓

Experiment

↓

Analysis of data

↓

Conclusions

Metric measurements are characteristic tools of scientific investigations. The English system of measurements is often used in the United States, so the investigator must make conversions from the English system to the metric system. Use table 1 for the conversion of English units of measure to metric units for length, mass, volume, time, and temperature.

Purpose of the Exercise

To become familiar with the scientific method of investigation, to learn how to formulate sound conclusions, and to provide opportunities to use the metric system of measurements.

Learning Outcomes

After completing this exercise, you should be able to

1. Convert English measurements to the metric system, and vice versa.
2. Measure and record upper limb lengths and heights of ten subjects.
3. Apply the scientific method to test the validity of a hypothesis concerning the direct, linear relationship between human upper limb length and height.
4. Design an experiment, formulate a hypothesis, and test it using the scientific method.

TABLE 1.1   Metric Measurement System and Conversions

Measurement	Unit & Abbreviation	Metric Equivalent	Conversion Factor Metric to English (approximate)	Conversion Factor English to Metric (approximate)
Length	1 kilometer (km)	1,000 (10^3) m	1 km = 0.62 mile	1 mile = 1.61 km
	1 meter (m)	100 (10^2) cm 1,000 mm	1 m = 1.1 yards = 3.3 feet = 39.4 inches	1 yard = 0.9 m 1 foot = 0.3 m
	1 decimeter (dm)	0.1 (10^{-1}) m	1 dm = 3.94 inches	1 inch = 0.25 dm
	1 centimeter (cm)	0.01 (10^{-2}) m	1 cm = 0.4 inches	1 foot = 30.5 cm 1 inch = 2.54 cm
	1 millimeter (mm)	0.001 (10^{-3}) m 0.1 cm	1 mm = 0.04 inches	
	1 micrometer (μm)	0.000001 (10^{-6}) m 0.001 mm		
Mass	1 metric ton (t)	1,000 kg	1 t = 1.1 ton	1 ton = 0.91 t
	1 kilogram (kg)	1,000 g	1 kg = 2.2 pounds	1 pound = 0.45 kg
	1 gram (g)	1,000 mg	1 g = 0.04 ounce	1 pound = 454 g 1 ounce = 28.35 g
	1 milligram (mg)	0.001 g		
Volume (liquids and gases)	1 liter (L)	1,000 mL	1 L = 1.06 quarts	1 gallon = 3.78 L 1 quart = 0.95 L
	1 milliliter (mL)	0.001 L 1 cubic centimeter (cc or cm^3)	1 mL = 0.03 fluid ounce 1 mL = 1/4 teaspoon 1 mL = 15–16 drops	1 quart = 946 mL 1 fluid ounce = 29.6 mL 1 teaspoon = 5 mL
Time	1 second (s)	1/60 minute	same	same
	1 millisecond (ms)	0.001 s	same	same
Temperature	Degrees Celsius (°C)		°F = 9/5 °C + 32	°C = 5/9 (°F − 32)

Procedure A—Using the Steps of the Scientific Method

1. Many people have observed a correlation between the length of the upper and lower limbs and the height (stature) of an individual. For example, a person who has long upper limbs (the arm, forearm, and hand combined) tends to be tall. Make some visual observations of other people in your class to observe a possible correlation.

2. From such observations, the following hypothesis is formulated: The length of a person's upper limb is equal to 0.4 (40%) of the height of the person. Test this hypothesis by performing the following experiment.

3. In this experiment, use a meterstick (fig. 1.1) to measure an upper limb length of ten subjects. For each measurement, place the meterstick in the axilla (armpit) and record the length in centimeters to the end of the longest finger (fig. 1.2). Obtain the height of each person in centimeters by measuring them without shoes against a wall (fig. 1.3). The height of each person can be calculated by multiplying each individual's height in inches by 2.54 to obtain his/her height in centimeters. Record all your measurements in Part A of Laboratory Report 1.

4. The data collected from all of the measurements can now be analyzed. The expected (predicted) correlation between upper limb length and height is determined using the following equation:

$$\text{Height} \times 0.4 = \text{expected upper limb length}$$

The observed (actual) correlation to be used to test the hypothesis is determined by

$$\text{Length of upper limb/height} = \text{actual \% of height}$$

5. A graph is an excellent way to display a visual representation of the data. Plot the subjects' data in Part A of the laboratory report. Plot the upper limb length of

---

FIGURE 1.1 Metric ruler with metric lengths indicated. A meterstick length would be 100 centimeters. (The image size is approximately to scale.)

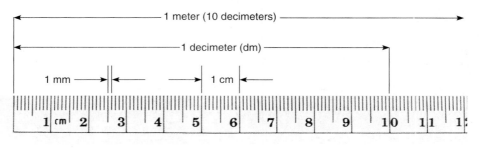

Metric ruler

FIGURE 1.2 Measurement of upper limb length.

FIGURE 1.3 Measurement of height.

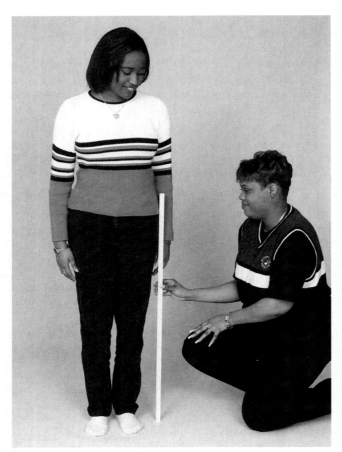

each subject on the x-axis and the height of each person on the y-axis. A line is already located on the graph that represents a hypothetical relationship of 0.4 (40%) upper limb length compared to height. This is a graphic representation of the original hypothesis.

6. Compare the distribution of all of the points (actual height and upper limb length) that you placed on the graph with the distribution of the expected correlation represented by the hypothesis.

7. Complete Part A of the laboratory report.

Procedure B—Design an Experiment

 Critical Thinking Activity

You have probably concluded that there is some correlation of the length of body parts to height. Often when a skeleton is found, it is not complete, especially when paleontologists discover a skeleton. It is occasionally feasible to use the length of a single bone to estimate the height of an individual. Observe human skeletons and locate the radius bone in the forearm. Use your observations to identify a mathematical relationship between the length of the radius and height. Formulate a hypothesis that can be tested. Make measurements, analyze data, and develop a conclusion from your experiment. Complete Part B of the laboratory report.

Name _____

Date _____

Section _____

The ⚠ corresponds to the Learning Outcome(s) listed at the beginning of the laboratory exercise.

Scientific Method and Measurements

Part A Assessments

1. Record measurements for the upper limb length and height of ten subjects. Use a calculator to determine the expected upper limb length and the actual percentage (as a decimal or a percentage) of the height for the ten subjects. Record your results in the following table. ⚠

Subject	Measured Upper Limb Length (cm)	Height* (cm)	Height × 0.4 = Expected Upper Limb Length (cm)	Actual % of Height = Upper Limb Length (cm)/ Height (cm)
1.				
2.				
3.				
4.				
5.				
6.				
7.				
8.				
9.				
10.				

*The height of each person can be calculated by multiplying each individual's height in inches by 2.54 to obtain his/her height in centimeters. ⚠

2. Plot the distribution of data (upper limb length and height) collected for the ten subjects on the following graph. The line located on the graph represents the *expected* 0.4 (40%) ratio of upper limb length to measured height (the original hypothesis). (The x-axis represents upper limb length and the y-axis represents height.) Draw a line of *best fit* through the distribution of points of the plotted data of the ten subjects. Compare the two distributions (expected line and the distribution line drawn for the ten subjects). **3**

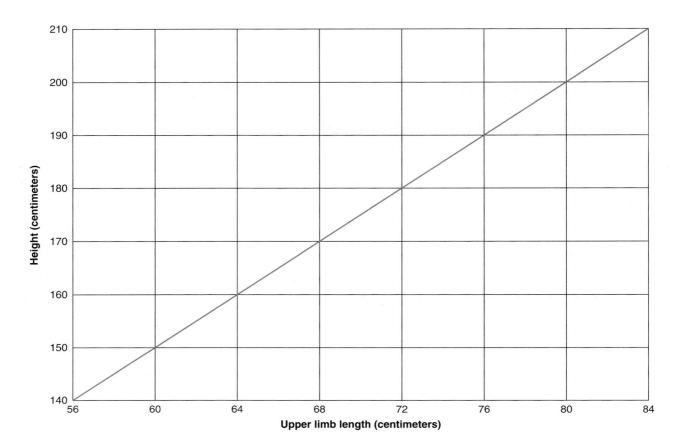

3. Does the distribution of the ten subjects' measured upper limb lengths support or reject the original hypothesis? _____ Explain your answer. **3**

Part B Assessments

1. Describe your observations of a possible correlation between the radius length and height. **4**

2. Write a hypothesis based on your observations. **4**

3. Describe the design of the experiment that you devised to test your hypothesis. **4**

4. Place your analysis of the data in this space in the form of a table and a graph. **4**

5. Based on an analysis of your data, what conclusions can you make? Did these conclusions confirm or refute your original hypothesis? ◢

6. Discuss your results and conclusions with classmates. What common conclusion can the class formulate about the correlation between radius length and height? ◢

Laboratory Exercise 2

Body Organization, Membranes, and Terminology

Pre-Lab

1. Carefully read the introductory material and examine the entire lab.
2. Be familiar with body cavities, membranes, organ systems, and body regions (from lecture or the textbook).
3. Visit www.mhhe.com/martinseriesl for pre-lab questions.

Materials Needed

Dissectible human torso model (manikin)
Variety of specimens or models sectioned along various planes

For Learning Extension Activity:
Colored pencils

The major features of the organization of the human body include certain body cavities. The dorsal body cavity includes a cranial cavity containing the brain and a vertebral canal (spinal cavity) containing the spinal cord. The ventral body cavity includes the thoracic cavity, which is subdivided into a mediastinum containing primarily the heart, esophagus, and trachea. The thoracic cavity also includes two pleural cavities, each surrounding a lung. Also included in the ventral body cavity is the abdominopelvic cavity, composed of an abdominal cavity and pelvic cavity. The entire abdominopelvic cavity is further subdivided into either nine regions or four quadrants. The large size of the abdominopelvic cavity, with its many visceral organs, warrants these further subdivisions into regions or quadrants for convenience and for accuracy in describing organ locations, injury sites, and pain locations. Because early anatomical references do not address dorsal and ventral body cavities, some anatomists prefer not to use those broader terms.

Located within the ventral body cavities are thin serous membranes containing small amounts of a lubricating serous fluid. The double-layered membranes that secrete this fluid include the pericardium, pleura, and peritoneum. The pericardium is associated with the heart; the pleura is associated with a lung; the peritoneum is associated with the viscera located in the abdominopelvic cavity. The inner portion of each membrane attached to the organ is the visceral component, whereas the outer parietal portion forms an outer cavity wall. Several abdominopelvic organs, such as the kidneys, are located just behind (are retroperitoneal to) the parietal peritoneum, thus lacking a mesentery.

Although the human body functions as one entire unit, it is customary to divide the body into eleven body organ systems. In order to communicate effectively with each other about the body, scientists have devised anatomical terminology. Foremost in this task we use *anatomical position* as our basis for communication, including directional terms, body regions, and planes of the body. A person in anatomical position is standing erect, facing forward, with upper limbs at the sides and palms forward. This standard position allows us to describe relative positions of various body parts using such directional terms as left-right, anterior (ventral)-posterior (dorsal), proximal-distal, and superior-inferior. Body regions include certain surface areas, portions of limbs, and portions of body cavities. In order to study internal structures, often the body is depicted as sectioned into a sagittal plane, frontal plane, or transverse plane.

Purpose of the Exercise

To review the organizational pattern of the human body, to review its organ systems and the organs included in each system, and to become acquainted with the terms used to describe the relative position of body parts, body sections, and body regions.

Learning Outcomes

After completing this exercise, you should be able to

1. Locate and name the major body cavities and identify the membranes associated with each cavity.
2. Associate the organs and functions included within each organ system and locate the organs in a dissectible human torso model.
3. Select the terms used to describe the relative positions of body parts.
4. Differentiate the terms used to identify body sections and identify the plane along which a particular specimen is cut.
5. Label body regions and associate the terms used to identify body regions.

Procedure A—Body Cavities and Membranes

1. Label figures 2.1, 2.2, and 2.3.
2. Locate the following features on the dissectible human torso model (fig. 2.4):

    body cavities
    cranial cavity
    vertebral canal (spinal cavity)
    thoracic cavity
        mediastinum (region between the lungs; includes pericardial cavity)
        pleural cavities (2)
    abdominopelvic cavity
        abdominal cavity
        pelvic cavity
    diaphragm
    smaller cavities within the head
    oral cavity
    nasal cavity with connected sinuses
    orbital cavity
    middle ear cavity
    membranes and cavities
    pleural cavity
        parietal pleura
        visceral pleura
    pericardial cavity
        parietal pericardium (covered by fibrous pericardium)
        visceral pericardium (epicardium)
    peritoneal cavity
        parietal peritoneum
        visceral peritoneum

3. Complete Part A of Laboratory Report 2.

FIGURE 2.1 Label these body cavities, left lateral view. ⚠1

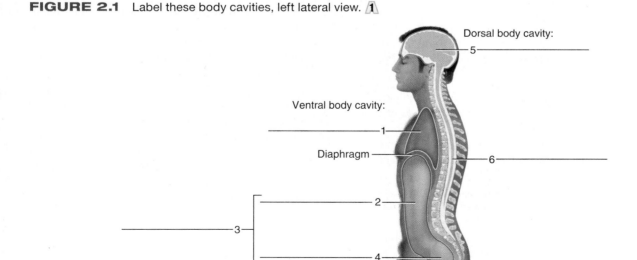

Dorsal body cavity:
5
Ventral body cavity:
1
Diaphragm
6
2
3
4

FIGURE 2.2 Label the thoracic membranes and cavities associated with (*a*) the lungs and (*b*) the heart, using the terms provided. 1

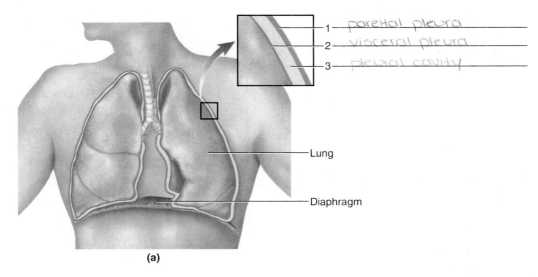

1 — parietal pleura
2 — visceral pleura
3 — pleural cavity

—Lung

—Diaphragm

(a)

Terms:
Parietal pleura (lines cavity wall)
Pleural cavity
Visceral pleura (covers lung)

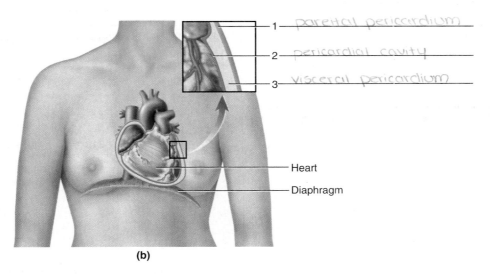

1 — parietal pericardium
2 — pericardial cavity
3 — visceral pericardium

—Heart
—Diaphragm

(b)

Terms:
Parietal pericardium (outer)
Pericardial cavity
Visceral pericardium (inner)

FIGURE 2.3 Label the abdominal cavity membranes as shown in this left lateral view, using the terms provided. ⚠️

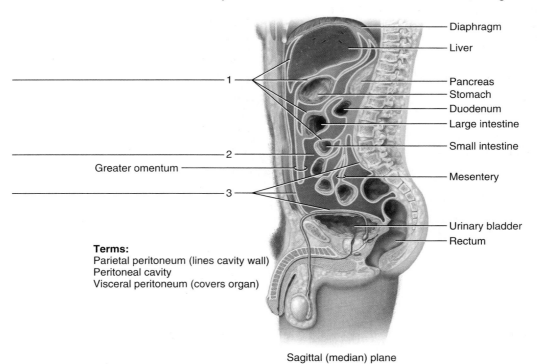

1

2

Greater omentum

3

Terms:
Parietal peritoneum (lines cavity wall)
Peritoneal cavity
Visceral peritoneum (covers organ)

Diaphragm
Liver
Pancreas
Stomach
Duodenum
Large intestine
Small intestine
Mesentery
Urinary bladder
Rectum

Sagittal (median) plane

FIGURE 2.4 Dissectible human torso model with body cavities, abdominopelvic quadrants, body planes, and major organs indicated.

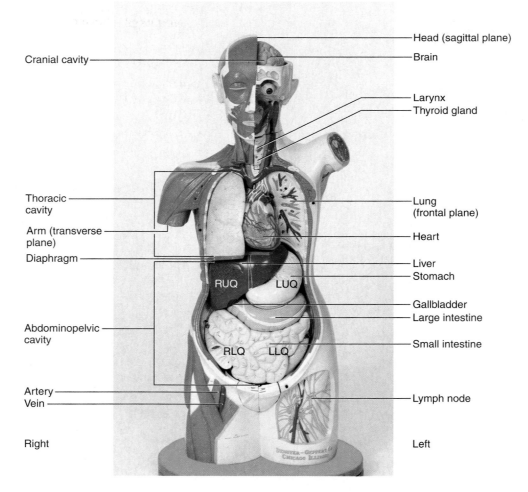

Cranial cavity

Thoracic cavity

Arm (transverse plane)

Diaphragm

Abdominopelvic cavity

Artery
Vein

Right

Head (sagittal plane)
Brain
Larynx
Thyroid gland
Lung (frontal plane)
Heart
Liver
Stomach
Gallbladder
Large intestine
Small intestine
Lymph node

RUQ   LUQ

RLQ   LLQ

Left

Procedure B—Organ Systems

1. Use the dissectible human torso model (fig. 2.4) to locate the following systems and their major organs:

   integumentary system
   - skin
   - accessory organs such as hair and nails

   skeletal system
   - bones
   - ligaments

   muscular system
   - skeletal muscles
   - tendons

   nervous system
   - brain
   - spinal cord
   - nerves

   endocrine system
   - pituitary gland
   - thyroid gland
   - parathyroid glands
   - adrenal glands
   - pancreas
   - ovaries
   - testes
   - pineal gland
   - thymus

   cardiovascular system
   - heart
   - arteries
   - veins

   lymphatic system
   - lymphatic vessels
   - lymph nodes
   - thymus
   - spleen

   respiratory system
   - nasal cavity
   - pharynx
   - larynx
   - trachea
   - bronchi
   - lungs

   digestive system
   - mouth
   - tongue
   - teeth
   - salivary glands
   - pharynx
   - esophagus
   - stomach
   - liver
   - gallbladder
   - pancreas
   - small intestine
   - large intestine

   urinary system
   - kidneys
   - ureters
   - urinary bladder
   - urethra

   male reproductive system
   - scrotum
   - testes
   - penis
   - urethra

   female reproductive system
   - ovaries
   - uterine tubes (oviducts; fallopian tubes)
   - uterus
   - vagina

2. Complete Part B of the laboratory report.

Procedure C—Relative Positions, Planes, Sections, and Regions

1. Observe the person standing in anatomical position (fig. 2.5). Anatomical terminology assumes the body is in anatomical position even though a person is often observed differently.
2. Label figures 2.6, 2.7, and 2.8.
3. Examine the sectioned specimens on the demonstration table and identify the plane along which each is cut.
4. Complete Parts C and D of the laboratory report.

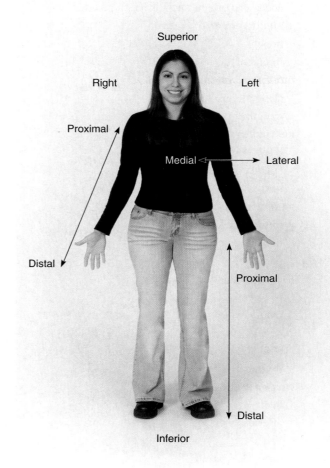

FIGURE 2.5 Anatomical position with directional terms indicated. The body is standing erect, face forward, with upper limbs at the sides and palms forward. This results in an anterior view of the body.

FIGURE 2.6 Label the planes represented in this illustration, using the terms provided. 4

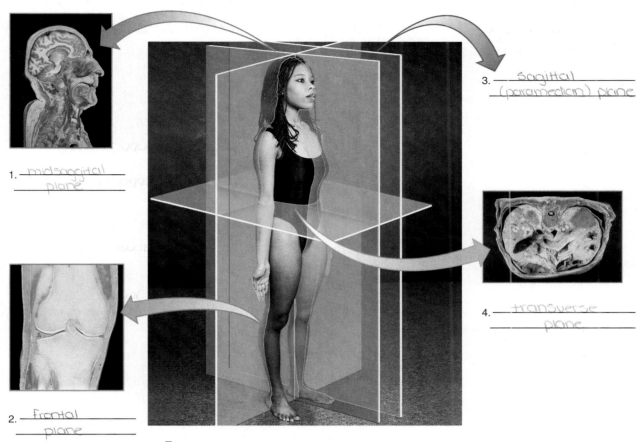

1. _midsagittal plane_

2. _Frontal plane_

3. _Sagittal (paramedian) plane_

4. _transverse plane_

Terms:
Frontal (coronal) plane (divides anterior and posterior)
Sagittal (median; midsagittal) plane (divides an equal left and right)
Sagittal (paramedian; parasagittal) plane (divides an unequal left and right)
Transverse (horizontal) plane (divides superior and inferior)

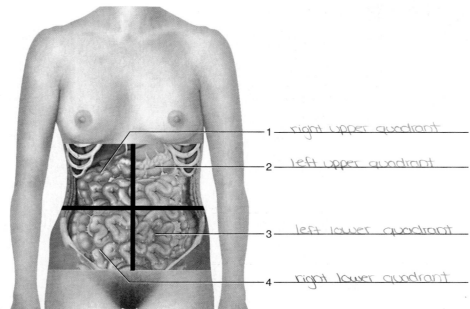

Terms:
Left lower (LLQ)
Left upper (LUQ)
Right lower (RLQ)
Right upper (RUQ)

1 ___ right upper quadrant
2 ___ left upper quadrant
3 ___ left lower quadrant
4 ___ right lower quadrant

(a) Quadrants (4)

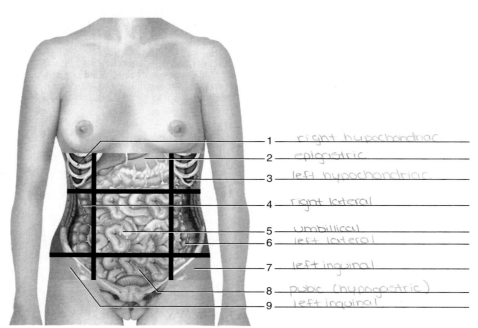

Terms:
Epigastric
Left hypochondriac
Left inguinal (iliac; groin)
Left lateral (lumbar; flank)
Pubic (hypogastric)
Right hypochondriac
Right inguinal (iliac; groin)
Right lateral (lumbar; flank)
Umbilical

1 ___ right hypochondriac
2 ___ epigastric
3 ___ left hypochondriac
4 ___ right lateral
5 ___ umbilical
6 ___ left lateral
7 ___ left inguinal
8 ___ pubic (hypogastric)
9 ___ left inguinal

(b) Regions (9)

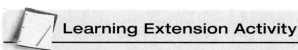

Learning Extension Activity

Use different colored pencils to distinguish body regions in figure 2.8.

FIGURE 2.8 Label these diagrams of the body regions using the terms provided: (*a*) anterior regions; (*b*) posterior regions. **5**

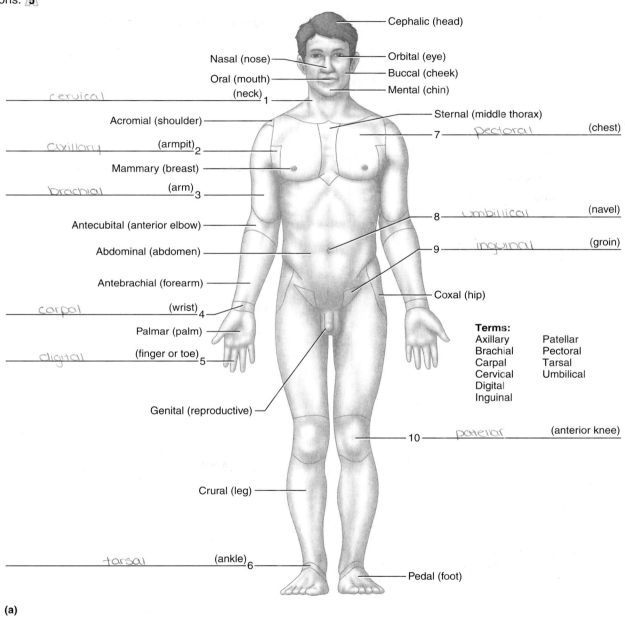

Cephalic (head)

Nasal (nose)

Oral (mouth)

(neck) 1 — cervical

Acromial (shoulder)

(armpit) 2 — axillary

Mammary (breast)

(arm) 3 — brachial

Antecubital (anterior elbow)

Abdominal (abdomen)

Antebrachial (forearm)

(wrist) 4 — carpal

Palmar (palm)

(finger or toe) 5 — digital

Genital (reproductive)

Crural (leg)

(ankle) 6 — tarsal

Orbital (eye)

Buccal (cheek)

Mental (chin)

Sternal (middle thorax)

7 — pectoral (chest)

8 — umbilical (navel)

9 — inguinal (groin)

Coxal (hip)

Terms:

Axillary	Patellar
Brachial	Pectoral
Carpal	Tarsal
Cervical	Umbilical
Digital	
Inguinal	

10 — posterior (anterior knee)

Pedal (foot)

(a)

FIGURE 2.8 *Continued.*

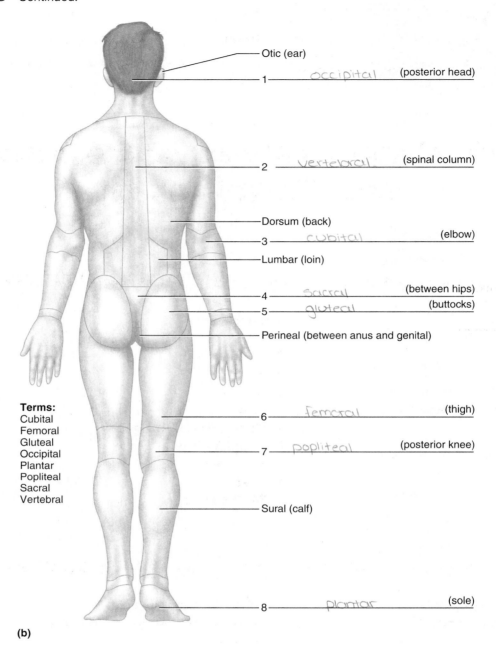

Otic (ear)

1 ___occipital___ (posterior head)

2 ___vertebral___ (spinal column)

Dorsum (back)

3 ___cubital___ (elbow)

Lumbar (loin)

4 ___sacral___ (between hips)

5 ___gluteal___ (buttocks)

Perineal (between anus and genital)

Terms:
Cubital
Femoral
Gluteal
Occipital
Plantar
Popliteal
Sacral
Vertebral

6 ___femoral___ (thigh)

7 ___popliteal___ (posterior knee)

Sural (calf)

8 ___plantar___ (sole)

(b)

Name _____

Date _____

Section _____

The ⬜ corresponds to the Learning Outcome(s) listed at the beginning of the laboratory exercise.

Body Organization, Membranes, and Terminology

Part A Assessments

Match the body cavities in column A with the organs contained in the cavities in column B. Place the letter of your choice in the space provided. ⬜1

Column A		Column B
a. Abdominal cavity	_____	1. Liver
b. Cranial cavity	_D_	2. Lungs
c. Pelvic cavity	_A_	3. Spleen
d. Thoracic cavity	_A_	4. Stomach
e. Vertebral canal (spinal cavity)	_B_	5. Brain
	_C_	6. Internal reproductive organs
	_C_	7. Urinary bladder
	_E_	8. Spinal cord
	_D_	9. Heart
	_A_	10. Small intestine

Part B Assessments

Match the organ systems in column A with the principal functions in column B. Place the letter of your choice in the space provided. ⬜2

Column A		Column B
a. Cardiovascular system	_C_	1. The main system that secretes hormones
b. Digestive system	_D_	2. Provides an outer covering of the body
c. Endocrine system	_H_	3. Produces gametes (eggs and sperm)
d. Integumentary system	_G_	4. Stimulates muscles to contract and interprets information from sensory organs
e. Lymphatic system		
f. Muscular system	_J_	5. Provides a framework and support for soft tissues and produces blood cells in red marrow
g. Nervous system		
h. Reproductive system	_I_	6. Exchanges gases between air and blood
i. Respiratory system	_E_	7. Transports excess fluid from tissues to blood
j. Skeletal system	_F-_	8. Movement via contractions and creates most body heat
k. Urinary system	_K_	9. Removes liquid and wastes from blood and transports them to the outside of the body
	_B_	10. Converts food molecules into forms that are absorbable
	_A_	11. Transports nutrients, wastes, and gases throughout the body

19

Part C Assessments

Indicate whether each of the following sentences makes correct or incorrect usage of the word in boldface type (assume that the body is in the anatomical position). If the sentence is incorrect, in the space provided supply a term that will make it correct. ◢3◣

1. The mouth is **superior** to the nose. _____

2. The stomach is **inferior** to the diaphragm. _____

3. The trachea is **anterior** to the spinal cord. _____

4. The larynx is **posterior** to the esophagus. _____

5. The heart is **medial** to the lungs. _____

6. The kidneys are **inferior** to the adrenal glands. _____

7. The hand is **proximal** to the elbow. _____

8. The knee is **proximal** to the ankle. _____

9. Blood in **deep** blood vessels gives color to the skin. _____

10. A **peripheral** nerve passes from the spinal cord into the limbs. _____

11. The spleen and gallbladder are **ipsilateral.** _____

12. The dermis is the **superficial** layer of the skin. _____

Part D Assessments

 Critical Thinking Activity

State the quadrant of the abdominopelvic cavity where the pain or sound would be located for each of the six conditions listed. In some cases, there may be more than one correct answer, and pain is sometimes referred to another region. This phenomenon, called *referred pain,* occurs when pain is interpreted as originating from some area other than the parts being stimulated. When referred pain is involved in the patient's interpretation of the pain location, the proper diagnosis of the ailment is more challenging. For the purpose of this exercise, assume the pain is interpreted as originating from the organ involved. ◢2◣

1. Stomach ulcer _____

2. Appendicitis _____

3. Bowel sounds _____

4. Gallbladder attack _____

5. Kidney stone in left ureter _____

6. Ruptured spleen _____

Chemistry of Life

Pre-Lab

1. Carefully read the introductory material and examine the entire lab content.
2. Be familiar with pH and organic molecules (from lecture or the textbook).
3. Visit www.mhhe.com/martinseries1 for pre-lab questions and LabCam videos.

10% glucose solution
Clear carbonated soft drink
10% starch solution
Potatoes for potato water
Distilled water
Vegetable oil
Brown paper
Numbered unknown organic samples

Materials Needed

For pH Tests:
Chopped fresh red cabbage
Beaker (250 mL)
Distilled water
Tap water
Vinegar
Baking soda
Laboratory scoop for measuring
Pipets for measuring
Full-range pH test papers
7 assorted common liquids clearly labeled in closed bottles on a tray or in a tub
Droppers labeled for each liquid

For Organic Tests:
Test tubes
Test-tube rack
Test-tube clamps
China marker
Hot plate
Beaker for hot water bath (500 mL)
Pipets for measuring
Benedict's solution
Biuret reagent (or 10% NaOH and 1% $CuSO_4$)
Iodine-potassium-iodide (IKI) solution
Sudan IV dye
Egg albumin

Safety

► Review all safety guidelines inside the front cover of your laboratory manual.
► Clean laboratory surfaces before and after laboratory procedures using soap and water.
► Use extreme caution when working with chemicals.
► Safety goggles must be worn at all times.
► Wear disposable gloves while working with the chemicals.
► Precautions should be taken to prevent chemicals from contacting your skin.
► Do not mix any of the chemicals together unless instructed to do so.
► Clean up any spills immediately and notify the instructor at once.
► Wash your hands before leaving the laboratory.

The complexities of the human body arise from the organization and interactions of chemicals. Organisms are made of matter, and the most basic unit of matter is the chemical element. The smallest unit of an element is an atom, and two or more of those can unite to form a molecule. All processes that occur within the body involve chemical reactions—interactions between atoms and molecules. We breathe to supply oxygen to our cells for energy. We eat and drink to bring chemicals into our bodies that our cells need. Water fills and bathes all of our cells and allows an amazing array of reactions to occur, all of which are designed to keep us alive. Chemistry forms the very basis of life and thus forms the foundation of anatomy and physiology.

The pH scale is from 0 to 14 with the midpoint of 7.0 (pure water) representing a neutral solution. If the solution has a pH of lower than 7.0, it represents an acidic solution with more hydrogen ions (H^+) than hydroxide ions (OH^-). The solution is a base if the pH is higher than 7.0 with more OH^- ions than H^+ ions. The weaker acids and bases are closer to 7, with the strongest acids near 0 and the strongest bases near 14. A seemingly minor change in the pH actually represents a more significant change as there is a tenfold difference in hydrogen ion concentrations with each whole number represented on the pH scale. Our cells can only function within minimal pH fluctuations.

The organic molecules (biomolecules) that are tested in this laboratory exercise include carbohydrates, lipids, and proteins. Carbohydrates are used to supply energy to our cells. The polysaccharide starch is composed of simple sugar (monosaccharide) building blocks. Lipids include fats, phospholipids, and steroids that provide some cellular structure and energy for cells. The building blocks of lipids include fatty acids and glycerol. Proteins compose important cellular structures, antibodies, and enzymes. Amino acids are the building blocks of proteins.

Purpose of the Exercise

To review pH and basic categories of organic compounds, and to differentiate between types of organic compounds.

Learning Outcomes

After completing this exercise, you should be able to

1. Demonstrate pH values of various substances through testing methods.

2. Determine categories of organic compounds with basic colorimetric tests.

3. Discover the organic composition of an unknown solution.

Procedure A—The pH Scale

1. Study figure 3.1, which shows the pH scale. Note the range of the scale and the pH value that is considered neutral.

2. **Cabbage water tests.** Many tests can determine a pH value, but among the more interesting are colorimetric tests in which an indicator chemical changes color when it reacts. Many plant pigments, especially anthocyanins (which give plants color, from blue to red), can be used as colorimetric pH indicators. One that works well is the red pigment in red cabbage.

    a. Prepare cabbage water to be used as a general pH indicator. Fill a 250 mL beaker to the 100 mL level with chopped red cabbage. Add water to make 150 mL. Place the beaker on a hot plate and simmer the mixture until the pigments come out of the cabbage and the water turns deep purple. Allow the water to cool. (You may proceed to step 3 while you wait for this to finish.)

    b. Label three clean test tubes: one for water, one for vinegar, and one for baking soda.

    c. Place 2 mL of cabbage water into each test tube.

    d. To the first test tube, add 2 mL of distilled water and swirl the mixture. Note and record the color in Part A of Laboratory Report 3.

    e. Repeat this procedure for test tube 2, adding 2 mL of vinegar and swirling the mixture. Note and record the results.

    f. Repeat this procedure for test tube 3, adding one laboratory scoop of baking soda. Swirl the mixture. Note and record the results.

3. **Testing with pH paper.** Many commercial pH indicators are available. A very simple one to use is pH paper, which comes in small strips. Don gloves for this procedure so your skin secretions do not contaminate the paper and to protect you from the chemicals you are testing.

FIGURE 3.1 As the concentration of hydrogen ions (H^+) increases, a solution becomes more acidic and the pH value decreases. As the concentration of ions that combine with hydrogen ions (such as hydroxide ions) increases, a solution becomes more basic (alkaline) and the pH value increases. The pH values of some common substances are shown.

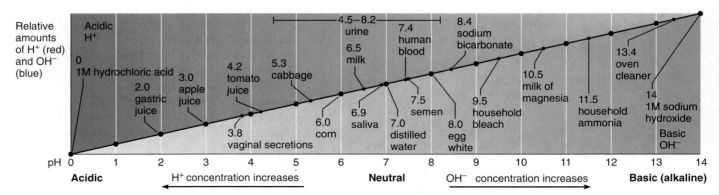

a. Test the pH of distilled water by dropping one or two drops of distilled water onto a strip of pH paper, and note the color. Compare the color to the color guide on the container. Record the pH value in Part A of the laboratory report.

b. Repeat this procedure for tap water. Record the results.

c. Individually test the various common substances found on your lab table. Record your results.

d. Complete Part A of the laboratory report.

Procedure B—Organic Molecules

In this section, you will perform some simple tests to check for the presence of the following categories of organic molecules (biomolecules): protein, sugar, starch, and lipid. Specific color changes will occur if the target compound is present. Please note the original color of the indicator being added so you can tell if the color really changed. For example, Biuret reagent and Benedict's solution are both initially blue, so an end color of blue would indicate no change. Use caution when working with these chemicals, and follow all directions. Carefully label the test tubes and avoid contaminating any of your samples. Only the Benedict's test requires heating and a time delay for accurate results. Do not heat any other tubes. To prepare for the Benedict's test, fill a 500 mL beaker half full with water and place it on the hot plate. Turn the hot plate on and bring the water to a boil. To save time, start the water bath before doing the Biuret test.

1. **Biuret test for protein.** In the presence of protein, Biuret reagent reacts with peptide bonds and changes to violet or purple. A pinkish color indicates that shorter polypeptides are present. The color intensity is proportional to the number of peptide bonds, thus the intensity reflects the length of polypeptides (amount of protein).

   a. Label six test tubes as follows: 1W, 1E, 1G, 1D, 1S, and 1P.

   b. To each test tube, add 2 mL of one of the samples as follows:
   1W—2 mL distilled water
   1E—2 mL egg albumin
   1G—2 mL 10% glucose solution
   1D—2 mL carbonated soft drink
   1S—2 mL 10% starch solution
   1P—2 mL potato juice

   c. To each, add 2 mL of Biuret reagent, swirl the tube to mix it, and note the final color. [*Note:* If Biuret reagent is not available, add 2 mL of 10% NaOH (sodium hydroxide) and about 10 drops of 1% $CuSO_4$ (copper sulfate)]. **Be careful—NaOH is very caustic.**

   d. Record your results in Part B of the laboratory report.

   e. Mark a "+" on any tubes with positive results and retain them for comparison while testing your unknown sample. Put remaining test tubes aside so they will not get confused with future trials.

2. **Benedict's test for sugar (monosaccharides).** In the presence of sugar, Benedict's solution changes from its initial blue to a green, yellow, orange, or reddish color, depending on the amount of sugar present. Orange and red indicate greater amounts of sugar.

   a. Label six test tubes as follows: 2W, 2E, 2G, 2D, 2S, and 2P.

   b. To each test tube, add 2 mL of one of the samples as follows:
   2W—2 mL distilled water
   2E—2 mL egg albumin
   2G—2 mL 10% glucose solution
   2D—2 mL carbonated soft drink
   2S—2 mL 10% starch solution
   2P—2 mL potato juice

   c. To each, add 2 mL of Benedict's solution, swirl the tube to mix it, and place all tubes into the boiling water bath for 3 to 5 minutes.

   d. Note the final color of each tube after heating and record the results in Part B of the laboratory report.

   e. Mark a "+" on any tubes with positive results and retain them for comparison while testing your unknown sample. Put remaining test tubes aside so they will not get confused with future trials.

3. **Iodine test for starch.** In the presence of starch, iodine turns a dark purple or blue-black color. Starch is a long chain formed by many glucose units linked together side by side. This regular organization traps the iodine molecules and produces the dark color.

   a. Label six test tubes as follows: 3W, 3E, 3G, 3D, 3S, and 3P.

   b. To each test tube, add 2 mL of one of the samples as follows:
   3W—2 mL distilled water
   3E—2 mL egg albumin
   3G—2 mL 10% glucose solution
   3D—2 mL carbonated soft drink
   3S—2 mL 10% starch solution
   3P—2 mL potato juice

   c. To each, add 0.5 mL of IKI (iodine solution) and swirl the tube to mix it.

   d. Record the final color of each tube in Part B of the laboratory report.

   e. Mark a "+" on any tubes with positive results and retain them for comparison while testing your unknown sample. Put remaining test tubes aside so they will not get confused with future trials.

4. **Tests for lipids.**

   a. Label two separate areas of a piece of brown paper as "water" or "oil."

   b. Place a drop of water on the area marked "water" and a drop of vegetable oil on the area marked "oil."

   c. Let the spots dry several minutes, then record your observations in Part C of the laboratory report. Upon drying, oil leaves a stain (grease spot) on brown paper; water does not leave such a spot.

d. In a test tube, add 2 mL of water and 2 mL of vegetable oil and observe. Shake the tube vigorously then let it sit for 5 minutes and observe again. Record your observations in Part B of the laboratory report.

e. Sudan IV is a dye that is lipid-soluble but not water-soluble. If lipids are present, the Sudan IV will stain them pink or red. Add a small amount of Sudan IV to the test tube that contains the oil and water. Swirl it then let it sit for a few minutes. Record your observations in Part B of the laboratory report.

Procedure C—Identifying Unknown Compounds

Now you will apply the information you gained with the tests in the previous section. You will retrieve an unknown sample that contains none, one, or any combination of the following types of organic compounds: protein, sugar, starch, or lipid. You will test your sample using each test from the previous section and record your results in Part C of the laboratory report.

1. Label four test tubes (1, 2, 3, 4).
2. Add 2 mL of your unknown sample to each test tube.
3. **Test for protein.** Add 2 mL of Biuret reagent to tube 1. Swirl the tube and observe the color. Record your observations in Part C of the laboratory report.
4. **Test for sugar.** Add 2 mL of Benedict's solution to tube 2. Swirl the tube and place it in a boiling water bath for 3 to 5 minutes. Record your observations in Part C of the laboratory report.
5. **Test for starch.** Add several drops of iodine solution to tube 3. Swirl the tube and observe the color. Record your observations in Part C of the laboratory report.
6. **Test for lipid.** Add 2 mL of water to tube 4. Swirl the tube and note if there is any separation.
7. Add a small amount of Sudan IV to test tube 4, swirl the tube and record your observations in Part C of the laboratory report.
8. Based on the results of these tests, determine if your unknown sample contains any organic compounds and, if so, what are they? Record and explain your identification in Part C of the laboratory report.

Name _____

Date _____

Section _____

The ⚠ corresponds to the Learning Outcome(s) listed at the beginning of the laboratory exercise.

Chemistry of Life

Part A—The pH Scale Assessments

1. Results from cabbage water test: ⚠

Substance	Cabbage Water	Distilled Water	Vinegar	Baking Soda
Color				
Acid, base, or neutral?				

2. Results from pH paper tests: ⚠

Substance Tested	Distilled Water	Tap Water	Sample 1: _____	Sample 2: _____	Sample 3: _____	Sample 4: _____	Sample 5: _____	Sample 6: _____	Sample 7: _____
pH Value									

3. Are the pH values the same for distilled water and tap water? _____

4. If not, what might explain this difference? _____

5. Draw the pH scale below, and indicate the following values: 0, 7 (neutral), and 14. Now label the scale with the names and indicate the location of the pH values for each substance you tested.

Part B—Testing for Organic Molecules Assessments

1. **Biuret test results for protein.** Enter your results from the Biuret test for protein. A color change to purple indicates that protein is present; pink indicates that short polypeptides are present. **2**

Tube	Contents	Color	Protein Present (+) or Absent (–)
1W	Distilled water		
1E	Egg albumin		
1G	Glucose solution		
1D	Soft drink		
1S	Starch solution		
1P	Potato juice		

2. **Benedict's test results for most sugars (monosaccharides).** Enter your results from the Benedict's test for sugars. A color change to green, yellow, orange, or red indicates that sugar is present. Note the color after the mixture has been heated for 3 to 5 minutes. **2**

Tube	Contents	Color	Sugar Present (+) or Absent (–)
2W	Distilled water		
2E	Egg albumin		
2G	Glucose solution		
2D	Soft drink		
2S	Starch solution		
2P	Potato juice		

3. **Iodine test results for starch.** Enter your results from the iodine test for starch. A color change to dark blue or black indicates that starch is present. **2**

Tube	Contents	Color	Sugar Present (+) or Absent (–)
3W	Distilled water		
3E	Egg albumin		
3G	Glucose solution		
3D	Soft drink		
3S	Starch solution		
3P	Potato juice		

4. **Lipid test results.** What did you observe when you allowed the drops of oil and water to dry on the brown paper? **2**

_____

What did you observe when you mixed the oil and water together? _____

_____

What did you observe when you added the Sudan IV dye? **2** _____

If a person were on a low-carbohydrate, high-protein diet, which of the six substances tested would the person want to increase? _____

Explain your answer.

_____

_____

_____

_____

Part C—Testing for Unknown Organic Compounds Assessments

1. What is the number of your sample? _____

2. Record the results of your tests on the unknown here: 🔺3

Test Performed	Results
Biuret test for protein	
Benedict's test for sugars	
Iodine test for starch	
Water test for lipid	
Sudan IV test for lipid	

3. Based on these results, what organic compound(s) does your unknown sample contain? 🔺3 _____

_____

4. Explain your answer. 🔺3 _____

_____

Care and Use of the Microscope

Pre-Lab

1. Carefully read the introductory material and examine the entire lab content.
2. Visit www.mhhe.com/martinseriesl for pre-lab questions and LabCam videos.

Materials Needed

Compound light microscope
Lens paper
Microscope slides
Coverslips
Transparent plastic millimeter ruler
Prepared slide of letter *e*
Slide of three colored threads
Medicine dropper
Dissecting needle (needle probe)
Specimen examples for wet mounts
Methylene blue (dilute) or iodine-potassium-iodide stain

For Demonstration Activities:
Micrometer scale
Stereomicroscope (dissecting microscope)

The human eye cannot perceive objects less than 0.1 mm in diameter so a microscope is an essential tool for the study of small structures such as cells. The microscope usually used for this purpose is the *compound light microscope*. It is called compound because it uses two sets of lenses: an eyepiece or ocular lens and an objective lens system. The eyepiece lens system magnifies, or compounds, the image reaching it after the image is magnified by the objective lens system. Such an instrument can magnify images of small objects up to about one thousand times.

Purpose of the Exercise

To become familiar with the major parts of a compound light microscope and their functions and to use the microscope to observe small objects.

Learning Outcomes

After completing this exercise, you should be able to

1. Differentiate the major functional parts of a compound light microscope.
2. Calculate the total magnification produced by various combinations of eyepiece and objective lenses.
3. Demonstrate proper use of the microscope to observe and measure small objects.
4. Prepare a simple microscope slide and sketch the objects you observed.

Procedure A—Microscope Basics

1. Familiarize yourself with the following list of rules for care of the microscope:
   a. Keep the microscope under its *dustcover* and in a cabinet when it is not being used.
   b. Handle the microscope with great care. It is an expensive and delicate instrument. To move it or carry it, hold it by its *arm* with one hand and support its *base* with the other hand (fig. 4.1).
   c. Always store the microscope with the scanning or lowest power objective in place. Always start with this objective when using the microscope.
   d. To clean the lenses, rub them gently with *lens paper* or a high-quality cotton swab. If the lenses need additional cleaning, follow the directions in the "Lens-Cleaning Technique" section that follows.
   e. If the microscope has a substage lamp, be sure the electric cord does not hang off the laboratory table where someone might trip over it. The bulb life can be extended if the lamp is cool before the microscope is moved.
   f. Never drag the microscope across the laboratory table after you have placed it.
   g. Never remove parts of the microscope or try to disassemble the eyepiece or objective lenses.
   h. If your microscope is not functioning properly, report the problem to your laboratory instructor immediately.
2. Observe a compound light microscope and study figure 4.1 to learn the names of its major parts. Your microscope could be equipped with a binocular body and two eyepieces (fig. 4.2). The lens system of a compound microscope includes three parts—the condenser, objective lens, and eyepiece or ocular.

FIGURE 4.1 Major parts of a compound light microscope with a monocular body and a mechanical stage. Some microscopes are equipped with a binocular body.

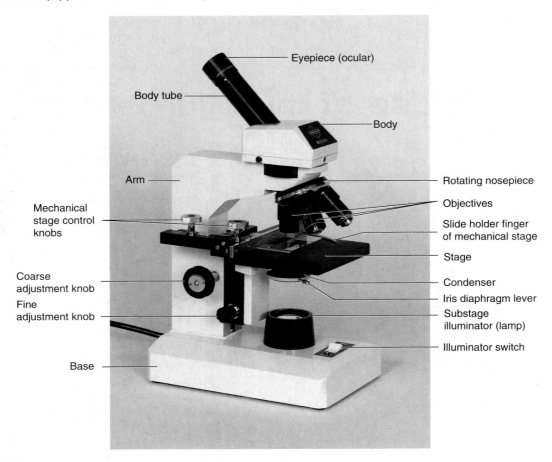

- Eyepiece (ocular)
- Body tube
- Body
- Arm
- Rotating nosepiece
- Objectives
- Mechanical stage control knobs
- Slide holder finger of mechanical stage
- Stage
- Coarse adjustment knob
- Condenser
- Iris diaphragm lever
- Fine adjustment knob
- Substage illuminator (lamp)
- Illuminator switch
- Base

FIGURE 4.2 Scientist using a compound light microscope equipped with a binocular body and two eyepieces.

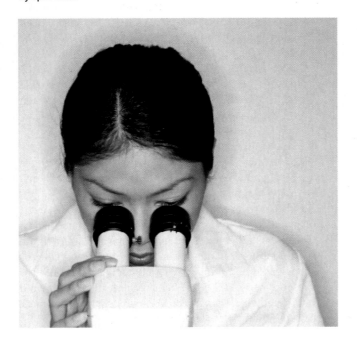

Lens-Cleaning Technique

1. Moisten one end of a high-quality cotton swab with one drop of Kodak lens cleaner. Keep the other end dry.
2. Clean the optical surface with the wet end. Dry it with the other end, using a circular motion.
3. Use a hand aspirator to remove lingering dust particles.
4. Start with the scanning objective and work upward in magnification, using a new cotton swab for each objective.
5. When cleaning the eyepiece, do not open the lens unless it is absolutely necessary.
6. Use alcohol for difficult cleaning and only as a last resort use xylene. Regular use of xylene will destroy lens coatings.

Light enters this system from a *substage illuminator (lamp)* or *mirror* and is concentrated and focused by a *condenser* onto a microscope slide or specimen placed on the *stage*. The condenser, which contains a set of lenses, usually is kept in its highest position possible.

The *iris diaphragm,* located between the light source and the condenser, can be used to increase or decrease the intensity of the light entering the condenser. Locate the lever that operates the iris diaphragm beneath the stage and move it back and forth. Note how this movement causes the size of the opening in the diaphragm to change. (Some microscopes have a revolving plate called a disc diaphragm beneath the stage instead of an iris diaphragm. Disc diaphragms have different-sized holes to admit varying amounts of light.) Which way do you move the diaphragm to increase the light intensity? _____ Which way to decrease it? _____

After light passes through a specimen mounted on a microscope slide, it enters an *objective lens system.* This lens projects the light upward into the *body tube,* where it produces a magnified image of the object being viewed.

The *eyepiece (ocular) lens* system then magnifies this image to produce another image seen by the eye. Typically, the eyepiece lens magnifies the image ten times (10×). Look for the number in the metal of the eyepiece that indicates its power (fig. 4.3). What is the eyepiece power of your microscope? _____

The objective lenses are mounted in a *rotating nosepiece* so that different magnifications can be achieved by rotating any one of several objective lenses into position above the specimen. Commonly, this set of lenses includes a scanning objective (4×), a low-power objective (10×), and a high-power objective, also called a high-dry-power objective (about 40×). Sometimes an oil immersion objective (about 100×) is present. Look for the number printed on each objective that indicates its power. What are the objective lens powers of your microscope? _____

To calculate the *total magnification* achieved when using a particular objective, multiply the power of the eyepiece by the power of the objective used. Thus, the 10× eyepiece and the 40× objective produce a total magnification of 10 × 40, or 400×. See table 4.1.

TABLE 4.1 Microscope Lenses

Objective Lens Name	Common Objective Lens Magnification	Common Eyepiece Lens Magnification	Total Magnification
Scan	4×	10×	40×
Low power (LP)	10×	10×	100×
High power (HP)	40×	10×	400×
Oil immersion	100×	10×	1,000×

Note: If you wish to observe an object under LP, HP, or oil immersion, locate and then center and focus the object first under scan magnification.

FIGURE 4.3 The powers of this 10× eyepiece (*a*) and this 40× objective (*b*) are marked in the metal. DIN is an international optical standard on quality optics. The 0.65 on the 40× objective is the numerical aperture, a measure of the light-gathering capabilities.

(a)

(b)

3. Complete Part A of Laboratory Report 4.
4. Turn on the substage illuminator and look through the eyepiece. You will see a lighted circular area called the *field of view.*

    You can measure the diameter of this field of view by focusing the lenses on the millimeter scale of a transparent plastic ruler. To do this, follow these steps:

   a. Place the ruler on the microscope stage in the spring clamp of a slide holder finger on a mechanical stage or under the stage (slide) clips. (*Note:* If your microscope is equipped with a mechanical stage, it may be necessary to use a short section cut from a transparent plastic ruler. The section should be several millimeters long and can be mounted on a microscope slide for viewing.)

   b. Center the millimeter scale in the beam of light coming up through the condenser, and rotate the scanning objective into position.

   c. While you watch from the side to prevent the lens from touching anything, raise the stage until the objective is as close to the ruler as possible, using the *coarse adjustment knob* and then using the *fine adjustment knob* (fig. 4.4). (*Note:* The adjustment

---

FIGURE 4.4 When you focus using a particular objective, you can prevent it from touching the specimen by watching from the side.

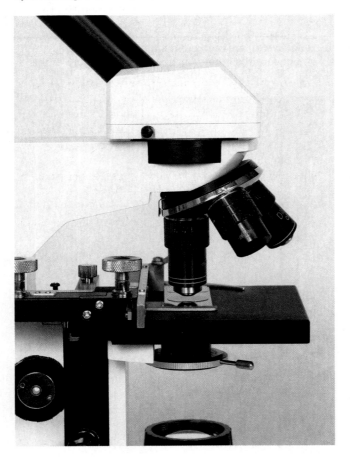

knobs on some microscopes move the body and objectives downward and upward for focusing.)

   d. Look into the eyepiece and use the coarse adjustment knob to raise the stage until the lines of the millimeter scale come into sharp focus.

   e. Adjust the light intensity by moving the *iris diaphragm lever* so that the field of view is brightly illuminated but comfortable to your eye. At the same time, take care not to over illuminate the field because transparent objects tend to disappear in bright light.

   f. Position the millimeter ruler so that its scale crosses the greatest diameter of the field of view. Also, move the ruler so that one of the millimeter marks is against the edge of the field of view.

   g. In millimeters, measure the distance across the field of view.

5. Complete Part B of the laboratory report.
6. Most microscopes are designed to be *parfocal.* This means that when a specimen is in focus with a lower-power objective, it will be in focus (or nearly so) when a higher-power objective is rotated into position. Always center the specimen in the field of view before changing to higher objectives.

    Rotate the low-power objective into position, and then look at the millimeter scale of the transparent plastic ruler. If you need to move the low-power objective to sharpen the focus, use the *fine adjustment knob.*

    Adjust the iris diaphragm so that the field of view is properly illuminated. Once again, adjust the millimeter ruler so that the scale crosses the field of view through its greatest diameter, and position the ruler so that a millimeter mark is against one edge of the field. Try to measure the distance across the field of view in millimeters.

7. Rotate the high-power objective into position, while you watch from the side, and then observe the millimeter scale on the plastic ruler. All focusing using high-power magnification should be done only with the fine adjustment knob. If you use the coarse adjustment knob with the high-power objective, you can accidently force the objective into the coverslip and break the slide. This is because the *working distance* (the distance from the objective lens to the slide on the stage) is much shorter when using higher magnifications.

    Adjust the iris diaphragm for proper illumination. Usually more illumination when using higher magnifications will help you to view the objects more clearly. Try to measure the distance across the field of view in millimeters.

8. Locate the numeral 4 (or 9) on the plastic ruler and focus on it using the scanning objective. Note how the number appears in the field of view. Move the plastic ruler to the right and note which way the image moves. Slide the ruler away from you and again note how the image moves.

9. Observe a letter *e* slide. Note the orientation of the letter *e* using the scan, LP, and HP objectives. As you

increase magnifications, note that the amount of the letter *e* shown is decreased. Move the slide to the left, and then away from you, and note the direction in which the observed image moves.

10. Examine the slide of the three colored threads using the low-power objective and then the high-power objective. Focus on the location where the three threads cross. By using the fine adjustment knob, determine the order from top to bottom by noting which color is in focus at different depths. The other colored threads will still be visible, but they will be blurred. Be sure to notice whether the stage or the body tube moves up and down with the adjustment knobs of the microscope being used for this depth determination. The vertical depth of the specimen clearly in focus is called the *depth of field (focus)*. Whenever specimens are examined, continue to use the fine adjustment focusing knob to determine relative depths of structures clearly in focus within cells, giving a three-dimensional perspective. The depth of field is less at higher magnifications.

Critical Thinking Activity

What was the sequence of the three colored threads from top to bottom? Explain how you came to that conclusion.

_____

_____

_____

11. Complete Parts C and D of the laboratory report.

Demonstration Activity

A compound light microscope is sometimes equipped with a micrometer scale mounted in the eyepiece. Such a scale is subdivided into fifty to one hundred equal divisions (fig. 4.5). These arbitrary divisions can be calibrated against the known divisions of a micrometer slide placed on the microscope stage. Once the values of the divisions are known, the length and width of a microscopic object can be measured by superimposing the scale over the magnified image of the object.

Observe the micrometer scale in the eyepiece of the demonstration microscope. Focus the low-power objective on the millimeter scale of a micrometer slide (or a plastic ruler) and measure the distance between the divisions on the micrometer scale in the eyepiece. What is the distance between the finest divisions of the scale in micrometers? _____

Procedure B—Slide Preparation

1. Prepare several temporary *wet mounts,* using any small, transparent objects of interest, and examine the specimens using the low-power objective and then a high-power objective to observe their details. To prepare a wet mount, follow these steps (fig. 4.6):
   a. Obtain a precleaned microscope slide.
   b. Place a tiny, thin piece of the specimen you want to observe in the center of the slide, and use a medicine dropper to put a drop of water over it. Consult with your instructor if a drop of stain might enhance the image of any cellular structures of your specimen. If the specimen is solid, you might want to tease some of it apart with dissecting needles. In any case, the specimen must be thin enough so that light can pass through it. Why is it necessary for the specimen to be so thin?

   _____

   _____

   c. Cover the specimen with a coverslip. Try to avoid trapping bubbles of air beneath the coverslip by slowly lowering it at an angle into the drop of water.
   d. Remove any excess water from the edge of the coverslip with absorbent paper. If your microscope has an inclination joint, do not tilt the microscope while observing wet mounts because the fluid will flow.
   e. Place the slide under the stage (slide) clips or in the slide holder on a mechanical stage, and position the slide so that the specimen is centered in the light beam passing up through the condenser.
   f. Focus on the specimen using the scanning objective first. Next focus using the low-power objective, and then examine it with the high-power objective.

2. If an oil immersion objective is available, use it to examine the specimen. To use the oil immersion objective, follow these steps:
   a. Center the object you want to study under the high-power field of view.
   b. Rotate the high-power objective away from the microscope slide, place a small drop of immersion oil on the coverslip, and swing the oil immersion objective into position. To achieve sharp focus, use the fine adjustment knob only.
   c. You will need to open the iris diaphragm more fully for proper illumination. More light is needed because the oil immersion objective covers a very small lighted area of the microscope slide.
   d. The oil immersion objective must be very close to the coverslip to achieve sharp focus, so care must be taken to avoid breaking the coverslip or damaging the objective lens. For this reason, never lower the objective when you are looking into the eyepiece. Instead, always raise the objective to achieve focus, or prevent the objective from touching the coverslip

FIGURE 4.5 The divisions of a micrometer scale in an eyepiece can be calibrated against the known divisions of a micrometer slide.

Source: Courtesy of Swift Instruments, Inc., San Jose, California

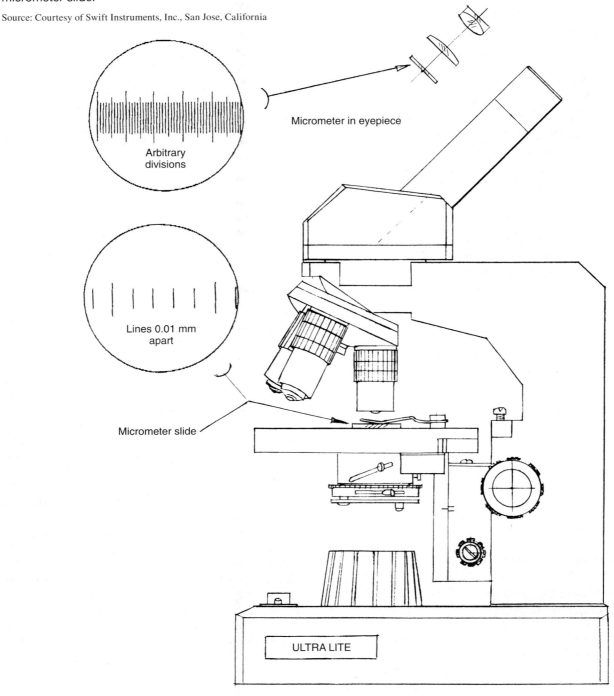

by watching the microscope slide and coverslip from the side if the objective needs to be lowered. Usually when using the oil immersion objective, only the fine adjustment knob needs to be used for focusing.

3. When you have finished working with the microscope, remove the microscope slide from the stage and wipe any oil from the objective lens with lens paper or a high-quality cotton swab. Swing the scanning objective or the low-power objective into position. Wrap the electric cord around the base of the microscope and replace the dustcover.

4. Complete Part E of the laboratory report.

FIGURE 4.6 Steps in the preparation of a wet mount.

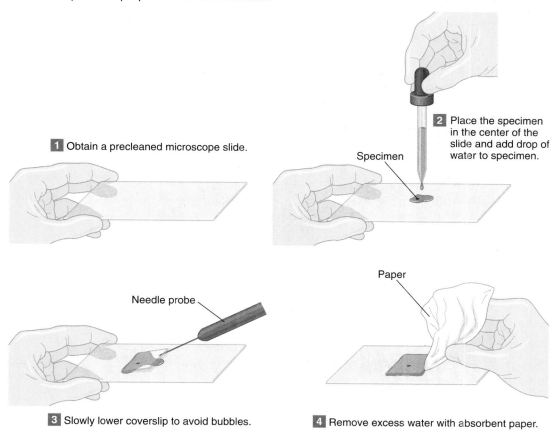

1 Obtain a precleaned microscope slide.

2 Place the specimen in the center of the slide and add drop of water to specimen.

Specimen

Needle probe

Paper

3 Slowly lower coverslip to avoid bubbles.

4 Remove excess water with absorbent paper.

Demonstration Activity

A stereomicroscope (dissecting microscope) (fig. 4.7) is useful for observing the details of relatively large, opaque specimens. Although this type of microscope achieves less magnification than a compound light microscope, it has the advantage of producing a three-dimensional image rather than the flat, two-dimensional image of the compound light microscope. In addition, the image produced by the stereomicroscope is positioned in the same manner as the specimen, rather than being reversed and inverted as it is by the compound light microscope.

Observe the stereomicroscope. The eyepieces can be pushed apart or together to fit the distance between your eyes. Focus the microscope on the end of your finger. Which way does the image move when you move your finger to the right? _____

When you move it away? _____

If the instrument has more than one objective, change the magnification to a higher power. Use the instrument to examine various small, opaque objects available in the laboratory.

FIGURE 4.7 A stereomicroscope, also called a dissecting microscope.

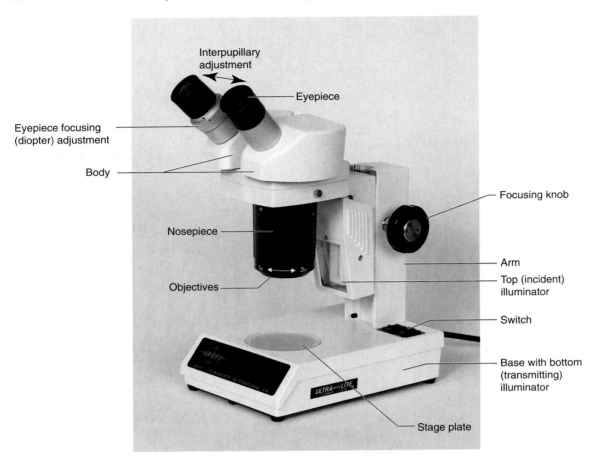

Interpupillary adjustment

Eyepiece

Eyepiece focusing (diopter) adjustment

Body

Nosepiece

Objectives

Focusing knob

Arm

Top (incident) illuminator

Switch

Base with bottom (transmitting) illuminator

Stage plate

Name _____

Date _____

Section _____

The ⬚ corresponds to the Learning Outcome(s) listed at the beginning of the laboratory exercise.

Care and Use of the Microscope

Part A Assessments

Revisit Procedure A, number 2, then complete the following: **2**

1. What total magnification will be achieved if the 10× eyepiece and the 10× objective are used? _____

2. What total magnification will be achieved if the 10× eyepiece and the 100× objective are used? _____

Part B Assessments

Revisit Procedure A, number 2, then complete the following:

1. Sketch the millimeter scale as it appears under the scanning objective magnification. (The circle represents the field of view through the microscope.)

2. In micrometers, what is the diameter of the scanning field of view? **3** _____

3. Microscopic objects often are measured in *micrometers*. A micrometer equals 1/1,000 of a millimeter and is symbolized by μm. In millimeters, what is the diameter of the scanning power field of view? **3** _____

4. If a circular object or specimen extends halfway across the scanning field, what is its diameter in millimeters? **3** _____

5. In micrometers, what is its diameter? **3** _____

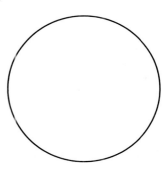

Part C Assessments

Complete the following:

1. Sketch the millimeter scale as it appears using the low-power objective.

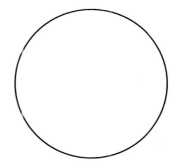

2. What do you estimate the diameter of this field of view to be in millimeters? **3**

   _____

3. How does the diameter of the scanning power field of view compare with that of the low-power field?

   _____

4. Why is it more difficult to measure the diameter of the high-power field of view than the low-power field?

   _____

   _____

5. What change occurred in the light intensity of the field of view when you exchanged the low-power objective for the high-power objective? _____

6. Sketch the numeral 4 (or 9) as it appears through the scanning objective of the compound microscope.

7. What has the lens system done to the image of the numeral? (Is it right side up, upside down, or what?) _____

8. When you moved the ruler to the right, which way did the image move? _____

9. When you moved the ruler away from you, which way did the image move? _____

Part D Assessments

Match the names of the microscope parts in column A with the descriptions in column B. Place the letter of your choice in the space provided. 1

Column A	Column B
a. Adjustment knob	_____ 1. Increases or decreases the light intensity
b. Arm	_____ 2. Platform that supports a microscope slide
c. Condenser	_____ 3. Concentrates light onto the specimen
d. Eyepiece (ocular)	_____ 4. Causes stage (or objective lens) to move upward or downward
e. Field of view	_____ 5. After light passes through the specimen, it next enters this lens system
f. Iris diaphragm	
g. Nosepiece	_____ 6. Holds a microscope slide in position
h. Objective lens system	_____ 7. Contains a lens at the top of the body tube
i. Stage	_____ 8. Serves as a handle for carrying the microscope
j. Stage (slide) clip	_____ 9. Part to which the objective lenses are attached
	_____ 10. Circular area seen through the eyepiece

Part E Assessments

Prepare sketches of the objects you observed using the microscope. For each sketch, include the name of the object, the magnification you used to observe it, and its estimated dimensions in millimeters and micrometers. 4

Cell Structure and Function

Pre-Lab

1. Carefully read the introductory material and examine the entire lab content.
2. Be familiar with the basic structures and functions of a cell (from lecture or the textbook).
3. Visit www.mhhe.com/martinseries1 for pre-lab questions.

Materials Needed

Animal cell model
Clean microscope slides
Coverslips
Flat toothpicks
Medicine dropper
Methylene blue (dilute) or iodine-potassium-iodide stain
Prepared microscope slides of human tissues
Compound light microscope
Ph.I.L.S. 3.0

For Learning Extension Activities:
Single-edged razor blade
Plant materials such as leaves, soft stems, fruits, onion peel, and vegetables
Cultures of *Amoeba* and *Paramecium*

Safety

▶ Review all the safety guidelines inside the front cover.
▶ Clean laboratory surfaces before and after laboratory procedures.
▶ Wear disposable gloves for the wet-mount procedures of the cells lining the inside of the cheek.
▶ Work only with your own materials when preparing the slide of the cheek cells. Observe the same precautions as with all body fluids.
▶ Dispose of laboratory gloves, slides, coverslips, and toothpicks as instructed.
▶ Precautions should be taken to prevent cellular stains from contacting your clothes and skin.
▶ Wash your hands before leaving the laboratory.

Cells are the "building blocks" from which all parts of the human body are formed. They account for the shape, organization, and construction of the body and are responsible for carrying on its life processes. Using a compound light microscope and the proper stain, one can easily see the **plasma (cell) membrane,** the **cytoplasm,** and the **nucleus.** The cytoplasm is composed of a clear fluid, the *cytosol,* and numerous *cytoplasmic organelles* that are suspended in the cytosol.

The plasma membrane, composed of lipids, carbohydrates, and proteins, represents the cell boundary and functions in various methods of membrane transport. The nucleus is surrounded by a double-layered nuclear envelope, and many of the cytoplasmic organelles have membrane boundaries similar to the plasma membrane. Movements of substances across these membranes can be by passive processes, involving kinetic energy or hydrostatic pressure, or active processes, using the cellular energy of ATP.

The nucleus contains fine strands of DNA and a protein called chromatin, which condenses to form chromosomes during cell divisions. The nucleolus of the nucleus is composed of RNA and protein. Cytoplasmic organelles provide for specialized metabolic functions. The endoplasmic reticulum (ER) has numerous canals that serve in transporting molecules throughout the cytoplasm. Ribosomes synthesize proteins and are located free in the cytoplasm or on the surface of the endoplasmic reticulum (rough ER). The Golgi apparatus, composed of flattened membranous sacs, is the site for packaging glycoproteins for secretion. Mitochondria provide the main location for the cellular energy production of ATP. Lysosomes contain intracellular digestive enzymes for destroying debris and worn-out organelles. Vesicles contain substances produced by the cell or form from pinching off pieces of the plasma membrane. A cytoskeleton contains microtubules, intermediate filaments, and microfilaments that support cellular structures and are involved in cellular movements.

Is the metabolic rate of cells the same in a man who is six feet tall as in one who is five foot nine? Surprisingly, an understanding of temperature regulation is required to answer this question. *Temperature regulation* is accomplished through a balance of heat production and heat loss. Heat is produced by cells through metabolism (second law of thermodynamics: For every chemical reaction, some energy is always lost as heat). The greater the number of cells or the more metabolically active the cells, the more heat that is produced. Heat is lost through the surface of the skin. The larger the surface area of the skin, the more that is lost. For a given volume of cells, the metabolic rate of cells is established and maintained to offset the heat that is lost by the surface of the skin (within limits). The more heat that is lost at the skin, the more metabolically active the cells must be to produce the heat required to keep the body warm.

Purpose of the Exercise

To review the structure and functions of major cellular components and to observe examples of human cells. To measure and compare the average cell's metabolic rate in individuals of different sizes (weight).

Learning Outcomes

After completing this exercise, you should be able to

1. Name and locate the components of a cell.
2. Differentiate the functions of cellular components.
3. Prepare a wet mount of cells lining the inside of the cheek; stain the cells; and identify the plasma (cell) membrane, nucleus, and cytoplasm of these cells.

FIGURE 5.1 Label the structures of this composite cell, using the terms provided. The structures are not drawn to scale. 1

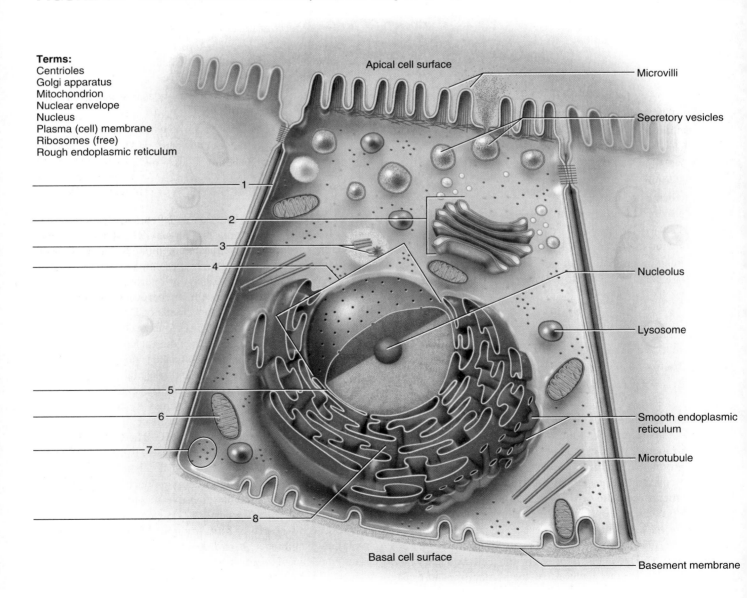

Terms:
Centrioles
Golgi apparatus
Mitochondrion
Nuclear envelope
Nucleus
Plasma (cell) membrane
Ribosomes (free)
Rough endoplasmic reticulum

Apical cell surface
Microvilli
Secretory vesicles
Nucleolus
Lysosome
Smooth endoplasmic reticulum
Microtubule
Basement membrane
Basal cell surface

FIGURE 5.2 Structure of the plasma (cell) membrane.

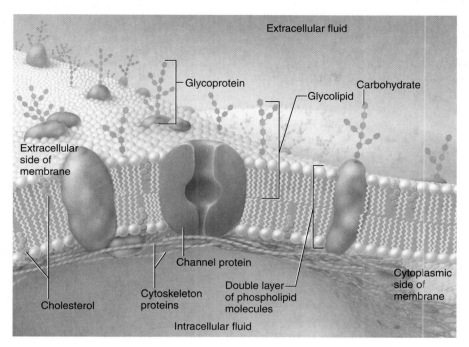

4 Examine cells on prepared slides of human tissues and identify their major components.

5 Explain the relationship of oxygen consumption with an animal's total weight.

6 Explain the relationship of oxygen consumption per gram of weight.

7 Explain the relationship of oxygen consumption and metabolic rate.

8 Calculate surface area–to–volume ratio and explain its significance to temperature regulation.

9 Integrate the concepts of weight, oxygen consumption, and metabolism and explain their application to the human body.

Procedure A—Cell Structure and Function

1. Observe the animal cell model and identify its major structures.
2. Label figure 5.1 and study figure 5.2.
3. Complete Part A of Laboratory Report 5.
4. Prepare a wet mount of cells lining the inside of the cheek. To do this, follow these steps:
   a. Gently scrape (force is not necessary and should be avoided) the inner lining of your cheek with the broad end of a flat toothpick.
   b. Stir the toothpick in a drop of water on a clean microscope slide and dispose of the toothpick as directed by your instructor.
   c. Cover the drop with a coverslip.
   d. Observe the cheek cells by using the microscope. Compare your image with figure 5.3. To report what

you observe, sketch a single cell in the space provided in Part B of the laboratory report.

5. Prepare a second wet mount of cheek cells, but this time, add a drop of dilute methylene blue or iodine-potassium-iodide stain to the cells. Cover the liquid with a coverslip and observe the cells with the microscope. Add to your sketch any additional structures you observe in the stained cells.

6. Answer the questions in Part B of the laboratory report.

7. Using the microscope, observe each of the prepared slides of human tissues. To report what you observe, sketch a single cell of each type in the space provided in Part C of the laboratory report.

8. Complete Parts C and D of the laboratory report.

FIGURE 5.3 Stained cell lining the inside of the cheek as viewed through the compound light microscope using the high-power objective (400×).

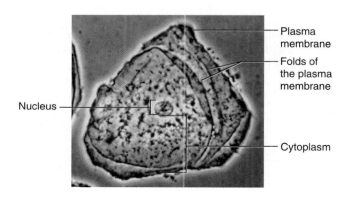

Critical Thinking Activity

The cells lining the inside of the cheek are frequently removed for making observations of basic cell structure. The cells are from stratified squamous epithelium. Explain why these cells are used instead of outer body surface tissue.

_____

_____

_____

Learning Extension Activity

Investigate the microscopic structure of various plant materials. To do this, prepare tiny, thin slices of plant specimens, using a single-edged razor blade. *(Take care not to injure yourself with the blade.)* Keep the slices in a container of water until you are ready to observe them. To observe a specimen, place it into a drop of water on a clean microscope slide and cover it with a coverslip. Use the microscope and view the specimen using low- and high-power magnifications. Observe near the edges where your section of tissue is most likely to be one cell thick. Add a drop of dilute methylene blue or iodine-potassium-iodide stain, and note if any additional structures become visible. How are the microscopic structures of the plant specimens similar to the human tissues you observed? _____

_____

How are they different? _____

_____

Learning Extension Activity

Prepare a wet mount of the *Amoeba* and *Paramecium* by putting a drop of culture on a clean glass slide. Gently cover with a clean coverslip. Observe the movements of the *Amoeba* with pseudopodia and the *Paramecium* with cilia. Try to locate cellular components such as the plasma (cell) membrane, nuclear envelope, nucleus, mitochondria, and contractile vacuoles. Describe the movement of the *Amoeba*.

_____

_____

Describe the movement of the *Paramecium*.

_____

_____

Procedure B—Ph.I.L.S. Lesson 2 Metabolism: Size and Basal Metabolic Rate

Hypothesis

Would you predict a larger animal will consume more total oxygen than a smaller animal? _____

Would you predict a larger animal will consume more oxygen *per gram of body weight* than a smaller animal? _____

Would you predict a larger animal will have a higher or lower metabolic rate than a smaller animal? _____

1. Open Exercise 2: Metabolism: Size and Basal Metabolic Rate.
2. Read the objectives and introduction and take the pre-lab quiz.
3. After completing the pre-lab quiz, read through the wet lab. Be sure to click open and view the videos that are indicated in red.
4. The lab exercise will open when you click Continue after completing the wet lab (fig. 5.4).

Weighing the Mouse

5. Click the power switch to turn on the scale.
6. Click tare to set the scale to zero.
7. Weigh one of the white mice by clicking on one mouse and dragging it to the scale.

Setting up the Chamber

8. After weighing the mouse, place it in the chamber by clicking and dragging on the mouse. (To "click and drag," move the mouse to position the arrow on the mouse; left click on mouse and hold; drag the mouse to the chamber; and when positioned release the left click).
9. Place bubbles on the end of the calibrated tube by clicking on the pipette and dragging the pipette to the calibration tube. Be sure to **touch** *pipette to the end of the calibration tube.* If a bubble does not appear try again.

Measuring the Bubbles

10. Measure the initial position of the bubbles (10 mm) and record at 0:00 time in the data table. Notice that the data table includes a change in time every 15 seconds between 0:00 and 2:00 minutes.
11. After clicking the Start button, measure the position of the soap bubble within the calibration tube at 15-second intervals (you will hear a beep) and record in the data table. (The timer will begin when you hit Start). You may find it more manageable to click Pause after each 15-second interval, measure, record, and then click on Start to begin again.
12. When finished click Pause.
13. Click "calc" to see graph (linear regression). Journal will open.

FIGURE 5.4 Opening Screen for the Laboratory Exercise on Metabolism: Size and Basal Metabolic Rate.

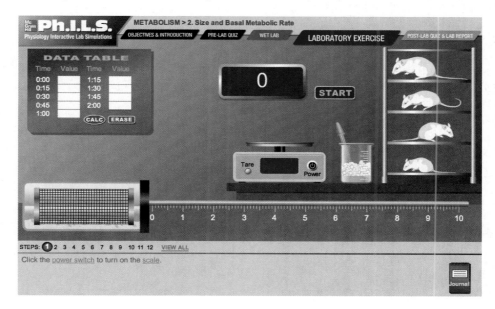

Examine the linear regression graph that relates time and oxygen consumption. Does the amount of oxygen consumed for this mouse increase or decrease over time (i.e., Does the amount of oxygen consumed for this mouse increase or decrease the longer the mouse is in the chamber)?_____ Why? _____
_____
_____

Repeating the Experiment

14. To complete with the same animal repeat steps 9–13.
15. To complete with a different animal, place the animal back in its cage by clicking and dragging the computer mouse. Repeat steps 5–13.

16. You can print your graphs by hitting Control (Ctrl) on your keyboard while pushing p.

Interpreting Results

17. *With the graph still on the screen,* answer the questions in Part E of the laboratory report in the laboratory manual. If you accidentally close the graph, click on the journal panel (red rectangle at bottom right of screen).
18. Click on post-lab quiz and lab report.
19. Complete the post-lab quiz by answering the ten questions on the computer screen.
20. Read the conclusion on the computer screen.
21. You may print the lab report for Metabolism: Size and Basal Metabolic Rate.

Name _____

Date _____

Section _____

The A corresponds to the Learning Outcome(s) listed at the beginning of the laboratory exercise.

Cell Structure and Function

Part A Assessments

Match the cellular components in column A with the descriptions in column B. Place the letter of your choice in the space provided. 2

Column A	Column B
a. Chromatin	_____ 1. Loosely coiled fibers containing protein and DNA within nucleus
b. Cytoplasm	_____ 2. Location of ATP production from digested food molecules
c. Endoplasmic reticulum	_____ 3. Small RNA-containing particles for the synthesis of proteins
d. Golgi apparatus (complex)	_____ 4. Membranous sac formed by the pinching off of pieces of cell membrane
e. Lysosome	
f. Microtubule	_____ 5. Dense body of RNA within the nucleus
g. Mitochondrion	_____ 6. Slender tubes that provide movement in cilia and flagella
h. Nuclear envelope	_____ 7. Composed of membrane-bound canals for tubular transport throughout the cytoplasm
i. Nucleolus	
j. Nucleus	_____ 8. Occupies space between plasma membrane and nucleus
k. Ribosome	_____ 9. Flattened membranous sacs that package a secretion
l. Vesicle	_____ 10. Membranous sac that contains digestive enzymes
	_____ 11. Separates nuclear contents from cytoplasm
	_____ 12. Spherical organelle that contains chromatin and nucleolus

Part B Assessments

Complete the following:

1. Sketch a single cheek cell. Label the cellular components you recognize. Add any additional structures observed to your sketch after staining was completed. (The circle represents the field of view through the microscope.) 3

Magnification _____ ×

2. After comparing the wet mount and the stained cheek cells, describe the advantage gained by staining cells.

_____

_____

3. Are the stained cheek cells nearly the same size and shape? _____ Propose an explanation for your answer.

_____

_____

Part C Assessments

Complete the following:

1. Sketch a single cell of each type you observed in the prepared slides of human tissues. Name the tissue, indicate the magnification used, and label the cellular components you recognize. ◢4

_____ ×
Tissue _____

_____ ×
Tissue _____

2. What do the various types of cells in these tissues have in common? _____

_____

3. What are the main differences you observed among these cells? _____

_____

_____

Part D Assessments

Electron micrographs represent extremely thin slices of cells. Each micrograph in figure 5.5 contains a section of a nucleus and some cytoplasm. Compare the organelles shown in these micrographs with organelles of the animal cell model and figure 5.1.

Identify the structures indicated by the arrows in figure 5.5. ◢1

1. _____   **6.** _____

2. _____   **7.** _____

3. _____   **8.** _____

4. _____   **9.** _____

5. _____   **10.** _____

46

FIGURE 5.5 Transmission electron micrographs of cellular components. The views are only portions of a cell. Magnifications: (a) 26,000×; (b) 10,000×. Identify the numbered cellular structures, using the terms provided. 🔎

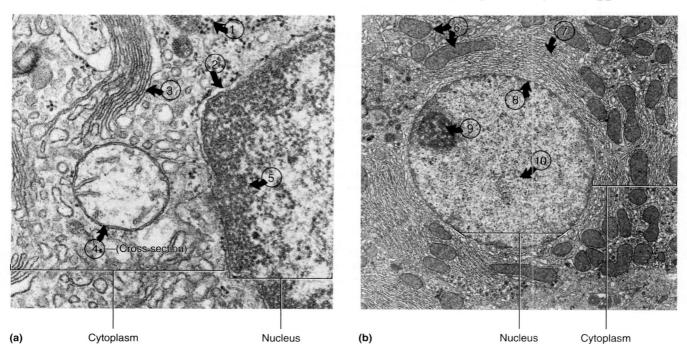

(a)   Cytoplasm   Nucleus   (b)   Nucleus   Cytoplasm

Terms:
Chromatin (use 2 times)
Endoplasmic reticulum
Golgi apparatus
Mitochondria
Mitochondrion (cross section)
Nuclear envelope (use 2 times)
Nucleolus
Ribosomes

Answer the following questions after observing the transmission electron micrographs in figure 5.5.

11. What cellular structures were visible in the transmission electron micrographs that were not apparent in the cells you observed using the microscope? _____

_____

_____

12. Before they can be observed by using a transmission electron microscope, cells are sliced into very thin sections. What disadvantage does this procedure present in the study of cellular parts? _____

_____

_____

Part E Ph.I.L.S. Lesson 2, Metabolism: Size and Basal Metabolic Rate Assessments

Interpreting Results

1. Examine the graph on the left of the computer screen. On the x-axis is weight (grams) and on the y-axis is average oxygen consumed per minute (av 0_2/m). As the weight of the mouse decreases, does the average oxygen consumed per minute decrease or increase? _____ Is the relationship between weight and oxygen consumed direct or inverse? _____ **5**

2. Examine the graph on the right. Weight (grams) is on the x-axis and average oxygen consumed per hour *per gram (of body weight)* (av 0_2/h/g) is on the y-axis. As the weight of the mouse decreases, does the average oxygen consumed per hour per gram decrease or increase? _____ Is the relationship between weight and oxygen consumed per gram of body weight direct or inverse? _____ **6**

3. Based on the results of the graph on the right *and* the fact that oxygen consumed is an indirect measure of metabolic rate, as the weight of the mouse decreases, does the metabolic rate decrease or increase? _____ Is the relationship between weight and metabolic rate direct or inverse? _____ **7**

Understanding the Results

4. **Surface Area to Volume Ratio:** As a (spherical) structure increases in radius, the surface area increases at a rate of r^2 and the volume increases at a rate of r^3. If the radius of the sphere is 2, the calculated surface area $2^2 = 2 \times 2 = 4$ and the calculated volume $2^3 = 2 \times 2 \times 2 = 8$ and the ratio of surface area/volume is $4/8 = 0.5$. If r is 4, calculate the surface area, volume, and surface area/volume ratio. Place your answers in the following box. **8**

$r = 4$ Surface area = _____   Volume = _____ Surface area/volume ratio = _____

5. As the (spherical) structure becomes larger, is the *ratio* becoming larger or smaller? (Hint: Compare the surface area/volume ratio between two spheres—one with a radius of 2 and the other with a radius of 4.) _____ Is the volume increasing at a faster rate than the surface area? _____ **8**

Relating to the Body

Use this information to answer question 6. If the surface area–to–volume ratio is applied to the human body, volume is the size (weight) of the body and surface area is the area of skin surface. **8** **9**

6. As a person increases in size:

   a. What happens to the ratio of surface area/volume? _____

   b. Is volume (size) increasing at a faster rate than surface area (skin)? _____

   c. Is this a more favorable ratio for keeping the body warm? _____

   d. Will each individual cell have to be as metabolically active as in a larger individual to keep the body warm? _____

   e. Is the metabolic rate of individual cells faster or slower? _____

Critical Thinking Activity

Walt, a six-foot man, walks with his friend Albert, who is five–foot–nine.

a. Would you predict that Walt, with the larger body, is consuming more oxygen than Albert? _____

b. Would you predict that Walt is consuming more oxygen *per gram of weight* than Albert? _____

c. Which man would have the higher *metabolic rate per gram of weight?* _____

Movements Through Membranes

 ## Pre-Lab

1. Carefully read the introductory material and examine the entire lab content.
2. Be familiar with diffusion, osmosis, tonicity, and filtration (from lecture or the textbook).
3. Visit www.mhhe.com/martinseries1 for pre-lab questions, Anatomy & Physiology Revealed animations, and LabCam videos.

 ## Materials Needed

For Procedure A—Diffusion:
Petri dish
White paper
Forceps
Potassium permanganate crystals
Millimeter ruler (thin and transparent)

For Procedure B—Osmosis:
Thistle tube
Molasses (or Karo dark corn syrup)
Selectively permeable (semipermeable) membrane (pre-soaked dialysis tubing of 1 5/16" or greater diameter)
Ring stand and clamp
Beaker
Rubber band
Millimeter ruler

For Procedure C—Hypertonic, Hypotonic, and Isotonic Solutions:
Test tubes
Marking pen
Test-tube rack
10 mL graduated cylinder
Medicine dropper
Uncoagulated animal blood
Distilled water
0.9% NaCl (aqueous solution)
3.0% NaCl (aqueous solution)

Clean microscope slides
Coverslips
Compound light microscope

For Procedure D—Filtration:
Glass funnel
Filter paper
Ring stand and ring
Beaker
Powdered charcoal or ground black pepper
1% glucose (aqueous solution)
1% starch (aqueous solution)
Test tubes
10 mL graduated cylinder
Water bath (boiling water)
Benedict's solution
Iodine-potassium-iodide solution
Medicine dropper

For Procedure E—Ph.I.L.S. Lesson 1 Osmosis and Diffusion: Varying Extracellular Concentrations
Ph.I.L.S. 3.0

For Alternative Osmosis Activity:
Fresh chicken egg
Beaker
Laboratory balance
Spoon
Vinegar
Corn syrup (Karo)

 ## Safety

▶ Clean laboratory surfaces before and after laboratory procedures.
▶ Wear disposable gloves when handling chemicals and animal blood.
▶ Wear safety glasses when using chemicals.
▶ Dispose of laboratory gloves and blood-contaminated items as instructed.
▶ Wash your hands before leaving the laboratory.

A plasma membrane functions as a gateway through which chemical substances and small particles may enter or leave a cell. These substances move through the membrane by physical processes such as simple diffusion, osmosis, and filtration, or by physiological processes such as active transport, phagocytosis, or pinocytosis.

This laboratory exercise contains examples of passive membrane transport through either living plasma membranes or artificial membranes. Simple diffusion is the random motion of molecules from an area of higher concentration toward an area of lower concentration. The rate of diffusion depends upon the concentration gradient, the molecular size, and the temperature. Facilitated diffusion is a carrier-mediated transport, which is not represented in this laboratory exercise.

A special case of diffusion, called osmosis, occurs when water molecules diffuse through a selectively permeable membrane from an area of higher water concentration toward an area of lower water concentration. Tonicity refers to the osmotic pressure of a solution in relation to a cell. A solution is isotonic if the osmotic pressure is the same as the cell. A 5% glucose solution and a 0.9% normal saline solution serve as isotonic solutions to human cells. A hypertonic solution has a higher concentration of solutes than the cell; a hypotonic solution has a lower concentration of solutes than a cell. The osmosis of water will occur out of a cell when immersed in a hypertonic solution, and the cell will shrink (crenate). Conversely, the osmosis of water will occur into a cell when immersed in a hypotonic solution, and the cell will swell, and may burst (lyse).

Filtration is the movement of water and solutes through a selectively permeable membrane by hydrostatic pressure upon the membrane. The pressure gradient forces the water and solutes (filtrate) from the higher hydrostatic pressure area to a lower hydrostatic pressure area. For example, blood pressure provides the hydrostatic pressure for water and dissolved substances to move through capillary walls.

Purpose of the Exercise

To demonstrate some of the physical processes by which substances move through membranes.

Learning Outcomes

After completing this exercise, you should be able to

1. Show *simple diffusion* and identify examples of diffusion.
2. Interpret diffusion by preparing and explaining a graph.
3. Show *osmosis* and identify examples of osmosis.
4. Distinguish among hypertonic, hypotonic, and isotonic solutions and examine the effects of these solutions on animal cells.
5. Show *filtration* and identify examples of filtration.
6. Identify the percent transmittance of light through a blood sample placed in varying concentrations of NaCl.
7. Interpret the results to explain the relationship of percent transmittance and condition of the red blood cells (normal, shriveled or ruptured) in the blood sample.

8. Interpret the results to explain the condition of the red blood cells (normal, shriveled, or ruptured) to the relative concentration of NaCl solutions (isotonic, hypertonic, or hypotonic).
9. Relate plasma concentration and red blood cell integrity with events that occur in the human body.

Procedure A—Diffusion

1. To demonstrate simple diffusion, follow these steps:
   a. Place a petri dish, half filled with water, on a piece of white paper that has a millimeter ruler positioned on the paper. Wait until the water surface is still. Allow approximately 3 minutes.
   Note: The petri dish should remain level. A second millimeter ruler may be needed under the petri dish as a shim to obtain a level amount of the water inside the petri dish.
   b. Using forceps, place one crystal of potassium permanganate near the center of the petri dish and near the millimeter ruler (fig. 6.1).
   c. Measure the radius of the purple circle at 1-minute intervals for 10 minutes and record the results in Part A of Laboratory Report 6.
2. Complete Part A of the report.

Learning Extension Activity

Repeat the demonstration of diffusion using a petri dish filled with ice-cold water and a second dish filled with very hot water. At the same moment, add a crystal of potassium permanganate to each dish and observe the circle as before. What difference do you note in the rate of diffusion in the two dishes? How do you explain this difference? _____

_____

_____

Procedure B—Osmosis

1. To demonstrate osmosis, refer to figure 6.2 as you follow these steps:
   a. One person plugs the tube end of a thistle tube with a finger.
   b. Another person then fills the bulb with molasses until it is about to overflow at the top of the bulb. Air remains trapped in the stem.
   c. Cover the bulb opening with a single-thickness piece of moist selectively permeable (semipermeable) membrane. Dialysis tubing that has been soaked for 30 minutes can easily be cut open because it becomes pliable.

FIGURE 6.1 To demonstrate diffusion, place one crystal of potassium permanganate in the center of a petri dish containing water. Place the crystal near the millimeter ruler (positioned under the petri dish).

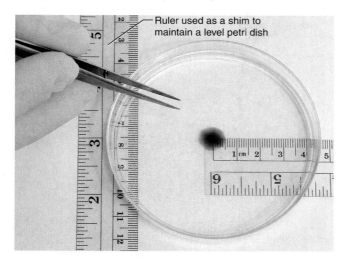

Ruler used as a shim to maintain a level petri dish

d. Tightly secure the membrane in place with several wrappings of a rubber band.
e. Immerse the bulb end of the tube in a beaker of water. If leaks are noted, repeat the procedures.
f. Support the upright portion of the tube with a clamp on a ring stand. Folded paper under the clamp will protect the thistle tube stem from breakage.
g. Mark the meniscus level of the molasses in the tube. *Note:* The best results will occur if the mark of the

molasses is a short distance up the stem of the thistle tube when the experiment starts.
h. Measure the level changes after 10 minutes and 30 minutes and record the results in Part B of the laboratory report.
2. Complete Part B of the laboratory report.

Alternative Activity

Eggshell membranes possess selectively permeable properties. To demonstrate osmosis using a natural membrane, soak a fresh chicken egg in vinegar for about 24 hours to remove the shell. Use a spoon to carefully handle the delicate egg. Place the egg in a hypertonic solution (corn syrup) for about 24 hours. Remove the egg, rinse it, and using a laboratory balance, weigh the egg to establish a baseline weight. Place the egg in a hypotonic solution (distilled water). Remove the egg and weigh it every 15 minutes for an elapsed time of 75 minutes. Explain any weight changes that were noted during this experiment.

_____ _____
_____ _____
_____ _____
_____ _____
_____ _____
_____ _____
_____ _____

FIGURE 6.2 (a) Fill the bulb of the thistle tube with molasses; (b) tightly secure a piece of selectively permeable (semipermeable) membrane over the bulb opening; and (c) immerse the bulb in a beaker of water. *Note:* These procedures require the participation of two people.

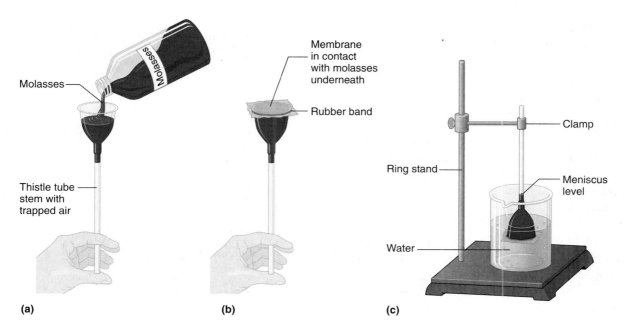

Molasses

Thistle tube stem with trapped air

Membrane in contact with molasses underneath

Rubber band

Clamp

Ring stand

Meniscus level

Water

(a)   (b)   (c)

Procedure C—Hypertonic, Hypotonic, and Isotonic Solutions

1. To demonstrate the effect of hypertonic, hypotonic, and isotonic solutions on animal cells, follow these steps:
   a. Place three test tubes in a rack and mark them *tube 1, tube 2,* and *tube 3.* (*Note:* One set of tubes can be used to supply test samples for the entire class.)
   b. Using 10 mL graduated cylinders, add 3 mL of distilled water to tube 1; add 3 mL of 0.9% NaCl to tube 2; and add 3 mL of 3.0% NaCl to tube 3.
   c. Place three drops of fresh, uncoagulated animal blood into each of the tubes, and gently mix the blood with the solutions. Wait 5 minutes.
   d. Using three separate medicine droppers, remove a drop from each tube and place the drops on three separate microscope slides marked *1, 2,* and *3.*
   e. Cover the drops with coverslips and observe the blood cells, using the high power of the microscope.
2. Complete Part C of the laboratory report.

Alternative Activity

Various substitutes for blood can be used for Procedure C. Onion, cucumber, or cells lining the inside of the cheek represent three possible options.

Procedure D—Filtration

1. To demonstrate filtration, follow these steps:
   a. Place a glass funnel in the ring of a ring stand over an empty beaker. Fold a piece of filter paper in half and then in half again. Open one thickness of the filter paper to form a cone. Wet the cone, and place it in the funnel. The filter paper is used to demonstrate how movement across membranes is limited by the size of the molecules, but it does not represent a working model of biological membranes.
   b. Prepare a mixture of 5 cc (approximately 1 teaspoon) powdered charcoal (or ground black pepper) and equal amounts of 1% glucose solution and 1% starch solution in a beaker. Pour some of the mixture into the funnel until it nearly reaches the top of the filter-paper cone. Care should be taken to prevent the mixture from spilling over the top of the filter paper. Collect the filtrate in the beaker below the funnel (fig. 6.3).
   c. Test some of the filtrate in the beaker for the presence of glucose. To do this, place 1 mL of filtrate in a clean test tube and add 1 mL of Benedict's solution. Place the test tube in a water bath of boiling water for 2 minutes and then allow the liquid to cool slowly. If the color of the solution changes to green, yellow, or red, glucose is present (fig. 6.4).

FIGURE 6.3 Apparatus used to illustrate filtration.

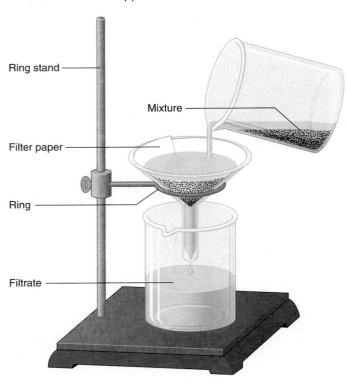

Ring stand

Mixture

Filter paper

Ring

Filtrate

FIGURE 6.4 Heat the filtrate and Benedict's solution in a boiling water bath for 2 minutes.

Water bath

Filtrate and Benedict's solution

OFF

Hot plate

   d. Test some of the filtrate in the beaker for the presence of starch. To do this, place a few drops of filtrate in a test tube and add one drop of iodine-potassium-iodide solution. If the color of the solution changes to blue-black, starch is present.
   e. Observe any charcoal in the filtrate.
2. Complete Part D of the laboratory report.

Procedure E—Ph.I.L.S. Lesson 1
Osmosis and Diffusion: Varying Extracellular Concentrations

Hypothesis

If red blood cells in a blood sample are placed in a HYPER-TONIC solution, water will move _____ (into or out of) the red blood cells and the cells might _____ (rupture open [hemolysis] or shrivel [crenate]). Would you predict that the sample of blood would transmit more or less light? _____ (Transmit = allowing more light through the sample; less light absorbed).

If red blood cells in a blood sample are placed in a HYPOTONIC solution, water will move _____ (into or out of) the red blood cells and the cells might _____ (rupture open [hemolysis] or shrivel [crenate]). Would you predict that the sample of blood would transmit more or less light? _____

1. Open Exercise 1: Osmosis and Diffusion: Varying Extracellular Concentration.
2. Read the objectives and introduction and take the pre-lab quiz.
3. After completing the pre-lab quiz, read through the wet lab. Be sure to click open and view the videos that are indicated in red in the wet lab.
4. The lab exercise will open when you click Continue after completing the wet lab (fig. 6.5).

Setup

5. Click the power switch to turn on the spectrophotometer.
6. Set the wavelength at 510 nm (wavelength at which cell membranes absorb light if intact).
7. Add 1 mL of blood to each of the tubes using the pipette. Lift the pipette (green) and drag over to the flask of blood. Using the up arrow, add 1 mL of blood to the 1 mL mark of the pipette.
8. Drag the pipette to an empty tube and release the blood by pressing the down arrow.
9. Repeat steps 7 and 8 until all of the test tubes are filled.

Calibrate Spectrophotometer

10. Set the transmittance value at 0 using the arrows at zero.
11. Open the lid of the "holder" of the spectrophotometer (If unsure, click on the word *spectrophotometer* in the instructions at the bottom of the screen) and click and drag the "blank" (tube on the far left of the rack) into the spectrophotometer. (To click and drag, move the mouse to position the arrow on the tube; left-click on mouse and hold; drag the tube to the holder of the spectrophotometer; and when positioned release the left click). Close the lid and set the transmittance at 100 using the arrows at Calibrate.
12. Click the lid of the spectrophotometer and then the tube (top of tube visible in the spectrophotometer) to remove it. The tube will come straight up out of the

---

FIGURE 6.5 Opening screen for the laboratory exercise on Osmosis and Diffusion: Varying Extracellular Concentration.

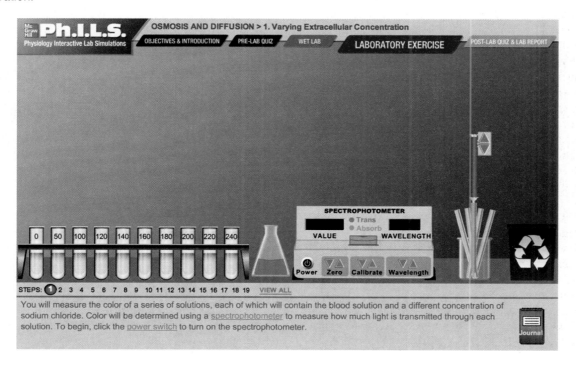

spectrophotometer. Then click on the tube to drag it back to the rack.

13. The journal will open automatically.

14. After viewing, close the journal by clicking on the X in the upper right corner of the graph.

Measuring Transmittance

15. Click and drag the next tube to the spectrophotometer and close the lid. (A reading is taken when the lid is closed). Open the lid. The journal will open automatically and you can view the data.

16. Click on the test tube to remove it. (The tube will come up out of the spectrophotometer and the journal will close automatically). Then drag it back to the rack.

17. Repeat steps 15 and 16 until all tubes have been tested.

Interpreting Results

18. *With the graph still on the screen,* answer the questions in Part E of the laboratory report in the lab manual. If you accidentally closed the graph, click on the journal panel (red rectangle at bottom right of screen). To help you to remember the relationship of % trans-

mittance and the state of the red blood cells, use the following:

a. *More* light is transmitted in samples where the red blood cells have taken on water and *ruptured open* (hemolysis). (Red blood cells are destroyed and no longer block the light as effectively, so more light is transmitted through the sample).

b. *Less* light is transmitted in samples where the red blood cells have lost water and *shriveled* (crenated). (Red blood cells are losing water and becoming more "compact," and they block the light more effectively, so less light is transmitted through the sample).

> Swelling cells: high transmittance
> Shrinking cells: low transmittance

19. Complete the post-lab quiz (click open post-lab quiz and lab report) by answering the ten questions on the computer screen.

20. Read the conclusion on the computer screen.

21. You may print the lab report for Osmosis and Diffusion: Varying Extracellular Concentration.

Name _____

Date _____

Section _____

The ⚠ corresponds to the Learning Outcome(s) listed at the beginning of the laboratory exercise.

Movements Through Membranes

Part A Assessments

Complete the following:

1. Enter data for changes in the movement of the potassium permanganate. ⚠

Elapsed Time	Radius of Purple Circle in Millimeters
Initial	_____
1 minute	_____
2 minutes	_____
3 minutes	_____
4 minutes	_____
5 minutes	_____
6 minutes	_____
7 minutes	_____
8 minutes	_____
9 minutes	_____
10 minutes	_____

2. Prepare a graph that illustrates the diffusion distance of potassium permanganate in 10 minutes. ⚠

3. Explain your graph. ⚠ _____

_____

_____

_____

_____

_____

_____

_____

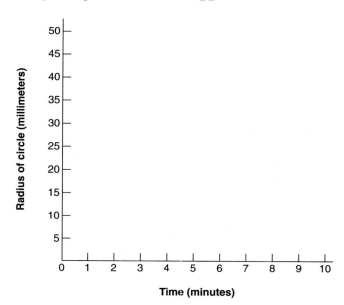

55

4. Define *simple diffusion.* _____

_____

_____

_____

Critical Thinking Activity

By answering yes or no, indicate which of the following provides an example of simple diffusion. ⟨**1**⟩

1. A perfume bottle is opened, and soon the odor can be sensed in all parts of the room. _____

2. A sugar cube is dropped into a cup of hot water, and, without being stirred, all of the liquid becomes sweet tasting. _____

3. Water molecules move from a faucet through a garden hose when the faucet is turned on. _____

4. A person blows air molecules into a balloon by forcefully exhaling. _____

5. A crystal of blue copper sulfate is placed in a test tube of water. The next day, the solid is gone, but the water is evenly colored. _____

Part B Assessments

Complete the following:

1. What was the change in the level of molasses in 10 minutes? _____

2. What was the change in the level of molasses in 30 minutes? _____

3. How do you explain this change? ⟨**3**⟩ _____

_____

4. Define *osmosis.* _____

_____

Critical Thinking Activity

By answering yes or no, indicate which of the following involves osmosis. ⟨**3**⟩

1. A fresh potato is peeled, weighed, and soaked in a strong salt solution. The next day, it is discovered that the potato has lost weight. _____

2. Garden grass wilts after being exposed to dry chemical fertilizer. _____

3. Air molecules escape from a punctured tire as a result of high pressure inside. _____

4. Plant seeds soaked in water swell and become several times as large as before soaking. _____

5. When the bulb of a thistle tube filled with water is sealed by a selectively permeable membrane and submerged in a beaker of molasses, the water level in the tube falls. _____

Part C Assessments

Complete the following:

1. In the spaces, sketch a few blood cells from each of the test tubes and indicate the magnification.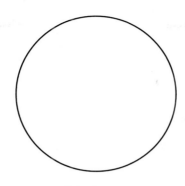

                                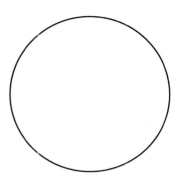

Tube 1   Tube 2   Tube 3
(distilled water)   (0.9% NaCl)   (3.0% NaCl)

_____×   _____×   _____×

2. Based on your results, which tube contained a solution hypertonic to the blood cells? **A** _____

Give the reason for your answer. **A** _____

_____

3. Which tube contained a solution hypotonic to the blood cells? **A** _____

Give the reason for your answer. **A** _____

_____

4. Which tube contained a solution isotonic to the blood cells? **A** _____

Give the reason for your answer. **A** _____

_____

5. Observe the RBCs shown in figure 6.6. Select the solutions in which the cells were placed to illustrate the effects of tonicity on cells.

FIGURE 6.6   Three blood cells placed in three different solutions: distilled water, 0.9% NaCl, and 3% NaCl. Select the solutions in which the cells were placed, using the terms provided. **A**

Terms:
Hypertonic
Hypotonic
Isotonic

(a) _____   (b) _____   (c) _____

Part D Assessments

Complete the following:

1. Which of the substances in the mixture you prepared passed through the filter paper into the filtrate? ◢5◣ _____

   _____

2. What evidence do you have for your answer to question 1? ◢5◣ _____

   _____

   _____

3. What force was responsible for the movement of substances through the filter paper? ◢5◣ _____

   _____

4. What substances did not pass through the filter paper? ◢5◣ _____

   _____

5. What factor prevented these substances from passing through? ◢5◣ _____

   _____

6. Define *filtration.* _____

   _____

Critical Thinking Activity

By answering yes or no, indicate which of the following involves filtration. ◢5◣

1. Oxygen molecules move into a cell and carbon dioxide molecules leave a cell because of differences in the concentrations of these substances on either side of the plasma membrane._____

2. Blood pressure forces water molecules from the blood outward through the thin wall of a blood capillary. _____

3. Urine is forced from the urinary bladder through the tubular urethra by muscular contractions._____

4. Air molecules enter the lungs through the airways when air pressure is greater outside these organs than inside._____

5. Coffee is made using a coffeemaker (not instant)._____

58

Part E Ph.I.L.S. Lesson 1, Osmosis and Diffusion: Varying Extracellular Concentrations Assessments

1. Use the data from the journal on the computer; diagram a graph of an NaCl concentration and % transmittance. (Include title. Label each axis, including units. Plot the points and connect.)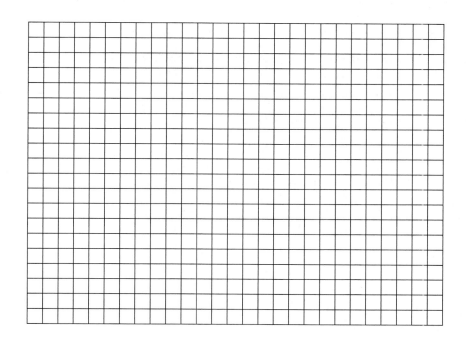

2. On the preceding graph, indicate the areas that correspond to the blood cells being in each of the following solutions: (A) isotonic solution; (B) hypertonic solution (C) hypotonic solution. 7 8

3. Physiological saline (0.9%) is isotonic. What concentration (mM) would be equivalent to physiological saline? 9

_____

4. Describe the net movement of water (into the cell, out of the cell, no net movement) when cells are placed in a(n): 9

   a. isotonic solution _____

   b. hypotonic solution _____

   c. hypertonic solution _____

Critical Thinking Activity 9

A patient is given an IV (intravenous) of deionized water (water with no solutes). If a large enough amount of fluid is administered, predict the effects. Plasma becomes

_____ (hypotonic/hypertonic) to the red blood cells.

_____ Water moves from the (plasma or red blood cells)

_____ to the (plasma or red blood cells)

_____ causing the red blood cells to (shrivel or swell).

A person drinks ocean salt water (3.5% salt). If the person drinks a large amount of the salt water (the salt water is absorbed into the blood), the plasma becomes

_____ (hypotonic/hypertonic) to the red blood cells.

_____ Water moves from the (plasma or red blood cells)

_____ to the (plasma or red blood cells)

_____ causing the red blood cells to (shrivel or swell).

What would you predict if a patient was administered an IV containing a physiological saline solution (0.9% sodium

chloride)? _____

_____

Cell Cycle

Pre-Lab

1. Carefully read the introductory material and examine the entire lab content.
2. Be familiar with the cell cycle (from lecture or the textbook).
3. Visit www.mhhe.com/martinseries1 for pre-lab questions.

Materials Needed

Models of animal mitosis
Microscope slides of whitefish mitosis (blastula)
Compound light microscope

For Demonstration Activity:
Microscope slide of human chromosomes from leukocytes in mitosis
Oil immersion objective

The cell cycle consists of the series of changes a cell undergoes from the time it is formed until it divides. Interphase, mitotic phase, and differentiation are stages of a cell cycle. Typically, a newly formed cell grows to a certain size and then divides to form two new cells (*daughter cells*).

Before the cell divides, it must synthesize biochemicals and other contents. This period of preparation is called *interphase.* The extensive period of interphase is divided into three phases. The S phase, when DNA synthesis occurs, is between two gap phases (G_1 and G_2), when cell growth occurs and cytoplasmic organelles duplicate. Interphase is followed by the *M (mitotic) phase,* which has two major portions: (1) division of the nucleus, called *mitosis,* and (2) division of the cell's cytoplasm, called *cytokinesis.* Mitosis includes four recognized phases: prophase, metaphase, anaphase, and telophase. Eventually, some specialized cells, such as skeletal muscle cells and most nerve cells, cease further cell division, but remain alive. A mature living cell that ceases cell division is in a state called G_0 (G-zero) for the remainder of its life span. This state can last from hours to decades. A special type of cell division, called *meiosis,*

occurs in the reproductive system to produce gametes and is not included in this laboratory exercise.

Purpose of the Exercise

To review the stages in the cell cycle and to observe cells in various stages of their life cycles.

Learning Outcomes

After completing this exercise, you should be able to

1. Describe the phases and structures of the cell cycle.
2. Identify and sketch the stages in the life cycle of a particular cell.
3. Arrange into a correct sequence a set of models or drawings of cells in various stages of their life cycles.

Procedure—Cell Cycle

1. Study the various stages of the cell's life cycle represented in figures 7.1 and 7.2 and label the structures indicated in figure 7.3.

FIGURE 7.1 The cell cycle: interphase (G_1, S, and G_2) and M (mitotic) phase (mitosis and cytokinesis).

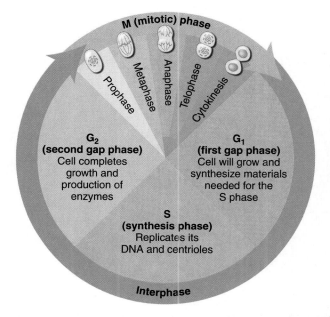

FIGURE 7.2 Mitosis drawings and photographs of the four phases and cytokinesis.

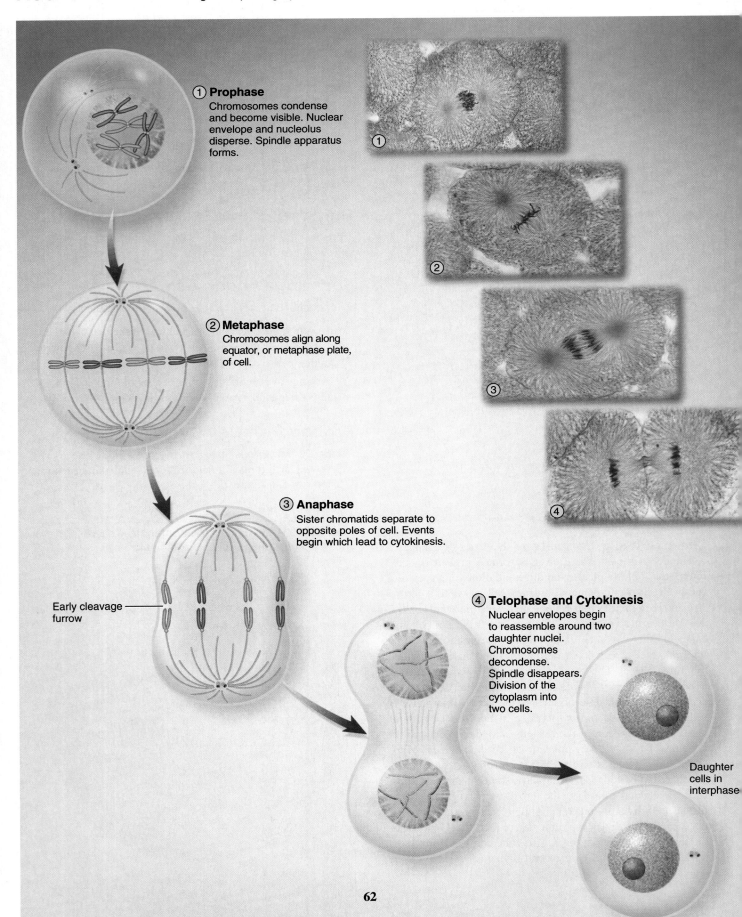

① **Prophase**
Chromosomes condense and become visible. Nuclear envelope and nucleolus disperse. Spindle apparatus forms.

② **Metaphase**
Chromosomes align along equator, or metaphase plate, of cell.

③ **Anaphase**
Sister chromatids separate to opposite poles of cell. Events begin which lead to cytokinesis.

Early cleavage furrow

④ **Telophase and Cytokinesis**
Nuclear envelopes begin to reassemble around two daughter nuclei. Chromosomes decondense. Spindle disappears. Division of the cytoplasm into two cells.

Daughter cells in interphase

FIGURE 7.3 Label the structures indicated in the dividing cell, using the terms provided. 🄰

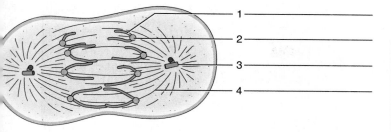

1 —————————

2 —————————

3 —————————

4 —————————

Terms:
Centrioles
Centromere
Chromatid
Spindle fiber (microtubules)

2. Observe the animal mitosis models and review the major events in a cell's life cycle represented by each of them. Be sure you can arrange these models in correct sequence if their positions are changed. The acronym IPMAT can help you arrange the correct order of phases in the cell cycle. This includes interphase followed by the four phases of mitosis. Cytokinesis overlaps anaphase and telophase.

3. Complete Part A of Laboratory Report 7.

4. Obtain a slide of the whitefish mitosis (blastula).

   a. Examine the slide using the high-power objective of a microscope. The tissue on this slide was obtained from a developing embryo (blastula) of a fish, and many of the embryonic cells are undergoing mitosis. The chromosomes of these dividing cells are darkly stained (fig. 7.4).

   b. Search the tissue for cells in various stages of cell division. There are several sections on the slide. If you cannot locate different stages in one section, examine the cells of another section because the stages occur randomly on the slide.

c. Each time you locate a cell in a different stage, sketch it in an appropriate circle in Part B of the laboratory report.

 Critical Thinking Activity

Which stage (phase) of the cell cycle was the most numerous in the blastula? _____

Explain your answer. _____

_____

_____

5. Complete Parts C, D, and E of the laboratory report.

Demonstration Activity

Using the oil immersion objective of a microscope, see if you can locate some human chromosomes by examining a prepared slide of human chromosomes from leukocytes. The cells on this slide were cultured in a special medium and were stimulated to undergo mitosis. The mitotic process was arrested in metaphase by exposing the cells to a chemical called colchicine, and the cells were caused to swell osmotically. As a result of this treatment, the chromosomes were spread apart. A complement of human chromosomes should be visible when they are magnified about 1,000×. Each chromosome is double-stranded and consists of two chromatids joined by a common centromere (fig. 7.5).

FIGURE 7.4 Cell in prophase (250× micrograph enlarged to 1,000×).

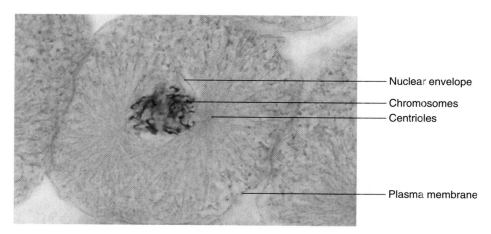

Nuclear envelope

Chromosomes

Centrioles

Plasma membrane

FIGURE 7.5 (*a*) A complement of human chromosomes (2,700×). (*b*) A *karyotype* can be constructed by arranging the homologous chromosome pairs together in a chart. The completed karyotype indicates a normal male.

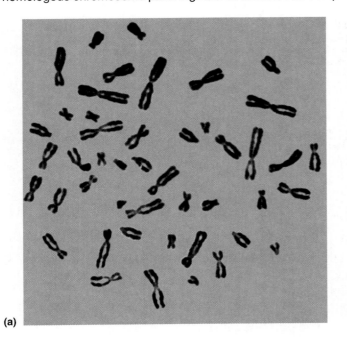

(a)

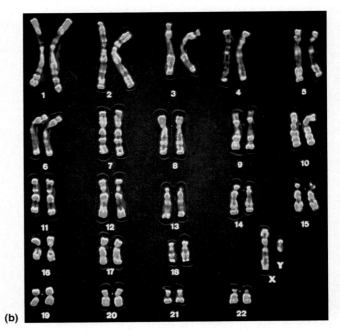

(b)

Name _____

Date _____

Section _____

The A corresponds to the Learning Outcome(s) listed at the beginning of the laboratory exercise.

Cell Cycle

Part A Assessments

Complete the table. 1

Stage	Major Events Occurring
Interphase — G_1, S, and G_2	
M (Mitotic) Phase — **Mitosis** Prophase	
Metaphase	
Anaphase	
Telophase	
Cytokinesis	

Part B Assessments

Sketch an interphase cell and cells in different stages of mitosis to illustrate the whitefish cell's life cycle. Label the major cellular structures represented in the sketches and indicate cytokinesis locations. (The circles represent fields of view through the microscope.) 2

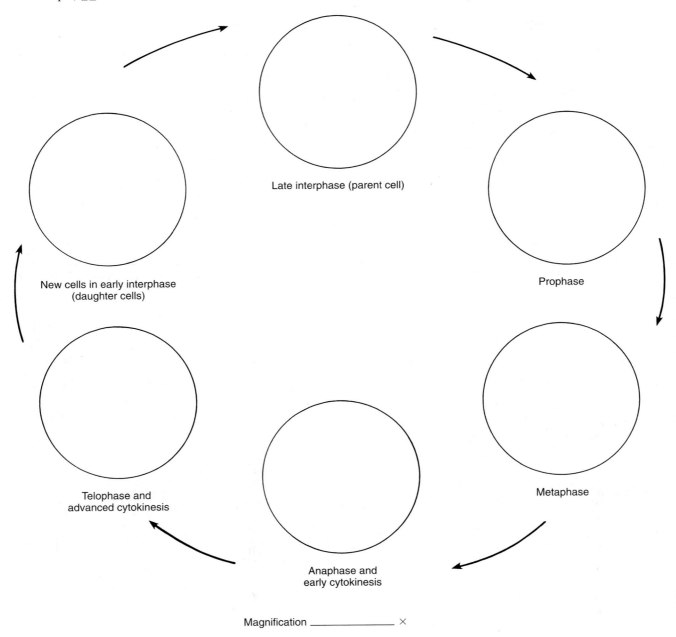

Late interphase (parent cell)

New cells in early interphase
(daughter cells)

Prophase

Telophase and
advanced cytokinesis

Metaphase

Anaphase and
early cytokinesis

Magnification _____ ×

Part C Assessments

Complete the following:

1. In what ways are the new cells (daughter cells), which result from a cell cycle, similar? _____

_____

2. How do the new cells slightly differ? _____

_____

3. Distinguish between mitosis and cytokinesis. ⚠ _____

_____

Part D Assessments

Identify the mitotic stage represented by each of the micrographs in figure 7.6 (a-d). ⚠

a. _____ c. _____

b. _____ d. _____

Part E Assessments

Identify the structures indicated by numbers in figure 7.6. ⚠

1. _____ 4. _____

2. _____ 5. _____

3. _____ 6. _____

FIGURE 7.6 Identify the mitotic stage and structures of the cell in each of these micrographs of the whitefish blastula (250×).

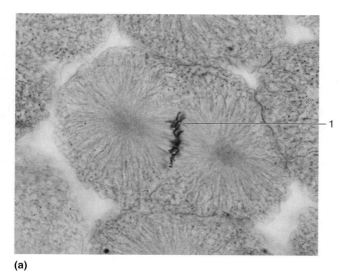

(a)

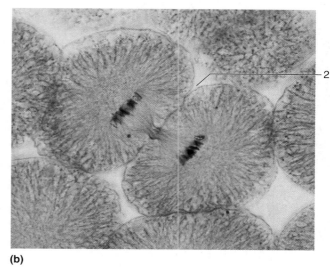

(b)

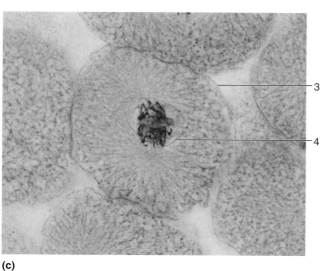

(c)

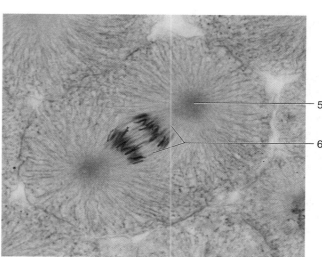

(d)

8

Epithelial Tissues

Pre-Lab

1. Carefully read the introductory material and examine the entire lab content.
2. Be familiar with epithelial tissues (from lecture or the textbook).
3. Visit www.mhhe.com/martinseriesl for pre-lab questions.

Materials Needed

Compound light microscope
Prepared slides of the following epithelial tissues:
    Simple squamous epithelium (lung)
    Simple cuboidal epithelium (kidney)
    Simple columnar epithelium (small intestine)
    Pseudostratified (ciliated) columnar epithelium
        (trachea)
    Stratified squamous epithelium (esophagus)
    Transitional epithelium (urinary bladder)

For Learning Extension Activity:
Colored pencils

A tissue is composed of a layer or group of cells similar in size, shape, and function. Within the human body, there are four major types of tissues: (1) epithelial, which cover the body's external and internal surfaces and most glands; (2) connective, which bind and support parts; (3) muscle, which make movement possible; and (4) nervous, which conduct impulses from one part of the body to another and help to control and coordinate body activities.

Epithelial tissues are tightly packed single (simple) to multiple (stratified) layers of cells that provide protective barriers. The underside of this tissue layer contains an acellular basement membrane layer of adhesives and collagen with which the epithelial cells anchor to an underlying connective tissue. A unique type of simple epithelium, pseudostratified columnar, appears to be multiple cells thick. However, this is a false appearance because all cells bind to the basement

membrane. The cells readily divide and lack blood vessels (are avascular).

Epithelial cells always have a free (apical) surface exposed to the outside or to an open space internally and a basal surface that attaches to the basement membrane. Many shapes of the cells, like squamous (flat), cuboidal, and columnar, exist that are used to name and identify the variations. Epithelial cell functions include protection, filtration, secretion, and absorption. Many of the prepared slides contain more than the tissue to be studied, so be certain that your view matches the correct tissue. Also be aware that stained colors of all tissues might vary.

The slides you observe might have representative tissue from more than one site, but the basic tissue structure will be similar to those represented in this laboratory exercise. Simple squamous epithelium is located in the air sacs (alveoli) of the lungs and inner linings of the heart and blood vessels. Simple cuboidal epithelium is situated in kidney tubules, thyroid gland, liver, and ducts of salivary glands. A nonciliated type of simple columnar epithelium is located in the linings of the uterus, stomach, and intestines, and a ciliated type lines the uterine tubes. A ciliated type of pseudostratified columnar epithelium is positioned in the linings of the upper respiratory tubes. A keratinized type of stratified squamous epithelium is found in the epidermis of the skin, and a nonkeratinized type in the linings of the oral cavity, esophagus, vagina, and anal canal. Transitional epithelium lines the urinary bladder, ureters, and part of the urethra.

Purpose of the Exercise

To review the characteristics of epithelial tissues and to observe examples.

Learning Outcomes

After completing this exercise, you should be able to

1. Identify and sketch six examples of epithelial tissues on microscope slides.
2. Differentiate the special characteristics of each type of epithelial tissue.
3. Indicate a location and function of each type of epithelial tissue.
4. Inventory the general characteristics of epithelial tissues that you were able to observe.

FIGURE 8.1 Micrographs of epithelial tissues. *Note:* The brackets to the right of each micrograph indicate the tissue.

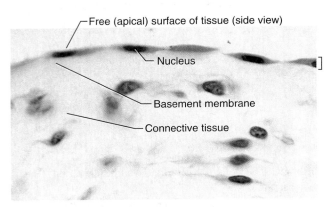

Free (apical) surface of tissue (side view)
Nucleus
Basement membrane
Connective tissue

(a) Simple squamous epithelium (side view)

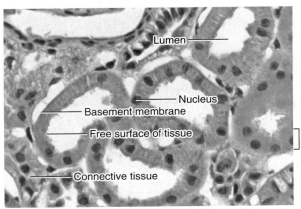

Lumen
Nucleus
Basement membrane
Free surface of tissue
Connective tissue

(b) Simple cuboidal epithelium (from kidney)

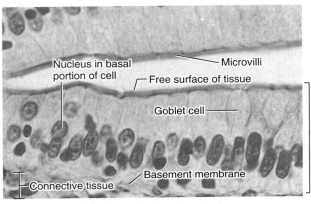

Nucleus in basal portion of cell
Microvilli
Free surface of tissue
Goblet cell
Basement membrane
Connective tissue

(c) Simple columnar epithelium (from intestine)

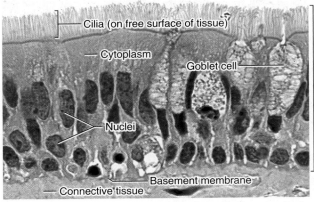

Cilia (on free surface of tissue)
Cytoplasm
Goblet cell
Nuclei
Basement membrane
Connective tissue

(d) Pseudostratified columnar epithelium with cilia (from trachea)

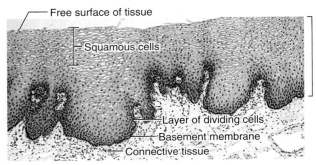

Free surface of tissue
Squamous cells
Layer of dividing cells
Basement membrane
Connective tissue

(e) Stratified squamous epithelium (nonkeratinized) (from esophagus)

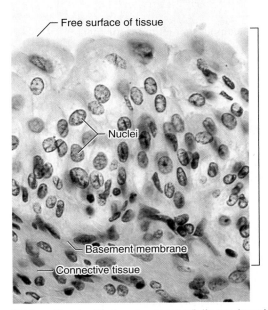

Free surface of tissue
Nuclei
Basement membrane
Connective tissue

(f) Transitional epithelium (unstretched) (from urinary bladder)

Procedure—Epithelial Tissues

1. Review the introductory material and complete Part A of Laboratory Report 8.
2. Use the microscope to observe the prepared slides of types of epithelial tissues. As you observe each tissue, look for its special distinguishing features such as cell size, shape, and arrangement. Compare your prepared slides of epithelial tissues to the micrographs in figure 8.1.

3. As you observe the tissues in figure 8.1 and the prepared slides, note characteristics epithelial tissues have in common. List characteristics you were able to observe in Part B of the laboratory report.
4. As you observe each type of epithelial tissue, prepare a labeled sketch of a representative portion of the tissue in Part C of the laboratory report.
5. Test your ability to recognize each type of epithelial tissue. To do this, have a laboratory partner select one of the prepared slides, cover its label, and focus the microscope on the tissue. Then see if you can correctly identify the tissue.
6. Complete Parts B and C of the laboratory report.

Learning Extension Activity

As you observe histology slides, be aware that tissues and some organs may have been sectioned in longitudinal (lengthwise cut), cross section (cut across), or oblique (angular cut) ways. Observe figure 8.2 for how this can be demonstrated on cuts of a banana. The direction in which the tissue or organ was cut will result in a certain perspective when it is sectioned, just as the banana was cut three different ways. Often a tissue slide has more than one tissue represented and has blood vessels or other structures cut in various ways. Locate an example of a longitudinal section, cross section, and oblique section of some structure on one of your tissue slides.

FIGURE 8.2 Three possible cuts of a banana: (a) longitudinal section; (b) cross section; (c) oblique section. Sections through an organ, as a body tube, frequently produce views similar to the cut banana.

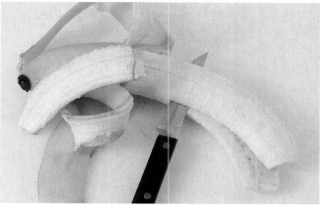

(a)

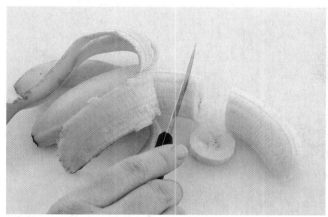

(b)

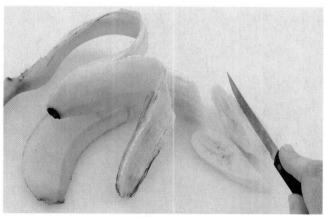

(c)

Name _____

Date _____

Section _____

The 🄰 corresponds to the Learning Outcome(s) listed at the beginning of the laboratory exercise.

Epithelial Tissues

Part A Assessments

Match the tissues in column A with the characteristics in column B. Place the letter of your choice in the space provided. (Some answers may be used more than once.) 🄰

Column A	Column B
a. Simple columnar epithelium	_____ 1. Consists of several layers of cube-shaped, elongated, and irregular cells
b. Simple cuboidal epithelium	_____ 2. Commonly possesses cilia that move dust and mucus out of the airways
c. Simple squamous epithelium	_____ 3. Single layer of flattened cells
d. Pseudostratified columnar epithelium	_____ 4. Nuclei located at different levels within cells
e. Stratified squamous epithelium	_____ 5. Forms walls of capillaries and air sacs of lungs
f. Transitional epithelium	_____ 6. Forms linings of trachea and bronchi
	_____ 7. Younger cells cuboidal, older cells flattened
	_____ 8. Forms inner lining of urinary bladder
	_____ 9. Lines kidney tubules and ducts of salivary glands
	_____ 10. Forms lining of stomach and intestines
	_____ 11. Nuclei located near basement membrane
	_____ 12. Forms lining of oral cavity, anal canal, and vagina

Part B Assessments

As you examined each specific epithelial tissue, you should have noted some of the general characteristics that they possess as described in the introduction of the laboratory exercise. List any of these you were able to observe. 🄰

_____

_____

_____

_____

Part C Assessments

In the space that follows, sketch a few cells of each type of epithelium you observed. For each sketch, label the major characteristics, indicate the magnification used, write an example of a location in the body, and provide a function. 1\ 2\ 3\

Simple squamous epithelium (____×) Location: _____ Function: _____	Simple cuboidal epithelium (____×) Location: _____ Function: _____
Simple columnar epithelium (____×) Location: _____ Function: _____	Pseudostratified columnar epithelium with cilia (____×) Location: _____ Function: _____
Stratified squamous epithelium (____×) Location: _____ Function: _____	Transitional epithelium (____×) Location: _____ Function: _____

Critical Thinking Activity

As a result of your observations of epithelial tissues, which one(s) provide(s) the best protection? Explain your answer. _____

_____

_____

Learning Extension Activity

Use colored pencils to differentiate various cellular structures in Part C. Select a different color for a nucleus, cytoplasm, plasma membrane, basement membrane, goblet cell, and cilia whenever visible.

Connective Tissues

Pre-Lab

1. Carefully read the introductory material and examine the entire lab content.
2. Be familiar with connective tissues (from lecture or the textbook).
3. Visit www.mhhe.com/martinseriesl for pre-lab questions and LabCam videos.

Materials Needed

Compound light microscope
Prepared slides of the following:
    Areolar connective tissue
    Adipose tissue
    Reticular connective tissue
    Dense regular connective tissue
    Dense irregular connective tissue
    Elastic connective tissue
    Hyaline cartilage
    Fibrocartilage
    Elastic cartilage
    Bone (compact, ground, cross section)
    Blood (human smear)

For Learning Extension Activity:
Colored pencils

Connective tissues contain a variety of cell types and occur in all regions of the body. They bind structures together, provide support and protection, fill spaces, store fat, and produce blood cells.

    Connective tissue cells are often widely scattered in an abundance of extracellular matrix. The matrix consists of fibers and a ground substance of various densities and consistencies. The protein fibers are among collagen (most abundant), reticular (collagen with glycoprotein), and elastic fibers. You might compare connective tissue to making gelatin: the gelatin of various densities represents the ground substance, added fruit represents cells, and added strands represent the fibers. Many of the prepared slides contain more than the tissue to be studied, so be certain that your view matches the correct tissue. Additional study of bone and blood will be found in Laboratory Exercises 12 and 41.

Purpose of the Exercise

To review the characteristics of connective tissues and to observe examples of the major types.

Learning Outcomes

After completing this exercise, you should be able to

1. Differentiate the special characteristics of each of the major types of connective tissue.
2. Sketch and label the characteristics of connective tissues that you were able to observe.
3. Indicate a location and function of each type of connective tissue.
4. Identify eleven types of connective tissues on microscope slides.

Procedure—Connective Tissues

1. Study table 9.1.
2. Complete Part A of Laboratory Report 9.
3. Use a microscope to observe the prepared slides of various connective tissues. As you observe each tissue, look for its special distinguishing features. Compare your prepared slides of connective tissues to the micrographs in figure 9.1.
4. As you observe each type of connective tissue, prepare a labeled sketch of a representative portion of the tissue in Part B of the laboratory report.
5. Complete Part B of the laboratory report.
6. Test your ability to recognize each of these connective tissues by having a laboratory partner select a slide, cover its label, and focus the microscope on this tissue. Then see if you correctly identify the tissue. 4

TABLE 9.1 Types of Mature Connective Tissues and Representative Locations*

Loose Connective	Dense Connective	Cartilage	Bone	Liquid Connective
Areolar Connective Locations: around body organs; binds skin to deeper organs	Dense Regular Connective Locations: tendons; ligaments	Hyaline Cartilage Locations: nasal septum; larynx; costal cartilage; ends of long bones; fetal skeleton	Compact Bone Locations: bone shafts; beneath periosteum	Blood Locations: lumens of blood vessels; heart chambers
Adipose Locations: subcutaneous layer; around kidneys and heart; yellow bone marrow; breasts	Dense Irregular Connective Locations: dermis; heart valves; periosteum on bone	Fibrocartilage Locations: between vertebrae; between pubic bones; pads (meniscus) in knee	Spongy (Cancellous) Bones Locations: ends of long bones; inside flat and irregular bones	Lymph Locations: lumens of lymphatic vessels
Reticular Connective Locations: spleen; thymus; lymph nodes; red bone marrow	Elastic Connective Locations: larger artery walls; vocal cords; ligaments between vertebrae	Elastic Cartilage Locations: outer ear; epiglottis		

*This table represents a scheme to organize connective tissue relationships. Spongy bone and lymph tissues are not examined microscopically in this laboratory exercise.

FIGURE 9.1 Micrographs of connective tissues.

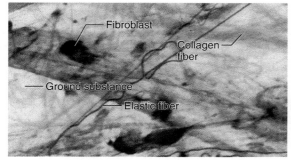

(a) Areolar connective (from fascia between muscles)

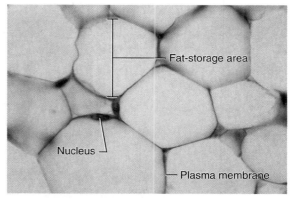

(b) Adipose (from subcutaneous layer)

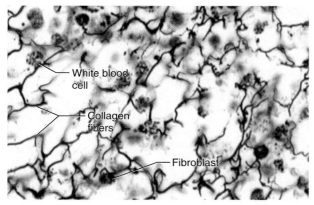

(c) Reticular connective (from spleen)

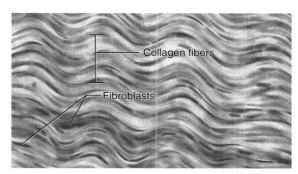

(d) Dense regular connective (from tendon)

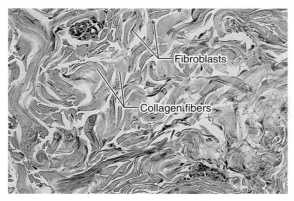

(e) Dense irregular connective (from dermis)

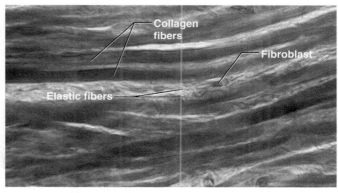

(f) Elastic connective (from artery wall)

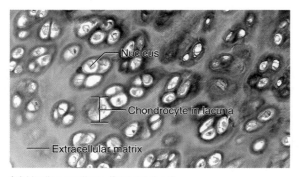

(g) Hyaline cartilage (from trachea)

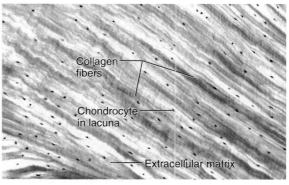

(h) Fibrocartilage (from pubic symphysis)

FIGURE 9.1 *Continued.*

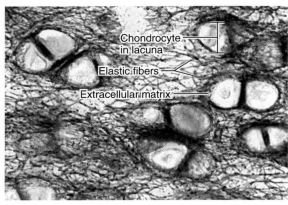

(i) Elastic cartilage (from ear)

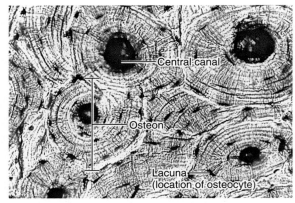

(j) Compact bone

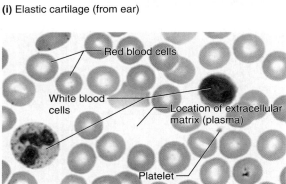

(k) Blood

Name _____

Date _____

Section _____

The A corresponds to the Learning Outcome(s) listed at the beginning of the laboratory exercise.

Connective Tissues

Part A Assessments

Match the tissues in column A with the characteristics in column B. Place the letter of your choice in the space provided. 1 3

Column A	Column B
a. Adipose	_____ 1. Forms framework of outer ear
b. Areolar connective	_____ 2. Functions as heat insulator beneath skin
c. Blood	_____ 3. Contains large amounts of fluid and lacks fibers
d. Bone (compact)	_____ 4. Cells arranged around central canal
e. Dense irregular connective	_____ 5. Binds skin to underlying organs
f. Dense regular connective	_____ 6. Main tissue of tendons and ligaments
g. Elastic cartilage	_____ 7. Forms the flexible part of the nasal septum
h. Elastic connective	_____ 8. Pads between vertebrae that are shock absorbers
i. Fibrocartilage	_____ 9. Main tissue of dermis
j. Hyaline cartilage	_____ 10. Occurs in ligament attachments between vertebrae and artery walls
k. Reticular connective	_____ 11. Forms supporting tissue in walls of thymus and spleen

Part B Assessments

In the space that follows, sketch a small section of each of the types of connective tissues you observed. For each sketch, label the major characteristics, indicate the magnification used, write an example of a location in the body, and provide a function. **1** **2** **3**

<table>
<tr>
<td>

Areolar connective (_____×)

Location: _____

Function: _____

</td>
<td>

Adipose (_____×)

Location: _____

Function: _____

</td>
</tr>
<tr>
<td>

Reticular connective (_____×)

Location: _____

Function: _____

</td>
<td>

Dense regular connective (_____×)

Location: _____

Function: _____

</td>
</tr>
<tr>
<td>

Dense irregular connective (_____×)

Location: _____

Function: _____

</td>
<td>

Elastic connective (_____×)

Location: _____

Function: _____

</td>
</tr>
</table>

Hyaline cartilage (_____×)

Location: _____

Function: _____

Fibrocartilage (_____×)

Location: _____

Function: _____

Elastic cartilage (_____×)

Location: _____

Function: _____

Compact bone (_____×)

Location: _____

Function: _____

Blood (_____×)

Location: _____

Function: _____

 Critical Thinking Activity

Abdominal impact injuries often involve the spleen. Explain the structural tissue characteristics that make the spleen so vulnerable to serious injury.

_____

_____

_____

_____

Learning Extension Activity

Use colored pencils to differentiate various cellular structures in Part B. Select a different color for the cells, fibers, and ground substance whenever visible.

Muscle and Nervous Tissues

Pre-Lab

1. Carefully read the introductory material and examine the entire lab content.
2. Be familiar with muscle tissues and nervous tissue (from lecture or the textbook).
3. Visit www.mhhe.com/martinseries1 for pre-lab questions.

Materials Needed

Compound light microscope
Prepared slides of the following:
   Skeletal muscle tissue
   Smooth muscle tissue
   Cardiac muscle tissue
   Nervous tissue (spinal cord smear and/or cerebellum)
For Learning Extension Activity:
Colored pencils

Muscle tissues are characterized by the presence of elongated cells, often called muscle fibers, that can contract to create movements. Many of our muscles are attached to the skeleton, but muscles are also components of many of our internal organs. During muscle contractions, considerable body heat is generated to help maintain our body temperature. Because more heat is generated than is needed to maintain our body temperature, much of the heat is dissipated from our body through the skin.

The three types of muscle tissues are skeletal, smooth, and cardiac. Skeletal muscles are under our conscious control and are considered voluntary. Although most skeletal muscles are attached to bones via tendons, other locations include the tongue, facial muscles, and voluntary sphincters. Functions include body movements, maintaining posture, breathing, speaking, controlling waste eliminations, and protection. Smooth muscle is considered involuntary and is located in many visceral organs, the iris, blood vessels,

respiratory tubes, and attached to hair follicles. Functions include the motions of visceral organs, controlling pupil size, blood flow, and airflow, and creating "goose bumps" if we are too cold or frightened. Cardiac muscle is located only in the heart wall. It is considered involuntary and functions to pump blood.

Nervous tissues occur in the brain, spinal cord, and peripheral nerves. The tissue consists of two cell types: neurons and neuroglia. Neurons, also called nerve cells, contain a cell body with the nucleus and most of the cytoplasm, and cellular processes that extend from the cell body. Cellular processes include one to many dendrites and a single axon (nerve fiber). Neurons are considered excitable cells because they can exhibit signals called action potentials (nerve impulses) along the neuron to another neuron or a muscle or gland. Neuroglia (glial cells) of various types are more abundant than neurons; they cannot conduct nerve impulses, but they have important supportive and protective functions for neurons.

Purpose of the Exercise

To review the characteristics of muscle and nervous tissues and to observe examples of these tissues.

Learning Outcomes

After completing this exercise, you should be able to

1. Differentiate the special characteristics of each type of muscle tissue and nervous tissue.
2. Sketch and label the characteristics of muscle tissues and nervous tissues that you were able to observe.
3. Indicate a location and function of each type of muscle tissue and nervous tissue.
4. Identify three types of muscle tissues and nervous tissue on microscope slides.

Procedure—Muscle and Nervous Tissues

1. Study the muscle characteristics in table 10.1.
2. Complete Part A of Laboratory Report 10.
3. Using the microscope, observe each of the types of muscle tissues on the prepared slides. Look for the special

TABLE 10.1 Muscle Tissue Characteristics

Characteristic	Skeletal Muscle	Smooth Muscle	Cardiac Muscle
Appearance of cells	Unbranched and relatively parallel	Spindle-shaped	Branched and connected in complex networks
Striations	Present and obvious	Absent	Present but faint
Nucleus	Multinucleated	Uninucleated	Uninucleated (usually)
Intercalated discs	Absent	Absent	Present
Control	Voluntary	Involuntary	Involuntary

features of each type. Compare your prepared slides of muscle tissues to the micrographs in figure 10.1.

4. As you observe each type of muscle tissue, prepare a labeled sketch of a representative portion of the tissue in the laboratory reoprt.

5. Observe the prepared slide of nervous tissue and identify neurons (nerve cells), neuron cellular processes, and neuroglia. Compare your prepared slide of nervous tissue to the micrograph in figure 10.1.

6. Complete Part B of the laboratory report.

7. Test your ability to recognize each of these muscle and nervous tissues by having your laboratory partner select a slide, cover its label, and focus the microscope on this tissue. Then see if you correctly identify the tissue. **4**

FIGURE 10.1 Micrographs of muscle and nervous tissues.

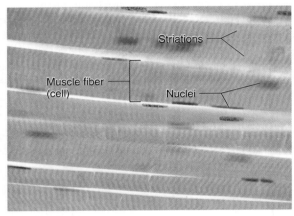

(a) Skeletal muscle

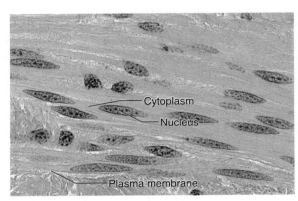

(b) Smooth muscle (from small intestine)

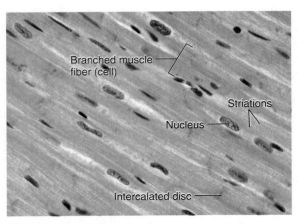

(c) Cardiac muscle (from heart)

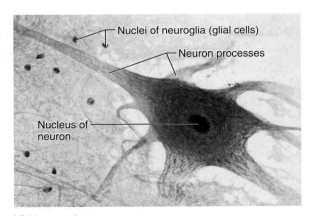

(d) Nervous tissue

Name _____

Date _____

Section _____

The ⃤ corresponds to the Learning Outcome(s) listed at the beginning of the laboratory exercise.

Muscle and Nervous Tissues

Part A Assessments

Match the tissues in column A with the characteristics in column B. Place the letter of your choice in the space provided. (Some answers may be used more than once.) ⃤1 ⃤3

Column A	Column B
a. Cardiac muscle	_____ **1.** Coordinates, regulates, and integrates body functions
b. Nervous tissue	_____ **2.** Contains intercalated discs
c. Skeletal muscle	_____ **3.** Muscle that lacks striations
d. Smooth muscle	_____ **4.** Striated and involuntary
	_____ **5.** Striated and voluntary
	_____ **6.** Contains neurons and neuroglia
	_____ **7.** Muscle attached to bones
	_____ **8.** Muscle that composes heart
	_____ **9.** Moves food through the digestive tract
	_____ **10.** Transmits impulses along cellular processes

Part B Assessments

In the space that follows, sketch a few cells or fibers of each of the three types of muscle tissues and of nervous tissue as they appear through the microscope. For each sketch, label the major structures of the cells or fibers, indicate the magnification used, write an example of a location in the body, and provide a function. A 1 A 2 A 3

<table>
<tr>
<td>

Skeletal muscle tissue (_____×)

Location: _____

Function: _____

</td>
<td>

Smooth muscle tissue (_____×)

Location: _____

Function: _____

</td>
</tr>
<tr>
<td>

Cardiac muscle tissue (_____×)

Location: _____

Function: _____

</td>
<td>

Nervous tissue (_____×)

Location: _____

Function: _____

</td>
</tr>
</table>

Learning Extension Activity

Use colored pencils to differentiate various cellular structures in Part B.

Laboratory Exercise 11

Integumentary System

Pre-Lab

1. Carefully read the introductory material and examine the entire lab content.
2. Be familiar with skin layers and accessory structures of the skin (from lecture or the textbook).
3. Visit www.mhhe.com/martinseriesl for pre-lab questions.

Materials Needed

Skin model
Hand magnifier or dissecting microscope
Forceps
Microscope slide and coverslip
Compound light microscope
Prepared microscope slide of human scalp or axilla
Prepared slide of dark (heavily pigmented) human skin
Prepared slide of thick skin (plantar or palmar)

For Learning Extension Activity:
Tattoo slide

The integumentary system includes the skin, hair, nails, sebaceous (oil) glands, and sweat (sudoriferous) glands. These structures provide a protective covering for deeper tissues, aid in regulating body temperature, retard water loss, house sensory receptors, synthesize various chemicals, and excrete small quantities of wastes.

The skin consists of two distinct layers. The outer layer, the *epidermis,* consists of stratified squamous epithelium. The inner layer, the *dermis,* consists of a superficial papillary region of areolar connective tissue and a thicker and deeper reticular region of dense irregular connective tissue. Beneath the dermis is the subcutaneous (hypodermis; superficial fascia) layer (not considered a true layer of the skin), composed of adipose and areolar connective tissues.

Accessory structures of the skin include those associated with the hair, which grows through a depression from the epidermis. A hair papilla at the base of the hair contains a network of capillaries that supply the nutrients for cell divisions for hair growth within the hair bulb. As the cells of the hair are forced toward the surface of the body, they become keratinized and pigmented and die. Attached to the follicle is the arrector pili muscle that can pull the hair to a more upright position, causing goose bumps when experiencing cold temperatures or fear. A sebaceous (oil) gland secretes sebum into the hair follicles, which keeps the hair and epidermal surface pliable and waterproof.

Sweat glands (sudoriferous glands) are distributed over most regions of the body and consist of two types of glands. The widespread eccrine sweat glands are most numerous on the palms, soles of the feet, and the forehead. Their ducts open to the surface at a sweat pore. Their secretions increase during hot days, physical exercise, and stress; they serve an excretory function and prevent overheating of our body temperature. The apocrine sweat glands are most abundant in the axillary and genital regions. Apocrine sweat ducts open into the hair follicles and become active at puberty. Their secretions increase during stress and pain and have little influence on thermoregulation.

Purpose of the Exercise

To observe the structures and tissues of the integumentary system and to review the functions of these parts.

Learning Outcomes

After completing this exercise, you should be able to

1. Locate the structures of the integumentary system.
2. Describe the major functions of these structures.
3. Distinguish the locations and tissues among epidermis, dermis, and the subcutaneous layer.
4. Sketch the layers of the skin and associated structures that you observed on the prepared slide.

Procedure—Integumentary System

1. Label figures 11.1 and 11.2. Locate as many of these structures as possible on a skin model.
2. Complete Part A of Laboratory Report 11.

FIGURE 11.1 Label this vertical section of the skin and subcutaneous layer, using the terms provided. 🔺

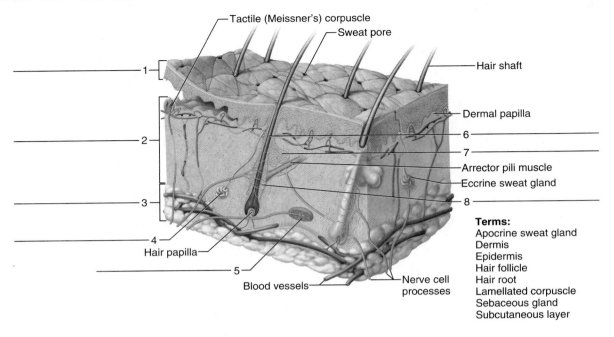

Tactile (Meissner's) corpuscle
Sweat pore
Hair shaft
Dermal papilla
Arrector pili muscle
Eccrine sweat gland
Hair papilla
Blood vessels
Nerve cell processes

1
2
3
4
5
6
7
8

Terms:
Apocrine sweat gland
Dermis
Epidermis
Hair follicle
Hair root
Lamellated corpuscle
Sebaceous gland
Subcutaneous layer

FIGURE 11.2 Label the epidermal layers in this section of thick skin from the fingertip, using the terms provided. 🔺

Terms:
Stratum basale
Stratum corneum
Stratum granulosum
Stratum spinosum

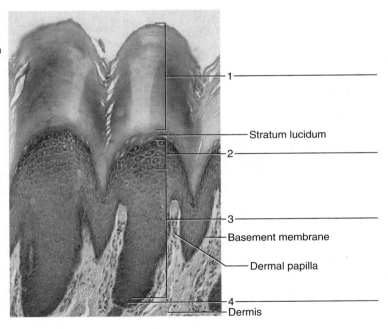

Stratum lucidum
Basement membrane
Dermal papilla
Dermis

1
2
3
4

3. Use the hand magnifier or dissecting microscope and proceed as follows:
   a. Observe the skin, hair, and nails on your hand.
   b. Compare the type and distribution of hairs on the front and back of your forearm.
4. Use low-power magnification of the compound light microscope and proceed as follows:
   a. Pull out a single hair with forceps and mount it on a microscope slide under a coverslip.

   b. Observe the root and shaft of the hair and note the scalelike parts that make up the shaft.
5. Complete Part B of the laboratory report.
6. As vertical sections of human skin are observed, remember that the lenses of the microscope invert and reverse images. It is important to orient the position of the epidermis, dermis, and subcutaneous layers using scan magnification before continuing with additional observations. Compare all of your skin observations to figure 11.3. Use

FIGURE 11.3 Features of human skin are indicated in these micrographs: (a) various structures; (b) epidermis of dark skin; (c) base of hair structures.

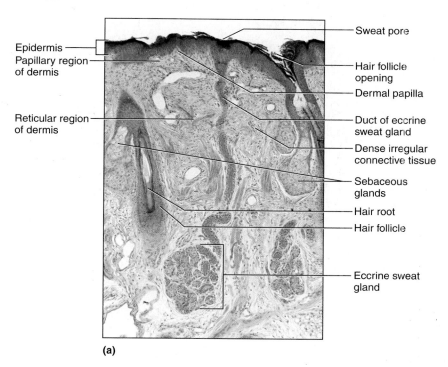

Epidermis
Papillary region of dermis
Reticular region of dermis

Sweat pore
Hair follicle opening
Dermal papilla
Duct of eccrine sweat gland
Dense irregular connective tissue
Sebaceous glands
Hair root
Hair follicle
Eccrine sweat gland

(a)

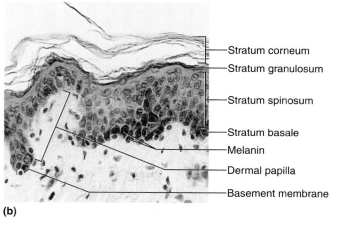

Stratum corneum
Stratum granulosum
Stratum spinosum
Stratum basale
Melanin
Dermal papilla
Basement membrane

(b)

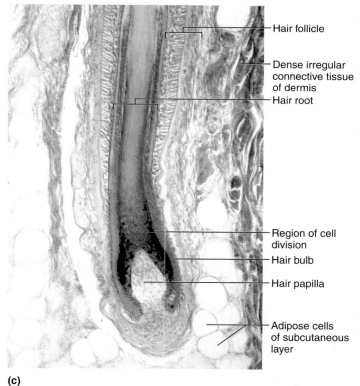

Hair follicle
Dense irregular connective tissue of dermis
Hair root
Region of cell division
Hair bulb
Hair papilla
Adipose cells of subcutaneous layer

(c)

low-power magnification of the compound light microscope and proceed as follows:

a. Observe the prepared slide of human scalp or axilla.

b. Locate the epidermis, dermis, and subcutaneous layer; a hair follicle; an arrector pili muscle; a sebaceous gland; and a sweat gland.

c. Focus on the epidermis with high power and locate the stratum corneum, stratum granulosum, stratum spinosum, and stratum basale. Note how the shapes of the cells in these layers differ.

d. Observe the dense irregular connective tissue that makes up the bulk of the dermis.

e. Observe the adipose tissue that composes most of the subcutaneous layer.

7. Observe the prepared slide of dark (heavily pigmented) human skin with low-power magnification. Note that the pigment is most abundant in the epidermis. Focus on this region with the high-power objective. The pigment-producing cells, or melanocytes, are located among the

89

Explain the advantage for melanin granules being located in the deep layer of the epidermis.

_____

_____

_____

_____

deeper layers of epidermal cells. Differences in skin color are primarily due to the amount of pigment (melanin) produced by these cells.

8. Observe the prepared slide of thick skin from the palm of a hand or the sole of a foot. Locate the stratum lucidum. Note how the stratum corneum compares to your observation of human scalp.

9. Complete Part C of the laboratory report.

10. Using low-power magnification, locate a hair follicle sectioned longitudinally through its bulblike base. Also locate a sebaceous gland close to the follicle and find a sweat gland (fig. 11.3). Observe the detailed structure of these parts with high-power magnification.

11. Complete Parts D and E of the laboratory report.

Learning Extension Activity

Observe a vertical section of human skin through a tattoo, using low-power magnification. Note the location of the dispersed ink granules within the upper portion of the dermis. From a thin vertical section of a tattoo, it is not possible to determine the figure or word of the entire tattoo as seen on the surface of the skin. Compare this to the location of melanin granules found in dark (heavily pigmented) skin. Suggest reasons why a tattoo is permanent and a suntan is not.

_____

_____

_____

_____

_____

Name _____

Date _____

Section _____

The ⚠ corresponds to the Learning Outcome(s) listed at the beginning of the laboratory exercise.

Integumentary System

Part A Assessments

Match the structures in column A with the description and functions in column B. Place the letter of your choice in the space provided. ⚠ ⚠

Column A	Column B
a. Apocrine sweat gland	_____ **1.** An oily secretion that helps to waterproof body surface
b. Arrector pili muscle	_____ **2.** Outermost layer of epidermis
c. Dermis	_____ **3.** Become active at puberty
d. Eccrine (merocrine) sweat gland	_____ **4.** Epidermal pigment
e. Epidermis	_____ **5.** Inner layer of skin
f. Hair follicle	_____ **6.** Responds to elevated body temperature
g. Keratin	_____ **7.** General name of entire superficial layer of the skin
h. Melanin	_____ **8.** Gland that secretes an oily substance
i. Sebaceous gland	_____ **9.** Hard protein of nails and hair
j. Sebum	_____ **10.** Cell division and deepest layer of epidermis
k. Stratum basale	_____ **11.** Tubelike part that contains the root of the hair
l. Stratum corneum	_____ **12.** Causes hair to stand on end and goose bumps to appear

Part B Assessments

Complete the following:

1. How does the skin of your palm differ from that on the back (posterior) of your hand? _____

_____

_____

2. Describe the differences you observed in the type and distribution of hair on the front (anterior) and back (posterior) of your forearm. _____

_____

3. Explain how a hair is formed. ⚠ _____

_____

4. What cells produce the pigment in hair? ⚠ _____

Part C Assessments

Complete the following:

1. Distinguish the locations and tissues among epidermis, dermis, and subcutaneous layer. **3** _____

    _____

    _____

2. How do the cells of stratum corneum and stratum basale differ? **3** _____

    _____

    _____

3. State the specific location of melanin observed in dark skin. **3** _____

    _____

4. What special qualities does the connective tissue of the dermis have? **3** _____

    _____

Part D Assessments

Complete the following:

1. What part of the hair extends from the hair papilla to the body surface? **1** _____

2. In which layer of skin are sebaceous glands found? **1** _____

3. How are sebaceous glands associated with hair follicles? **1** _____

    _____

4. In which layer of skin are sweat glands usually located? **1** _____

Part E Assessments

Sketch a vertical section of human skin, using the scanning objective. Label the skin layers and a hair follicle, a sebaceous gland, and a sweat gland. **4**

Laboratory Exercise 12

Bone Structure and Classification

Pre-Lab

1. Carefully read the introductory material and examine the entire lab content.
2. Be familiar with bone structure and bone tissue (from lecture or the textbook).
3. Visit www.mhhe.com/martinseries1 for pre-lab questions, Anatomy & Physiology Revealed animations list, and LabCam videos.

Materials Needed

Prepared microscope slide of ground compact bone
Human bone specimens including long, short, flat, irregular, and sesamoid types
Human long bone, sectioned longitudinally
Fresh animal bones, sectioned longitudinally and transversely
Compound light microscope
Dissecting microscope

For Demonstration Activity:
Fresh chicken bones (radius and ulna from wings)
Vinegar or dilute hydrochloric acid

Safety

▶ Wear disposable gloves for handling fresh bones and for the demonstration of a bone soaked in vinegar or dilute hydrochloric acid.
▶ Wash your hands before leaving the laboratory.

A bone represents an organ of the skeletal system. As such, it is composed of a variety of tissues including bone tissue, cartilage, dense connective tissue, blood, and nervous tissue. Bones are not only alive, but also multifunctional.

They support and protect softer tissues, provide points of attachment for muscles, house blood-producing cells, and store inorganic salts.

Living bone is a combination of about one-third organic matter and two-thirds inorganic matter. The organic matter consists mostly of embedded cells and collagen fibers. The inorganic matter is mostly complex salt crystals, hydroxyapatite, consisting of calcium phosphate. Lesser amounts of calcium carbonate and ions of magnesium, fluoride, and sodium become incorporated into the inorganic crystals.

Bones are classified according to their shapes as long, short, flat, irregular, or sesamoid (round). Long bones are much longer than they are wide and have expanded ends. Short bones are somewhat cube shaped, with similar lengths and widths. Flat bones have wide surfaces, but they are sometimes curved, such as those of the cranium. Irregular bones have numerous shapes and often have articulations with more than one other bone. Sesasmoid bones are small and embedded within a tendon near joints where compression often occurs. The patella is a sesamoid bone that is included in the total skeletal number of 206. Any additional sesamoid bones that may develop in compression areas of the hand or foot are not considered among the 206 bones of the adult skeleton. Although bones of the skeleton vary greatly in size and shape, they have much in common structurally and functionally.

Purpose of the Exercise

To review the way bones are classified and to examine the structure of a long bone.

Learning Outcomes

After completing this exercise, you should be able to

1 Locate the major structures of a long bone.
2 Distinguish between compact and spongy bone.
3 Differentiate the special characteristics of compact bone tissue.
4 Arrange five groups of bones based on their shapes and identify an example for each group.
5 Describe the functions of various structures of a bone.

Procedure—Bone Structure and Classification

1. Label figures 12.1 and 12.2.

2. Reexamine the microscopic structure of bone tissue by observing a prepared microscope slide of ground compact bone. Use figure 12.3 of bone tissue to locate the following features:

   osteon (Haversian system)—cylinder-shaped unit

   central canal (Haversian canal; osteonic canal)—contains blood vessels and nerves

   lamella—concentric ring of matrix around central canal

   lacuna—small chamber for an osteocyte

   bone extracellular matrix—collagen and calcium phosphate

   canaliculus—minute tube containing cellular process

Critical Thinking Activity

Explain how bone cells embedded in a solid ground substance obtain nutrients and eliminate wastes.

_____

_____

_____

_____

FIGURE 12.1 Label the major structures of this long bone (living femur), using the terms provided. 1 2

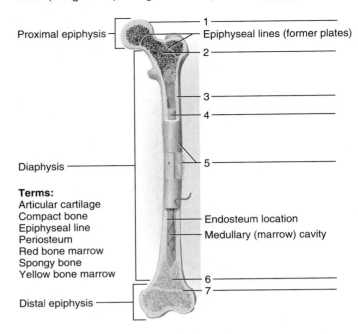

Proximal epiphysis

1 —
Epiphyseal lines (former plates)
2 —
3 —
4 —
5 —

Diaphysis —

Terms:
Articular cartilage
Compact bone
Epiphyseal line
Periosteum
Red bone marrow
Spongy bone
Yellow bone marrow

Endosteum location
Medullary (marrow) cavity

6 —
7 —

Distal epiphysis —

3. Observe the individual bone specimens and arrange them into groups, according to the following shapes and examples (fig. 12.4): 4

   long—femur; humerus; phalanges

   short—carpals; tarsals

   flat—ribs; scapula; most cranial bones

   irregular—vertebra; some facial bones as sphenoid

   sesamoid (round)—patella

FIGURE 12.2 Label the structures associated with a bone, using the terms provided. 2 3

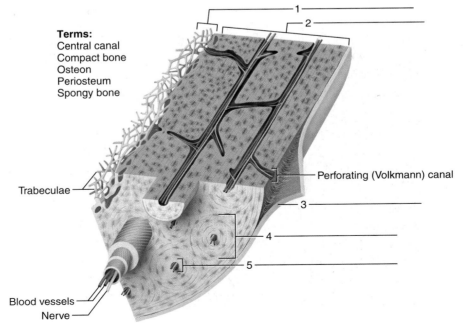

Terms:
Central canal
Compact bone
Osteon
Periosteum
Spongy bone

1 —
2 —

Perforating (Volkmann) canal

Trabeculae —

3 —
4 —
5 —

Blood vessels —
Nerve —

FIGURE 12.3 Micrograph of ground compact bone tissue (160×).

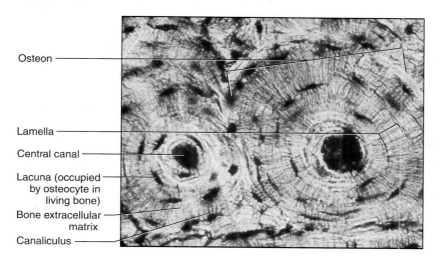

Osteon

Lamella

Central canal

Lacuna (occupied
by osteocyte in
living bone)

Bone extracellular
matrix

Canaliculus

4. Complete Part A of Laboratory Report 12.
5. Examine the sectioned long bones and locate the following (fig. 12.4):

epiphysis
proximal—nearest torso
distal—farthest from torso
epiphyseal plate—growth zone of hyaline cartilage
articular cartilage—on ends of epiphyses
diaphysis—shaft between epiphyses
periosteum—membrane around bone (except articular cartilage) of dense irregular connective tissue
compact (dense) bone—forms diaphysis and epiphyseal surfaces

spongy (cancellous) bone—within epiphyses
trabeculae—a structural lattice of plates in spongy bone
medullary (marrow) cavity—hollow chamber
endosteum—thin membrane lining medullary cavity of reticular connective tissue
yellow bone marrow—occupies medullary cavity
red bone marrow—occupies spongy bone in some epiphyses and flat bones

6. Use the dissecting microscope to observe the compact bone and spongy bone of the sectioned specimens. Also examine the marrow in the medullary cavity and the spaces within the spongy bone of the fresh specimen.
7. Complete Parts B and C of the laboratory report.

FIGURE 12.4 Radiograph (X ray) of a child's hand showing numerous epiphyseal plates. Only single epiphyseal plates develop in the hand and fingers.

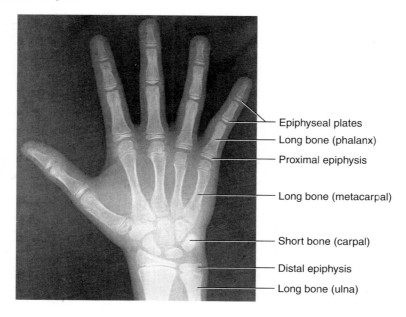

Epiphyseal plates

Long bone (phalanx)

Proximal epiphysis

Long bone (metacarpal)

Short bone (carpal)

Distal epiphysis

Long bone (ulna)

Demonstration Activity

Examine a fresh chicken bone and a chicken bone that has been soaked for several days in vinegar or overnight in dilute hydrochloric acid. Wear disposable gloves for handling these bones. This acid treatment removes the inorganic salts from the bone extracellular matrix. Rinse the bones in water and note the texture and flexibility of each (fig. 12.5a). Based on your observations, what quality of the fresh bone seems to be due to the inorganic salts removed by the acid treatment? 3

_____

_____

Examine the specimen of chicken bone that has been exposed to high temperature (baked at 121°C/250°F for 2 hours). This treatment removes the protein and other organic substances from the bone extracellular matrix (fig. 12.5b). What quality of the fresh bone seems to be due to these organic materials? 3

_____

_____

FIGURE 12.5 Results of fresh chicken bone demonstration: (*a*) soaked in vinegar; (*b*) baked in oven.

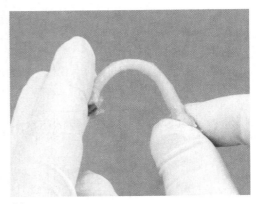

(a)

(b)

Name _____

Date _____

Section _____

The A corresponds to the Learning Outcome(s) listed at the beginning of the
laboratory exercise.

Bone Structure and Classification

Part A Assessments

Complete the following statements: (*Note:* Questions 1–6 pertain to bone classification by shape.)

1. A bone that has a wide surface is classified as a(an) _____ bone. 4

2. The bones of the wrist are examples of _____ bones. 4

3. The bone of the thigh is an example of a(an) _____ bone. 4

4. Vertebrae are examples of _____ bones. 4

5. The patella (kneecap) is an example of a large _____ bone. 4

6. The bones of the skull that form a protective covering for the brain are examples of _____ bones. 4

7. Distinguish between the epiphysis and the diaphysis of a long bone. 1 _____

_____

8. Describe where cartilage is found on the surface of a long bone. 1 _____

_____

9. Describe where dense irregular connective tissue is found on the surface of a long bone. 1 _____

_____

10. Distinguish the locations and tissues between the periosteum and the endosteum. 1 _____

_____

_____

Part B Assessments

Complete the following:

1. What structural differences did you note between compact bone and spongy bone? 2 _____

_____

_____

_____

2. How are these structural differences related to the locations and functions of these two types of bone? **5** _____

_____

_____

3. From your observations, how does the marrow in the medullary cavity compare with the marrow in the spaces of the

spongy bone? **5** _____

_____

_____

Part C Assessments

Identify the structures indicated in figure 12.6.

FIGURE 12.6   Identify the structures indicated in (a) the unsectioned long bone (fifth metatarsal) and (b) the partially sectioned long bone, using the terms provided. **1** **2**

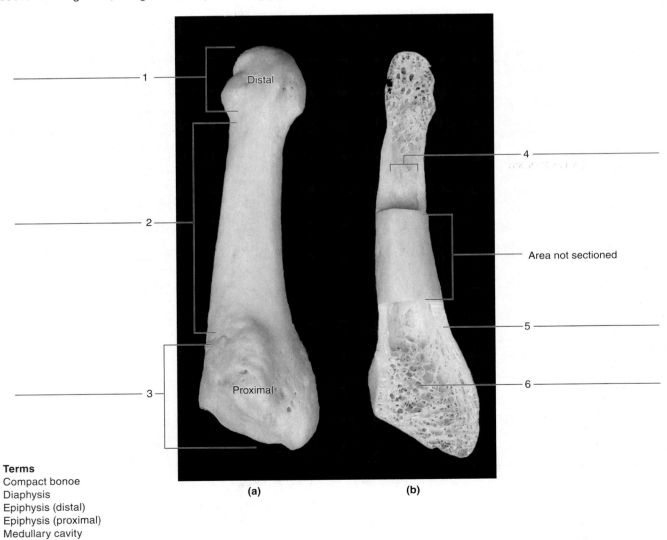

Terms
Compact bonoe
Diaphysis
Epiphysis (distal)
Epiphysis (proximal)
Medullary cavity
Spongy bone

(a)   (b)

Organization of the Skeleton

Pre-Lab

1. Carefully read the introductory material and examine the entire lab content.
2. Be familiar with the axial and appendicular skeleton (from lecture or the textbook).
3. Visit www.mhhe.com/martinseriesl for pre-lab questions.

Materials Needed

Articulated human skeleton

For Learning Extension Activity:
Colored pencils

For Demonstration Activity:
Radiographs (X rays) of skeletal structures

The skeleton can be divided into two major portions: (1) the axial skeleton, which consists of the bones and cartilages of the head, neck, and trunk, and (2) the appendicular skeleton, which consists of the bones of the limbs and those that anchor the limbs to the axial skeleton. The bones that anchor the limbs include the pectoral and pelvic girdles.

The number of bones in the adult skeleton is often reported to be 206 for both males and females. However, at birth the number of bones is closer to 275 as many ossification centers are still composed of cartilage, and some bones form during childhood. For example, each hip bone includes an ilium, ischium, and pubis, and many long bones have three ossification centers separated by epiphyseal plates. The sternum, composed of a manubrium, body, and xiphoid process, becomes a single bone much later than when we reach our full height. Some people have additional bones not considered in the total number. Sesamoid bones other than the the patellae may develop in the hand or the foot. Also, extra bones sometimes form in the skull within the sutures; these are called sutural (wormian) bones.

Special terminology is used to describe the features of a bone. The term used depends on whether the feature is a type of projection, articulation, depression, or opening. Many of these features can be noted when viewing radiographs. Some of the features can be palpated if they are located near the surface of the body.

Purpose of the Exercise

To review the organization of the skeleton, the major bones of the skeleton, and the terms used to describe skeletal structures.

Learning Outcomes

After completing this exercise, you should be able to

1. Distinguish between the axial skeleton and the appendicular skeleton.
2. Locate and label the major bones of the human skeleton.
3. Associate the terms used to describe skeletal structures and locate examples of such structures on the human skeleton.

Procedure—Organization of the Skeleton

1. Label figure 13.1.
2. Examine the human skeleton and locate the following parts. As you locate the following bones, note the number of each in the skeleton. Palpate as many of the corresponding bones in your skeleton as possible.

axial skeleton	
skull	
cranium	(8)
face	(14)
middle ear bone	(6)
hyoid bone	(1)
vertebral column	
vertebra	(24)
sacrum	(1)
coccyx	(1)
thoracic cage	
rib	(24)
sternum	(1)

FIGURE 13.1 Label the major bones of the skeleton: (*a*) anterior view; (*b*) posterior view, using the terms provided. 🔺 2️⃣

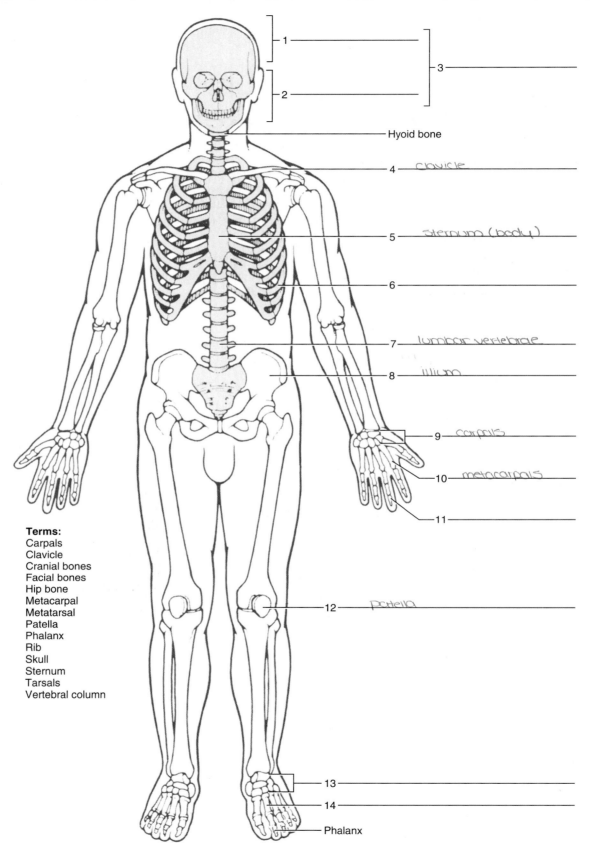

1

2

3

Hyoid bone

4 _clavicle_

5 _sternum (body)_

6

7 _lumbar vertebrae_

8 _illium_

9 _carpals_

10 _metacarpals_

11

Terms:
Carpals
Clavicle
Cranial bones
Facial bones
Hip bone
Metacarpal
Metatarsal
Patella
Phalanx
Rib
Skull
Sternum
Tarsals
Vertebral column

12 _patella_

13

14

Phalanx

FIGURE 13.1 *Continued.*

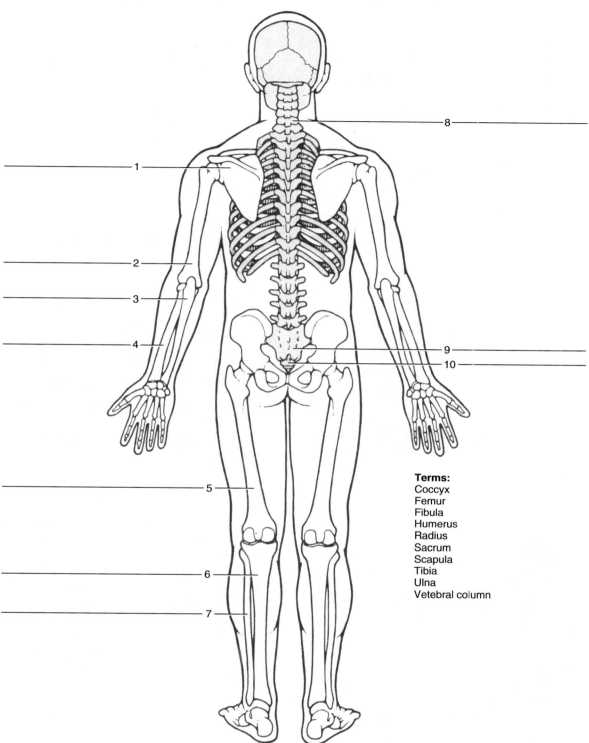

Terms:
Coccyx
Femur
Fibula
Humerus
Radius
Sacrum
Scapula
Tibia
Ulna
Vetebral column

appendicular skeleton

    pectoral girdle

        scapula   (2)

        clavicle   (2)

    upper limbs

        humerus   (2)

        radius   (2)

        ulna   (2)

        carpal   (16)

        metacarpal   (10)

        phalanx   (28)

    pelvic girdle

        hip bone (coxa; pelvic   (2)
           bone; innominate)

    lower limbs

        femur   (2)

        tibia   (2)

        fibula   (2)

        patella   (2)

        tarsal   (14)

        metatarsal   (10)

        phalanx   (28)

    Total   **206 bones**

Learning Extension Activity

Use colored pencils to distinguish the individual bones in figure 13.1.

3. Bone features (bone markings) can be grouped together in a category of projections, articulations, depressions, or openings. Within each category more specific examples occur. The bones listed in this section only represent an example of a location in the human body. Locate each of the following features on the example bone listed, noting the size, shape, and location in the human skeleton.

Projections:

  crest (ridgelike)—hip bone

  epicondyle (superior to condyle)—femur

  line (linea) (slightly raised ridge)—femur

  process (prominent)—vertebra

  protuberance (outgrowth)—skull (occipital)

  ramus (extension)—hip bone

  spine (thornlike)—scapula

  trochanter (large)—femur

  tubercle (knoblike)—humerus

  tuberosity (rough elevation)—tibia

Articulations:

  condyle (rounded process)—skull (occipital)

  facet (nearly flat)—vertebra

  head (expanded end)—femur

Depressions:

  fossa (shallow basin)—humerus

  fovea (tiny pit)—femur

Openings:

  canal (tubular passage)—skull (occipital)

  fissure (slit)—skull (orbit)

  foramen (hole)—vertebra

  meatus (tubelike)—skull (temporal)

  sinus (cavity)—skull (maxilla)

Critical Thinking Activity

Locate and name the largest foramen in the skull.

foramen magnum

Locate and name the largest foramen in the skeleton.

4. Complete Parts A, B, C, and D of Laboratory Report 13.

Demonstration Activity

Images on radiographs (X rays) are produced by allowing X rays from an X-ray tube to pass through a body part and to expose photographic film positioned on the opposite side of the part. The image that appears on the film after it is developed reveals the presence of parts with different densities. Bone, for example, is very dense tissue and is a good absorber of X rays. Thus, bone generally appears light on the film. Air-filled spaces, on the other hand, absorb almost no X rays and appear as dark areas on the film. Liquids and soft tissues absorb intermediate quantities of X rays, so they usually appear in various shades of gray.

Examine the available radiographs of skeletal structures by holding each film in front of a light source. Identify as many of the bones and features as you can.

Name _____

Date _____

Section _____

The ⚠ corresponds to the Learning Outcome(s) listed at the beginning of the laboratory exercise.

Organization of the Skeleton

Part A Assessments

Complete the following statements:

1. The extra bones that sometimes develop between the flat bones of the skull are called _____.

2. Small bones occurring in some tendons are called _____ bones.

3. The cranium and facial bones compose the _____.

4. The _____ bone supports the tongue. ⚠

5. The _____ at the inferior end of the sacrum is composed of several fused vertebrae. ⚠

6. Most ribs are attached anteriorly to the _____. ⚠

7. The thoracic cage is composed of _____ pairs of ribs. ⚠

8. The scapulae and clavicles together form the _____.

9. Which of the following bones is not part of the appendicular skeleton: clavicle, femur, scapula, sternum? _____ ⚠

10. The wrist is composed of eight bones called _____. ⚠

11. The hip bones are attached posteriorly to the _____. ⚠

12. The pelvic girdle (hip bones), sacrum, and coccyx together form the _____. ⚠

13. The _____ covers the anterior surface of the knee. ⚠

14. The bones that articulate with the distal ends of the tibia and fibula are called _____. ⚠

15. All finger and toe bones are called _____. ⚠

Part B Assessments

Match the terms in column A with the definitions in column B. Place the letter of your choice in the space provided. 🔼3

Column A	Column B
a. Condyle	_____ **1.** Small, nearly flat articular surface
b. Crest	__F___ **2.** Deep depression or shallow basin
c. Facet	_____ **3.** Rounded process
d. Fontanel	__E___ **4.** Opening or passageway
e. Foramen	__G___ **5.** Interlocking line of union
f. Fossa	__B___ **6.** Narrow, ridgelike projection
g. Suture	_____ **7.** Soft region between bones of skull

Part C Assessments

Match the terms in column A with the definitions in column B. Place the letter of your choice in the space provided. 🔼3

Column A	Column B
a. Fovea	_____ **1.** Tubelike passageway
b. Head	_____ **2.** Tiny pit or depression
c. Meatus	_____ **3.** Small, knoblike process
d. Sinus	_____ **4.** Thornlike projection
e. Spine	_____ **5.** Rounded enlargement at end of bone
f. Trochanter	__D___ **6.** Air-filled cavity within bone
g. Tubercle	_____ **7.** Relatively large process

Part D Assessments

Identify the bones indicated in figure 13.2.

FIGURE 13.2 Identify the bones in this random arrangement, using the terms provided. **2**

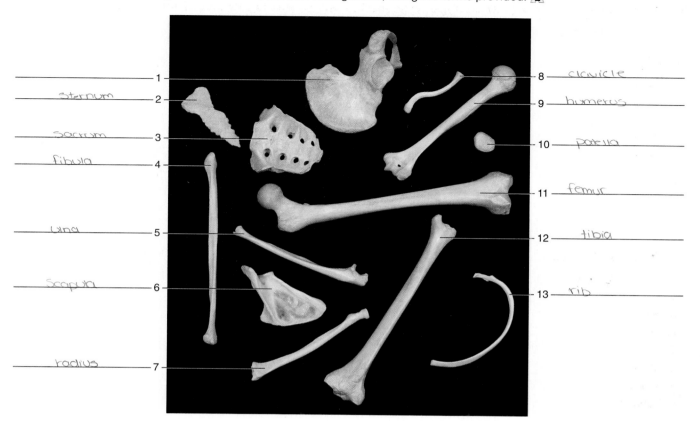

2 _Sternum_

3 _Sacrum_

4 _fibula_

5 _ulna_

6 _Scapula_

7 _radius_

8 _clavicle_

9 _humerus_

10 _patella_

11 _femur_

12 _tibia_

13 _rib_

Terms:

Clavicle	Patella	Scapula
Femur	Radius	Sternum
Fibula	Rib	Tibia
Hip bone	Sacrum	Ulna
Humerus		

Skull

 **Pre-Lab**

1. Carefully read the introductory material and examine the entire lab content.
2. Be familiar with the skull (from lecture or the textbook).
3. Visit www.mhhe.com/martinseriesl for pre-lab questions and Anatomy & Physiology Revealed animations.

 Materials Needed

Human skull, articulated
Human skull, disarticulated (Beauchene)
Human skull, sagittal section
For Learning Extension Activity:
Colored pencils

A human skull consists of twenty-two bones that, except for the lower jaw, are firmly interlocked along sutures. Eight of these immovable bones make up the braincase, or cranium, and thirteen more immovable bones and the mandible form the facial skeleton.

Associated bones with the skull include the auditory ossicles (middle ear bones) located within the temporal bones. They include the malleus, incus, and stapes that are examined in Laboratory Exercise 37. The stapes is considered the smallest bone in the human body among the total number of 206. The hyoid bone is suspended from the temporal bones and serves for tongue and larynx muscle attachments. A fractured hyoid bone is often used as an indication of strangulation as the cause of death.

After intramembranous ossification of the flat bones of the skull is complete, a fibrous type of joint remains between them. These joints are considered immovable and are known as sutures. The suture between the two parietal bones is the sagittal suture. The lambdoid suture is between the occipital and parietal bones. The coronal suture is bordered by the frontal bone and the parietal bones. The squamous suture is primarily along the superior border of the temporal bone with the parietal bone.

Sinus cavities are found in the frontal bone, ethmoid bone, sphenoid bone, and both maxillary bones. Because they are associated with the nasal passages, they are called paranasal sinuses. These air-filled sinuses are lined with a mucous membrane that is continuous with the nasal passages. The sinuses function to lighten the skull, assist in warming and humidifying the air during breathing, and provide resonance to our voice. If infected, the large maxillary sinuses are often misinterpreted to be a toothache.

Purpose of the Exercise

To examine the structure of the human skull and to identify the bones and major features of the skull.

Learning Outcomes

After completing this exercise, you should be able to

(1) Locate and label the bones of the skull and their major features.

(2) Locate and label the major sutures of the cranium.

(3) Locate and label the sinuses of the skull.

Procedure—Skull

1. Label figures 14.1, 14.2, 14.3, 14.4, and 14.5.
2. Examine the **cranial bones** of the articulated human skull and the sectioned skull. Also observe the corresponding disarticulated bones. Locate the following bones and features in the laboratory specimens and, at the same time, palpate as many of these bones and features in your skull as possible.

frontal bone	**(1)**
supraorbital foramen	
frontal sinus	
parietal bone	**(2)**
sagittal suture	
coronal suture	

occipital bone (1)
  lambdoid suture
  external occipital protuberance
  foramen magnum
  occipital condyle
temporal bone (2)
  squamous suture
  external acoustic meatus
  mandibular fossa
  mastoid process
  styloid process
  carotid canal
  jugular foramen
  internal acoustic meatus
  zygomatic process
sphenoid bone (1)
  sella turcica
  greater and lesser wings
  sphenoidal sinus
ethmoid bone (1)
  cribriform plate
  perpendicular plate
  superior nasal concha
  middle nasal concha
  ethmoidal sinus
  crista galli

3. Complete Parts A and B of Laboratory Report 14.
4. Examine the **facial bones** of the articulated and sectioned skulls and the corresponding disarticulated bones. Locate the following:

maxilla (2)
  maxillary sinus
  palatine process
  alveolar process
  alveolar arch
palatine bone (2)
zygomatic bone (2)
  temporal process
  zygomatic arch
lacrimal bone (2)
nasal bone (2)
vomer bone (1)
inferior nasal concha (2)
mandible (1)
  ramus
  mandibular condyle
  coronoid process
  alveolar border
  mandibular foramen
  mental foramen

5. Complete Part C of the laboratory report.

FIGURE 14.1 Label the anterior bones and features of the skull, using the terms provided. (If the line lacks the word bone, label the particular feature of that bone.) 1 2

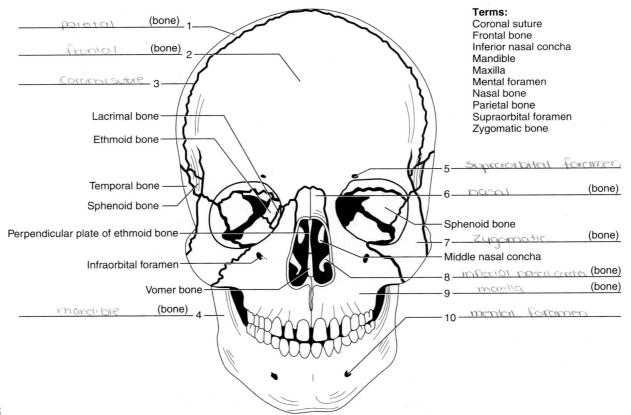

Terms:
Coronal suture
Frontal bone
Inferior nasal concha
Mandible
Maxilla
Mental foramen
Nasal bone
Parietal bone
Supraorbital foramen
Zygomatic bone

1. parietal (bone)
2. frontal (bone)
3. coronal suture
Lacrimal bone
Ethmoid bone
Temporal bone
Sphenoid bone
Perpendicular plate of ethmoid bone
Infraorbital foramen
Vomer bone
4. mandible (bone)
5. Supraorbital foramen
6. nasal (bone)
Sphenoid bone
7. zygomatic (bone)
Middle nasal concha
8. inferior nasal concha (bone)
9. maxilla (bone)
10. mental foramen

FIGURE 14.2 Label the lateral bones and features of the skull, using the terms provided. ⚊1 ⚊2

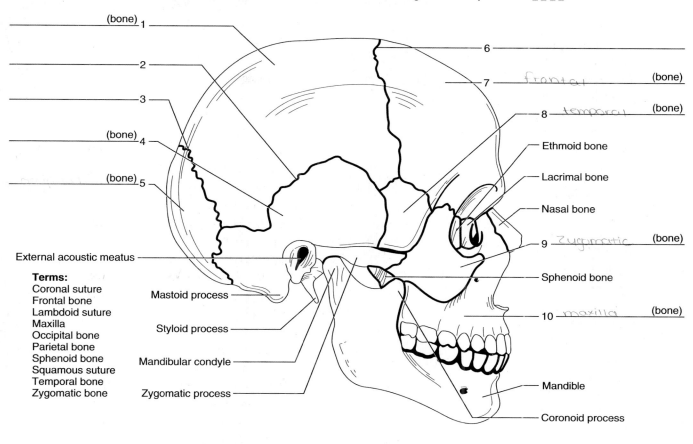

(bone) 1
2
3
(bone) 4
(bone) 5

6
7 _frontal_ (bone)
8 _temporal_ (bone)
Ethmoid bone
Lacrimal bone
Nasal bone
9 _zygomatic_ (bone)
Sphenoid bone
10 _maxilla_ (bone)

External acoustic meatus

Terms:
Coronal suture
Frontal bone
Lambdoid suture
Maxilla
Occipital bone
Parietal bone
Sphenoid bone
Squamous suture
Temporal bone
Zygomatic bone

Mastoid process
Styloid process
Mandibular condyle
Zygomatic process

Mandible
Coronoid process

FIGURE 14.3 Label the inferior bones and features of the skull, using the terms provided. ⚊1

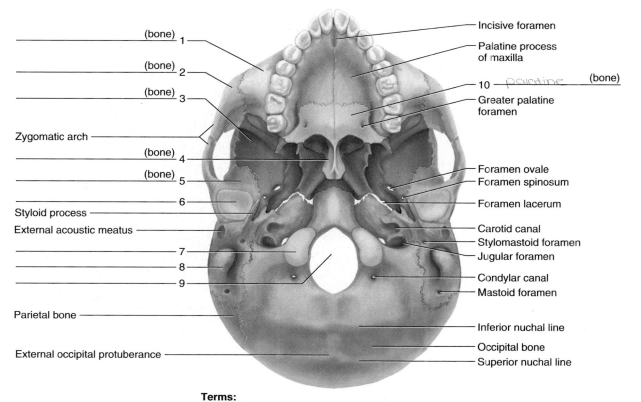

(bone) 1
(bone) 2
(bone) 3
Zygomatic arch
(bone) 4
(bone) 5
6
Styloid process
External acoustic meatus
7
8
9
Parietal bone
External occipital protuberance

Incisive foramen
Palatine process of maxilla
10 _palatine_ (bone)
Greater palatine foramen
Foramen ovale
Foramen spinosum
Foramen lacerum
Carotid canal
Stylomastoid foramen
Jugular foramen
Condylar canal
Mastoid foramen
Inferior nuchal line
Occipital bone
Superior nuchal line

Terms:
Foramen magnum   Occipital condyle   Temporal bone
Mandibular fossa   Palatine bone   Vomer bone
Mastoid process   Sphenoid bone   Zygomatic bone
Maxilla

109

FIGURE 14.4 Label the bones and features of the floor of the cranial cavity in this superior view, using the terms provided. 1

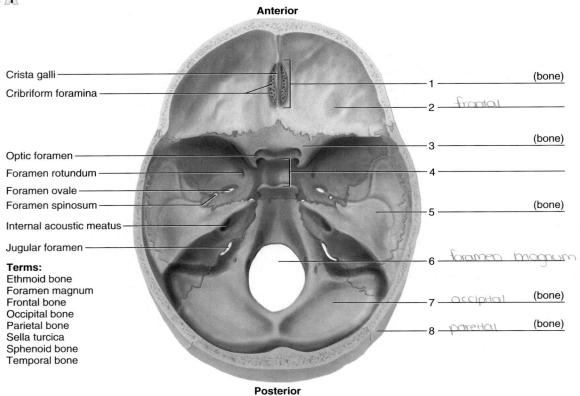

Anterior

Crista galli

Cribriform foramina

1 _____ (bone)

2 _frontal_

3 _____ (bone)

Optic foramen

Foramen rotundum

Foramen ovale

Foramen spinosum

Internal acoustic meatus

Jugular foramen

4 _____

5 _____ (bone)

6 _foramen magnum_

7 _occipital_ (bone)

8 _parietal_ (bone)

Terms:
Ethmoid bone
Foramen magnum
Frontal bone
Occipital bone
Parietal bone
Sella turcica
Sphenoid bone
Temporal bone

Posterior

FIGURE 14.5 Label the bones and features of the sagittal section of the skull, using the terms provided. 1 2 3

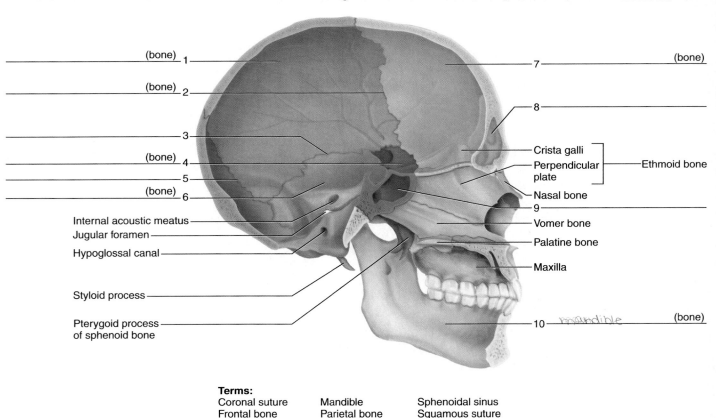

(bone) 1 _____

(bone) 2 _____

_____ 3

(bone) 4 _____

_____ 5

(bone) 6 _____

Internal acoustic meatus

Jugular foramen

Hypoglossal canal

Styloid process

Pterygoid process
of sphenoid bone

7 _____ (bone)

8 _____

Crista galli

Perpendicular plate } Ethmoid bone

Nasal bone

9 _____

Vomer bone

Palatine bone

Maxilla

10 _mandible_ (bone)

Terms:

Coronal suture	Mandible	Sphenoidal sinus
Frontal bone	Parietal bone	Squamous suture
Frontal sinus	Sphenoid bone	Temporal bone
Lambdoid suture		

Learning Extension Activity

Use colored pencils to differentiate the bones illustrated in figures 14.1 and 14.2. Select a different color for each bone in the series. This activity should help you locate various bones shown in different views in the figures.

6. Reexamine figures 14.3 to 14.5 and locate the following features of the human skull:

carotid canal
condylar canal
foramen lacerum
foramen magnum
foramen ovale
foramen rotundum
foramen spinosum
greater palatine foramen
hypoglossal canal
incisive foramen
inferior orbital fissure
infraorbital foramen

internal acoustic meatus
jugular foramen
mandibular foramen
mastoid foramen
mental foramen
optic canal
stylomastoid foramen
superior orbital fissure
supraorbital foramen

8. Complete Part D of the laboratory report.

Critical Thinking Activity

Examine the inside of the cranium on a sectioned skull. What appears to be the weakest area? Explain your answer.

_____

_____

_____

_____

Name _____

Date _____

Section _____

The 🄰 corresponds to the Learning Outcome(s) listed at the beginning of the laboratory exercise.

Skull

Part A Assessments

Match the bones in column A with the features in column B. Place the letter of your choice in the space provided. (Some answers are used more than once.) 🄰

Column A	Column B
a. Ethmoid bone	_____ **1.** Forms sagittal, coronal, squamous, and lambdoid sutures
b. Frontal bone	_____ **2.** Cribriform plate
c. Occipital bone	_____ **3.** Crista galli
d. Parietal bone	_____ **4.** External acoustic meatus
e. Sphenoid bone	_____ **5.** Foramen magnum
f. Temporal bone	_____ **6.** Mandibular fossa
	_____ **7.** Mastoid process
	_____ **8.** Middle nasal concha
	_____ **9.** Occipital condyle
	_____ **10.** Sella turcica
	_____ **11.** Styloid process
	_____ **12.** Supraorbital foramen

Part B Assessments

Complete the following statements:

1. The _____ suture joins the frontal bone to the parietal bones. 🄖

2. The parietal bones are firmly interlocked along the midline by the _____ suture. 🄖

3. The _____ suture joins the parietal bones to the occipital bone. 🄖

4. The temporal bones are joined to the parietal bones along the _____. sutures. 🄖

5. Name the three cranial bones that contain sinuses. 🄗 _____

_____

6. Name a facial bone that contains a sinus. 🄗 _____

Part C Assessments

Match the bones in column A with the characteristics in column B. Place the letter of your choice in the space provided. △

Column A	Column B
a. Inferior nasal concha	_____ **1.** Forms bridge of nose
b. Lacrimal bone	_____ **2.** Only movable bone in the facial skeleton
c. Mandible	_____ **3.** Contains coronoid process
d. Maxilla	_____ **4.** Creates prominence of cheek inferior and lateral to the eye
e. Nasal bone	_____ **5.** Contains sockets of upper teeth
f. Palatine bone	_____ **6.** Forms inferior portion of nasal septum
g. Vomer bone	_____ **7.** Forms anterior portion of zygomatic arch
h. Zygomatic bone	_____ **8.** Scroll-shaped bone
	_____ **9.** Forms anterior roof of mouth
	_____ **10.** Contains mental foramen
	_____ **11.** Forms posterior roof of mouth
	_____ **12.** Scalelike part in medial wall of orbit

Part D Assessments

Identify the numbered bones and features of the skulls indicated in figures 14.6, 14.7, 14.8, 14.9, and 14.10.

FIGURE 14.6 Identify the bones and features indicated on this anterior view of the skull, using the terms provided. /1\

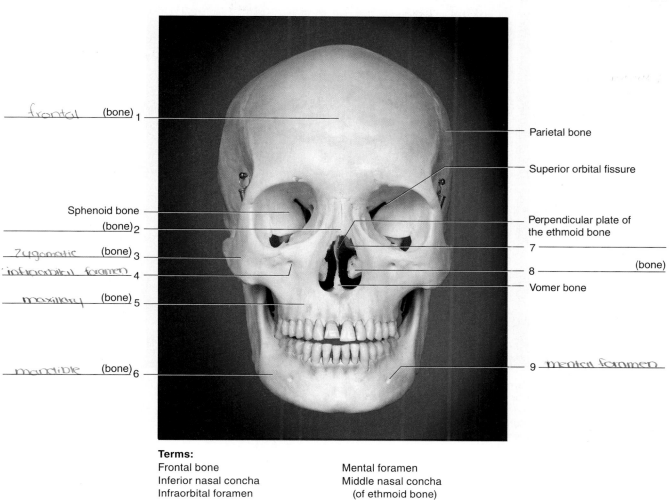

frontal (bone) 1

Sphenoid bone
(bone) 2

Zygomatic (bone) 3

infraorbital foramen 4

maxillary (bone) 5

mandible (bone) 6

Parietal bone

Superior orbital fissure

Perpendicular plate of the ethmoid bone

7

8 (bone)

Vomer bone

9 mental foramen

Terms:

Frontal bone

Inferior nasal concha

Infraorbital foramen

Mandible

Maxilla

Mental foramen

Middle nasal concha (of ethmoid bone)

Nasal bone

Zygomatic bone

FIGURE 14.7 Identify the bones and features indicated on this lateral view of the skull, using the terms provided. ⚠ 2

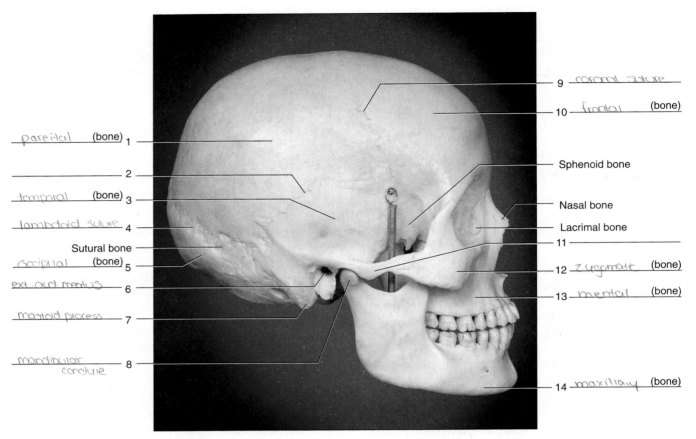

pareital ___ (bone) 1

___ 2

tempral ___ (bone) 3

lambdoid suture ___ 4

Sutural bone ___

occipital ___ (bone) 5

ext. aud meatus ___ 6

mastoid process ___ 7

mandibular ___ 8
condule

9 ___ coronal suture

10 ___ frontal ___ (bone)

Sphenoid bone

Nasal bone

Lacrimal bone

11 ___

12 ___ Zygomatic ___ (bone)

13 ___ mental ___ (bone)

14 ___ maxillary ___ (bone)

Terms:

Coronal suture
External acoustic meatus
Frontal bone
Lambdoid suture
Mandible
Mandibular condyle
Mastoid process
Maxilla

Occipital bone
Parietal bone
Squamous suture
Temporal bone
Zygomatic bone
Zygomatic process
   (of temporal bone)

FIGURE 14.8   Identify the bones and features indicated on this inferior view of the skull, using the terms provided. 🔺

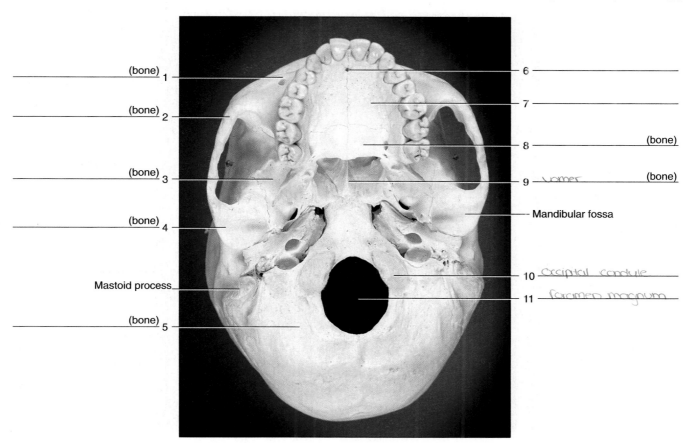

(bone) 1
(bone) 2
(bone) 3
(bone) 4
Mastoid process
(bone) 5

6
7
8 (bone)
9 _Vomer_ (bone)
Mandibular fossa
10 _Occipital condyle_
11 _foramen magnum_

Terms:

Foramen magnum
Incisive foramen
Maxilla
Occipital bone
Occipital condyle
Palatine bone

Palatine process of maxilla
Sphenoid bone
Temporal bone
Vomer bone
Zygomatic bone

FIGURE 14.9 Identify the bones and features on this floor of the cranial cavity of a skull, using the terms provided. 🄰

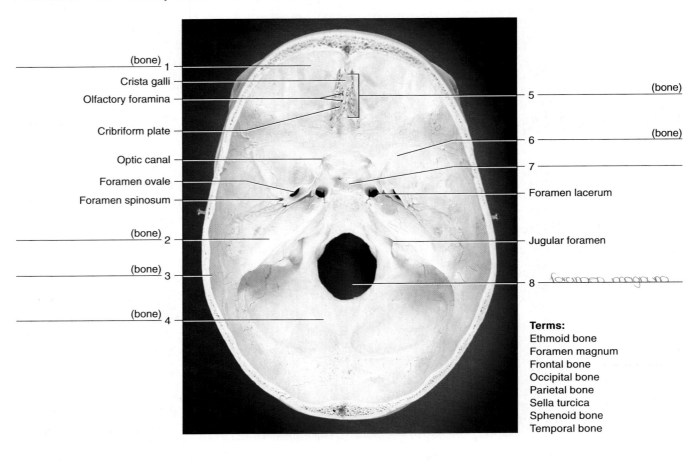

(bone) 1

Crista galli

Olfactory foramina

Cribriform plate

Optic canal

Foramen ovale

Foramen spinosum

(bone) 2

(bone) 3

(bone) 4

(bone) 5

(bone) 6

7

Foramen lacerum

Jugular foramen

8 — foramen magnum

Terms:
Ethmoid bone
Foramen magnum
Frontal bone
Occipital bone
Parietal bone
Sella turcica
Sphenoid bone
Temporal bone

FIGURE 14.10 Identify the bones on this disarticulated skull, using the terms provided. 🄰

1

2

3

4

5

6

7

Ethmoid bone

Inferior nasal concha

Vomer bone

Terms:

Frontal bone	Sphenoid bone
Mandible	Temporal bone
Maxilla	Zygomatic
Parietal bone	bone

Vertebral Column and Thoracic Cage

Pre-Lab

1. Carefully read the introductory material and examine the entire lab content.
2. Be familiar with the vertebral column and the thoracic cage (from lecture or the textbook).
3. Visit www.mhhe.com/martinseriesl for pre-lab questions.

Materials Needed

Human skeleton, articulated
Samples of cervical, thoracic, and lumbar vertebrae
Human skeleton, disarticulated

The vertebral column, consisting of twenty-six bones, extends from the skull to the pelvis and forms the vertical axis of the human skeleton. The vertebral column includes seven cervical vertebrae, twelve thoracic vertebrae, five lumbar vertebrae, one sacrum of five fused vertebrae, and one coccyx of usually four fused vertebrae. To help you to remember the number of cervical, thoracic, and lumbar vertebrae from superior to inferior, consider this saying: breakfast at 7, lunch at 12, and dinner at 5. These vertebrae are separated from one another by cartilaginous intervertebral discs and are held together by ligaments.

The thoracic cage surrounds the thoracic and upper abdominal cavities. It includes the ribs, the thoracic vertebrae, the sternum, and the costal cartilages. Men and women, although variations can exist, possess the same total bone number of 206.

Purpose of the Exercise

To examine the vertebral column and the thoracic cage of the human skeleton and to identify the bones and major features of these parts.

Learning Outcomes

After completing this exercise, you should be able to

1. Identify the major features of the vertebral column.
2. Locate the features of a vertebra.
3. Distinguish among a cervical, thoracic, and lumbar vertebra and the sacrum and coccyx.
4. Identify the structures of the thoracic cage.
5. Distinguish between true and false ribs.

Procedure A—The Vertebral Column

1. Label figures 15.1, 15.2, 15.3, and 15.4.
2. Examine the vertebral column of the human skeleton and locate the following bones and features. At the same time, locate as many of the corresponding bones and features in your skeleton as possible.

atlas (C1)	**(1)**
axis (C2)	**(1)**
vertebra prominens (C7)	**(1)**
cervical vertebrae (includes atlas, axis, and vertebra prominens)	**(7)**
thoracic vertebrae	**(12)**
lumbar vertebrae	**(5)**
intervertebral discs (fibrocartilage)	
vertebral canal (contains spinal cord)	
sacrum	**(1)**
coccyx	**(1)**
cervical curvature	
thoracic curvature	
lumbar curvature	
sacral (pelvic) curvature	
intervertebral foramina (passageway for spinal nerves)	

 Critical Thinking Activity

Note the four curvatures of the vertebral column. What functional advantages exist with curvatures for skeletal structure instead of a straight vertebral column?

_____

_____

_____

_____

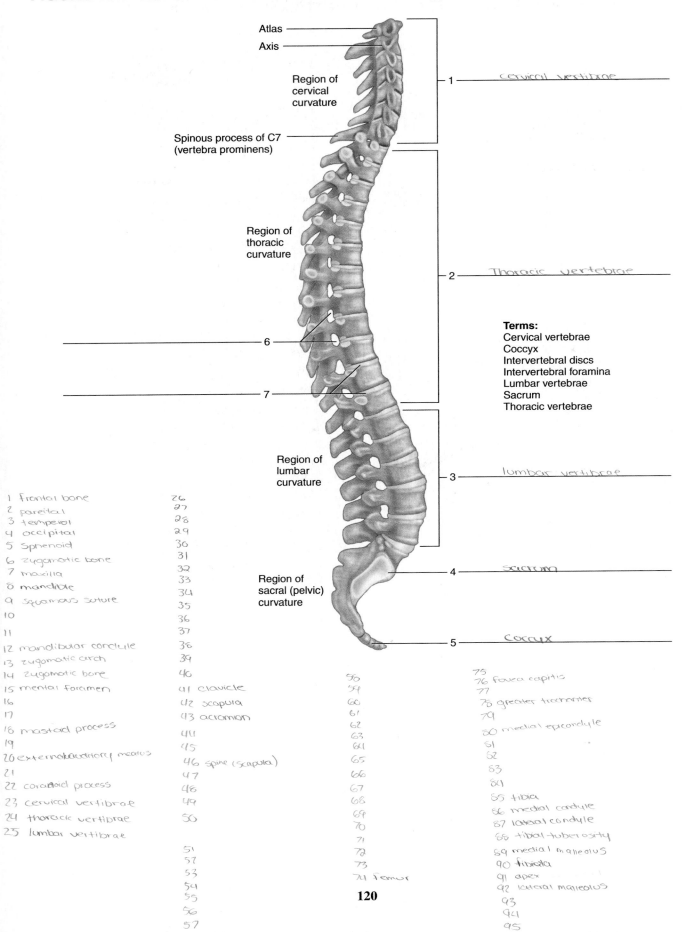

Atlas

Axis

Region of
cervical
curvature

Spinous process of C7
(vertebra prominens)

Region of
thoracic
curvature

1 — cervical vertibrae

2 — Thoracic vertebrae

6

7

Terms:
Cervical vertebrae
Coccyx
Intervertebral discs
Intervertebral foramina
Lumbar vertebrae
Sacrum
Thoracic vertebrae

Region of
lumbar
curvature

3 — lumbar vertibrae

Region of
sacral (pelvic)
curvature

4 — sacrum

5 — coccyx

1 frontal bone
2 pareital
3 tempeval
4 occipital
5 sphenoid
6 zygomatic bone
7 maxilla
8 mandible
9 squamous suture
10
11
12 mandibular condyle
13 zugomatic arch
14 zugomatic bone
15 mental foramen
16
17
18 mastoid process
19
20 external auditory meatus
21
22 coracoid process
23 cervical vertibrae
24 thoracic vertibrae
25 lumbar vertibrae

26
27
28
29
30
31
32
33
34
35
36
37
38
39
40
41 clavicle
42 scapula
43 acromion
44
45
46 spine (scapula)
47
48
49
50

51
52
53
54
55
56
57

58
59
60
61
62
63
64
65
66
67
68
69
70
71
72
73
74 femur

75
76 fovea capitis
77
78 greater trochanter
79
80 medial epicondyle
81
82
83
84
85 tibia
86 medial condyle
87 lateral condyle
88 tibial tuberosity
89 medial malleolus
90 fibiula
91 apex
92 lateral malleolus
93
94
95

120

FIGURE 15.2 Label the superior features of (*a*) the atlas and the superior and right lateral features of (*b*) the axis by placing the correct numbers in the spaces provided. (The broken arrow indicates a transverse foramen.) 2 3

_____ Body
_____ Dens
__4__ Facet that articulates with occipital condyle
_____ Spinous process
_____ Superior articular facet
__6__ Transverse foramen
__5__ Transverse process
_____ Vertebral foramen

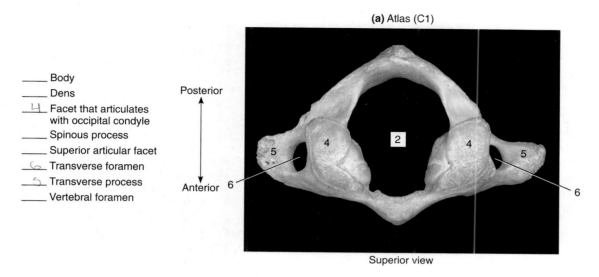

(a) Atlas (C1)

Posterior
Anterior

Superior view

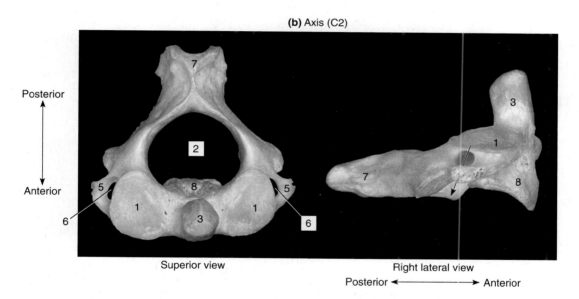

(b) Axis (C2)

Posterior
Anterior

Superior view

Right lateral view

Posterior ◀——▶ Anterior

FIGURE 15.3 Label the superior and right lateral features of the (a) cervical, (b) thoracic, and (c) lumbar vertebrae by placing the correct numbers in the spaces provided. (The broken arrow indicates a transverse foramen.) 🄰 🄱

Superior views   Right lateral views

_____ Body
_____ Inferior vertebral notch
_____ Lamina
_____ Pedicle
_____ Spinous process
_____ Superior articular process
_____ Transverse foramen
_____ Transverse process
_____ Vertebral foramen

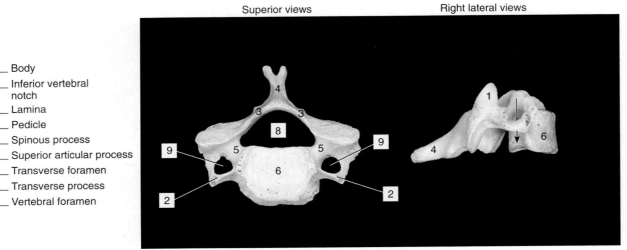

(a) Cervical vertebra

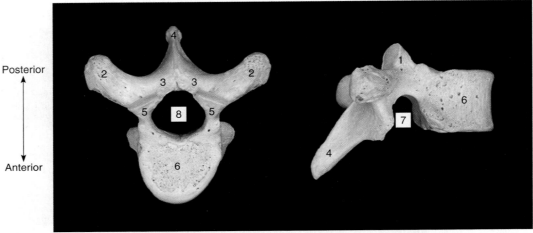

Posterior

Anterior

(b) Thoracic vertebra

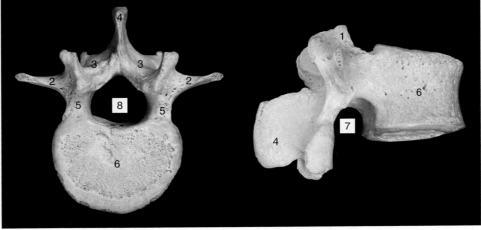

(c) Lumbar vertebra

Posterior ⟷ Anterior

FIGURE 15.4 Label this diagram of the sacrum by placing the correct numbers in the spaces provided. 2 3

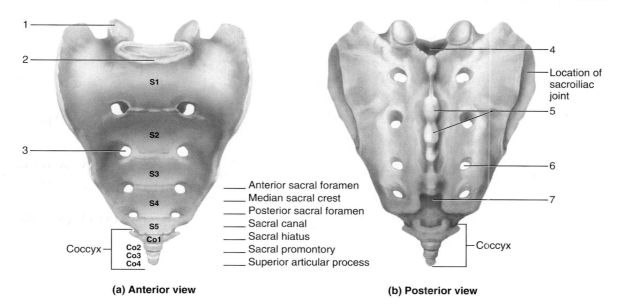

1
2
S1
S2
3
S3
S4
S5
Co1
Coccyx — Co2
Co3
Co4

_____ Anterior sacral foramen
_____ Median sacral crest
_____ Posterior sacral foramen
_____ Sacral canal
_____ Sacral hiatus
_____ Sacral promontory
_____ Superior articular process

4
Location of sacroiliac joint
5
6
7
— Coccyx

(a) Anterior view   **(b) Posterior view**

3. Compare the available samples of cervical, thoracic, and lumbar vertebrae by noting differences in size and shapes and by locating the following features:

    vertebral foramen
    body
    pedicles
    laminae
    spinous process
    transverse processes
    facets
    superior articular processes
    inferior articular processes
    inferior vertebral notch
    transverse foramina
    dens of axis

4. Examine the sacrum and coccyx. Locate the following features:

    sacrum
      superior articular process
      posterior sacral foramen
      anterior sacral foramen
      sacral promontory
      sacral canal
      median sacral crest
      sacral hiatus
    coccyx

5. Complete Parts A and B of Laboratory Report 15.

Procedure B—The Thoracic Cage

1. Label figure 15.5.
2. Examine the thoracic cage of the human skeleton and locate the following bones and features:

    rib
      head
      tubercle
      neck
      shaft
      anterior (sternal) end
      facets
      true ribs (pairs 1–7)
      false ribs (pairs 8–12) (includes floating ribs)
        floating ribs (pairs 11–12)
    costal cartilages (hyaline cartilage)
    sternum
      jugular (suprasternal) notch
      clavicular notch
      manubrium
      sternal angle
      body
      xiphoid process

3. Complete Parts C and D of the laboratory report.

FIGURE 15.5 Label the bones and features of the thoracic cage (anterior view), using the terms provided. 4 5

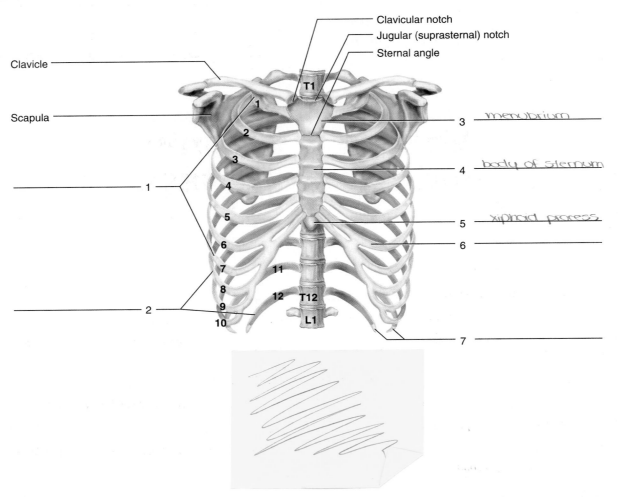

Clavicular notch

Jugular (suprasternal) notch

Sternal angle

Clavicle

Scapula

T1

1

2

3

4

5

6

7

8

9

10

11

12

T12

L1

1

2

3 — menubrium

4 — body of sternum

5 — xiphoid process

6

7

Name _____

Date _____

Section _____

The ⚠ corresponds to the Learning Outcome(s) listed at the beginning of the laboratory exercise.

Vertebral Column and Thoracic Cage

Part A Assessments

Complete the following statements:

1. The vertebral column encloses and protects the _____. ⚠

2. The number of separate bones in the vertebral column of an adult is _____. ⚠

3. The seventh cervical vertebra is called the _____ and has an obvious spinous process surface feature that can be palpated. ⚠

4. The _____ of the vertebrae support the weight of the head and trunk. ⚠

5. The _____ separate adjacent vertebrae, and they soften the forces created by walking. ⚠

6. The intervertebral foramina provide passageways for _____. ⚠

7. Transverse foramina of _____ vertebrae serve as passageways for blood vessels leading to the brain. ⚠

8. The first vertebra also is called the _____. ⚠

9. When the head is moved from side to side, the first vertebra pivots around the _____ of the second vertebra. ⚠

10. The _____ vertebrae have the largest and strongest bodies. ⚠

11. The number of vertebrae that fuse to form the sacrum is _____. ⚠

Part B Assessments

Based on your observations, compare typical cervical, thoracic, and lumbar vertebrae in relation to the characteristics indicated in the table. For your responses, consider characteristics such as size, shape, presence or absence, and unique features. ⚠ ⚠

Vertebra	Number	Size	Body	Spinous Process	Transverse Foramina
Cervical					
Thoracic					
Lumbar					

Part C Assessments

Complete the following statements:

1. The adult skeleton of most men and women contains a total of _____ (number) bones.

2. The last two pairs of ribs that have no cartilaginous attachments to the sternum are sometimes called _____ ribs. ⑤

3. There are _____ pairs of true ribs. ⑤

4. Costal cartilages are composed of _____ tissue. ④

5. The manubrium articulates with the _____ on its superior border. ④

6. List three general functions of the thoracic cage. _____

_____

Part D Assessments

Identify the bones and features indicated in the radiograph of the neck in figure 15.6.

FIGURE 15.6   Identify the bones and features indicated in this radiograph of the neck (lateral view), using the terms provided. ① ②

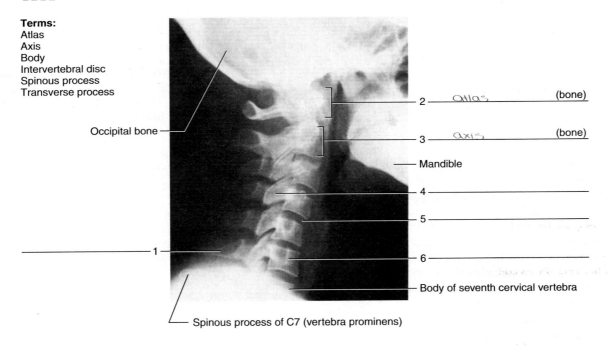

Terms:
Atlas
Axis
Body
Intervertebral disc
Spinous process
Transverse process

Occipital bone

2 ___atlas___ (bone)

3 ___axis___ (bone)

Mandible

4 _____

5 _____

1 _____

6 _____

Body of seventh cervical vertebra

Spinous process of C7 (vertebra prominens)

Pectoral Girdle and Upper Limb

Pre-Lab

1. Carefully read the introductory material and examine the entire lab content.
2. Be familiar with the pectoral girdle and upper limb bones (from lecture or the textbook).
3. Visit www.mhhe.com/martinseriesl for pre-lab questions.

Materials Needed

Human skeleton, articulated
Human skeleton, disarticulated

For Learning Extension Activity:
Colored pencils

A pectoral girdle (shoulder girdle) consists of an anterior clavicle and a posterior scapula. A pectoral girdle represents an incomplete ring (girdle) of bones as the posterior scapulae do not meet each other, but muscles extend from their medial borders to the vertebral column. The clavicles on their medial ends form a joint with the manubrium of the sternum. The pectoral girdle supports the upper limb and serves as attachments for various muscles that move the upper limb. This allows considerable flexibility of the shoulder. Relatively loose attachments of the pectoral girdle with the humerus allow a wide range of movements, but shoulder joint injuries are somewhat common. Additionally, the clavicle is a frequently broken bone when one reaches with an upper limb to break a fall.

Each upper limb includes a humerus in the arm, a radius and ulna in the forearm, and eight carpals, five metacarpals, and fourteen phalanges in the hand. (Anatomically, arm represents the region from shoulder to elbow, forearm is elbow to the wrist, and hand includes the wrist to the end of digits.) These bones form the framework of the upper limb. They also function as parts of levers when muscles contract.

Purpose of the Exercise

To examine the bones of the pectoral girdle and upper limb and to identify the major features of these bones.

Learning Outcomes

After completing this exercise, you should be able to

1. Locate and identify the bones of the pectoral girdle and their major features.
2. Locate and identify the bones of the upper limb and their major features.

Procedure A—The Pectoral Girdle

1. Label figures 16.1 and 16.2.
2. Examine the bones of the pectoral girdle and locate the following features. At the same time, locate as many of the corresponding surface bones and features of your own skeleton as possible.

clavicle
   sternal (medial) end
   acromial (lateral) end
scapula
   spine
   acromion
   glenoid cavity
   coracoid process
   borders
      superior border
      medial (vertebral) border
      lateral (axillary) border
   fossae
      supraspinous fossa
      infraspinous fossa
      subscapular fossa
   angles
      superior angle
      inferior angle

FIGURE 16.1 Label the bones and features of the right shoulder and upper limb (anterior view), using the terms provided. 🅰 🅰

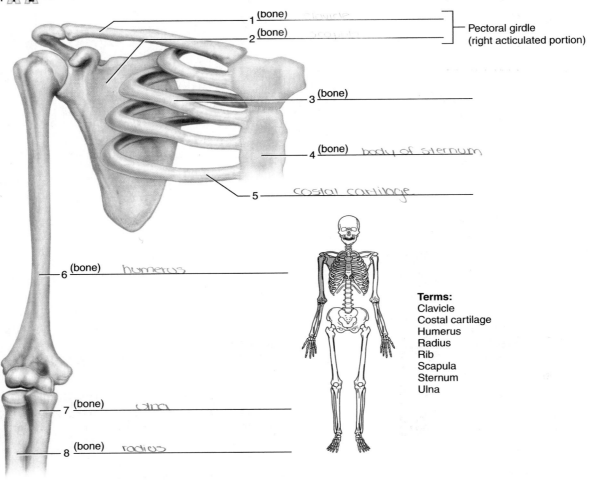

1 (bone) _clavicle_
2 (bone) _scapula_

— Pectoral girdle (right acticulated portion)

3 (bone)

4 (bone) _body of sternum_

5 _costal cartilage_

6 (bone) _humerus_

Terms:
Clavicle
Costal cartilage
Humerus
Radius
Rib
Scapula
Sternum
Ulna

7 (bone) _ulna_

8 (bone) _radius_

FIGURE 16.2 Label (a) the anterior surface and (b) the posterior surface of the right scapula, using the terms provided. 🅰

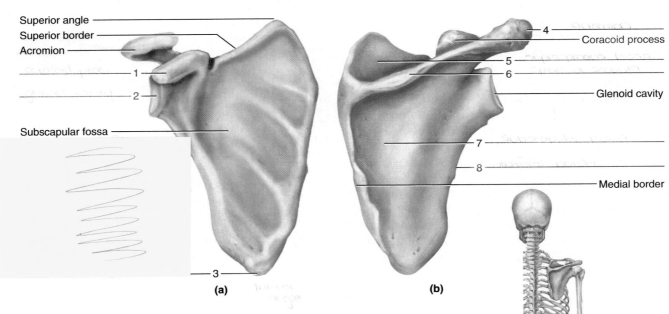

Superior angle
Superior border
Acromion

1

2

Subscapular fossa

3

(a)

4
Coracoid process

5
6

Glenoid cavity

7

8

Medial border

(b)

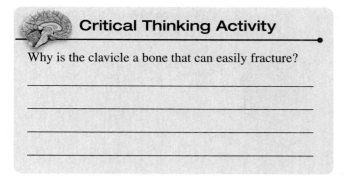

Critical Thinking Activity

Why is the clavicle a bone that can easily fracture?

_____

_____

_____

_____

3. Complete Part A of Laboratory Report 16.

Procedure B—The Upper Limb

1. Label figures 16.3, 16.4, and 16.5.

2. Examine the following bones and features of the upper limb:

humerus
- proximal features
  - head
  - greater tubercle
  - lesser tubercle
  - anatomical neck
  - surgical neck
  - intertubercular sulcus (groove)
- shaft
  - deltoid tuberosity
- distal features
  - capitulum
  - trochlea
  - medial epicondyle
  - lateral epicondyle
  - coronoid fossa
  - olecranon fossa

radius
- head of radius
- radial tuberosity
- styloid process of radius
- ulnar notch of radius

ulna
- trochlear notch (semilunar notch)
- radial notch of ulna
- olecranon process
- coronoid process
- styloid process of ulna
- head of ulna

carpal bones
- proximal row (listed lateral to medial)
  - scaphoid
  - lunate
  - triquetrum
  - pisiform
- distal row (listed medial to lateral)
  - hamate
  - capitate
  - trapezoid
  - trapezium

The following mnemonic device will help you learn the eight carpals:

**So Long Top Part
Here Comes The Thumb**

The first letter of each word corresponds to the first letter of a carpal. This device arranges the carpals in order for the proximal, transverse row of four bones from lateral to medial, followed by the distal, transverse row from medial to lateral, which ends nearest the thumb. This arrangement assumes the hand is in the anatomical position.

---

metacarpals (I–V)
phalanges
- proximal phalanx
- middle phalanx
- distal phalanx

3. Complete Parts B, C, and D of the laboratory report.

FIGURE 16.3 Label the (a) anterior features and (b) posterior features of the right humerus, using the terms provided.

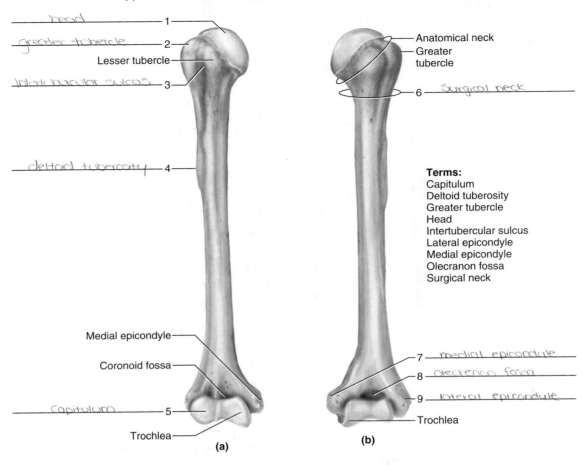

head — 1

greater tubercle — 2

Lesser tubercle —

Intertubucular sulcus — 3

deltoid tuberosity — 4

Medial epicondyle —

Coronoid fossa —

Capitulum — 5

Trochlea —

(a)

Anatomical neck

Greater tubercle

6 — _Surgical neck_

Terms:
Capitulum
Deltoid tuberosity
Greater tubercle
Head
Intertubercular sulcus
Lateral epicondyle
Medial epicondyle
Olecranon fossa
Surgical neck

7 — _medial epicondyle_

8 — _olecranon fossa_

9 — _lateral epicondyle_

Trochlea

(b)

FIGURE 16.4 Label the major anterior features of the right radius and ulna, using the terms provided.

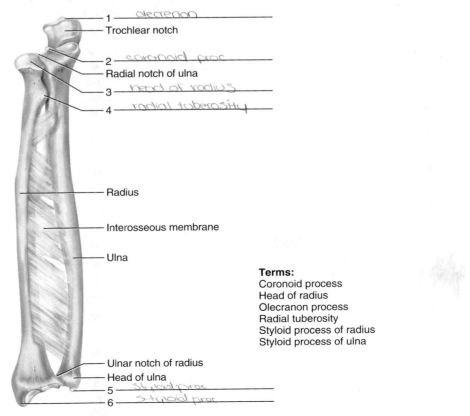

1 — _olecranon_

Trochlear notch

2 — _coronoid proc_

Radial notch of ulna

3 — _head of radius_

4 — _radial tuberosity_

Radius

Interosseous membrane

Ulna

Terms:
Coronoid process
Head of radius
Olecranon process
Radial tuberosity
Styloid process of radius
Styloid process of ulna

Ulnar notch of radius

Head of ulna

5 — _styloid proc_

6 — _styloid proc_

130

Learning Extension Activity

Use different colored pencils to distinguish the individual bones in figure 16.5.

FIGURE 16.5 Label the bones and groups of bones in this anterior view of the right hand, using the terms provided. **2**

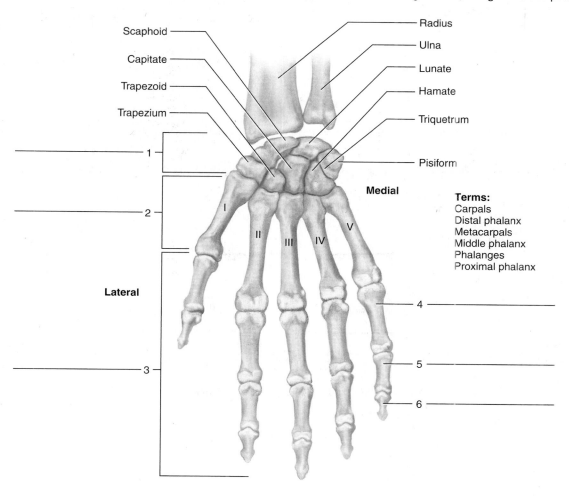

Scaphoid

Capitate

Trapezoid

Trapezium

Radius

Ulna

Lunate

Hamate

Triquetrum

Pisiform

1

2

3

Lateral

Medial

I II III IV V

Terms:
Carpals
Distal phalanx
Metacarpals
Middle phalanx
Phalanges
Proximal phalanx

4

5

6

Name _____

Date _____

Section _____

The ⚠ corresponds to the Learning Outcome(s) listed at the beginning of the laboratory exercise.

Pectoral Girdle and Upper Limb

Part A Assessments

Complete the following statements:

1. The pectoral girdle is an incomplete ring because it is open in the back between the _____. ⚠

2. The medial ends of the clavicles articulate with the _____ of the sternum. ⚠

3. The lateral ends of the clavicles articulate with the _____ of the scapulae. ⚠

4. The _____ is a bone that serves as a brace between the sternum and the scapula. ⚠

5. The _____ divides the scapula into unequal portions. ⚠

6. The tip of the shoulder is the _____ of the scapula. ⚠

7. Near the lateral end of the scapula, the _____ curves anteriorly and inferiorly from the clavicle. ⚠

8. The glenoid cavity of the scapula articulates with the _____ of the humerus. ⚠

Part B Assessments

Match the bones in column A with the bones and features in column B. Place the letter of your choice in the space provided. ②

Column A	Column B
a. Carpals	_____ 1. Capitate
b. Humerus	_____ 2. Coronoid fossa
c. Metacarpals	_____ 3. Deltoid tuberosity
d. Phalanges	_____ 4. Greater tubercle
e. Radius	_____ 5. Five palmar bones
f. Ulna	_____ 6. Fourteen bones in digits
	_____ 7. Intertubercular sulcus
	_____ 8. Lunate
	_____ 9. Olecranon fossa
	_____ 10. Radial tuberosity
	_____ 11. Trapezium
	_____ 12. Trochlear notch

Part C Assessments

Identify the bones and features indicated in the radiographs of figures 16.6, 16.7, and 16.8.

FIGURE 16.6 Identify the bones and features indicated on this radiograph of the right elbow (anterior view), using the terms provided. [2]

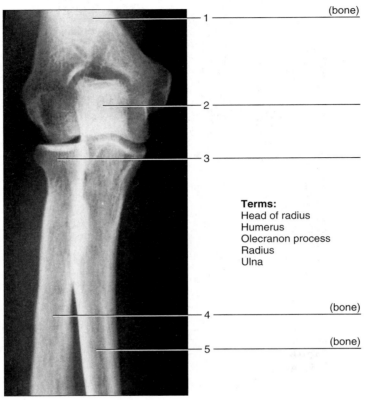

1 _____ (bone)

2 _____

3 _____

Terms:
Head of radius
Humerus
Olecranon process
Radius
Ulna

4 _____ (bone)

5 _____ (bone)

FIGURE 16.7 Identify the bones and features indicated on this radiograph of the anterior view of the right shoulder, using the terms provided. [1] [2]

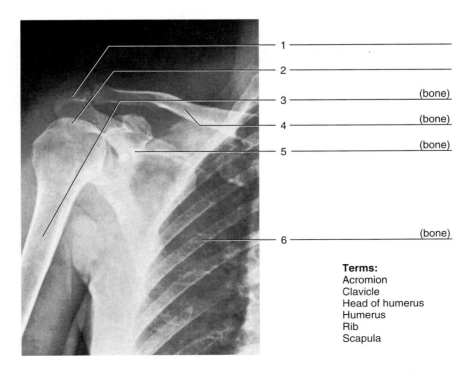

1 _____

2 _____

3 _____ (bone)

4 _____ (bone)

5 _____ (bone)

6 _____ (bone)

Terms:
Acromion
Clavicle
Head of humerus
Humerus
Rib
Scapula

FIGURE 16.8 Identify the bones indicated on this radiograph of the right hand (anterior view), using the terms provided. **2**

Terms:
Carpals
Distal phalanx
Metacarpals
Phalanges
Proximal phalanx

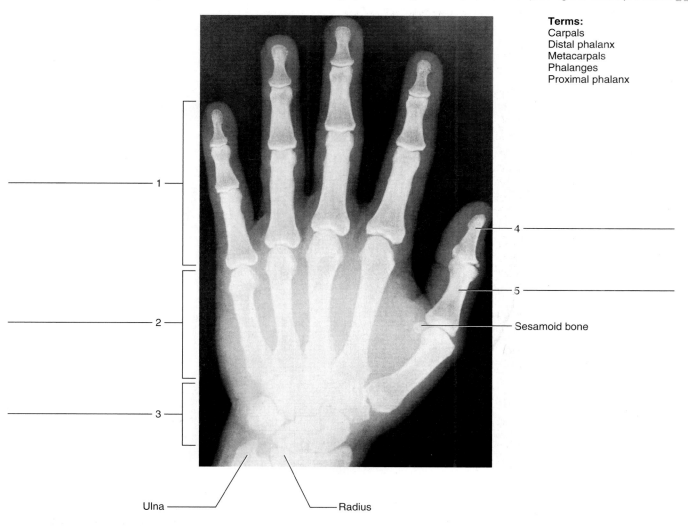

1

4

5

Sesamoid bone

2

3

Ulna

Radius

Part D Assessments

Identify the bones of the hand in figure 16.9.

FIGURE 16.9   Label the bones numbered on this anterior view of the right hand by placing the correct numbers in the spaces provided. 2

_____ Capitate

_____ Distal phalanges

_____ Hamate

_____ Lunate

_____ Metacarpals

_____ Middle phalanges

_____ Pisiform

_____ Proximal phalanges

_____ Scaphoid

_____ Trapezium

_____ Trapezoid

_____ Triquetrum

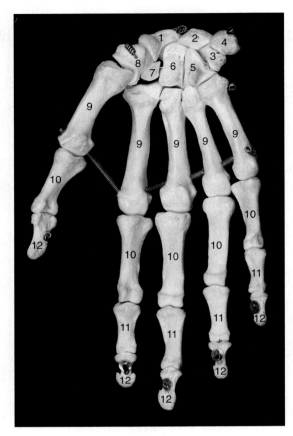

Pelvic Girdle and Lower Limb

Pre-Lab

1. Carefully read the introductory material and examine the entire lab content.
2. Be familiar with the pelvic girdle and lower limb bones (from lecture or the textbook).
3. Visit www.mhhe.com/martinseries1 for pre-lab questions.

Materials Needed

Human skeleton, articulated
Human skeleton, disarticulated
Male and female pelves

For Learning Extension Activity:
Colored pencils

The pelvic girdle includes two hip bones that articulate with each other anteriorly at the pubic symphysis. Posteriorly, each hip bone articulates with a sacrum at a sacroiliac joint. Together, the pelvic girdle, sacrum, and coccyx comprise the pelvis. The pelvis, in turn, provides support for the trunk of the body and provides attachments for the lower limbs. The pelvis supports and protects the viscera in the pelvic region of the abdominopelvic cavity. The pelvic outlet, with boundaries of the coccyx, inferior border of the pubic symphysis, and between the ischial tuberosities, is clinically important in females. The pelvic outlet must be large enough to successfully accommodate the fetal head during a vaginal delivery. Each acetabulum of a hip bone articulates with the head of the femur of a lower limb. The hip joint structures provide a more stable joint compared to a shoulder joint.

The bones of the lower limb form the framework of the thigh, leg, and foot. (Anatomically, thigh represents the region from hip to knee, leg is from knee to ankle, and foot includes the ankle to the end of the toes.) Each limb includes a femur in the thigh, a patella in the knee, a tibia and fibula in the leg, and seven tarsals, five metatarsals, and fourteen phalanges in the foot. These bones and large muscles are for weight-bearing support and locomotion and thus are considerably larger and possess more stable joints than those of an upper limb.

Purpose of the Exercise

To examine the bones of the pelvic girdle and lower limb, and to identify the major features of these bones.

Learning Outcomes

After completing this exercise, you should be able to

1. Locate and identify the bones of the pelvic girdle and their major features.
2. Locate and identify the bones of the lower limb and their major features.

Procedure A—The Pelvic Girdle

1. Label figures 17.1 and 17.2.
2. Examine the bones of the pelvic girdle and locate the following:

   hip bone (coxa; pelvic bone; innominate)
   ilium
       iliac crest
       anterior superior iliac spine
       posterior superior iliac spine
       greater sciatic notch (portion in ischium)
       iliac fossa
   ischium
       ischial tuberosity
       ischial spine
       lesser sciatic notch
   pubis
       pubic symphysis (joint between pubic bones)
       pubic arch (subpubic angle) (between pubic bones of pelvis)
   acetabulum (formed by ilium, ischium, and pubis)
   obturator foramen (formed by ischium and pubis)

3. Complete Part A of Laboratory Report 17.

FIGURE 17.1 Label the bones of the pelvis (anterosuperior view). ◢

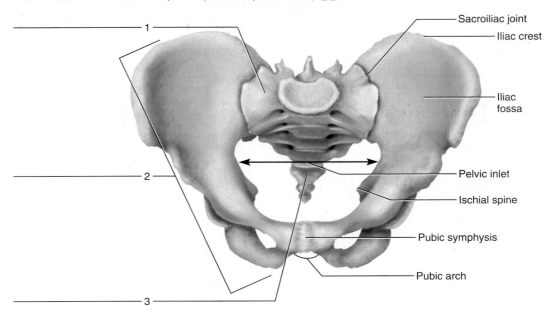

1

2

3

Sacroiliac joint

Iliac crest

Iliac fossa

Pelvic inlet

Ischial spine

Pubic symphysis

Pubic arch

FIGURE 17.2 Label the lateral features of the right hip bone, using the terms provided. ◢

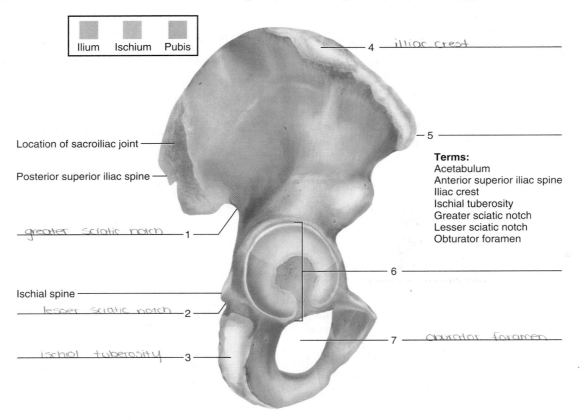

Ilium   Ischium   Pubis

Location of sacroiliac joint

Posterior superior iliac spine

greater sciatic notch — 1

Ischial spine

lesser sciatic notch — 2

ischial tuberosity — 3

4 — illiac crest

5

Terms:
Acetabulum
Anterior superior iliac spine
Iliac crest
Ischial tuberosity
Greater sciatic notch
Lesser sciatic notch
Obturator foramen

6

7 — oburator foramen

Examine the male and female pelves. Look for major differences between them. Note especially the flare of the iliac bones, the angle of the pubic arch, the distance between the ischial spines and ischial tuberosities, and the curve and width of the sacrum. In what ways are the differences you observed related to the function of the female pelvis as a birth canal?

_____

_____

_____

_____

Procedure B—The Lower Limb

1. Label figures 17.3, 17.4, and 17.5.
2. Examine the bones of the lower limb and locate each of the following:

  femur
    proximal features
      head
      fovea capitis
      neck
      greater trochanter
      lesser trochanter
    shaft
      gluteal tuberosity
      linea aspera

FIGURE 17.3   Label the features of (a) the anterior surface and (b) the posterior surface of the right femur, using the terms provided. The anterior and posterior views of the patella are included in the figure. **2**

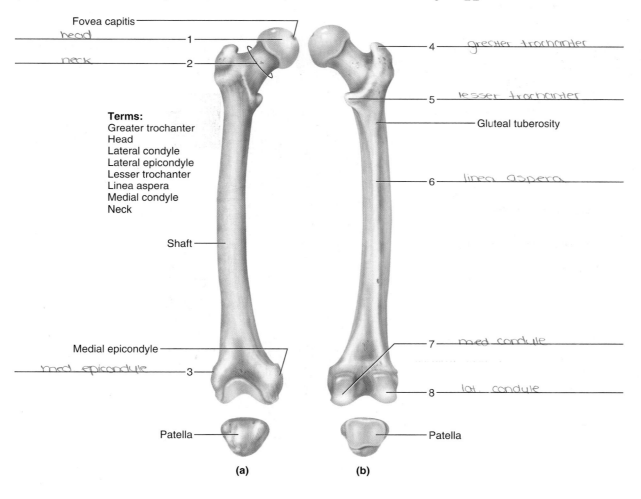

Fovea capitis

head

neck

1

2

Terms:
Greater trochanter
Head
Lateral condyle
Lateral epicondyle
Lesser trochanter
Linea aspera
Medial condyle
Neck

Shaft

Medial epicondyle

med epicondyle

3

Patella

(a)

4 grecter trochanter

5 lesser trochanter

Gluteal tuberosity

6 linea aspera

7 med condyle

8 lat. condyle

Patella

(b)

distal features
    lateral epicondyle
    medial epicondyle
    lateral condyle
    medial condyle
patella
tibia
  medial condyle
  lateral condyle
  tibial tuberosity
  anterior border (crest; margin)
  medial malleolus
fibula
  head
  lateral malleolus

tarsal bones
  talus
  calcaneus
  navicular
  cuboid
  lateral cuneiform
  intermediate (middle) cuneiform
  medial cuneiform
metatarsal bones
phalanges
  proximal phalanx
  middle phalanx
  distal phalanx

3. Complete Parts B, C, and D of the laboratory report.

FIGURE 17.4 Label the features of the right tibia and fibula in this anterior view, using the terms provided. **2**

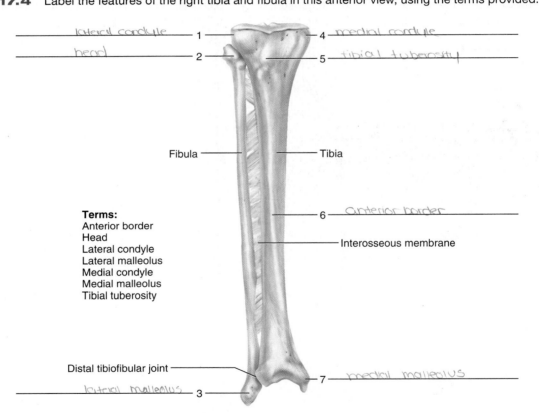

1. lateral condyle
2. head
3. lateral malleolus
4. medial condyle
5. tibial tuberosity
6. anterior border
7. medial malleolus

Fibula

Tibia

Interosseous membrane

Distal tibiofibular joint

Terms:
Anterior border
Head
Lateral condyle
Lateral malleolus
Medial condyle
Medial malleolus
Tibial tuberosity

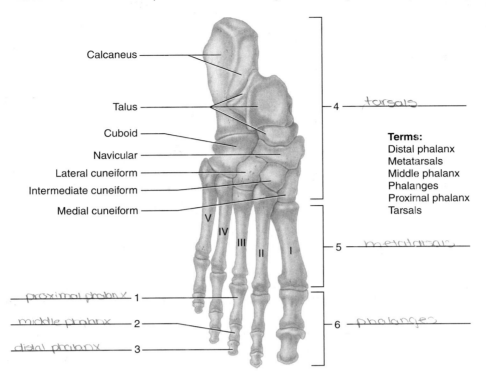

Calcaneus

Talus ———— 4 — tarsals

Cuboid

Navicular

Lateral cuneiform

Intermediate cuneiform

Medial cuneiform

V

IV

III

II

I

Terms:
Distal phalanx
Metatarsals
Middle phalanx
Phalanges
Proximal phalanx
Tarsals

5 — metatarsals

proximal phalanx — 1

middle phalanx — 2

distal phalanx — 3

6 — phalanges

Learning Extension Activity

Use different colored pencils to distinguish the individual bones in figure 17.5.

Laboratory Report

17

Name _____

Date _____

Section _____

The ◮ corresponds to the Learning Outcome(s) listed at the beginning of the laboratory exercise.

Pelvic Girdle and Lower Limb

Part A Assessments

Complete the following statements:

1. The pelvic girdle consists of two _____. ◮

2. The head of the femur articulates with the _____ of the hip bone. ◮

3. The _____ is the largest portion of the hip bone. ◮

4. The distance between the _____ represents the shortest diameter of the pelvic outlet. ◮

5. The pubic bones come together anteriorly to form the joint called the _____. ◮

6. The _____ is the superior margin of the ilium that causes the prominence of the hip. ◮

7. When a person sits, the _____ of the ischium supports the weight of the body. ◮

8. The angle formed by the pubic bones below the pubic symphysis is called the _____. ◮

9. The _____ is the largest foramen in the skeleton. ◮

10. The ilium joins the sacrum at the _____ joint. ◮

Part B Assessments

Match the bones in column A with the features in column B. Place the letter of your choice in the space provided. ◭

Column A	Column B
a. Femur	_____ 1. Middle phalanx
b. Fibula	_____ 2. Lesser trochanter
c. Metatarsals	_____ 3. Medial malleolus
d. Patella	_____ 4. Fovea capitis
e. Phalanges	_____ 5. Calcaneus
f. Tarsals	_____ 6. Lateral cuneiform
g. Tibia	_____ 7. Tibial tuberosity
	_____ 8. Talus
	_____ 9. Linea aspera
	_____ 10. Lateral malleolus
	_____ 11. Sesamoid bone
	_____ 12. Five bones that form the instep

Part C Assessments

Identify the bones and features indicated in the radiographs of figures 17.6, 17.7, and 17.8.

FIGURE 17.6 Identify the bones and features indicated on this radiograph of the anterior view of the pelvic region, using the terms provided. 🔺1 🔺2

Terms:
Head of femur
Ilium
Obturator foramen
Pubic symphysis
Pubis
Sacrum

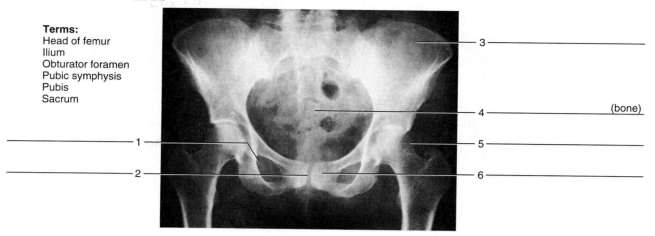

3

4 _____ (bone)

1

5

2

6

FIGURE 17.7 Identify the bones and features indicated in this radiograph of the right knee (anterior view), using the terms provided. 🔺2

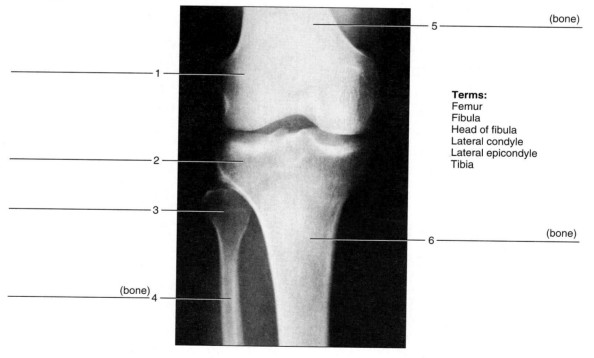

5 _____ (bone)

1

Terms:
Femur
Fibula
Head of fibula
Lateral condyle
Lateral epicondyle
Tibia

2

3

6 _____ (bone)

(bone) 4

FIGURE 17.8 Identify the bones indicated in this radiograph of the right foot (medial side), using the terms provided. 2

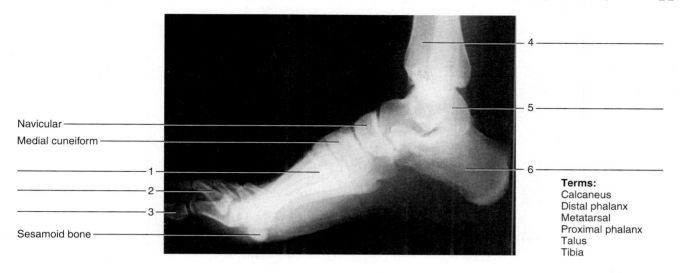

Navicular
Medial cuneiform
1
2
3
Sesamoid bone

4
5
6

Terms:
Calcaneus
Distal phalanx
Metatarsal
Proximal phalanx
Talus
Tibia

Part D Assessments

Identify the bones of the foot in figure 17.9.

FIGURE 17.9 Identify the bones indicated on this superior view of the right foot, using the terms provided. 2

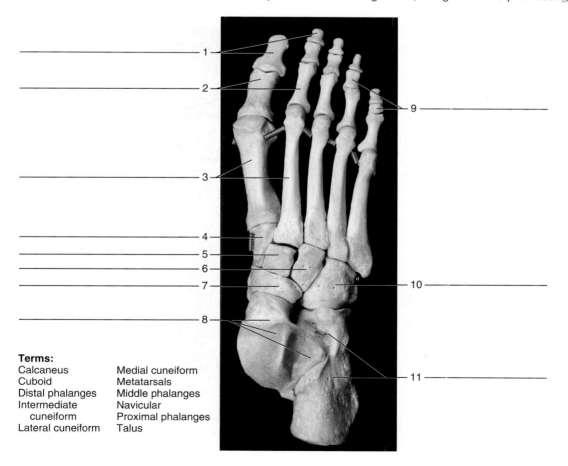

1
2
3
4
5
6
7
8

9
10
11

Terms:
Calcaneus   Medial cuneiform
Cuboid   Metatarsals
Distal phalanges Middle phalanges
Intermediate   Navicular
  cuneiform   Proximal phalanges
Lateral cuneiform Talus

Fetal Skeleton

Pre-Lab

1. Carefully read the introductory material and examine the entire lab.
2. Be familiar with bone development and a fetal skull (from lecture or the textbook).
3. Visit www.mhhe.com/martinseriesl for pre-lab questions and LabCam videos.

Materials Needed

Human skull
Human adult skeleton
Human fetal skeleton
Meterstick
Metric ruler
Note: If a fetal skeleton is not available, use the figures represented in this laboratory exercise with the metersticks included in the photographs.

A human adult skeleton consists of 206 bones. However, the skeleton of a newborn child may have nearly 275 bones due to the numerous ossification centers and epiphyseal plates present for many bones. Development of the embryo and the fetus progresses from cephalic toward caudal regions of the body. As a result, there is a noticeable difference between the proportions of the head, the entire body, and the upper and lower limbs between a fetus and an adult. Additionally, the bones most important for protection are among the early ossification sites.

Purpose of the Exercise

To examine and make comparisons between fetal skeletons and adult skeletons.

Learning Outcomes

After completing this exercise, you should be able to

1. Locate and describe the function of the fontanels of a fetal skull.
2. Measure and compare the size of the cranial bones and the facial bones between a fetal skull and the adult skull.
3. Describe the ossification of the frontal and temporal bones.
4. Measure and compare the total height and the upper and lower limb lengths for a fetal skeleton and an adult skeleton.
5. Locate and describe six fetal bones or body regions that are not completely ossified, but become single bones or groups of bones in the adult skeleton.
6. Estimate the gestational age of a fetus from measured lengths.

Procedure A—Skulls

1. Examine the fetal skull from three views (figures 18.1, 18.2, and 18.3). Fontanels (sometimes called "soft spots") are fibrous membranes that have not completed ossification. Fontanels allow for compression of the cranium during a vaginal delivery and for some additional brain development after birth. The posterior and lateral fontanels close during the first year after birth, but the anterior fontanel does not close until nearly two years of age. Locate the following fontanels:

Mastoid fontanel	**(2)**
Sphenoidal fontanel	**(2)**
Posterior fontanel	**(1)**
Anterior fontanel	**(1)**

2. Complete Part A of Laboratory Report 18.

FIGURE 18.1 Superior view of the fetal skull.

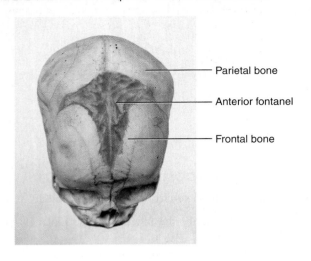

- Parietal bone
- Anterior fontanel
- Frontal bone

FIGURE 18.2 Lateral view of the fetal skull.

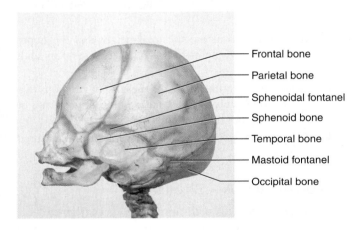

- Frontal bone
- Parietal bone
- Sphenoidal fontanel
- Sphenoid bone
- Temporal bone
- Mastoid fontanel
- Occipital bone

FIGURE 18.3 Posterior view of fetal skull.

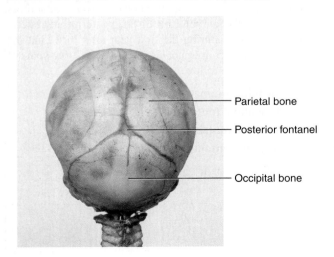

- Parietal bone
- Posterior fontanel
- Occipital bone

3. Examine the anterior views of a fetal skull next to an adult skull (fig. 18.4). Use a meterstick and measure the height of the entire cranium (braincase) and the height of the face of the fetal skull (fig. 18.4*a*). The brain is part of the early development and a large cranium, which gives a perspective of a large head. However, the small face is related to undersized maxillae, maxillary sinuses, mandible, lack of teeth eruptions, and nasal cavities. The facial portion of a fetal skull at birth represents about one-eighth of the entire skull, whereas the adult facial portion represents nearly one-half of the complete skull.

4. As growth of the fetus advances, the relative size of the cranium becomes smaller as the size of the face increases. Use a meterstick and measure the height of the cranium and face of an adult skull (fig. 18.4*b*).

5. Record your measurements in Part B of the laboratory report.

6. Make careful observations of the ossification of the frontal bone and the temporal bone.

7. Complete Part B of the laboratory report.

Procedure B—Entire Skeleton

1. Observe and then measure the total height (crown-to-heel) of the entire fetal skeleton (fig. 18.5) and an adult skeleton. Compare the height of the fetal skeleton with the data in table 18.1. Observe and then measure the total height (crown-to-heel) and the sitting height (crown-to-rump) of the human fetus shown in figure 18.6. Figure 18.7 illustrates the proportionate changes of bodies during development from an embryo to the adult. Record the measurements in Part C of the laboratory report.

2. Much of the development of the upper limb occurs during fetal development, especially during the second trimester. The rapid development of the lower limb is much greater after birth (figs. 18.5 and 18.7).

3. Observe and then measure the total length of the upper limb and the lower limb of the fetus (fig. 18.5). Record the measurements in Part C of the laboratory report.

4. Observe the adult skeleton and measure the total length of the upper limb and the lower limb. Record the measurements in Part C of the laboratory report.

5. Observe very carefully a hip bone, sternum, and sacrum of a fetal skeleton. Record your observations in Part C of the laboratory report.

6. Make careful observations of the anterior knee, thoracic cage, wrist, and ankles of the fetal skeleton (fig. 18.5). Record your observations in Part C of the laboratory report.

7. Complete Part C of the laboratory report.

FIGURE 18.4 Anterior views of skulls next to metric scales: (*a*) fetal skull; (*b*) adult skull. The dotted lines indicate the division between the cranium and the face.

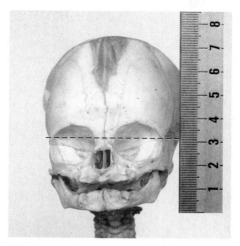

(a)

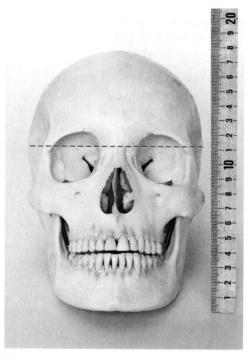

(b)

FIGURE 18.5 Anterior view of a fetal skeleton next to a metric scale. To estimate the gestational age of this fetus, measure the total height (crown-to-heel).

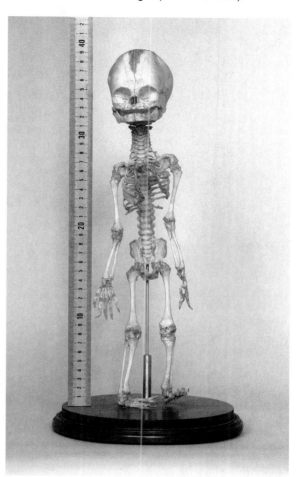

TABLE 18.1 Embryonic and Fetal Height (Length)

Gestational Age (Months)	Sitting Height (crown-to-rump)*	Total Height (crown-to-heel)*
1	0.6 cm	
2	3 cm	
3	7–9 cm	9–10 cm
4	12–14 cm	16–20 cm
5	18–19 cm	25–27 cm
6		30–32 cm
7		35–37 cm
8		40–45 cm
9		50–53 cm

*Measurements of height until about 20 weeks are more often from crown-to-rump; measurements after 20 weeks are more often from crown-to-heel.

FIGURE 18.6 Human fetus next to a metric scale. To estimate the gestational age of this young fetus, measure the total height (crown-to-heel) and the sitting height (crown-to-rump).

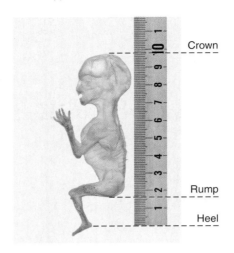

Crown

Rump

Heel

FIGURE 18.7 During development, proportions of the head, entire body, and upper and lower limbs change considerably.

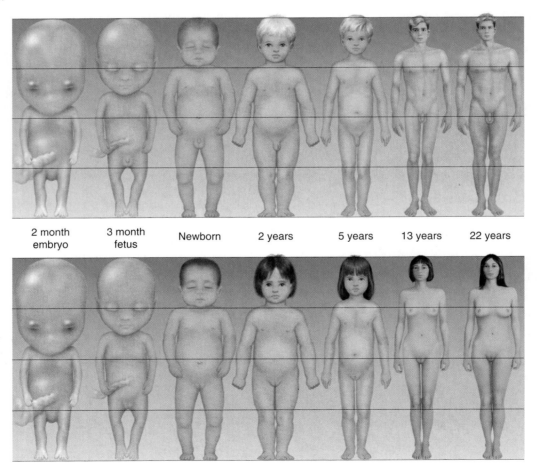

| 2 month embryo | 3 month fetus | Newborn | 2 years | 5 years | 13 years | 22 years |

Laboratory Report

18

Name _____

Date _____

Section _____

The ▲ corresponds to the Learning Outcome(s) listed at the beginning of the laboratory exercise.

Fetal Skeleton

Part A Assessments

Complete the following:

1. List all of the bones that form part of the border with the mastoid fontanel. ▲1

_____

2. List all of the bones that form part of the border with the sphenoidal fontanel. ▲1

_____

3. List all of the bones that form part of the border with the posterior fontanel. ▲1

_____

4. List all of the bones that form part of the border with the anterior fontanel. ▲1

_____

5. Describe the function of fontanels. ▲1

_____

6. What is the final outcome of the fontanels? ▲1

_____

Part B Assessments

1. Enter the measurements of the fetal and adult skulls in the table. ▲2

Skull	Total Height of Skull (cm)	Cranial Height (cm)	Facial Height (cm)
Fetus			
Adult			

2. Compare the frontal bone of the fetal skull with the adult skull. Describe some unusual aspects of the fetal frontal bone as compared to the adult structure. ▲3

_____

_____

3. Compare the temporal bone of the fetal skull with the adult skull. Describe some unusual aspects of the fetal temporal bone as compared to the adult structure. ⁄3\

_____

_____

4. Summarize the measured and observed relationships between a fetal skull and an adult skull. ⁄2\

_____

_____

_____

_____

Critical Thinking Activity

Why is it difficult to match the identifications of baby pictures with those of adults? ⁄2\

_____

_____

_____

_____

Part C Assessments

1. Enter the measurements of the fetal and adult skeletons in the table. ⁄4\

Skeleton	Total Height (cm)	Upper Limb Length (cm)	Lower Limb Length (cm)
Fetus			
Adult			

2. Summarize the relationship of total body height to limb lengths for the fetal skeleton and the adult skeleton. ⁄4\

_____

_____

_____

_____

3. Locate and describe the ossification of the following fetal bones and body regions and compare them with the adult skeleton. ⁄5\

    a. Hip bone:

_____

_____

b. Sternum:

_____

_____

c. Sacrum:

_____

_____

d. Anterior knee:

_____

_____

e. Thoracic cage:

_____

_____

f. Wrist and ankle:

_____

_____

4. Estimate the age of fetal development of the fetal skeleton in figure 18.5 and the one that is represented at your school. Estimate the age of the human fetus shown in figure 18.6. Use the data provided in table 18.1 for your determination of the estimates. **6**

   a. Fetal skeleton in figure 18.5: _____

   b. Fetal skeleton at your school: _____

   c. Fetus in figure 18.6: _____

Joint Structure and Movements

ARTICULATIONS

Pre-Lab

1. Carefully read the introductory material and examine the entire lab content.
2. Be familiar with joint structures and movements (from lecture or the textbook).
3. Visit www.mhhe.com/martinseriesl for pre-lab questions, Anatomy & Physiology Revealed animations list, and LabCam videos.

Materials Needed

Human skull
Human skeleton, articulated
Models of synovial joints (shoulder, elbow, hip, and knee)

For Demonstration Activities:
Fresh animal joint (knee joint preferred)
Radiographs of major joints

 ## Safety

▶ Wear disposable gloves when handling the fresh animal joint.
▶ Wash your hands before leaving the laboratory.

Joints are junctions between bones. Although they vary considerably in structure, they can be classified according to the type of tissue that binds the bones together. Thus, the three groups of structural joints can be identified as fibrous joints, cartilaginous joints, and synovial joints. Fibrous joints are filled with dense fibrous connective tissue, cartilaginous joints are filled with a type of cartilage, and synovial joints contain synovial fluid inside a joint cavity.

Joints can also be classified by the degree of functional movement allowed: synarthroses are immovable, amphiarthroses allow slight movement, and diarthroses allow free

movement. Movements occurring at freely movable synovial joints are due to the contractions of skeletal muscles. In each case, the type of movement depends on the type of joint involved and the way in which the muscles are attached to the bones on either side of the joint.

Purpose of the Exercise

To examine examples of the three types of structural and functional joints, to identify the major features of these joints, and to review the types of movements produced at synovial joints.

Learning Outcomes

After completing this exercise, you should be able to

1. Distinguish structural features among fibrous, cartilaginous, and synovial joints.
2. Distinguish functional characteristics of synarthroses, amphiarthroses, and diarthroses.
3. Locate examples of each type of structural and functional joint.
4. Examine the structure and types of movements of the shoulder, elbow, hip, and knee joints.
5. Demonstrate the types of movements that occur at synovial joints.

Procedure A—Types of Joints

1. Study table 19.1 and figure 19.1.
2. Examine the human skull and articulated skeleton to locate examples of the following types of structural and functional joints:

   fibrous joint
   suture—synarthrosis
   gomphosis—synarthrosis
   syndesmosis—amphiarthrosis
   cartilaginous joint
   synchondrosis—synarthrosis
   symphysis—amphiarthrosis
   synovial joints—diarthrosis

3. Complete Part A of Laboratory Report 19.

FIGURE 19.1 Examples of all types of synovial joints and the possible movements of each type of joint.

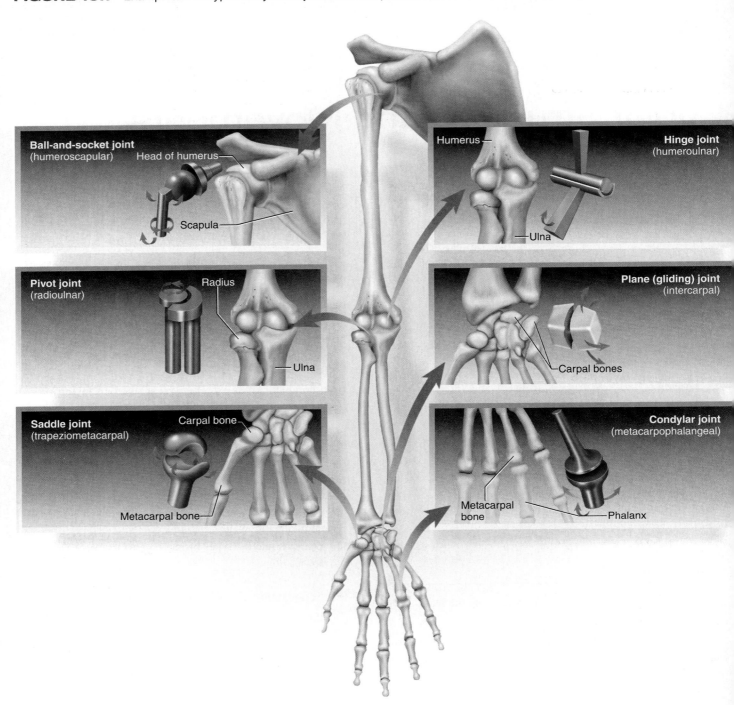

Ball-and-socket joint (humeroscapular)
Head of humerus
Scapula

Pivot joint (radioulnar)
Radius
Ulna

Saddle joint (trapeziometacarpal)
Carpal bone
Metacarpal bone

Hinge joint (humeroulnar)
Humerus
Ulna

Plane (gliding) joint (intercarpal)
Carpal bones

Condylar joint (metacarpophalangeal)
Metacarpal bone
Phalanx

4. Using table 19.1 and figure 19.1 as references, locate examples of the following types of synovial joints in the skeleton. At the same time, examine the corresponding joints in the models and in your body. Experiment with each joint to experience its range of movements. 🄰 🄱

ball-and-socket joint	hinge joint
condylar (ellipsoid) joint	pivot joint
plane (gliding) joint	saddle joint

5. Complete Parts B and C of the laboratory report.

Procedure B—Examples of Synovial Joints

1. Study the major features of a synovial joint in figure 19.2.
2. Examine models of the shoulder, elbow, hip, and knee joints.

TABLE 19.1 Classification of Joints

Structural Classification	Structural Features	Functional Classification	Location Examples
Fibrous			
Suture	Dense fibrous connective tissue	Synarthrosis—immovable	Sutures of skull
Gomphosis	Periodontal ligament	Synarthrosis—immovable	Tooth sockets
Syndesmosis	Dense fibrous connective tissue	Amphiarthrosis—slightly movable	Tibiofibular joint
Cartilaginous			
Synchondrosis	Hyaline cartilage	Synarthrosis—immovable	Epiphyseal plates
Symphysis	Fibrocartilage	Amphiarthrosis—slightly movable	Pubic symphysis; intervertebral discs
Synovial			
Ball-and-socket	All synovial joints contain articular cartilage, a synovial membrane, and a joint cavity filled with synovial fluid	Diarthrosis—movements in all planes	Hip; shoulder
Hinge		Diarthrosis—flexion and extension	Elbow; knee; interphalangeal
Condylar (ellipsoid)		Diarthrosis—flexion, extension, abduction, and adduction	Radiocarpal; metacarpophalangeal
Pivot		Diarthrosis—rotation	Radioulnar at elbow; dens at atlas
Saddle		Diarthrosis—variety of movements mainly in two planes	Base of thumb with trapezium
Plane (gliding)		Diarthrosis—sliding or twisting	Intercarpal; intertarsal

FIGURE 19.2 Basic structure of a synovial joint.

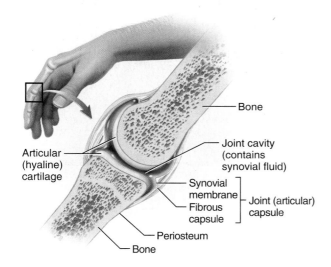

- Bone
- Joint cavity (contains synovial fluid)
- Articular (hyaline) cartilage
- Synovial membrane
- Fibrous capsule
- Joint (articular) capsule
- Periosteum
- Bone

 Demonstration Activity

Examine a longitudinal section of a fresh synovial animal joint. Locate the dense connective tissue that forms the joint capsule and the hyaline cartilage that forms the articular cartilage on the ends of the bones. Locate the synovial membrane on the inside of the joint capsule. Does the joint have any semilunar cartilages (menisci)? ⁴

_____

_____

What is the function of such cartilages? _____

_____

_____

Demonstration Activity

Study the available radiographs of joints by holding the films in front of a light source. Identify the type of joint and the bones incorporated in the joint. Also identify other major visible features.

3. Locate the major knee joint structures of figure 19.3 that are visible on the knee joint model.
4. Complete Part D of the laboratory report.

Procedure C—Joint Movements

1. Examine figures 19.4 and 19.5.
2. When the body is in anatomical position, most joints are extended and/or adducted. Skeletal muscle action

157

FIGURE 19.3 Right knee joint (*a*) anterior view and (*b*) superior view of tibial end.

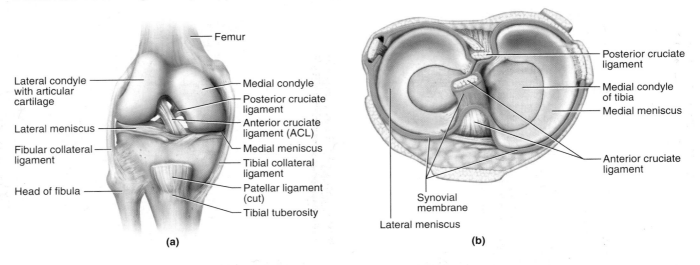

(a)

(b)

FIGURE 19.4 Examples of angular, rotational, and gliding movements of synovial joints.

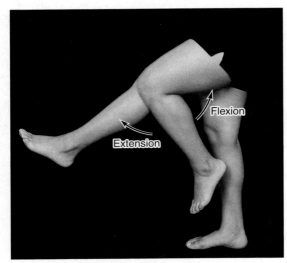

(a) Flexion and extension of the knee joint

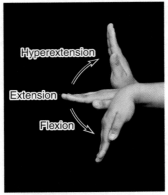

(b) Flexion, extension, and hyper-extension of the wrist joint

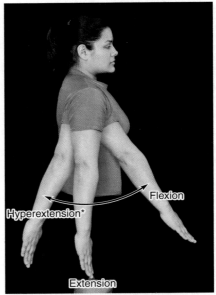

(c) Flexion, extension, and hyperextension* of the shoulder joint (*Hyperextension of the shoulder and hip joints is considered normal extension in some health professions.)

FIGURE 19.4 *Continued.*

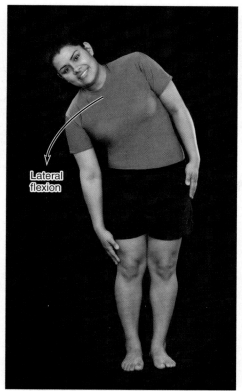

(d) Lateral flexion of the trunk (vertebral column)

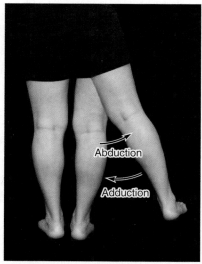

(e) Abduction and adduction of the hip joint

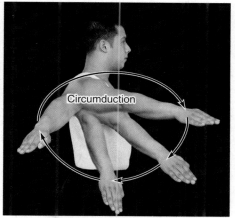

(f) Circumduction of the shoulder joint

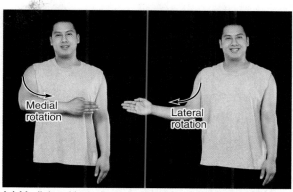

(g) Medial and lateral rotation of the arm at the shoulder joint

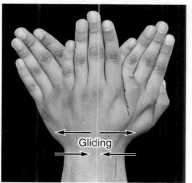

(h) Gliding movements among the carpals of the wrist

FIGURE 19.5 Special movements of synovial joints.

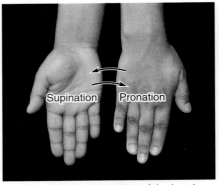

(a) Supination and pronation of the hand involving movement at the radioulnar joint

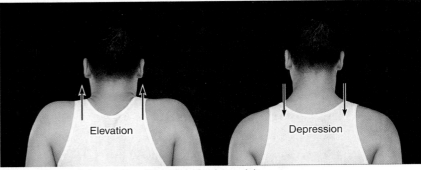

(b) Elevation and depression of the shoulder (scapula)

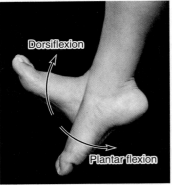

(c) Dorsiflexion and plantar flexion of the foot at the ankle joint

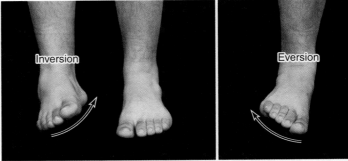

(d) Inversion and eversion of the right foot at the ankle joint (The left foot is unchanged for comparison.)

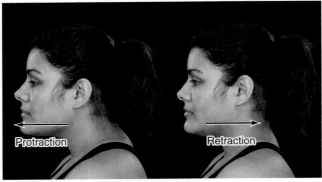

(e) Protraction and retraction of the head

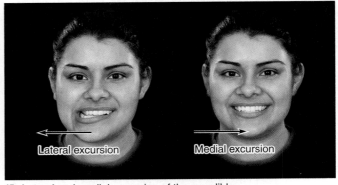

(f) Lateral and medial excursion of the mandible

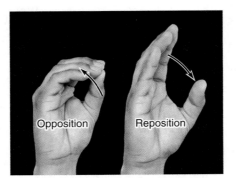

(g) Opposition and reposition of the thumb at the carpometacarpal (saddle) joint

involves the movable end (*insertion*) being pulled toward the stationary end (*origin*). In the limbs, the origin is usually proximal to the insertion; in the trunk, the origin is usually medial to the insertion. Use these concepts as reference points as you move joints. Move various parts of your body to demonstrate the joint movements described in tables 19.2 and 19.3. ⑤

3. Have your laboratory partner do some of the preceding movements and see if you can correctly identify the movements made. ⑤

4. Complete Part E of the laboratory report.

TABLE 19.2   Angular, Rotational, and Gliding Movements

Movement	Description
Flexion	Decrease of an angle (usually in the sagittal plane)
Lateral flexion	Bending the trunk (vertebral column) to the side
Extension	Increase of an angle (usually in the sagittal plane)
Hyperextension	Extension beyond anatomical position
Abduction	Movement away from the midline (usually in the frontal plane)
Adduction	Movement toward the midline (usually in the frontal plane)
Circumduction	Circular movement (combines flexion, abduction, extension, and adduction)
Rotation	Movement of part around its long axis
Medial (internal)	Inward rotation
Lateral (external)	Outward rotation
Gliding	Back-and-forth and side-to-side sliding movements of plane joints

TABLE 19.3   Special Movements (pertain to specific joints)

Movement	Description
Supination	Movement of palm of hand anteriorly or upward
Pronation	Movement of palm of hand posteriorly or downward
Elevation	Movement of body part upward
Depression	Movement of body part downward
Dorsiflexion	Movement of ankle joint so dorsum (superior) of foot becomes closer to anterior surface of leg (as standing on heels)
Plantar flexion	Movement of ankle joint so the plantar surface of foot becomes closer to the posterior surface of leg (as standing on toes)
Inversion (called supination in some health professions)	Medial movement of sole of foot at ankle joint
Eversion (called pronation in some health professions)	Lateral movement of sole of foot at ankle joint
Protraction	Anterior movement in the transverse plane
Retraction	Posterior movement in the transverse plane
Excursion	Special movements of mandible when grinding food
Lateral	Movement of mandible laterally
Medial	Movement of mandible medially
Opposition	Movement of the thumb to touch another finger of the same hand
Reposition	Return of thumb to anatomical position

Critical Thinking Activity

Describe a body position that can exist when all major body parts are flexed.

_____

_____

_____

_____

Name _____

Date _____

Section _____

The ⒶAcorresponds to the Learning Outcome(s) listed at the beginning of the laboratory exercise.

Joint Structure and Movements

Part A Assessments

Match the terms in column A with the descriptions in column B. Place the letter of your choice in the space provided. Ⓐ1️⃣ 2️⃣

Column A	Column B
a. Gomphosis	_____ **1.** Immovable joint between flat bones of the skull united by a thin layer of dense connective tissue
b. Suture	
c. Symphysis	_____ **2.** Fibrocartilage fills the slightly movable joint
d. Synchondrosis	_____ **3.** Temporary joint in which bones are united by bands of hyaline cartilage
e. Syndesmosis	
	_____ **4.** Slightly movable joint in which bones are united by interosseous membrane
	_____ **5.** Joint formed by union of tooth root in bony socket

Part B Assessments

Identify the types of structural and functional joints numbered in figure 19.6. 3️⃣

1. _____

2. _____

3. _____

4. _____

5. _____

6. _____

7. _____

8. _____

9. _____

FIGURE 19.6 Identify the types of structural and functional joints numbered in these illustrations.

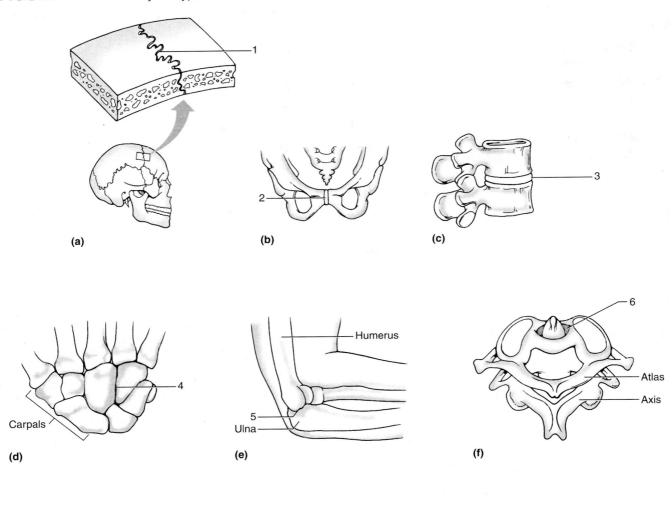

(a)

(b)

(c)

Humerus

Carpals

(d)

(e)

Ulna

6

Atlas

Axis

(f)

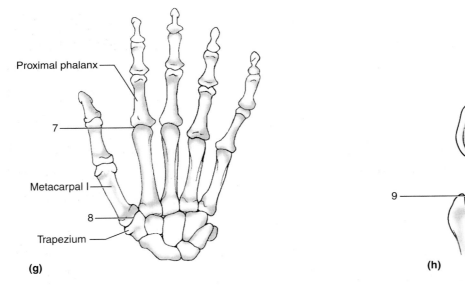

Proximal phalanx

7

Metacarpal I

8

Trapezium

(g)

9

(h)

Part C Assessments

Match the types of synovial joints in column A with the examples in column B. Place the letter of your choice in the space provided. ▲3

Column A	Column B
a. Ball-and-socket	_____ 1. Hip joint
b. Condylar (ellipsoid)	_____ 2. Metacarpal-phalanx
c. Hinge	_____ 3. Proximal radius-ulna
d. Pivot	_____ 4. Humerus-ulna of the elbow joint
e. Plane (gliding)	_____ 5. Phalanx-phalanx
f. Saddle	_____ 6. Shoulder joint
	_____ 7. Knee joint
	_____ 8. Carpal-metacarpal of the thumb
	_____ 9. Carpal-carpal
	_____ 10. Tarsal-tarsal

Part D Assessments

Complete the following table: ▲4 ▲5

Name of Joint	Type of Joint	Bones Included	Types of Movement Possible
Shoulder joint			
Elbow joint			
Hip joint			
Knee joint			

Part E Assessments

Identify the types of joint movements numbered in figure 19.7. ▲5

1. _____ (of head)
2. _____ (of shoulder)
3. _____ (of shoulder)
4. _____ (of hand)
5. _____ (of hand)
6. _____ (of arm at shoulder)
7. _____ (of arm at shoulder)
8. _____ (of hand at wrist)
9. _____ (of hand at wrist)
10. _____ (of thigh at hip)
11. _____ (of thigh at hip)
12. _____ (of lower limb at hip)
13. _____ (of chin/mandible)
14. _____ (of chin/mandible)
15. _____ (of vertebral column/trunk)
16. _____ (of vertebral column/trunk)
17. _____ (of head and neck)
18. _____ (of head and neck)
19. _____ (of arm at shoulder)
20. _____ (of arm at shoulder)
21. _____ (of forearm at elbow)
22. _____ (of forearm at elbow)
23. _____ (of thigh at hip)
24. _____ (of thigh at hip)
25. _____ (of leg at knee)
26. _____ (of leg at knee)
27. _____ (of foot at ankle)
28. _____ (of foot at ankle)

Skeletal Muscle Structure and Function

Pre-Lab

1. Carefully read the introductory material and examine the entire lab content.
2. Be familiar with skeletal muscle tissue and muscle actions (from lecture or the textbook).
3. Visit www.mhhe.com/martinseriesl for pre-lab questions and Anatomy & Physiology Revealed animations list.

Materials Needed

Compound light microscope
Prepared microscope slide of skeletal muscle tissue
Human torso model with musculature
Model of skeletal muscle fiber
Ph.I.L.S. 3.0

For Demonstration Activity:
Fresh round beefsteak

Safety

▶ Wear disposable gloves when handling the fresh beefsteak.
▶ Wash your hands before leaving the laboratory.

A skeletal muscle represents an organ of the muscular system and is composed of several types of tissues. These tissues include skeletal muscle tissue, nervous tissue, blood, and various connective tissues.

Each skeletal muscle is encased and permeated with connective tissue sheaths. The connective tissues extend into the structure of a muscle and separate it into compartments. The outer layer, called *epimysium,* extends deeper into the muscle as *perimysium,* which covers bundles of cells (fascicles), and then farther inward extends around each muscle cell (fiber) as a thin *endomysium.* The connective tissues provide support and reinforcement during muscular contractions and allow portions of a muscle to contract somewhat independently. The connective tissue often extends beyond the end of a muscle, providing an attachment to other muscles or to bones. Some collagen fibers of the connective tissue are continuous with the tendon and the periosteum, making for a strong structural continuity.

Muscles are named according to their location, size, shape, action, attachments, or the direction of the fibers. Examples of how muscles are named include: gluteus maximus (location and size); adductor longus (action and shape); sternocleidomastoid (attachments); and orbicularis oculi (direction of fibers and location).

Skeletal muscles, such as the biceps brachii, are composed of many muscle fibers (cells). These muscle fibers are innervated by motor neurons of the nervous system. Anatomically, each motor neuron is arranged to extend to a specific number of muscle fibers. (Note that each muscle fiber is innervated by only one motor neuron.) The motor neuron and the collection of muscle fibers it innervates is called a *motor unit.* When the motor neuron relays a neural impulse to the muscle, all the muscle fibers that it innervates (i.e., the motor unit) will be stimulated to contract. If only a few motor units are stimulated, the muscle as a whole exerts little force. If a larger number of motor units are stimulated, the muscle will exert a greater force. Thus, to lift a heavy object like a suitcase requires the activation of more motor units in the biceps brachii than the lifting of a lighter object like a book.

Purpose of the Exercise

To study the structure and function of skeletal muscles as cells and as organs.

Learning Outcomes

After completing this exercise, you should be able to

(1) Locate the major structures of a skeletal muscle fiber (cell).

(2) Describe how connective tissue is associated with muscle tissue within a skeletal muscle.

(3) Distinguish between the origin and insertion of a muscle.

(4) Describe and demonstrate the general actions of prime movers (agonists), synergists, fixators, and antagonists.

(5) Diagram and label a muscle twitch.

(6) Demonstrate *threshold* (the minimum voltage required to observe the appearance of muscle contraction).

(7) Illustrate *recruitment* (the stimulation of additional motor units in the muscle) by observing the change in amplitude of the contraction.

(8) Determine maximum contraction (the stimulation of all motor units in the muscle) by observing no further increase in the amplitude of the contraction.

(9) Measure amplitude of muscle contraction when the muscle is stimulated with varying degrees of intensity (voltage).

(10) Integrate the concepts of recruitment and maximum stimulation as applied to the muscles of the body.

Procedure A—Skeletal Muscle Structure

1. Reexamine the microscopic structure of skeletal muscle by observing a prepared microscope slide of this tissue. Use figure 20.1 of skeletal muscle tissue to locate the following features:

    skeletal muscle fiber (cell)
    nuclei
    striations (alternating light and dark)

2. Study figures 20.2 and 20.3.

3. Examine the human torso model and locate examples of fascia, tendons, and aponeuroses. An origin tendon is attached to a fixed location, while the insertion tendon is attached to a more movable location. Sheets of connective tissue, called aponeuroses, also serve for some muscle attachments. Locate examples of tendons in your body.

4. Complete Part A of Laboratory Report 20.

 Demonstration Activity

Examine the fresh round beefsteak. It represents a cross section through the beef thigh muscles. Note the white lines of connective tissue that separate the individual skeletal muscles. Also note how the connective tissue extends into the structure of a muscle and separates it into small compartments of muscle tissue. Locate the epimysium and the perimysium of an individual muscle.

5. Examine the model of the skeletal muscle fiber and locate the following:

    sarcolemma
    sarcoplasm
    myofibril
      thick (myosin) filament
      thin (actin, tropomyosin, and troponin) filament
    sarcomere (functional contractile unit within muscle fiber)
      A band (anisotropic; dark)
      I band (isotropic; light)
      H zone (band)
      M line
      Z line (disc)
    sarcoplasmic reticulum
      terminal cisternae
    transverse (T) tubules

FIGURE 20.1 Structures found in skeletal muscle fibers (cells) (250× micrograph enlarged to 700×).

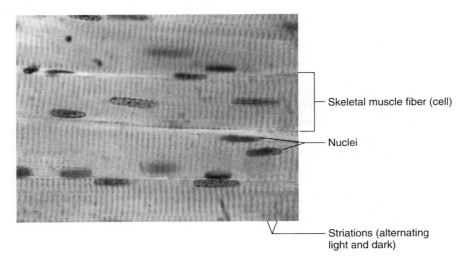

Skeletal muscle fiber (cell)

Nuclei

Striations (alternating light and dark)

FIGURE 20.2 Skeletal muscle structure from the gross anatomy to the microscopic arrangement. Note the distribution pattern of the epimysium, perimysium, and endomysium.

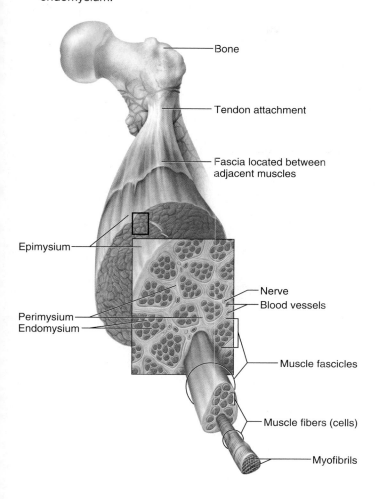

- Bone
- Tendon attachment
- Fascia located between adjacent muscles
- Epimysium
- Perimysium
- Endomysium
- Nerve
- Blood vessels
- Muscle fascicles
- Muscle fibers (cells)
- Myofibrils

6. Complete Part B of the laboratory report.
7. Study figure 20.4.
8. Locate the biceps brachii, brachialis, and triceps brachii and their origins and insertions in the human torso model and in your body.
9. Make various movements with your upper limb at the shoulder and elbow. For each movement, determine the location of the muscles functioning as prime movers (agonists) and as antagonists. **Remember, when a prime mover contracts (shortens) and a joint moves, its antagonist relaxes (lengthens).** Synergistic muscles often supplement the contraction force of a prime mover, or by acting as fixators, they might also stabilize nearby joints. Antagonistic muscle pairs pull from opposite sides of the same body region. The role of a muscle as a prime mover, antagonist, or synergist depends upon the movement under consideration, as their roles change.
10. Complete Part C of the laboratory report.

FIGURE 20.3 Structures of a segment of a muscle fiber (cell).

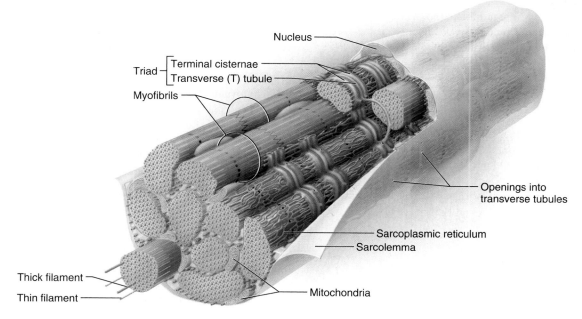

- Nucleus
- Triad
  - Terminal cisternae
  - Transverse (T) tubule
- Myofibrils
- Openings into transverse tubules
- Sarcoplasmic reticulum
- Sarcolemma
- Thick filament
- Thin filament
- Mitochondria

FIGURE 20.4 Antagonistic muscle pairs are located on opposite sides of the same body region as shown in an arm. The muscle acting as the prime mover depends upon the movement that occurs.

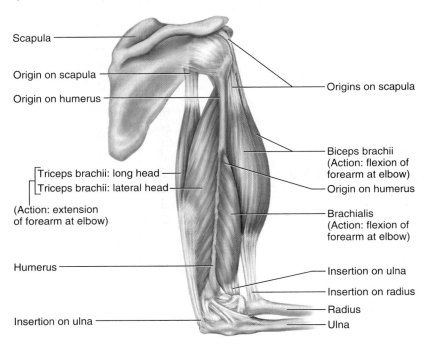

Procedure B—Ph.I.L.S. Lesson 4 Skeletal Muscle Function: Stimulus Dependent Force Generation

Hypothesis

Would you predict that there would be a minimum voltage for stimulating a muscle to contract? _____

Would you predict a stronger contraction as the voltage stimulating the muscle is increased? _____

Would you predict that there would be a voltage so high that all muscle fibers are stimulated to contract? _____

What would you predict would happen to the force of contraction as the voltage is increased past this point? _____

1. Open Exercise 4: Skeletal Muscle Function—Stimulus Dependent Force Generation.
2. Read the objectives and introduction and take the pre-lab quiz. See figure 20.5.
3. After completing the pre-lab quiz, read through the wet lab.
4. The lab exercise will open when you have completed the wet lab (fig. 20.6).

5. Follow the instructions at the bottom of the screen to turn on the power of the data acquisition unit and to connect the transducer and electrodes.
6. Set the stimulus voltage to at least 1.0 volts. (This voltage will elicit a muscle contraction.)
7. To stimulate the muscle, click the shock button in the control panel. (A graph will appear with a red line and a blue line.)
8. Measure the tension (degree of contraction) by positioning the crosshairs (using the mouse) at the top of the wave. Click to make a black arrow appear. If not in the correct location, reposition by dragging to a new position. Now position the crosshairs at the bottom of the deflection (after the wave or 0 volts) and click. The result will display in the (yellow) data panel. Amplitude (amp) is the change in tension of the muscle representing the degree of movement of the transducer when the muscle contracted.
9. Click the journal panel (red rectangle at bottom right of screen) to enter your value into the journal. A table with volts and amplitude (amp) and a graph will appear. Note the range in values for volts from 0 to 1.6 volts.
10. Close the journal window by clicking on the X in the right-hand corner.

FIGURE 20.5 Motor unit.

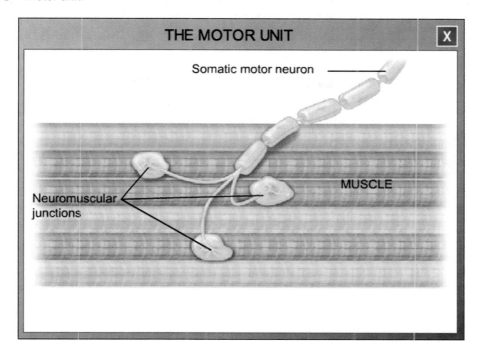

FIGURE 20.6 Opening screen for the laboratory exercise on Skeletal Muscle Function: Stimulus Dependent Force Generation.

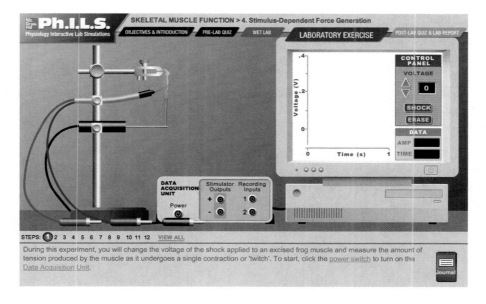

11. To complete the table and produce the graph,
    a. Click Erase.
    b. Set voltage (1.1 volts).
    c. Click Shock.
    d. Measure.
    e. Click Journal.
    f. Repeat all steps except set voltage at new value. Complete all values in the table (range from 0 to 1.6 in increments of 0.1 volts).
12. After finishing the laboratory exercise, you can print the line tracing by clicking on the P in the bottom left of the screen (of the laboratory experiment).

13. Click on Post-lab quiz and lab report.
14. Complete the post-lab quiz by answering the ten questions on the computer screen.
15. Read the conclusion on the computer screen.
16. You may print the lab report for the Skeletal Muscle Function: Stimulus Dependent Force Generation computer simulation.
17. Complete Part D of the laboratory report in your laboratory manual.

Laboratory Report

20

Name _____

Date _____

Section _____

The ⚠ corresponds to the Learning Outcome(s) listed at the beginning of the laboratory exercise.

Skeletal Muscle Structure and Function

Part A Assessments

Match the terms in column A with the definitions in column B. Place the letter of your choice in the space provided. ⚠ ⚠

Column A	Column B
a. Endomysium	_____ **1.** Membranous channel extending inward from muscle fiber membrane
b. Epimysium	_____ **2.** Cytoplasm of a muscle fiber
c. Fascia	_____ **3.** Connective tissue located between adjacent muscles
d. Fascicle	_____ **4.** Layer of connective tissue that separates a muscle into small bundles called fascicles
e. Myosin	
f. Perimysium	_____ **5.** Plasma membrane of a muscle fiber
g. Sarcolemma	_____ **6.** Layer of connective tissue that surrounds a skeletal muscle
h. Sarcomere	_____ **7.** Unit of alternating light and dark striations between Z lines
i. Sarcoplasm	
j. Sarcoplasmic reticulum	_____ **8.** Layer of connective tissue that surrounds an individual muscle fiber
k. Tendon	_____ **9.** Cellular organelle in muscle fiber corresponding to the endoplasmic reticulum
l. Transverse (T) tubule	
	_____ **10.** Cordlike part that attaches a muscle to a bone
	_____ **11.** Protein found within thick filament
	_____ **12.** A small bundle of muscle fibers

Part B Assessments

Provide the labels for the electron micrograph in figure 20.7.

FIGURE 20.7 Label this transmission electron micrograph (16,000×) of a relaxed sarcomere by placing the correct numbers in the spaces provided. ⚠

____ A band (dark)
____ I band (light)
____ Sarcomere
____ Z line

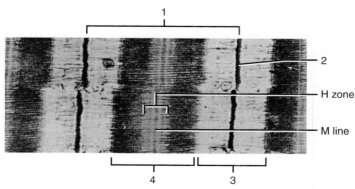

173

Part C Assessments

Complete the following statements:

1. The _____ of a muscle is usually attached to a fixed location. /3\

2. The _____ of a muscle is usually attached to a movable location. /3\

3. A muscle responsible for most of a movement is called a(n) _____. /4\

4. Assisting muscles are called _____. /4\

5. Antagonists are muscles that resist the actions of _____ and cause movement in the opposite direction. /4\

6. When the forearm is extended at the elbow joint, the _____ muscle acts as the prime mover. /4\

7. When the biceps brachii acts as the prime mover, the _____ muscle assists as a synergist. /4\

Part D Ph.I.L.S. Lesson 4, Skeletal Muscle Function: Stimulus Dependent Force Generation Assessments

1. Diagram a graph of a twitch and label: latent period, contraction, and relaxation of a twitch. /5\

2. What was the threshold voltage for stimulation of the frog's gastrocnemius muscle? /6\ /9\ _____

3. What voltage produced maximal contraction? /8\ /9\ _____

4. What is occurring between the threshold and the maximum contraction? /7\ /9\ _____

5. Explain the relationship of increasing voltage stimulation to the recruitment of motor units. /6\ /7\ /8\ _____

_____

_____

Critical Thinking Activity

You are getting ready to leave the lab. You lift your pencil to put it in your book bag. Then you lift your lab book and place it in your book bag. Which scenario resulted from the stimulation of more motor units? /10\

_____

Which would require more neural signals being sent by the nervous system to the muscles of your arm? /10\

_____

What concept covered in this lab (threshold, recruitment, or maximum contraction) explains why there are certain objects that are too heavy for you to lift? /10\

_____

_____

Electromyography: BIOPAC© Exercise

This is the first of a series of laboratory exercises that will utilize the BIOPAC computer-based data acquisition and analysis system. This is a complete system consisting of hardware and software. Once the software package has been loaded onto your computer, each lesson will use a different set of hardware. Connections and how to use the instrumentation will be explained in each lesson.

The hardware includes the MP30 Acquisition Unit, connection cables, transformers, transducers, electrodes and electrode cables, and other accessories. The electrodes and other accessories will often contain sensors that pick up a signal and send it to the MP30 Acquisition Unit. The software picks up electrical signals coming into the MP30 Acquisition Unit, which are then displayed on the screen as a waveform. The software then guides the user through the lesson and manages data saving and review.

The BIOPAC computer-based data acquisition and analysis system works in a manner similar to other recording equipment in sensing electrical signals in the body. The major advantage to the BIOPAC system is that it has dynamic calibration. With other equipment, the instructor or student needs to play with dials and settings to get a good signal for the recording. With the BIOPAC Student Lab, the adjustments are made automatically for each subject during the calibration procedure in each lesson.

Before you start any lesson, the software should already be installed on the computer and you should read through each lesson completely. The BIOPAC Student Laboratory Manual that comes with the system has a full description of the lessons and some troubleshooting information. The lessons as described in this manual are shortened versions of these. Figure 21.1 and Table 21.1 have been included here as a quick reference guide to the Display Tools for Analysis as you proceed through the lessons.

In this laboratory exercise you will investigate the electrical properties of skeletal muscle using *electromyography*. This is a recording of skin-surface voltage that is produced by underlying skeletal muscle contraction. The voltage produced by contraction is weak, but it can be picked up by surface sensors to produce the recording, referred to as an *electromyogram (EMG)*.

FIGURE 21.1 Display window icons.

Source: Courtesy of and © *BIOPAC* Systems, Inc.

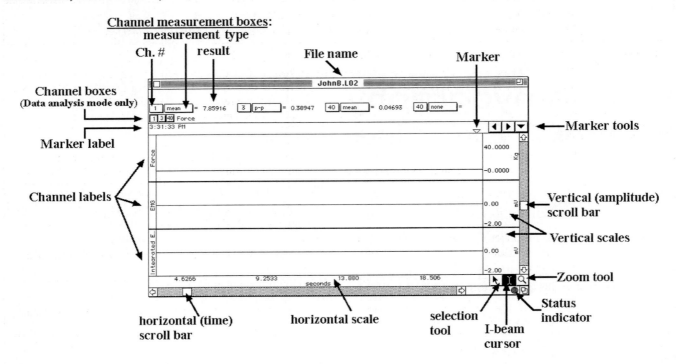

Part 1—Electromyography: Standard and Integrated EMG

Purpose of the Exercise

To measure and calculate fist clench strength from electromyograms obtained using the BIOPAC system and to listen to the EMG "sounds."

Learning Outcomes

After completing this exercise, you should be able to

1. Record the maximum clench strength for dominant and nondominant hands.

2. Examine, record, and correlate motor unit recruitment with increased power of skeletal muscle contraction.

3. Listen to EMG "sounds" and correlate sound intensity with motor unit recruitment.

Procedure A—Setup

1. With your computer turned **ON** and the BIOPAC MP30 unit turned **OFF,** plug the electrode lead set (SSL2) into Channel 3 and BIOPAC SS25 Hand Dyanamometer into Channel 1. Plug the headphones into the back of the unit if desired. (The dynamometer will be used in Part 2.)

Human skeletal muscle consists of hundreds of cylindrical skeletal muscle cells (also called fibers). Each fiber must be stimulated to contract by a somatic motor neuron. Although each muscle fiber is stimulated by only one motor neuron, each motor neuron can stimulate a number of muscle fibers. The functional unit formed by a single somatic motor neuron and all the muscle fibers that it stimulates is referred to as a *motor unit.* As a result, each skeletal muscle consists of numerous motor units. The strength of muscle contraction can vary depending upon how many motor units are stimulated: the more strength that is needed, the more motor units that are stimulated. This phenomenon is referred to as *motor unit recruitment.*

The primary function of muscle is to convert chemical energy (ATP) into mechanical energy in order to contract. Skeletal muscles that are required to contract repetitively or at their maximum strength will eventually run out of energy and *fatigue.*

This laboratory exercise has been divided into two parts. In Part 1 you will observe a standard EMG using a simple fist clench while electrodes are attached to the forearm. In Part 2 you will use a *hand dynamometer* equipped with an electronic transducer for recording. This will produce a special type of recording called a *dynagram* that measures the force exerted during the fist clench.

TABLE 21.1 Display Tools for Analysis

Viewing Tool	Overview
Selection tool	The **selection** icon is located in the lower right corner of the display window. The **selection tool** is a general-purpose cursor, used for selecting waveforms and scrolling through data.
I-Beam tool	The **I-beam** icon is located in the lower right corner of the display window. The **I-beam tool** is used to select an area for measurement. To activate it, click on it and move the mouse such that the cursor is positioned at the beginning of the region that you want to select. The cursor *must* be positioned within the data window. Hold the mouse button down and move (drag) the cursor to the second position. Release the mouse button to complete the selection. The selected area should remain darkened. Measurements will apply to the selected area.
Zoom tool	The **zoom** icon is in the lower right corner of the display window. Once selected, it allows you to define an area of the waveform by click-hold-drag-releasing the mouse over the desired section. The horizontal (time) and vertical (amplitude) scales will expand to display just the selected section.
Zoom previous	Click on the **Display** menu to access this option. After you have used the Zoom tool, this selection will revert the display to the previous scale settings.
Horizontal (time) scroll bar	Located below the time division scale. To move the display's time position, click on the scroll box arrows or hold the mouse button down on the scroll box and drag the box left (earlier) or right (later).
Vertical (amplitude) scroll bar	Located to the right of the amplitude (mV) division scale. To move the selected channel's vertical position, click on the scroll box or hold the mouse button down on the scroll box and drag the box up or down.
Autoscale waveforms	Click on the **Display** menu to access this option. Optimizes the vertical (amplitude) scale so that all the data will be shown on the screen.
Autoscale horizontal	Click on the **Display** menu to access this option. Compresses or expands the horizontal (time) scale so the entire recording will fit on the screen.
Grids	Click on the **File** menu and choose **Preferences** to access this option. Grids provide a quick visual guide to determine amplitude and duration. To turn the grids display on or off, change the Preferences setting.
Overlap button	Emulates an oscilloscope display.
Split button	Emulates a chart recorder display. This is the default display mode.
Adjust Baseline button	The Adjust Baseline button only appears in "Lesson 3 ECG I." Allows you to position the waveform up or down in small increments so that the baseline can be set to exactly zero. This is not needed to get accurate amplitude measurements, but may be desired before making a printout or when using grids. When the **Adjust Baseline** button is pressed, **Up** and **Down** buttons will appear. Simply click on these to move the waveform up or down.

2. Turn on the MP30 Data Acquisition Unit.
3. Attach the electrodes to the dominant forearm (the right arm if the subject is right-handed, the left if he or she is left-handed) of the subject as shown in figure 21.2. This will be **Forearm I.** To help to insure a good contact, make sure the area to be in contact with the electrodes is clean by wiping with the abrasive pad (ELPAD) or alcohol wipe. A small amount of BIOPAC electrode gel (GEL1) may also be used to make better contact between the sensor in the electrode and the skin. For optimal adhesion, the electrodes should be placed on the skin at least 5 minutes before the calibration procedure.
4. Attach the electrode lead set by the pinch connectors as shown in figure 21.2. Make sure the proper color is attached to the appropriate electrode.
5. Start the BIOPAC Student Lab Program for Electromyography I, lesson L01-EMG-1.
6. Designate a unique filename to be used to save the subject's data.

FIGURE 21.2 Electrode lead attachment.

Source: Courtesy of and © *BIOPAC* Systems, Inc.

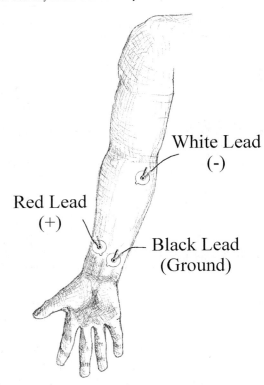

White Lead
(-)

Red Lead
(+)

Black Lead
(Ground)

Procedure B—Calibration

1. Click on **Calibrate.** A dialog box will direct you through the calibration procedure. After reading the box, click **OK.**
2. Wait about 2 seconds, clench the fist as hard as possible, then release. Wait for the calibration to stop (this will take about 8 seconds).

3. The calibration recording should look similar to figure 21.3. If not, click **Redo Calibration.** If the baseline is not at zero, the subject did not wait 2 seconds before clenching.

Procedure C—Recording

1. Fill out the subject profile on Laboratory Report 21. Click on **Record** and have the subject:
   a. Clench-Release-Wait, holding for 2 seconds.
   b. Repeat this cycle 4 times. During the first cycle the clench should be gentle, then increase the strength with each successive cycle so that the fourth clench is the maximum force.
2. Click **Suspend.** The recording should be similar to figure 21.4. If it is not, click **Redo** and repeat step 1.
3. If desired, you can remove the electrodes from the dominant forearm and place the electrodes and the lead set on the nondominant forearm of the same subject. Click **Resume** and repeat from step 1 of Procedure B—Calibration.
4. Click Stop when you have completed the procedure.
5. If you want to listen to the EMG signal, put on the headphones and click **Listen.** Have the subject experiment by changing the clench force as you watch the screen. When finished, click **Stop.**

Procedure D—Data Analysis

1. You can either analyze the current file or save it and do the analysis later by using the **Review Saved Data mode** and selecting the correct file.
2. Note the channel number designations: **CH 3** displays **Raw EMG** (the actual voltage recording) and **CH 40** the **Integrated EMG** (which reflects the absolute intensity of the voltage).

FIGURE 21.3 Calibration recording part 1 (single wave).

Source: Courtesy of and © *BIOPAC* Systems, Inc.

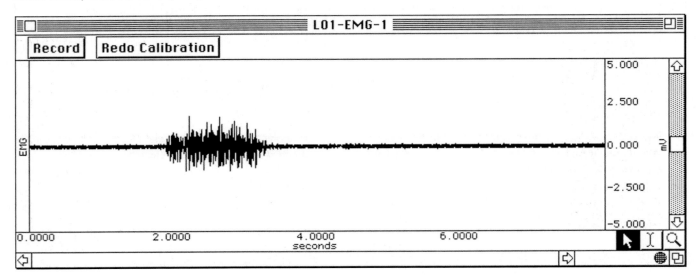

FIGURE 21.4 Clench-release recording.

Source: Courtesy of and © *BIOPAC* Systems, Inc.

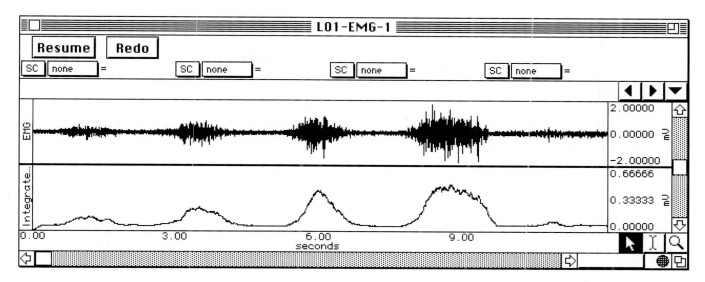

3. Set up the first four pairs of channel/measurement boxes as follows:

   CH 3/min, CH 3/max, CH 3/p-p, CH 40/mean.

   This will give you these measurements:

   min—the minimum value in the selected area

   max—the maximum value in the selected area

   p-p—subtracts the minimum from the maximum value in the selected area

   mean—the average value in the selected area

4. After clicking the I-beam box (this will activate the select function), select the first EMG cluster as shown in figure 21.5.

5. The values in the measurement boxes should be recorded in the table in Part IA of Laboratory Report 21.

6. Repeat for the other three clusters and record the data in Part IA of the laboratory report.

7. Complete Part I of the laboratory report.

FIGURE 21.5 I-beam measurement.

Source: Courtesy of and © *BIOPAC* Systems, Inc.

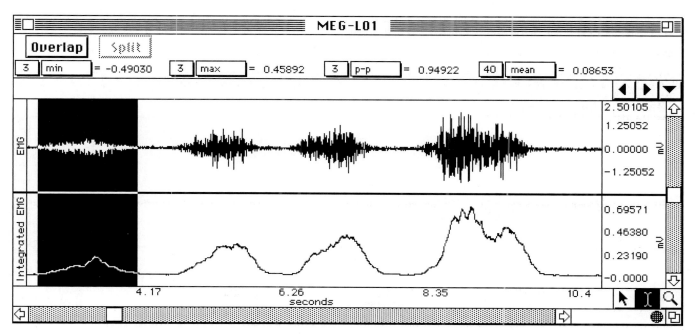

Part 2—Electromyography: Motor Unit Recruitment and Fatigue

Purpose of the Exercise

To measure and calculate the force generated during motor unit recruitment.

Learning Outcomes

After completing this exercise, you should be able to

4. Examine, record, and calculate motor unit recruitment with increased power of skeletal muscle contraction.

5. Determine the maximum clench force between dominant and nondominant hands (and male and female, if possible).

6. Record the force produced by clenched muscles during fatigue.

7. Calculate the time it takes clenched muscles to fatigue.

Procedure A—Setup

If continuing with the same subject used in Part 1, move to setup step 5.

1. With your computer turned **ON** and the BIOPAC MP30 unit turned **OFF,** plug the electrode lead set (SSL2) into Channel 3 and BIOPAC SS25 Hand Dynamometer into Channel 1. Plug the headphones into the back of the unit if desired.

2. Turn on the MP30 Data Acquisition Unit.

3. Attach the electrodes to the dominant forearm (the right arm if the subject is right-handed, the left if he or she is left-handed) of the subject as shown in figure 21.2. This will be **Forearm I.** To help to insure a good contact, make sure the area to be in contact with the electrodes is clean by wiping with the abrasive pad (ELPAD) or alcohol wipe. A small amount of BIOPAC electrode gel (GEL1) may also be used to make better contact between the sensor in the electrode and the skin. For optimal adhesion, the electrodes should be placed on the skin at least 5 minutes before the calibration procedure.

4. Attach the electrode lead set by the pinch connectors as shown in figure 21.2. Make sure the proper color is attached to the appropriate electrode.

5. Start the BIOPAC Student Lab Program for Electromyography 1, lesson L02-EMG-2.

6. Designate a unique filename to be used to save the subject's data, then click **OK.**

Procedure B—Calibration

1. Click **Calibrate.** Set the hand dynamometer on the table and click **OK.** This establishes a zero-force calibration before you continue the calibration sequence.

2. Grasp the hand dynamometer with the dominant hand (Forearm 1) as close to the dynagrip crossbar as possible without touching the crossbar. Note the position of your hand and try to replicate this during the recording phase.

3. Follow the instructions on the two successive pop-up windows and click **OK** when ready for each.

4. Wait 2 seconds, then clench the hand dynamometer as hard as possible, then release.

5. Wait for the calibration to stop. This will take about 8 seconds.

6. If the calibration recording does not resemble figure 21.6, click **Redo Calibration.**

7. Before proceeding to the recording procedure, you need to determine the force increments that you will use. Using the maximum force from the calibration, the increments should be as follows: 0–25 kg use 5 kg increments; 25–50 kg use 10 kg increments; > 50 kg use 20 kg increments.

FIGURE 21.6 Calibration recording part 2 (single wave).

Source: Courtesy of and © *BIOPAC* Systems, Inc.

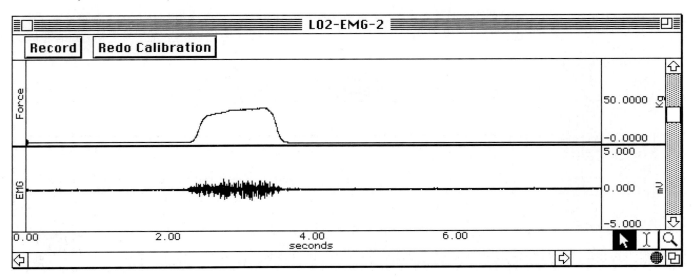

FIGURE 21.7 Motor unit recruitment.

Source: Courtesy of and © *BIOPAC* Systems, Inc.

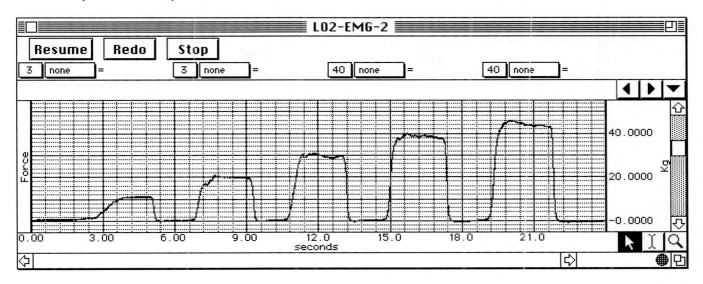

Procedure C—Recording

1. Click on **Record** and have the subject:
   - Clench-Release-Wait and repeat with increasing clench force by the increment determined in step 7 above.
   - There should be a 2-second interval between clenches.
2. Click **Suspend.** The recording should resemble figure 21.7 for Motor Unit Recruitment. If it does not, click **Redo** and repeat step 1.
3. Click **Resume.**
4. Clench the hand dynamometer with your maximum force. Note the force and try to maintain it.
5. When the maximum clench force has decreased by 50%, click **Suspend.** The recording should resemble

figure 21.8 for Fatigue. If it does not, click **Redo** and repeat steps 3–5.
6. If correct, click **Stop.**
7. If you want to listen to the EMG signal, put on the headphones and click **Listen.** Have the subject experiment by changing the clench force as you watch the screen. When finished, click **Stop.**
8. Remove the electrodes from the dominant forearm and place new electrodes and the lead set on the nondominant forearm of the same subject. Click **Forearm 2.** This will return you to the Calibration sequence. Proceed as with the dominant arm for Calibration and Recording.
9. When finished, click **Done.**

FIGURE 21.8 Fatigue.

Source: Courtesy of and © *BIOPAC* Systems, Inc.

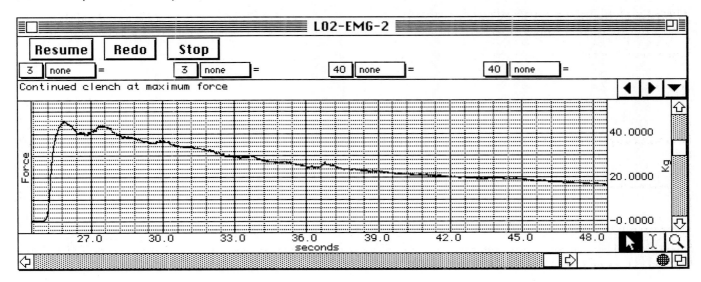

Procedure D—Data Analysis

1. You can either analyze the current file or save it and do the analysis later using the Review Saved Data mode and selecting the correct file. The file with the extension "1-L02" is Forearm 1 and "2-L02" is Forearm 2.
2. Note the channel number designations: **CH 1** displays **Force** (kg), **CH 3** displays **Raw EMG** (mV) and **CH 40** displays the **Integrated EMG** (mV).
3. Set up the channel/measurement boxes as follows:

CH 1/mean, CH 3/p-p, CH 40/mean

4. Use the I-beam cursor to select an area on the plateau phase of the first clench. Record the measurements on Laboratory Report 21 on the table for Motor Unit recruitment (Part II A1).
5. Repeat step 4 on the plateau for each successive clench.
6. Scroll to the second recording segment (for Fatigue).
7. Set up the channel/measurement boxes as follows:

CH 1/value (the highest value in the selected area)
CH 40/Δ T (the amount of time in the selected area)

8. Use the I-beam cursor to select a point of maximal clench force at the start of Segment 2. Record the measurements on the laboratory report on the table for Fatigue (Part II A2).
9. Calculate 50% of the maximum clench force from step 8. Record the calculation on the laboratory report on the table for Fatigue (Part II A2).
10. Select the area from the point of 50% clench force to the maximum clench force by dragging the cursor. Note the time to fatigue (CH 40/Δ T) and record on the laboratory report on the table for Fatigue (Part II A2).
11. Repeat from step 2 for the second forearm.
12. Complete Part II of the laboratory report.
13. Exit the program.

Name _____

Date _____

Section _____

The A corresponds to the Learning Outcome(s) listed at the beginning of the laboratory exercise.

Electromyography: BIOPAC© Exercise

Subject Profile

Name _____   Height _____

Age_____   Weight _____

Gender: Male/Female   Dominant Forearm: right/left

I. Standard and Integrated EMG

Part A Data and Calculations Assessments

1. EMG Measurements A1 A2

Cluster #	Forearm 1 (Dominant)				Forearm 2			
	min	max	p-p	mean	min	max	p-p	mean
1								
2								
3								
4								

2. Use the mean measurement from the preceding table to compute the percentage increase in EMG activity recorded between the weakest clench and the strongest clench of Forearm 1.

Calculation:

Answer: _____ %

Part B Assessments

Complete the following:

1. If you did both forearms on the same subject, are the mean measurements for the right and left maximum grip EMG cluster the same? Which one suggests the greater grip strength? Explain. A1

_____

_____

_____

2. What factors in addition to gender contribute to observed differences in clench strength? **2**

_____

_____

_____

3. Explain the source of signals detected by the EMG electrodes. **3**

_____

_____

_____

4. What is meant by the term motor unit recruitment?

_____

_____

_____

5. Define electromyography.

_____

_____

_____

II. Motor Unit Recruitment and Fatigue

Part A Data and Calculations Assessments

1. Motor Unit Recruitment Data: **4 5**

Peak #	Force (kg) Increments Assigned	Forearm 1 (Dominant)			Forearm 2		
		Force at Peak [CH 1] mean (kg)	Raw EMG [CH 3] p-p (mV)	Int. EMG [CH 40] mean (mV)	Force at Peak [CH 1] mean (kg)	Raw EMG [CH 3] p-p (mV)	Int. EMG [CH 40] mean (mV)
1	kg						
2	kg						
3	kg						
4	kg						
5	kg						
6	kg						
7	kg						
8	kg						
9	kg						

2. Fatigue Data: 6 7

Forearm 1 (Dominant)			Forearm 2		
Maximum Clench Force	50% of Max Clench Force	Time to Fatigue	Maximum Clench Force	50% of Max Clench Force	Time to Fatigue
CH 1 Value	Calculate	CH 40 Δ T	CH 1 Value	Calculate	CH 40 Δ T

Part B Assessments

Complete the following:

1. Describe the strength of the dominant forearm and the nondominant forearm. 5

_____

_____

_____

2. Is there a difference in the absolute values of force generated by males and females in your class? What might explain the difference? 5

_____

_____

_____

3. Is there a difference in fatigue times between the two forearms? Why might this happen?

_____

_____

_____

4. Define fatigue.

_____

_____

_____

5. Define dynamometry.

_____

_____

_____

22

Muscles of the Head and Neck

Pre-Lab

1. Carefully read the introductory material and examine the entire lab.
2. Be familiar with the locations, actions, origins, and insertions of the muscles of the head and neck (from lecture or the textbook).
3. Visit www.mhhe.com/martinseriesl for pre-lab questions and Anatomy & Physiology Revealed animations list.

Materials Needed

Human torso model with musculature
Human skull
Human skeleton, articulated

For Learning Extension Activity:
Long rubber bands

The skeletal muscles of the head include the muscles of facial expression. Human facial expressions are more variable than those of other mammals because many of the muscles have insertions in the skin, rather than on bones, allowing great versatility of movements. The extrinsic muscles of eye movements will be covered in a later laboratory exercise.

The muscles of mastication (chewing) are all inserted on the mandible. Muscles for chewing allow the movements of depression and elevation of the mandible to open and close the jaw. However, jaw movements also involve protraction, retraction, lateral excursion, and medial excursion needed to bite off pieces of food and grind the food.

The muscles that move the head are located in the neck. The actions of flexion, extension, and hyperextension of the head and the cervical vertebral column occur when right and left paired muscles work simultaneously. Additional actions, such as lateral flexion and rotation, result from muscles on one side only contracting or alternating muscles contracting.

A group of neck muscles that have attachments on the hyoid bone and larynx assist in swallowing and speech. Some of the neck muscles that move the head and neck have their attachments in the thorax.

Purpose of the Exercise

To review the locations, actions, origins, and insertions of the muscles of the head and neck.

Learning Outcomes

After completing this exercise, you should be able to

1. Locate and identify the muscles of facial expression, the muscles of mastication, the muscles that move the head and neck, and the muscles that move the hyoid bone and larynx.
2. Describe and demonstrate the action of each of these muscles.
3. Locate the origin and insertion of each of these muscles in a human skeleton and the musculature of the human torso model.

Procedure—Muscles of the Head and Neck

1. Study and label figures 22.1, 22.2, 22.3, 22.4, and 22.5.
2. Locate the following muscles in the human torso model and in your body whenever possible: 1

muscles of facial expression
epicranius (occipitofrontalis)
 frontalis
 occipitalis
nasalis
orbicularis oculi
orbicularis oris
risorius
zygomaticus major
zygomaticus minor
buccinator
platysma

muscles of mastication
- masseter
- temporalis
- lateral pterygoid
- medial pterygoid

muscles that move head and neck
- sternocleidomastoid
- trapezius (superior part)
- scalenes (anterior, middle, and posterior)
- splenius capitis
- semispinalis capitis

muscles that move hyoid bone and larynx
- suprahyoid muscles
  - digastric (2 parts)
  - stylohyoid
  - mylohyoid
- infrahyoid muscles
  - sternohyoid
  - omohyoid (2 parts)
  - sternothyroid
  - thyrohyoid

FIGURE 22.1 Label the muscles of facial expression and mastication for (a) anterior view and (b) lateral view, using the terms provided. ⚠

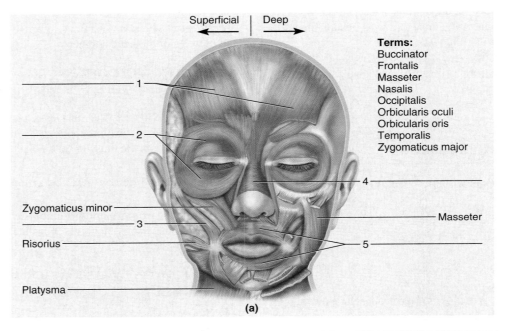

Superficial | Deep

Terms:
Buccinator
Frontalis
Masseter
Nasalis
Occipitalis
Orbicularis oculi
Orbicularis oris
Temporalis
Zygomaticus major

Zygomaticus minor
Risorius
Platysma
Masseter

(a)

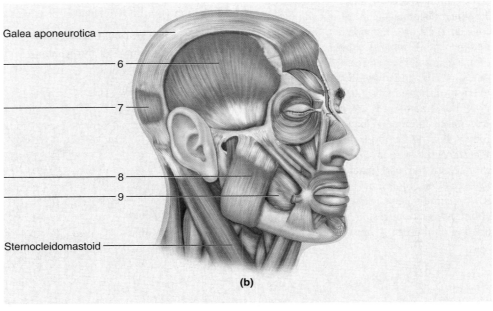

Galea aponeurotica
Sternocleidomastoid

(b)

FIGURE 22.2 Deep muscles of mastication.

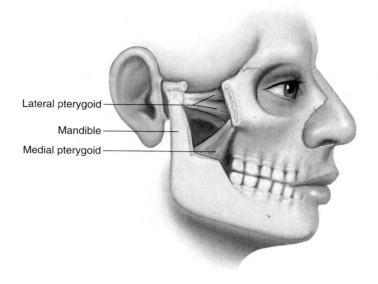

Lateral pterygoid

Mandible

Medial pterygoid

FIGURE 22.3 Label the posterior muscles that move the head and neck, using the terms provided. ⚠

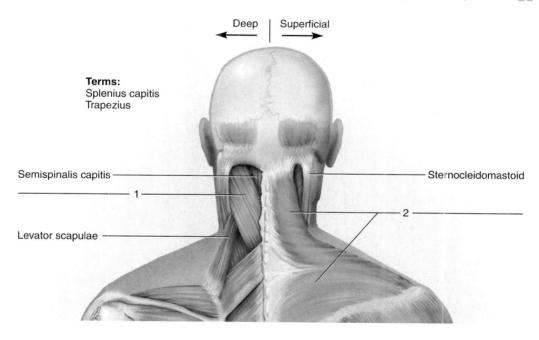

Deep | Superficial

Terms:
Splenius capitis
Trapezius

Semispinalis capitis

1

Levator scapulae

Sternocleidomastoid

2

3. Demonstrate the actions of these muscles in your body.
4. Locate the origins and insertions of these muscles in the human skull and skeleton.
5. Complete Parts A, B, C, and D of Laboratory Report 22.

 Learning Extension Activity

A long rubber band can be used to simulate muscle locations, origins, insertions, and actions on the human torso model, the skeleton, or a laboratory partner. Hold one end of the rubber band firmly on the origin location of a muscle, then slightly stretch the rubber band and hold the other end on the insertion site. Allow the insertion end to slowly move toward the origin end to simulate the contraction and action of the muscle. ⚠ ⚠ ⚠

FIGURE 22.4 Label the muscles that move the head and neck in this lateral view, using the terms provided.

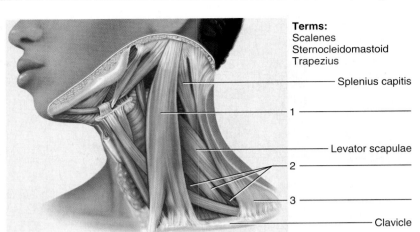

Terms:
Scalenes
Sternocleidomastoid
Trapezius

— Splenius capitis

1 —————————

— Levator scapulae

2 —————————

3 —————————

— Clavicle

FIGURE 22.5 Muscles that move the hyoid bone and larynx assist in swallowing and speech and are grouped into suprahyoid and infrahyoid muscles.

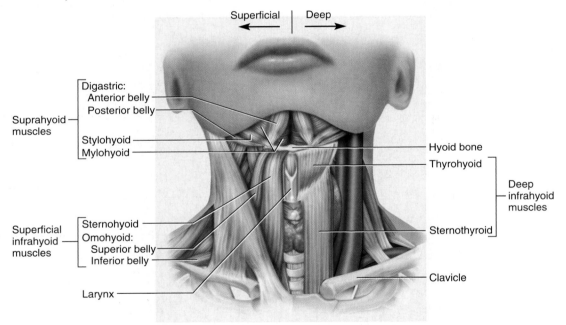

Superficial | Deep

Suprahyoid muscles
Digastric:
  Anterior belly
  Posterior belly
Stylohyoid
Mylohyoid

Superficial infrahyoid muscles
Sternohyoid
Omohyoid:
  Superior belly
  Inferior belly
Larynx

Hyoid bone
Thyrohyoid

Deep infrahyoid muscles

Sternothyroid

Clavicle

Name _____

Date _____

Section _____

The Ⓐ corresponds to the Learning Outcome(s) listed at the beginning of the laboratory exercise.

Muscles of the Head and Neck

Part A Assessments

Identify the muscles indicated in the head and neck of a cadaver in figure 22.6.

FIGURE 22.6   Identify the muscles of the head and neck in this lateral view of a cadaver, using the terms provided. Ⓐ

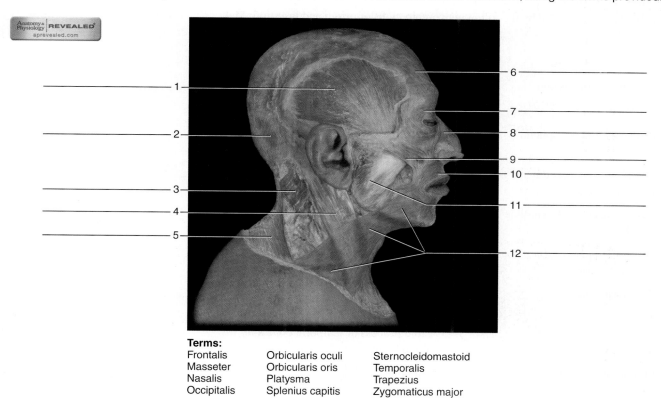

Terms:

Frontalis	Orbicularis oculi	Sternocleidomastoid
Masseter	Orbicularis oris	Temporalis
Nasalis	Platysma	Trapezius
Occipitalis	Splenius capitis	Zygomaticus major

Part B Assessments

Complete the following statements:

1. When the _____ contracts, the corner of the mouth is drawn upward and laterally when laughng. **2**

2. The _____ acts to compress the wall of the cheeks when air is blown out of the mouth. **2**

3. The _____ causes the lips to close and pucker during kissing, whistling, and speaking. **2**

4. The temporalis acts to _____. **2**

5. The _____ pterygoid can close the jaw and can pull it sideways. **2**

6. The _____ pterygoid can protrude the jaw, pull the jaw sideways, and open the mouth. **2**

7. The _____ can close the eye, as in blinking. **2**

8. The _____ can pull the head toward the chest. **2**

9. The muscle used for pouting and to express horror is the _____. **2**

10. The muscle used to widen the nostrils is the _____. **2**

Part C Assessments

Name the muscle indicated by the following combinations of origin and insertion. Make each selection from the list of muscles provided. **3**

Origin	Insertion	Muscle
1. Manubrium of sternum	Thyroid cartilage of larynx	_____
2. Zygomatic arch	Lateral surface of mandible	_____
3. Pterygoid plate of sphenoid bone	Anterior surface below mandibular condyle	_____
4. Manubrium of sternum and medial clavicle	Mastoid process of temporal bone	_____
5. Lateral surfaces of mandible and maxilla	Orbicularis oris	_____
6. Fascia in upper chest	Lower border of mandible and skin around corner of mouth	_____
7. Temporal bone	Coronoid process and anterior ramus of mandible	_____
8. Spinous processes of cervical and thoracic vertebrae (C7–T6)	Mastoid process of temporal bone and occipital bone	_____
9. Styloid process of temporal bone	Hyoid bone	_____

Muscles:
Buccinator
Lateral pterygoid
Masseter
Platysma
Splenius capitis
Sternocleidomastoid
Sternothyroid
Stylohyoid
Temporalis

Part D Assessments

Identify the muscles of various facial expressions in the photographs of figure 22.7.

1. _____

2. _____

3. _____

4. _____

5. _____

FIGURE 22.7 Identify the muscles of facial expression being contracted in each of these photographs (*a–c*), using the terms provided. 〈1〉 〈2〉

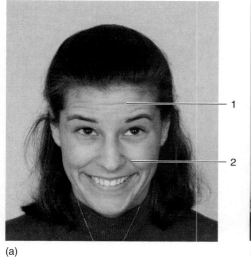

(a)

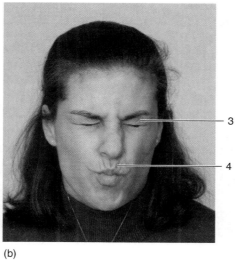

(b)

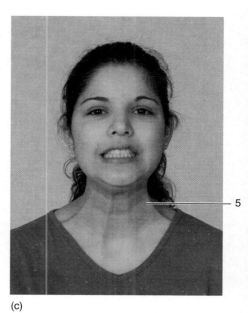

(c)

Terms:
Epicranius/frontalis
Orbicularis oculi
Orbicularis oris
Platysma
Zygomaticus major

Muscles of the Chest, Shoulder, and Upper Limb

Pre-Lab

1. Carefully read the introductory material and examine the entire lab content.
2. Be familiar with the locations, actions, origins, and insertions of the muscles of the chest, shoulder, and upper limb (from lecture or the textbook).
3. Visit www.mhhe.com/martinseriesl for pre-lab questions, Anatomy & Physiology Revealed animations list, and LabCam videos.

Materials Needed

Human torso model
Human skeleton, articulated
Muscular models of the upper limb

For Learning Extension Activity:
Long rubber bands

Many of the chest and shoulder muscles have their insertions on the scapula and provide various movements of the pectoral girdle. Some chest and shoulder muscles crossing the shoulder joint have insertions on the humerus and are responsible for moving the arm in various ways at the shoulder joint. Muscles located in the arm having insertions on the radius and ulna are responsible for forearm movements at the elbow joint. Muscles located in the forearm with insertions on carpals, metacarpals, or phalanges are responsible for various movements of the hand.

The muscles of the upper limb vary from large sizes that provide powerful actions to numerous smaller muscles in the forearm and hand that allow finer motor skills. The human hand, with its opposable thumb, enables us to grasp objects and provides great dexterity for tasks such as writing, drawing, and playing musical instruments.

As you demonstrate the actions of muscles in an upper limb, use the standard anatomical regions to describe them. The arm refers to shoulder to elbow; the forearm refers to elbow to wrist; and the hand includes the wrist, palm, and fingers. Also remember that the belly of a muscle responsible for a joint movement is usually proximal to the region being pulled. If a muscle crosses two joints, the actions allowed are even more diverse.

Purpose of the Exercise

To review the locations, actions, origins, and insertions of the muscles in the chest, shoulder, and upper limb.

Learning Outcomes

After completing this exercise, you should be able to

1. Locate and identify the muscles of the chest, shoulder, and upper limb.
2. Describe and demonstrate the action of each of these muscles.
3. Locate the origin and insertion of each of these muscles in a human skeleton and on the muscular models.

Procedure—Muscles of the Chest, Shoulder, and Upper Limb

1. Study and label figures 23.1, 23.2, 23.3, 23.4, and 23.5.
2. Locate the following muscles in the human torso model and models of the upper limb. Also locate in your body as many of the muscles as you can.

muscles that move the pectoral girdle
anterior muscles
   pectoralis minor
   serratus anterior
posterior muscles
   trapezius
   rhomboid major
   rhomboid minor
   levator scapulae

muscles that move the arm
  origins on axial skeleton
    pectoralis major
    latissimus dorsi
  origins on scapula
    deltoid
    teres major
    coracobrachialis
  rotator cuff (SITS) muscles (origins also on scapula)
    supraspinatus
    infraspinatus
    teres minor
    subscapularis

muscles that move the forearm
  muscle bellies in arm
    biceps brachii
    brachialis
    triceps brachii
  muscle bellies in forearm
    brachioradialis
    anconeus

  supinator
  pronator teres
  pronator quadratus

muscles that move the hand
  anterior superficial flexor muscles
    palmaris longus (absent sometimes)
    flexor carpi radialis
    flexor capri ulnaris
    flexor digitorum superficialis (intermediate depth)
  anterior deep flexor muscles
    flexor digitorum profundus
    flexor pollicis longus
  posterior superficial extensor muscles
    extensor digitorum
    extensor carpi radialis longus
    extensor carpi radialis brevis
    extensor carpi ulnaris
  posterior deep extensor muscles
    abductor pollicis longus
    extensor pollicis longus

FIGURE 23.1   Label the anterior muscles of the chest, shoulder, and arm, using the terms provided.

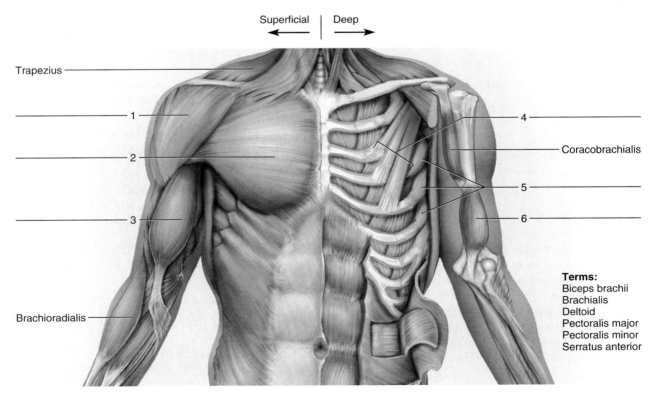

Superficial | Deep

Trapezius

1
2
3
Brachioradialis

4
Coracobrachialis
5
6

Terms:
Biceps brachii
Brachialis
Deltoid
Pectoralis major
Pectoralis minor
Serratus anterior

FIGURE 23.2 Label the posterior muscles of the shoulder, back, and arm, using the terms provided.

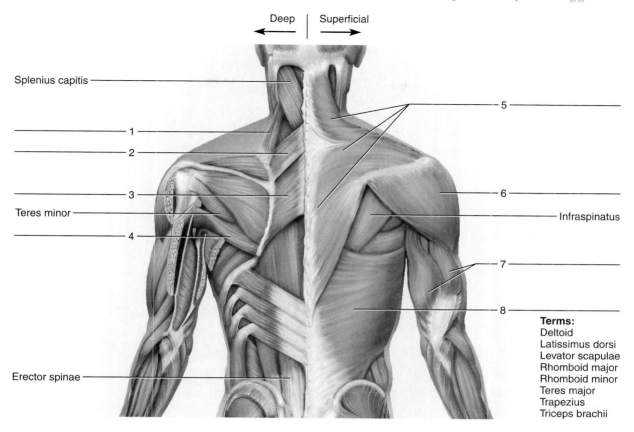

Deep | Superficial

Splenius capitis

Teres minor

Erector spinae

Infraspinatus

Terms:
Deltoid
Latissimus dorsi
Levator scapulae
Rhomboid major
Rhomboid minor
Teres major
Trapezius
Triceps brachii

3. Demonstrate the action of these muscles in your body.
4. Locate the origins and insertions of these muscles in the human skeleton.
5. Complete Parts A, B, C, and D of Laboratory Report 23.

Learning Extension Activity

A long rubber band can be used to simulate muscle locations, origins, insertions, and actions on muscular models, the skeleton, or a laboratory partner. Hold one end of the rubber band firmly on the origin location of a muscle, then slightly stretch the rubber band and hold the other end on the insertion site. Allow the insertion end to slowly move toward the origin end to simulate the contraction and action of the muscle.

FIGURE 23.3 Rotator cuff muscles: (*a*) anterior view; (*b*) posterior view; (*c*) lateral view. Some regional bones, features, and muscles are also labeled.

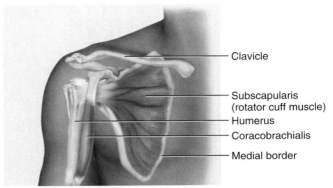

Clavicle

Subscapularis (rotator cuff muscle)

Humerus

Coracobrachialis

Medial border

(a) Anterior view

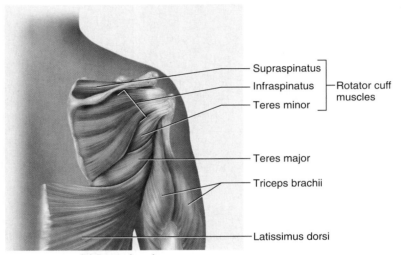

Supraspinatus

Infraspinatus

Teres minor

Rotator cuff muscles

Teres major

Triceps brachii

Latissimus dorsi

(b) Posterior view

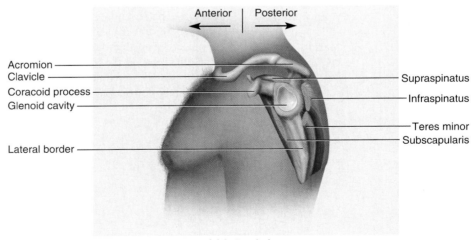

Anterior   Posterior

Acromion
Clavicle
Coracoid process
Glenoid cavity

Lateral border

Supraspinatus

Infraspinatus

Teres minor
Subscapularis

(c) Lateral view

FIGURE 23.4 Label the superficial and deep anterior forearm muscles (*a, b,* and *c*), using the terms provided. ⚠

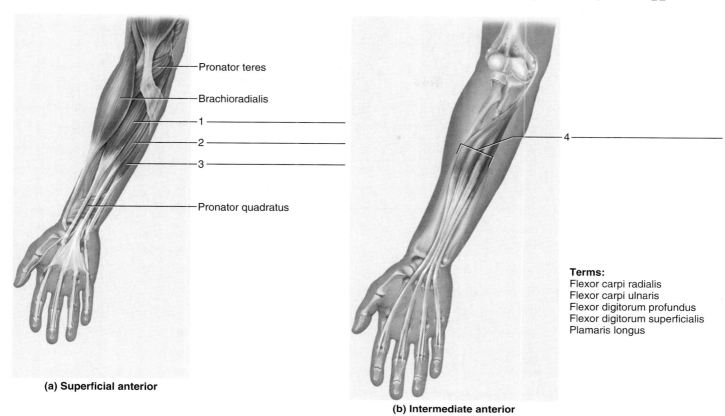

Pronator teres

Brachioradialis

1 ————————

2 ————————

3 ————————

Pronator quadratus

(a) Superficial anterior

4 ——————————

Terms:
Flexor carpi radialis
Flexor carpi ulnaris
Flexor digitorum profundus
Flexor digitorum superficialis
Plamaris longus

(b) Intermediate anterior

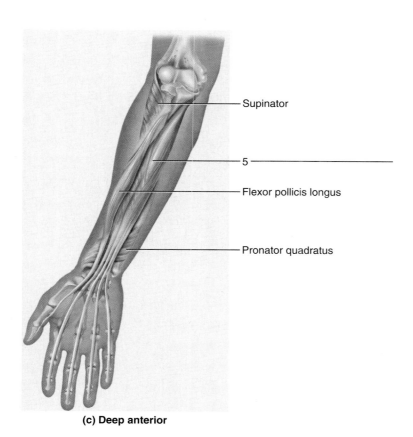

Supinator

5 ————————

Flexor pollicis longus

Pronator quadratus

(c) Deep anterior

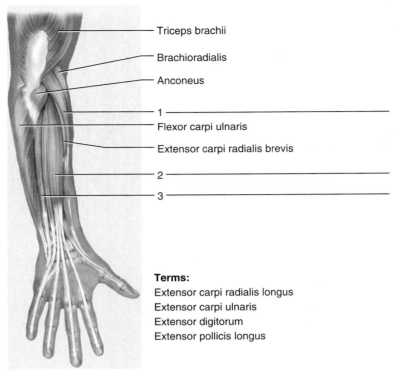

- Triceps brachii
- Brachioradialis
- Anconeus
- 1 ———————————
- Flexor carpi ulnaris
- Extensor carpi radialis brevis
- 2 ———————————
- 3 ———————————

Terms:
Extensor carpi radialis longus
Extensor carpi ulnaris
Extensor digitorum
Extensor pollicis longus

(a) Superficial posterior

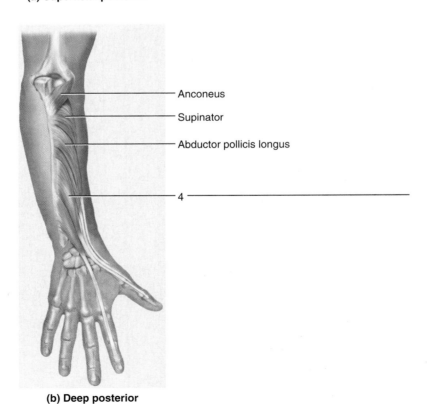

- Anconeus
- Supinator
- Abductor pollicis longus
- 4 ———————————

(b) Deep posterior

Name _____

Date _____

Section _____

The ⒶＲ corresponds to the Learning Outcome(s) listed at the beginning of the laboratory exercise.

Muscles of the Chest, Shoulder, and Upper Limb

Part A Assessments

Identify the muscles indicated in the shoulder, arm, and forearm of a cadaver in figures 23.6 and 23.7.

FIGURE 23.6   Identify the posterior muscles of the left shoulder and arm of a cadaver, using the terms provided. The deltoid and trapezius muscles have been removed. ⒶＲ

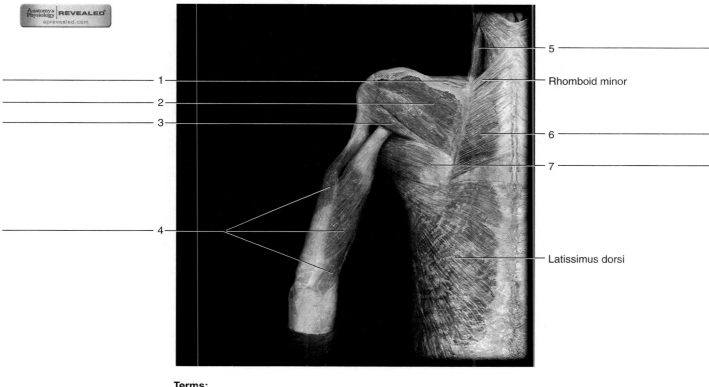

Terms:

Infraspinatus	Teres major
Levator scapulae	Teres minor
Rhomboid major	Triceps brachii
Supraspinatus	

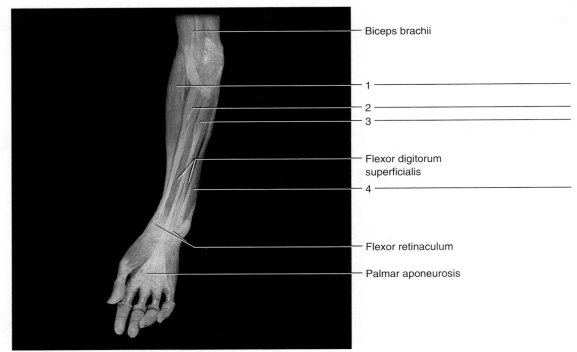

Biceps brachii

1

2

3

Flexor digitorum superficialis

4

Flexor retinaculum

Palmar aponeurosis

Terms:
Brachioradialis
Flexor carpi radialis
Flexor carpi ulnaris
Palmaris longus

Part B Assessments

Match the muscles in column A with the actions in column B. Place the letter of your choice in the space provided. **2**

Column A	Column B
a. Brachialis	_____ **1.** Abducts arm
b. Coracobrachialis	_____ **2.** Pulls arm forward (flexion) and across chest (adduction) and rotates arm medially
c. Deltoid	
d. Extensor carpi ulnaris	_____ **3.** Flexes and adducts hand at the wrist
e. Flexor carpi ulnaris	_____ **4.** Raises and adducts (retracts) scapula
f. Infraspinatus	_____ **5.** Raises ribs in forceful inhalation or pulls scapula forward and downward
g. Pectoralis major	
h. Pectoralis minor	_____ **6.** Used to thrust shoulder anteriorly (protraction), as when pushing something
i. Rhomboid major	
j. Serratus anterior	_____ **7.** Flexes the forearm at the elbow
k. Teres major	_____ **8.** Flexes and adducts arm at the shoulder along with pectoralis major
l. Triceps brachii	
	_____ **9.** Extends the forearm at the elbow
	_____ **10.** Extends, adducts, and rotates arm medially
	_____ **11.** Extends and adducts hand at the wrist
	_____ **12.** Rotates arm laterally

Part C Assessments

Name the muscle indicated by the following combinations of origin and insertion. Make each selection from the list of muscles provided. **3**

	Origin	Insertion	Muscle
1.	Spines of upper thoracic vertebrae	Medial border of scapula	_____
2.	Outer surfaces of upper ribs	Anterior surface of scapula	_____
3.	Sternal ends of upper ribs	Coracoid process of scapula	_____
4.	Coracoid process of scapula	Shaft of humerus	_____
5.	Inferior angle of scapula	Intertubercular sulcus of humerus	_____
6.	Anterior surface of scapula	Lesser tubercle of humerus	_____
7.	Lateral border of scapula	Greater tubercle of humerus	_____
8.	Anterior shaft of humerus	Coronoid process of ulna	_____
9.	Medial epicondyle of humerus	Fascia of palm	_____

Muscles:

Brachialis

Coracobrachialis

Palmaris longus

Pectoralis minor

Rhomboid major

Serratus anterior

Subscapularis

Teres major

Teres minor

Part D Assessments

 Critical Thinking Activity

Identify the muscles indicated in figure 23.8.

FIGURE 23.8 Label these muscles that appear as body surface features in these photographs (*a, b,* and *c*) by placing the correct numbers in the spaces provided. **1**

_____ Biceps brachii

_____ Deltoid

_____ External oblique

_____ Pectoralis major

_____ Rectus abdominis

_____ Serratus anterior

_____ Sternocleidomastoid

_____ Trapezius

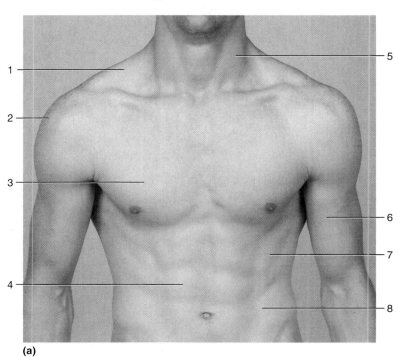

(a)

FIGURE 23.8 *Continued.*

_____ Biceps brachii
_____ Deltoid
_____ Infraspinatus
_____ Latissimus dorsi
_____ Trapezius
_____ Triceps brachii

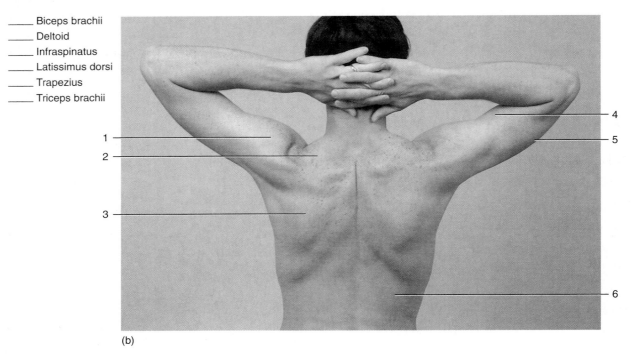

(b)

_____ Biceps brachii
_____ Brachioradialis
_____ Deltoid
_____ Pectoralis major
_____ Serratus anterior
_____ Trapezius
_____ Triceps brachii

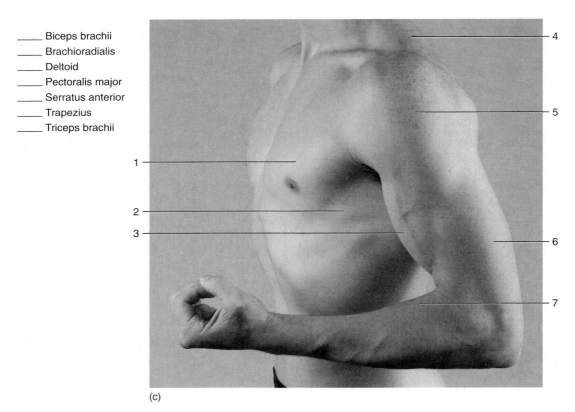

(c)

Muscles of the Deep Back, Abdominal Wall, and Pelvic Outlet

Pre-Lab

1. Carefully read the introductory material and examine the entire lab content.
2. Be familiar with the locations, actions, origins, and insertions of muscles of the deep back, abdominal wall, and pelvic outlet (from lecture or the textbook).
3. Visit www.mhhe.com/martinseriesl for pre-lab questions and Anatomy & Physiology Revealed animations.

Materials Needed

Human torso model with musculature
Human skelton, articulated
Muscular models of male and female pelves

The deep muscles of the back extend the vertebral column. Because the muscles have many origins, insertions, and subgroups, the muscles overlap each other. The deep back muscles can extend the spine when contracting as a group but also help to maintain posture and normal spine curvatures.

The anterior and lateral walls of the abdomen contain broad, flattened muscles arranged in layers. These muscles connect the rib cage and vertebral column to the pelvic girdle. The abdominal wall muscles compress the abdominal visceral organs, help maintain posture, assist in forceful exhalation, and contribute to trunk flexion and waist rotation.

The muscles in the region of the perineum form the pelvic floor (outlet) and can be divided into two triangles: a urogenital and an anal triangle. The anterior urogenital triangle contains superficial and deeper muscle layers associated with the external genitalia. The posterior anal triangle contains the pelvic diaphragm and the anal sphincter. The pelvic floor is penetrated by the urethra, vagina, and anus in a female and thus are important in obstetrics.

Purpose of the Exercise

To review the actions, origins, and insertions of the muscles of the deep back, abdominal wall, and pelvic outlet.

Learning Outcomes

After completing this exercise, you should be able to

1. Locate and identify the muscles of the deep back, abdominal wall, and pelvic outlet.
2. Describe the action of each of these muscles.
3. Locate the origin and insertion of each of these muscles in a human skeleton or on muscular models.

Procedure—Muscles of the Deep Back, Abdominal Wall, and Pelvic Outlet

1. Study and label figures 24.1, 24.2, 24.3, and 24.4.
2. Locate the following muscles in the human torso model:

   erector spinae group
   iliocostalis (lateral group)
   longissimus (intermediate group)
   spinalis (medial group)
   quadratus lumborum
   external oblique
   internal oblique
   transversus abdominis (transverse abdominal)
   rectus abdominis

3. Demonstrate the actions of these muscles in your body.
4. Locate the origin and insertion of each of these muscles in the human skeleton.
5. Complete Parts A and B of Laboratory Report 24.
6. Locate the following muscles from figure 24.4 in the models of the male and female pelves:

   levator ani
   coccygeus
   superficial transverse perineal muscle
   bulbospongiosus
   ischiocavernosus

FIGURE 24.1 Deep back muscles.

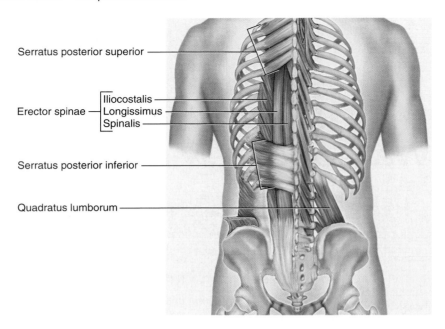

Serratus posterior superior

Erector spinae — Iliocostalis
Longissimus
Spinalis

Serratus posterior inferior

Quadratus lumborum

FIGURE 24.2 Label the muscles of the abdominal wall, using the terms provided.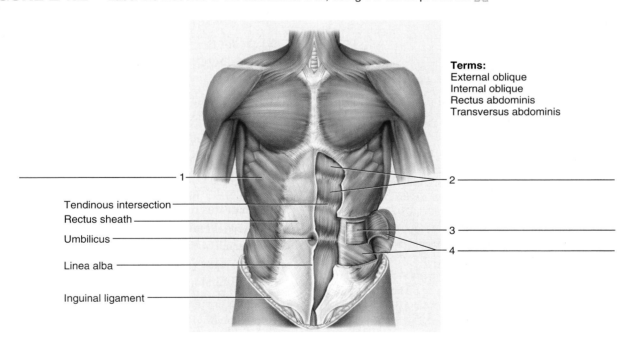

Terms:
External oblique
Internal oblique
Rectus abdominis
Transversus abdominis

1

2

Tendinous intersection
Rectus sheath
Umbilicus
Linea alba

3

4

Inguinal ligament

7. Locate the origin and insertion of each of these muscles in the human skeleton. **3**

8. Complete Part C of the laboratory report.

FIGURE 24.3   Transverse view of the midsection of the abdominal wall.

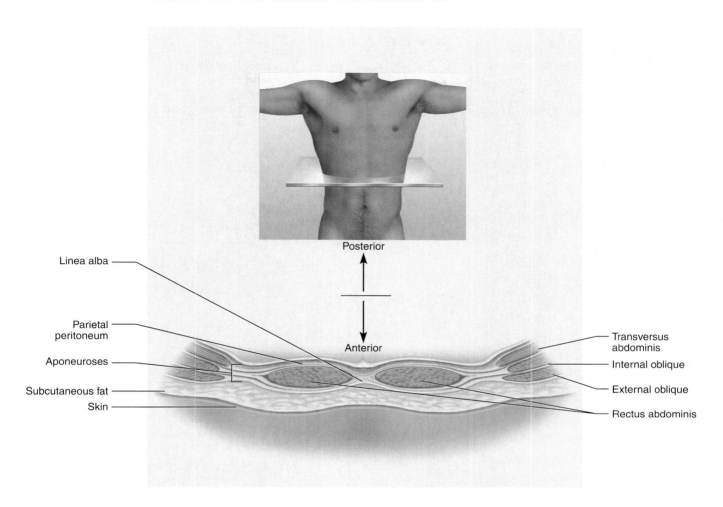

Posterior

Anterior

Linea alba

Parietal peritoneum

Aponeuroses

Subcutaneous fat

Skin

Transversus abdominis

Internal oblique

External oblique

Rectus abdominis

Critical Thinking Activity

List the muscles from superficial to deep for an appendectomy incision. ⚠1

_____ _____

_____ _____

_____ _____

_____ _____

FIGURE 24.4 Muscles of the pelvic floor (outlet): (*a*) inferior view of superficial perineal muscles of male and female; (*b*) super view of pelvic diaphragm of female.

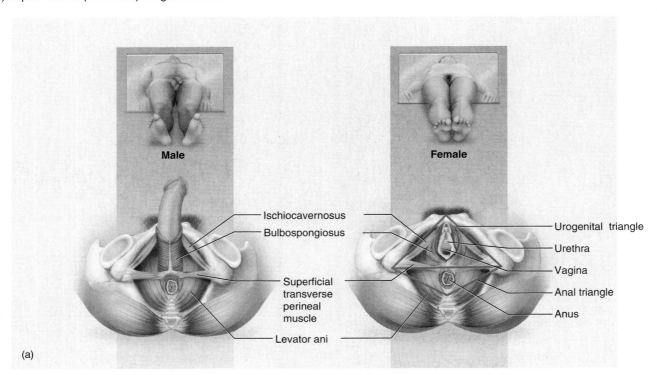

Male

Female

- Ischiocavernosus
- Bulbospongiosus
- Superficial transverse perineal muscle
- Levator ani

- Urogenital triangle
- Urethra
- Vagina
- Anal triangle
- Anus

(a)

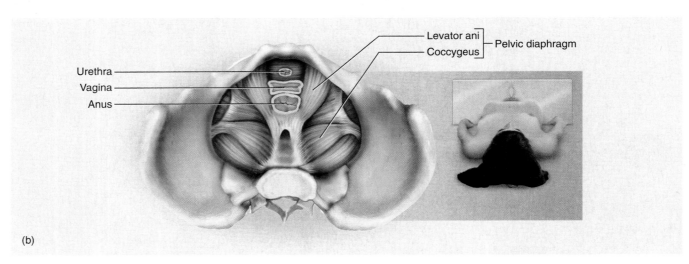

- Levator ani
- Coccygeus
- Pelvic diaphragm

Urethra
Vagina
Anus

(b)

Name _____

Date _____

Section _____

The △ corresponds to the Learning Outcome(s) listed at the beginning of the laboratory exercise.

Muscles of the Deep Back, Abdominal Wall, and Pelvic Outlet

Part A Assessments

Identify the muscles indicated in the deep back of a cadaver in figure 24.5.

FIGURE 24.5 Label the deep back muscles of a cadaver, using the terms provided. The trapezius, latissimus dorsi, rhomboids, and serratus posterior have been removed. △

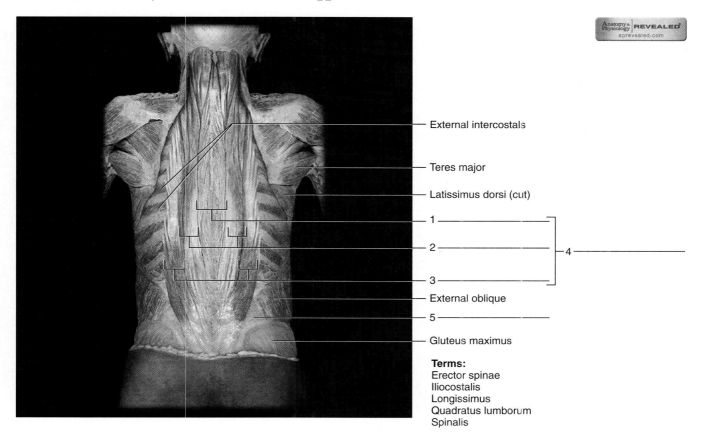

External intercostals

Teres major

Latissimus dorsi (cut)

1 ———————

2 ——————— 4 ———————

3 ———————

External oblique

5 ———————

Gluteus maximus

Terms:
Erector spinae
Iliocostalis
Longissimus
Quadratus lumborum
Spinalis

Part B Assessments

Complete the following statements:

1. A band of tough connective tissue in the midline of the anterior abdominal wall called the _____ serves as a muscle attachment. /1\

2. The _____ muscle spans from the costal cartilages and sternum to the pubic bones. /3\

3. The _____ forms the third layer (deepest layer) of the abdominal wall muscles. /1\

4. The action of the external oblique muscle is to _____. /2\

5. The action of the rectus abdominis is to _____. /2\

6. The iliocostalis, longissimus, and spinalis muscles together form the _____. /1\

7. The _____ has the origin on the iliac crest and the insertion on rib 12 and L1–L4 vertebrae. /3\

Part C Assessments

Complete the following statements:

1. The levator ani and coccygeus together form the _____ diaphragm. /1\

2. The posterior portion of the perineum is called the _____ triangle. /1\

3. The _____ surrounds the base of the penis. /1\

4. In females, the bulbospongiosus acts to constrict the _____. /2\

5. In the female, the pelvic floor (outlet) is penetrated by the urethra, vagina, and _____. /1\

6. In the female, the _____ muscles are separated by the vagina, urethra, and anal canal. /1\

7. Name five muscles of the pelvic outlet that males and females both have in common. /1\ (There are more than five correct answers.)

_____

_____

_____

_____

_____

25

Muscles of the Hip and Lower Limb

Pre-Lab

1. Carefully read the introductory material and examine the entire lab content.
2. Be familiar with the locations, actions, origins, and insertions of the muscles of the hip and lower limb.
3. Visit www.mhhe.com/martinseriesl for pre-lab questions, Anatomy & Physiology Revealed animations list, and LabCam videos.

Materials Needed

Human torso model with musculature
Human skeleton, articulated
Muscular models of the lower limb

For Learning Extension Activity:
Long rubber bands

The muscles that move the thigh at the hip joint have their origins on the pelvis and insertions usually on the femur. Those muscles attached on the anterior pelvis act to flex the thigh at the hip joint; those attached on the posterior pelvis act to extend the thigh at the hip joint; those attached on the lateral side of the pelvis act as abductors and rotators of the thigh at the hip joint; and those attached on the medial pelvis act as adductors of the thigh at the hip joint.

The muscles that move the leg at the knee joint have their origins on the femur or the hip bone. The anterior thigh muscles have their insertions on the proximal tibia and serve as prime movers of extension of the leg at the knee joint. The posterior thigh muscles have their insertions on the tibia or fibula and act as prime movers of flexion of the leg at the knee joint.

The muscles that move the foot have their origins on the tibia, fibula, or femur. The anterior leg muscles have actions of dorsiflexion and extension of the toes; the posterior leg muscles have actions of plantar flexion and flexion of the toes; the lateral leg muscles have actions of eversion of the foot.

In contrast to the upper limb muscles, those in the lower limb tend to be much larger and are involved in powerful contractions for walking or running as well as isometric contractions in standing. As compared to the forearm, the leg has fewer muscles because the foot is not as involved in actions that require precision movements as the hand.

As you demonstrate the actions of muscles in the lower limb, refer to standard anatomical regions of the limb. The thigh refers to hip to knee; the leg refers to knee to ankle; and the foot refers to ankle, metatarsal area, and toes. Several muscles in the hip and lower limb cross more than one joint and involve movements of both joints. Examples of muscles crossing more than one joint include the sartorius and gastrocnemius.

Purpose of the Exercise

To review the actions, origins, and insertions of the muscles that move the thigh, leg, and foot.

Learning Outcomes

After completing this exercise, you should be able to

① Locate and identify the muscles that move the thigh, leg, and foot.

② Describe and demonstrate the actions of each of these muscles.

③ Locate the origin and insertion of each of these muscles in a human skeleton and on muscular models.

Procedure—Muscles of the Hip and Lower Limb

1. Label figures 25.1, 25.2, 25.3, 25.4, 25.5, and 25.6.
2. Locate the following muscles in the human torso model and in the lower limb models. Also locate as many of them as possible in your body.

   muscles that move the thigh
      anterior hip muscles
         iliopsoas group
            iliacus
            psoas major

posterior and lateral hip muscles
  gluteus maximus
  gluteus medius
  gluteus minimus
  tensor fasciae latae
  piriformis
medial adductor muscles
  pectineus
  adductor longus
  adductor magnus
  adductor brevis
  gracilis

muscles that move the leg
  anterior thigh muscles
    sartorius
    quadriceps femoris group
      rectus femoris
      vastus lateralis
      vastus medialis
      vastus intermedius
  posterior thigh muscles
    hamstring group
      biceps femoris
      semitendinosus
      semimembranosus

FIGURE 25.1 Label the hip and thigh muscles in this right anterior view, using the terms provided. /1\

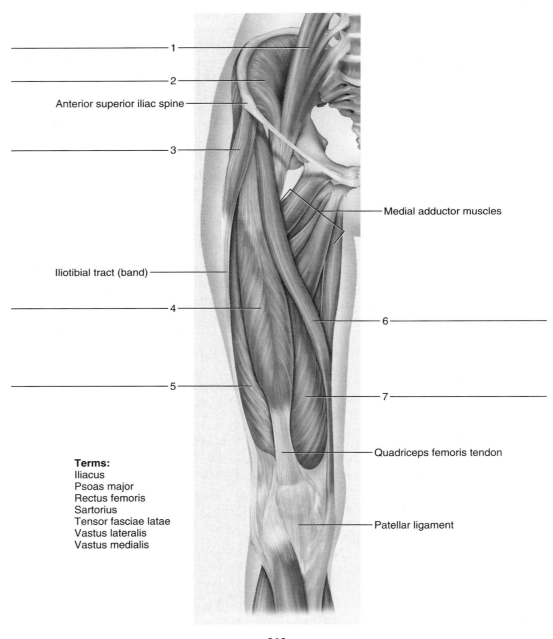

Anterior superior iliac spine

Medial adductor muscles

Iliotibial tract (band)

Quadriceps femoris tendon

Patellar ligament

Terms:
Iliacus
Psoas major
Rectus femoris
Sartorius
Tensor fasciae latae
Vastus lateralis
Vastus medialis

muscles that move the foot
    anterior leg muscles
        tibialis anterior
        extensor digitorum longus
        fibularis (peroneus) tertius
    posterior leg muscles
        superficial group
            gastrocnemius
            soleus
        deep group
            flexor digitorum longus
            tibialis posterior
            flexor hallucis longus
    lateral leg muscles
        fibularis (peroneus) longus
        fibularis (peroneus) brevis

3. Demonstrate the action of each of these muscles in your body.
4. Locate the origin and insertion of each of these muscles in the human skeleton.
5. Complete Parts A, B, C, and D of Laboratory Report 25.

Learning Extension Activity

A long rubber band can be used to simulate muscle locations, origins, insertions, and actions on muscular models, the skeleton, or on a laboratory partner. Hold one end of the rubber band firmly on the origin location of a muscle, then slightly stretch the rubber band and hold the other end on the insertion site. Allow the insertion end to slowly move toward the origin end to simulate the contraction and action of the muscle. 1 2 3

FIGURE 25.2 Label the anterior hip muscles and the medial adductor muscles on the right side, using the terms provided. 1

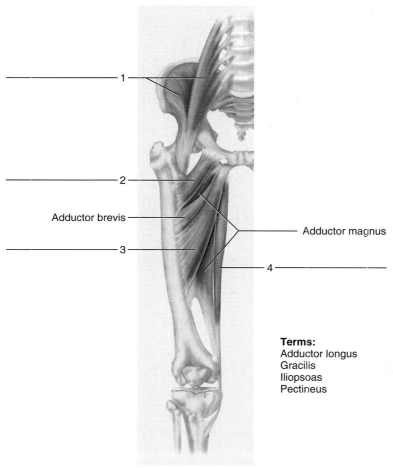

Adductor brevis

Adductor magnus

Terms:
Adductor longus
Gracilis
Iliopsoas
Pectineus

FIGURE 25.3 Label the hip and thigh muscles in this right posterior view, using the terms provided. 🔺

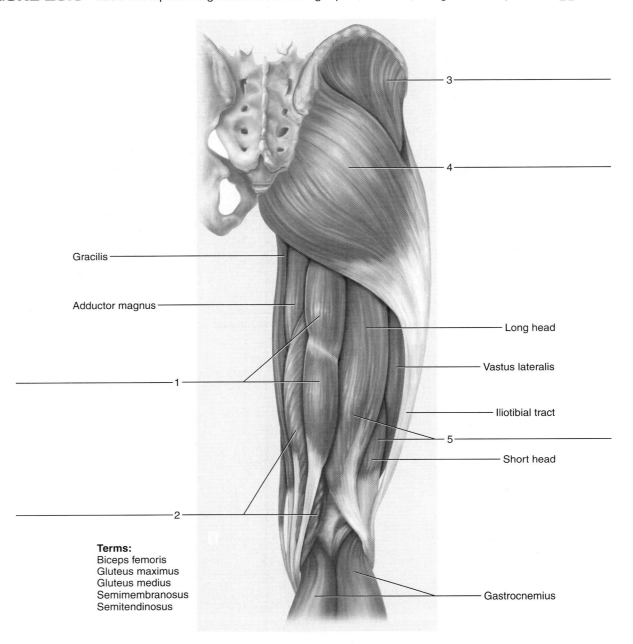

Gracilis

Adductor magnus

1

2

3

4

Long head

Vastus lateralis

Iliotibial tract

5

Short head

Gastrocnemius

Terms:
Biceps femoris
Gluteus maximus
Gluteus medius
Semimembranosus
Semitendinosus

FIGURE 25.4 Label the superficial and deep hip muscles in this posterior view, using the terms provided. ⚠

Superficial | Deep

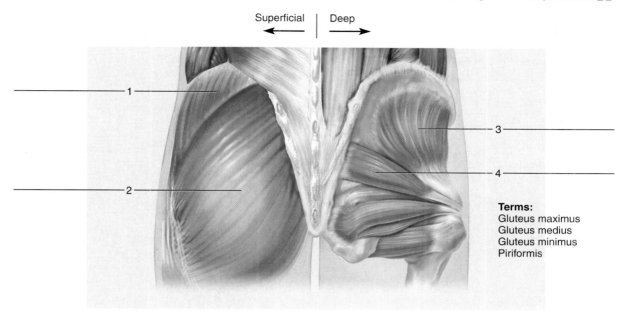

Terms:
Gluteus maximus
Gluteus medius
Gluteus minimus
Piriformis

FIGURE 25.5 Label the leg muscles in this right side anterior view, using the terms provided. ⚠

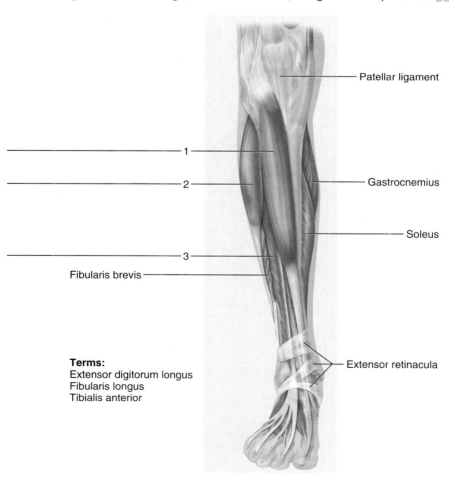

Patellar ligament

Gastrocnemius

Soleus

Fibularis brevis

Extensor retinacula

Terms:
Extensor digitorum longus
Fibularis longus
Tibialis anterior

FIGURE 25.6 Label the superficial and deep right leg muscles (a and b), using the terms provided.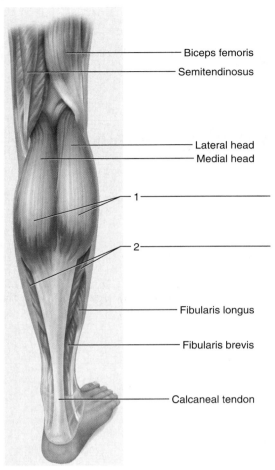

Biceps femoris
Semitendinosus

Lateral head
Medial head

1

2

Fibularis longus

Fibularis brevis

Calcaneal tendon

(a) Superficial

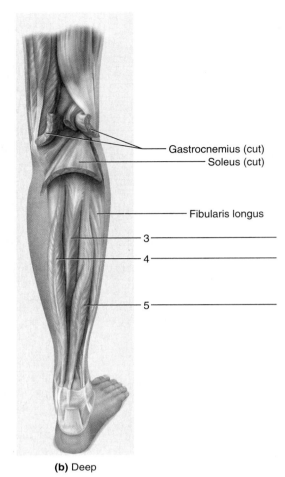

Gastrocnemius (cut)
Soleus (cut)

Fibularis longus

3

4

5

(b) Deep

Terms:
Flexor digitorum longus
Flexor hallucis longus
Gastrocnemius
Soleus
Tibialis posterior

216

Name _____

Date _____

Section _____

The ⚠ corresponds to the Learning Outcome(s) listed at the beginning of the laboratory exercise.

Muscles of the Hip and Lower Limb

Part A Assessments

Identify the muscles indicated in the posterior hip and thigh of a cadaver in figure 25.7.

FIGURE 25.7 Label the right posterior hip and thigh muscles of a cadaver, using the terms provided. The gluteus maximus has been removed. ⚠

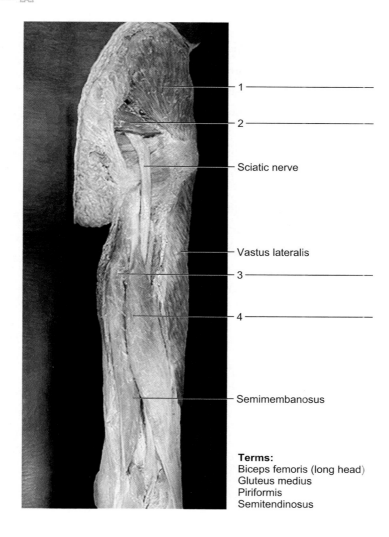

1 _____

2 _____

Sciatic nerve

Vastus lateralis

3 _____

4 _____

Semimembanosus

Terms:
Biceps femoris (long head)
Gluteus medius
Piriformis
Semitendinosus

Part B Assessments

Match the muscles in column A with the actions in column B. Place the letter of your choice in the space provided. ◢2◣

Column A

a. Biceps femoris
b. Fibularis (peroneus) longus
c. Gluteus medius
d. Gracilis
e. Psoas major and iliacus
f. Quadriceps femoris group
g. Sartorius
h. Tibialis anterior
i. Tibialis posterior

Column B

_____ 1. Adducts thigh

_____ 2. Plantar flexion and eversion of foot

_____ 3. Flexes thigh at the hip

_____ 4. Abducts thigh and rotates it laterally

_____ 5. Abducts thigh and rotates it medially

_____ 6. Plantar flexion and inversion of foot

_____ 7. Flexes leg at the knee

_____ 8. Extends leg at the knee

_____ 9. Dorsiflexion and inversion of foot

Part C Assessments

Name the muscle indicated by the following combinations of origin and insertion. Make each selection from the list of muscles provided. ◢3◣

Origin	Insertion	Muscle
1. Lateral surface of ilium	Greater trochanter of femur	_____
2. Anterior superior iliac spine	Medial surface of proximal tibia	_____
3. Lateral and medial condyles of femur	Posterior surface of calcaneus	_____
4. Anterior iliac crest and anterior superior iliac spine	Iliotibial tract to lateral condyle of tibia	_____
5. Greater trochanter and posterior surface of femur	Patella to tibial tuberosity	_____
6. Ischial tuberosity	Medial surface of tibia	_____
7. Medial surface of femur	Patella to tibial tuberosity	_____
8. Posterior surface of tibia	Distal phalanges of four lateral toes	_____
9. Lateral condyle and lateral surface of tibia	Medial cuneiform and first metatarsal	_____

Muscles:
Flexor digitorum longus
Gastrocnemius
Gluteus medius
Sartorius
Semitendinosus
Tensor fasciae latae
Tibialis anterior
Vastus lateralis
Vastus medialis

Part D Assessments

Critical Thinking Activity

Identify the muscles indicated in figure 25.8.

FIGURE 25.8 Label these muscles that appear as lower limb surface features in these photographs (*a* and *b*), by placing the correct numbers in the spaces provided.

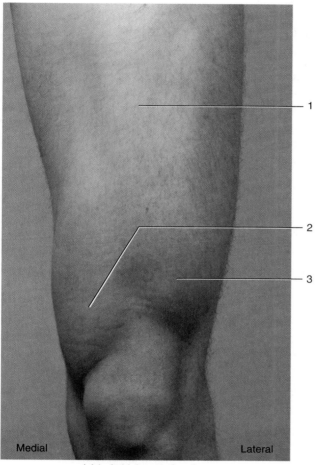

(a) Left thigh, anterior view

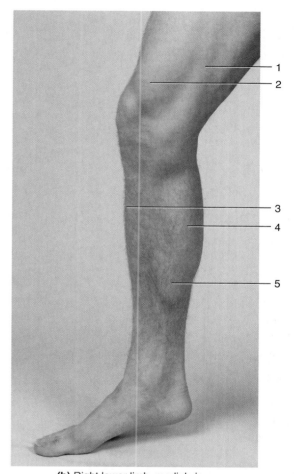

(b) Right lower limb, medial view

_____ Rectus femoris
_____ Vastus lateralis
_____ Vastus medialis

_____ Gastrocnemius
_____ Sartorius
_____ Soleus
_____ Tibialis anterior
_____ Vastus medialis

Laboratory Exercise  26

Surface Anatomy

Pre-Lab

1. Carefully read the introductory material and examine the entire lab content.
2. Be familiar with body regions, the skeletal system, and the muscular system (from lecture or the textbook).
3. Visit www.mhhe.com/martinseries1 for pre-lab questions.

Materials Needed

Small round stickers
Colored pencils (black and red)

External landmarks, called surface anatomy, located on the human body provide an opportunity to examine surface features and to help locate other internal structures. This exercise will focus on bony landmarks and superficial soft tissue structures, primarily related to the skeletal and muscular systems. The technique used for the examination of surface features is called *palpation,* accomplished by touching with the hands or fingers. This enables us to determine the location, size, and texture of the structure. Some of the surface features can be visible without the need of palpation, especially on body builders and very thin individuals. If the individual involved has considerable subcutaneous adipose tissue, additional pressure might be necessary to palpate some of the structures.

These surface features will help us to locate additional surface features and deeper structures during the study of the additional systems. Some of the respiratory structures in the anterior neck can be palpated. Surface features will help us to determine the pulse locations of superficial arteries, to assess injection sites, to describe pain and injury sites,

to insert examination tubes, and to listen to heart, lung, and bowel sounds with a stethoscope. For those who choose careers in emergency medical services, nurses, physicians, physical educators, physical therapists, chiropractors, and massage therapists, surface anatomy is especially valuable.

Purpose of the Exercise

To examine the surface features of the human body and the terms used to describe them.

Learning Outcomes

After completing this exercise you should be able to

1. Locate and identify major surface features of the human body.
2. Arrange surface features by body region.
3. Distinguish between surface features that are bony landmarks and soft tissue.

Procedure—Surface Anatomy

1. Review laboratory manual figures on body regions, skeletal system, and muscular system.
2. Use the laboratory manual figures and the other figures provided to complete figures 26.1 through 26.8. Many features have already been labeled. Features in boldface are bony features. Other labeled features (not boldface) are soft tissue features. Bony features are a part of a bone and the skeletal system. Soft tissue features are comprised of tissue other than bone such as muscle, connective tissue, or skin. The missing features are listed by each figure.
3. Complete Parts A and B of Laboratory Report 26.
4. Use figures 26.1 through 26.8 to determine which items listed in table 26.1 are bony features and which items are soft tissue features. Examine the list of bony and soft tissue surface features in the first column. If the feature listed is a bony surface feature, place a ✓ mark in the second column; if the feature listed is a soft tissue feature, place a ✓ mark in the third column. **3**
5. Complete Part C of the laboratory report.

FIGURE 26.1 Using the terms provided, label the surface features of (a) the anterior head and neck, (b) the lateral head and neck, and (c) the posterior head and neck. (**Boldface** indicates bony features; not boldface indicates soft tissue features.) ⊿3

Terms:
Hyoid bone
Mandible
Sternocleidomastoid
Trapezius (upper)

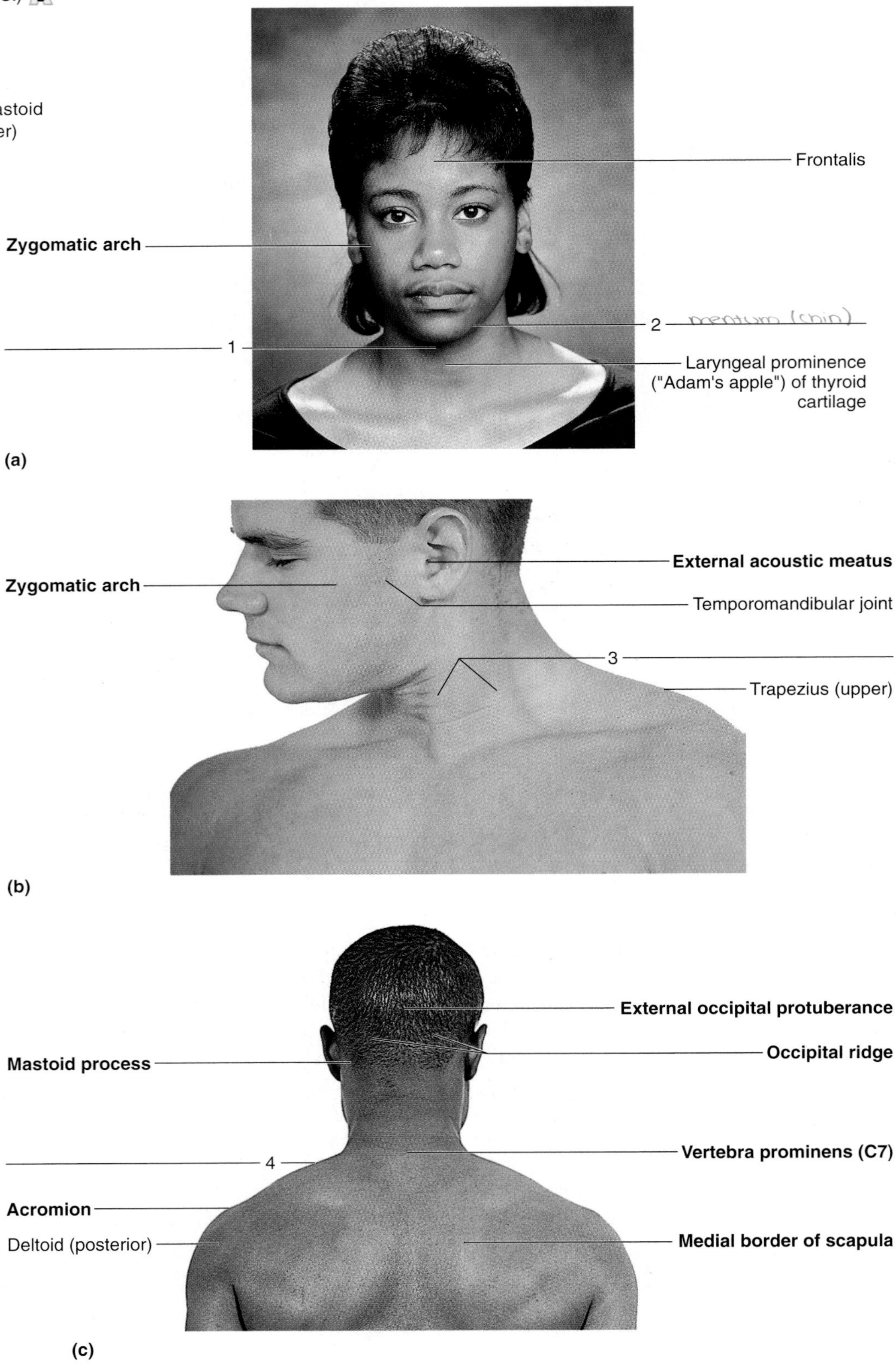

—————————————— Frontalis

Zygomatic arch ————————

2 ———— mentum (chin)

1 ——————

————— Laryngeal prominence ("Adam's apple") of thyroid cartilage

(a)

Zygomatic arch ————————

External acoustic meatus

Temporomandibular joint

3 ——————

Trapezius (upper)

(b)

External occipital protuberance

Occipital ridge

Mastoid process ————————

Vertebra prominens (C7)

4 ——————

Acromion ————————

Deltoid (posterior) ————————

Medial border of scapula

(c)

222

FIGURE 26.2 Surface features of the posterior shoulder and torso. (**Boldface** indicates bony features; not boldface indicates soft tissue features.)

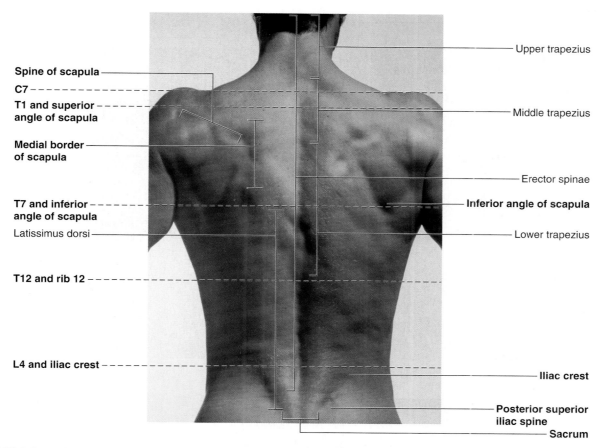

Spine of scapula
C7
T1 and superior angle of scapula
Medial border of scapula
T7 and inferior angle of scapula
Latissimus dorsi
T12 and rib 12
L4 and iliac crest

Upper trapezius
Middle trapezius
Erector spinae
Inferior angle of scapula
Lower trapezius
Iliac crest
Posterior superior iliac spine
Sacrum

TABLE 26.1 Representative Bony and Soft Tissue Surface Features

Bony and Soft Tissues	Bony Features	Soft Tissue Features
Biceps brachii		
Calcaneal tendon		
Deltoid		
Gastrocnemius		
Greater trochanter		
Head of fibula		
Iliac crest		
Iliotibial tract		
Lateral malleolus		
Mandible		
Mastoid process		
Medial border of scapula		
Rectus abdominis		
Sacrum		
Serratus anterior		
Sternocleidomastoid		
Sternum		
Thenar eminence		
Tibialis anterior		
Zygomatic arch		

FIGURE 26.3 Using the terms provided, label the anterior surface features of (a) upper torso and (b) lower torso. (**Boldface** indicates bony features; not boldface indicates soft tissue features.) **3**

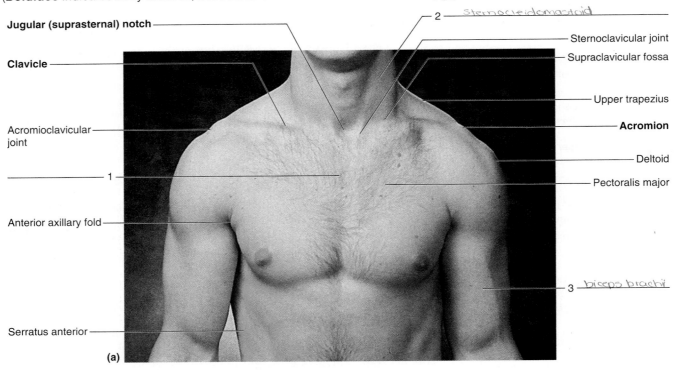

Jugular (suprasternal) notch

Clavicle

Acromioclavicular joint

1

Anterior axillary fold

Serratus anterior

(a)

2 — Sternocleidomastoid

Sternoclavicular joint

Supraclavicular fossa

Upper trapezius

Acromion

Deltoid

Pectoralis major

3 — biceps brachii

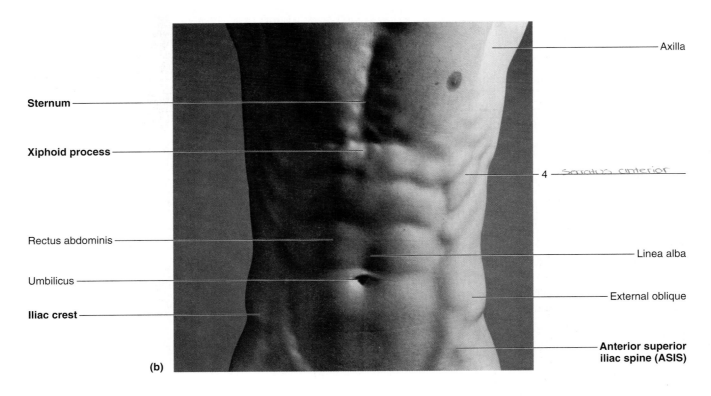

Sternum

Xiphoid process

Rectus abdominis

Umbilicus

Iliac crest

(b)

Axilla

4 — Serratus anterior

Linea alba

External oblique

Anterior superior iliac spine (ASIS)

Terms:
Biceps brachii
Serratus anterior
Sternocleidomastoid
Sternum

FIGURE 26.4 Using the terms provided, label the surface features of the lateral shoulder and upper limb. (**Boldface** indicates bony features; not boldface indicates soft tissue features.) ⚠

Terms:
Acromion
Olecranon process

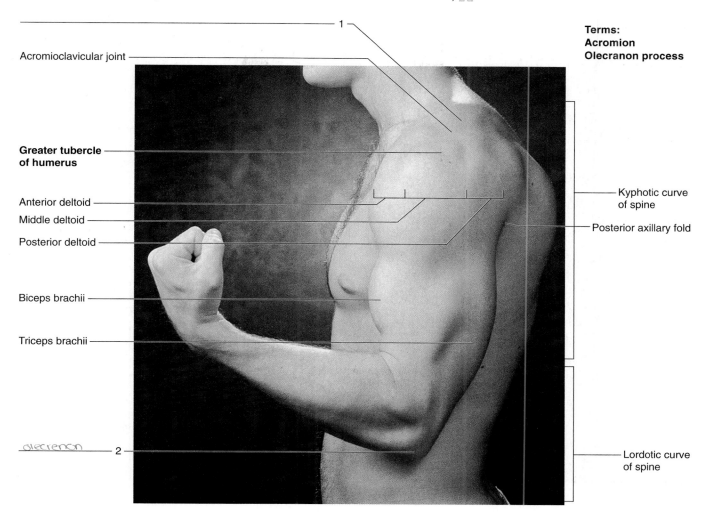

1

Acromioclavicular joint

Greater tubercle of humerus

Anterior deltoid

Middle deltoid

Posterior deltoid

Biceps brachii

Triceps brachii

olecrenon

2

Kyphotic curve of spine

Posterior axillary fold

Lordotic curve of spine

Learning Extension Activity

Select fifteen surface features presented in the laboratory exercise figures. Using small, round stickers, write the name of a surface feature on each sticker. Working with a lab partner, locate the selected surface features on yourself and on your lab partner. Use the stickers to accurately "label" your partner. ⚠

• Were you able to find all the features on both of you? _____

• Was it easier to find the surface features on yourself or your lab partner? _____

FIGURE 26.5 Surface features of the upper limb (*a*) anterior view and (*b*) posterior view. (**Boldface** indicates bony features; not boldface indicates soft tissue features.)

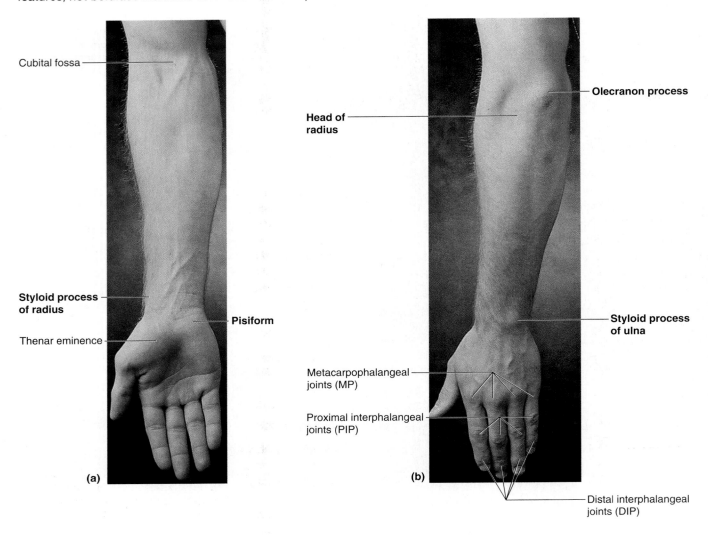

Cubital fossa

**Head of
radius**

Olecranon process

**Styloid process
of radius**

Thenar eminence

Pisiform

**Styloid process
of ulna**

Metacarpophalangeal
joints (MP)

Proximal interphalangeal
joints (PIP)

Distal interphalangeal
joints (DIP)

(a)

(b)

FIGURE 26.6 Using the terms provided, label the surface features of the anterior view of the body. (**Boldface** indicates bony features; not boldface indicates soft tissue features.) ⚠

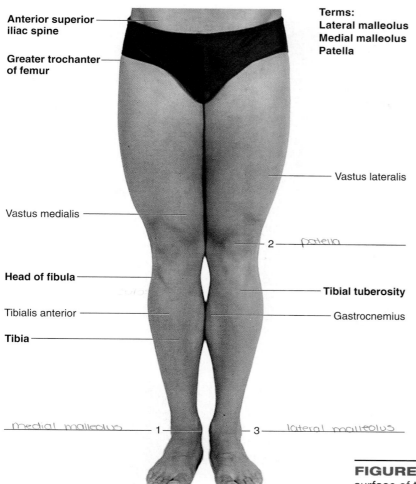

Anterior superior iliac spine

Greater trochanter of femur

Vastus medialis

Head of fibula

Tibialis anterior

Tibia

Terms:
Lateral malleolus
Medial malleolus
Patella

Vastus lateralis

2 ___ patella

Tibial tuberosity

Gastrocnemius

medial malleolus ___ 1 ___ 3 ___ lateral malleolus

FIGURE 26.7 Surface features of the plantar surface of the foot. (**Boldface** indicates bony features; not boldface indicates soft tissue features.)

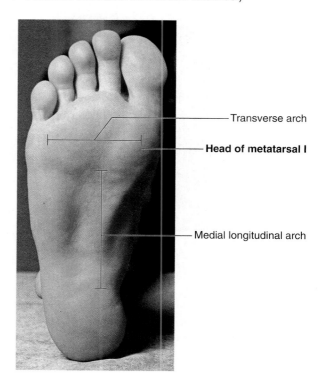

Transverse arch

Head of metatarsal I

Medial longitudinal arch

227

FIGURE 26.8 Using the terms provided, label the surface features of the (a) posterior lower limb, (b) lateral lower limb, and (c) medial lower limb. (**Boldface** indicates bony features; not boldface indicates soft tissue features.) 3

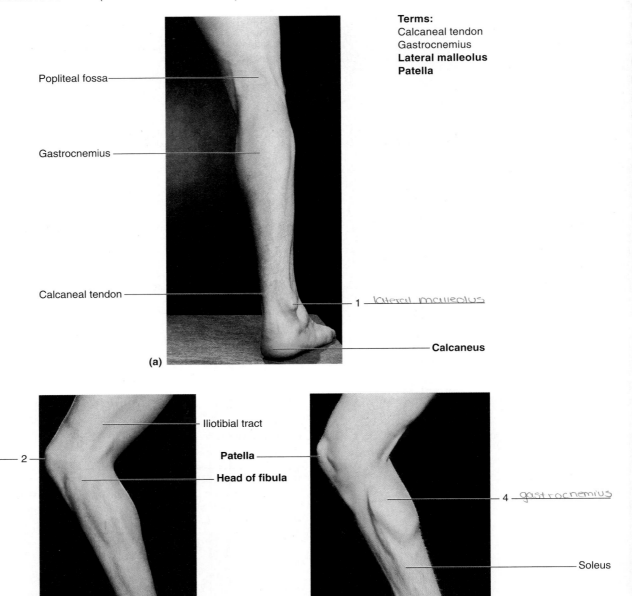

Terms:
Calcaneal tendon
Gastrocnemius
Lateral malleolus
Patella

Popliteal fossa

Gastrocnemius

Calcaneal tendon

1 _lateral malleolus_

Calcaneus

(a)

Iliotibial tract

2

Patella

Head of fibula

4 _gastrocnemius_

Soleus

3 _calcaneal tendon_

Medial malleolus
Lateral malleolus

(b)

(c)

Name _____

Date _____

Section _____

The A corresponds to the Learning Outcome(s) listed at the beginning of the laboratory exercise.

Surface Anatomy

Part A Assessments

Match the body region in column A with the surface feature in column B. Place the letter of your choice in the space provided. 2

Column A	Column B
a. Head	_____ **1.** Umbilicus
b. Upper limb	_____ **2.** Medial malleolus
c. Lower limb	_____ **3.** Iliac crest
d. Trunk	_____ **4.** Transverse arch
	_____ **5.** Spine of scapula
	_____ **6.** External occipital protuberance
	_____ **7.** Proximal interphalangeal joints
	_____ **8.** Sternum
	_____ **9.** Olecranon process
	_____ **10.** Zygomatic arch
	_____ **11.** Mastoid process
	_____ **12.** Thenar eminence

Part B Assessments

Label figure 26.9 with the surface features provided.

FIGURE 26.9 Using the terms provided, label the surface features of the (a) anterior view of body and (b) posterior view of the body. (**Boldface** indicates bony features; not boldface indicates soft tissue features.) 🅰

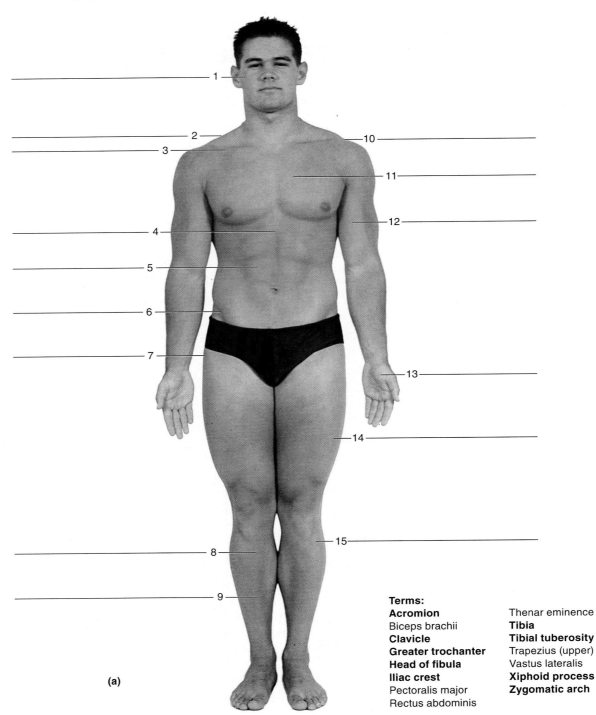

(a)

Terms:

Acromion	Thenar eminence
Biceps brachii	**Tibia**
Clavicle	**Tibial tuberosity**
Greater trochanter	Trapezius (upper)
Head of fibula	Vastus lateralis
Iliac crest	**Xiphoid process**
Pectoralis major	**Zygomatic arch**
Rectus abdominis	

FIGURE 26.9 *Continued.*

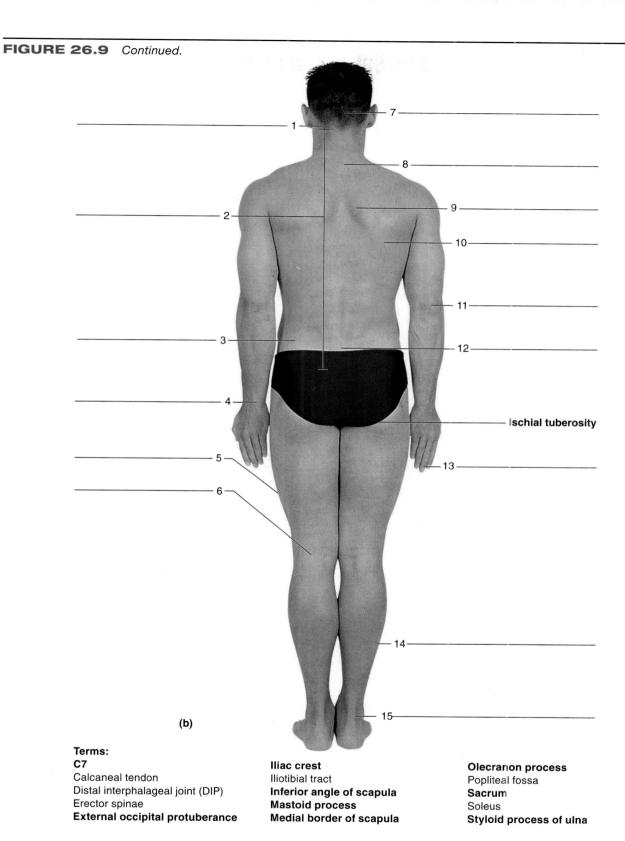

Ischial tuberosity

(b)

Terms:
C7
Calcaneal tendon
Distal interphalageal joint (DIP)
Erector spinae
External occipital protuberance

Iliac crest
Iliotibial tract
Inferior angle of scapula
Mastoid process
Medial border of scapula

Olecranon process
Popliteal fossa
Sacrum
Soleus
Styloid process of ulna

Part C Assessments

Using table 26.1 from Laboratory Exercise 26, indicate the locations of the surface features with an X on figure 26.10. Use a black X for the bony features and a red X for the soft tissue features.

FIGURE 26.10 Indicate bony surface features with a black X and soft tissue surface features with a red X using the results from table 26.1, on the (a) anterior and (b) posterior diagrams of the body. **3**

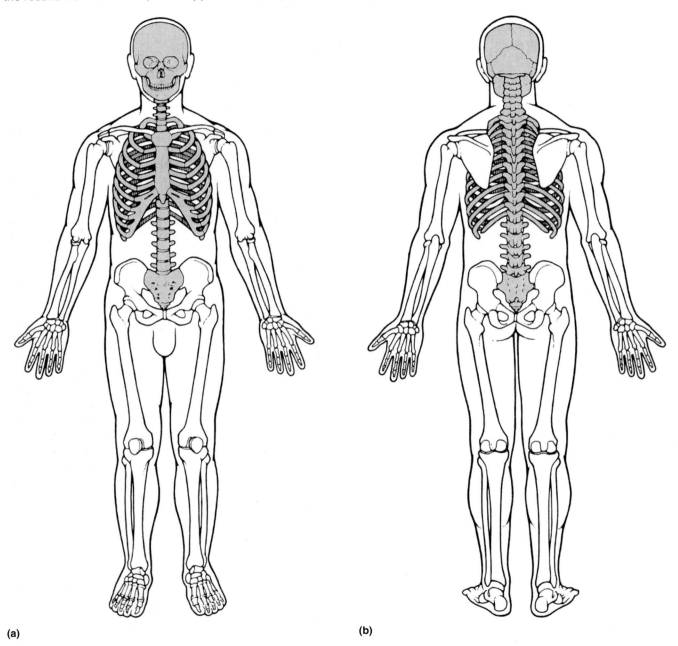

(a)   (b)

Nervous Tissue and Nerves

Pre-Lab

1. Carefully read the introductory material and examine the entire lab content.
2. Be familiar with the structures and functions of nervous tissue and nerves (from lecture or the textbook).
3. Visit www.mhhe.com/martinseriesl for pre-lab questions and Anatomy & Physiology Revealed animations list.

Materials Needed

Compound light microscope
Prepared microscope slides of the following:
   Spinal cord (smear)
   Dorsal root ganglion (section)
   Neuroglial cells (astrocytes)
   Peripheral nerve (cross section and longitudinal section)
Neuron model

For Learning Extension Activity:
Prepared microscope slide of Purkinje cells from cerebellum

Nervous tissue, which occurs in the brain, spinal cord, and peripheral nerves, contains *neurons* (nerve cells) and *neuroglia* (neuroglial cells; glial cells). Neurons are irritable (excitable) and easily respond to stimuli, resulting in the conduction of an action potential (nerve impulse). A neuron contains a cell body with a nucleus and most of the cytoplasm, and elongated cell processes (dendrites and axons) along which impulse conductions occur. Neurons can be classified according to structural variations of their cell processes: multipolar with many dendrites and one axon (nerve fiber), bipolar with one dendrite and one axon, and unipolar with a single process where the dendrite leads

directly into the axon. Functional classifications are used according to the direction of the impulse conduction: sensory (afferent) neurons conduct impulses from receptors to the central nervous system (CNS), interneurons (association neurons) conduct impulses within the CNS, and motor (efferent) neurons conduct impulses away from the CNS to effectors (muscles or glands).

Neuroglial cells (supportive cells) are located in close association with neurons. Four types of neuroglia are in the CNS: astrocytes, microglia, oligodendrocytes, and ependymal cells. Astrocytes are numerous, and their branches support neurons and blood vessels. Microglia can phagocytize microorganisms and nerve tissue debris. Oligodendrocytes produce the myelin insulation in the CNS. Ependymal cells line the brain ventricles and secrete CSF (cerebrospinal fluid), and the cilia on their apical surfaces aid the circulation of the CSF.

Two types of neuroglia are located in the peripheral nervous system (PNS): Schwann cells and satellite cells. Schwann cells surround nerve fibers numerous times. The wrappings nearest the nerve fiber represent the myelin sheath and serve to insulate and increase the impulse speed, while the last wrapping, called the neurilemma, contains most of the cytoplasm and the nucleus and functions in nerve fiber regeneration in PNS neurons. Satellite cells surround and support the cell body regions, called ganglia, of peripheral neurons.

The PNS contains 12 pairs of cranial nerves and 31 pairs of spinal nerves, all containing parallel axons of neurons and neuroglia representing the nervous tissue components. Most nerves are mixed in that they contain both sensory and motor neurons, but some contain only sensory or motor components. The nerves represent an organ structure as they also contain small blood vessels and fibrous connective tissue. The fibrous connective tissue around each nerve fiber and the Schwann cell is the endoneurium; a bundle of nerve fibers, representing a fascicle, is surrounded by a perineurium; and the entire nerve is surrounded by an epineurium. The fibrous connective components of a nerve provide protection and stretching during body movements.

Purpose of the Exercise

To review the characteristics of nervous tissue and to observe neurons, neuroglia, and various features of the nerves.

Learning Outcomes

After completing this exercise, you should be able to

(1) Describe and locate the general characteristics of nervous tissue.

(2) Distinguish structural and functional characteristics between neurons and neuroglia.

(3) Identify and sketch the major structures of a neuron and a nerve.

Procedure—Nervous Tissue and Nerves

1. Label figures 27.1 and 27.2.
2. Complete Parts A and B of Laboratory Report 27.
3. Obtain a prepared microscope slide of a spinal cord smear. Using low-power magnification, search the slide and locate the relatively large, deeply stained cell bodies of motor neurons (multipolar neurons).

FIGURE 27.1 Label this diagram of a motor neuron, using the terms provided. 3

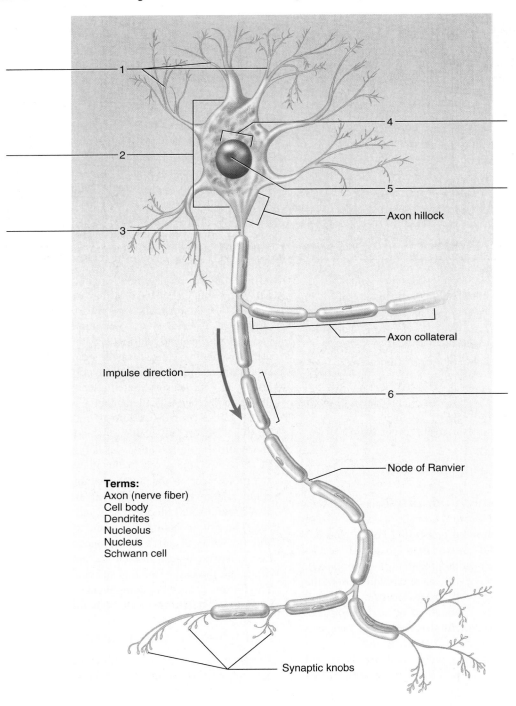

Axon hillock

Axon collateral

Impulse direction

Node of Ranvier

Terms:
Axon (nerve fiber)
Cell body
Dendrites
Nucleolus
Nucleus
Schwann cell

Synaptic knobs

FIGURE 27.2 Label the features of the myelinated axon (nerve fiber), using the terms provided. **3**

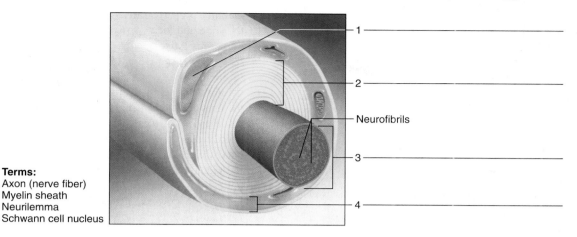

1 _____

2 _____

Neurofibrils

3 _____

4 _____

Terms:
Axon (nerve fiber)
Myelin sheath
Neurilemma
Schwann cell nucleus

4. Observe a single motor neuron, using high-power magnification, and note the following features:

cell body (soma)
   nucleus
   nucleolus
   Nissl bodies (chromatophilic substance)

neurofibrils (threadlike structures extending into nerve fibers)

axon hillock (origin region of axon)

dendrites

axon (nerve fiber)

Compare the slide to the neuron model and to figure 27.3. You also may note small, darkly stained nuclei of neuroglia around the motor neuron.

FIGURE 27.3 Micrograph of a multipolar neuron and neuroglia from a spinal cord smear (100× micrograph enlarged to 600×).

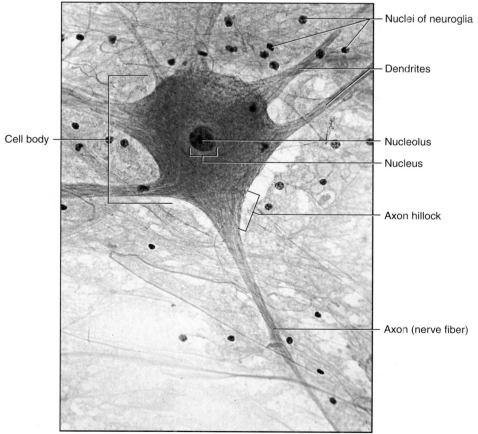

Nuclei of neuroglia

Dendrites

Cell body

Nucleolus

Nucleus

Axon hillock

Axon (nerve fiber)

5. Sketch and label a motor (efferent) neuron in the space provided in Part C of the laboratory report.
6. Obtain a prepared microscope slide of a dorsal root ganglion. Search the slide and locate a cluster of sensory neuron cell bodies. You also may note bundles of nerve fibers passing among groups of neuron cell bodies (fig. 27.4).

FIGURE 27.4 Micrograph of a dorsal root ganglion (50× micrograph enlarged to 100×).

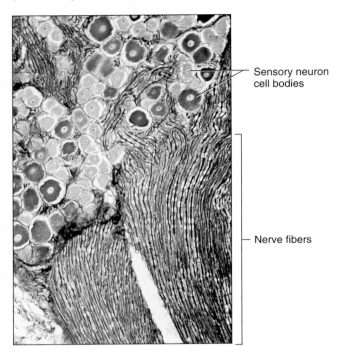

Sensory neuron cell bodies

Nerve fibers

FIGURE 27.5 Micrograph of astrocytes (250× micrograph enlarged to 1,000×).

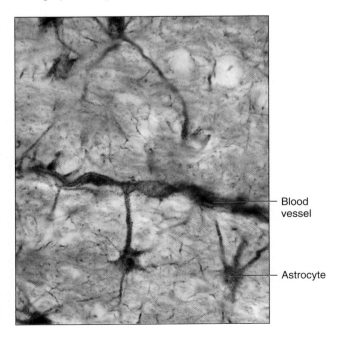

Blood vessel

Astrocyte

7. Sketch and label a sensory (afferent) neuron cell body in the space provided in Part C of the laboratory report.
8. Obtain a prepared microscope slide of neuroglia. Search the slide and locate some darkly stained astrocytes with numerous long, slender processes (fig. 27.5).
9. Sketch a neuroglial cell in the space provided in Part C of the laboratory report.
10. Obtain a prepared microscope slide of a nerve. Locate the cross section of the nerve and note the many round nerve fibers inside. Also note the dense layer of connective tissue (perineurium) that encircles a fascicle of nerve fibers and holds them together in a bundle. The individual nerve fibers are surrounded by a layer of more delicate connective tissue (endoneurium) (fig. 27.6).
11. Using high-power magnification, observe a single nerve fiber and note the following features:

   central axon

   myelin sheath around the axon of Schwann cell (most of the myelin may have been dissolved and lost during the slide preparation)

   neurilemma of Schwann cell

12. Sketch and label a nerve fiber with Schwann cell (cross section) in the space provided in Part D of the laboratory report.
13. Locate the longitudinal section of the nerve on the slide (fig. 27.7). Note the following:

   central axons

   myelin sheath of Schwann cells

   neurilemma of Schwann cells

   nodes of Ranvier

FIGURE 27.6 Cross section of a bundle of neurons within a nerve (400×).

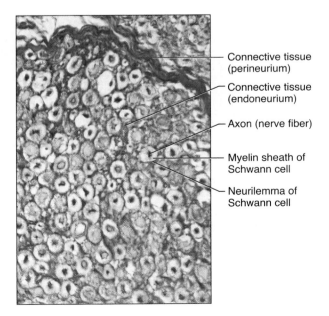

Connective tissue (perineurium)

Connective tissue (endoneurium)

Axon (nerve fiber)

Myelin sheath of Schwann cell

Neurilemma of Schwann cell

236

FIGURE 27.7 Longitudinal section of a nerve (250× micrograph enlarged to 2,000×).

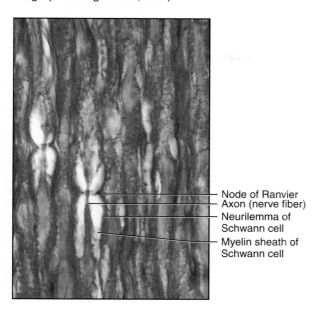

- Node of Ranvier
- Axon (nerve fiber)
- Neurilemma of Schwann cell
- Myelin sheath of Schwann cell

14. Sketch and label a nerve fiber with Schwann cell (longitudinal section) in the space provided in Part D of the laboratory report.

Learning Extension Activity

Obtain a prepared microscope slide of Purkinje cells. To locate these neurons, search the slide for large, flask-shaped cell bodies. Each cell body has one or two large, thick dendrites that give rise to branching networks of fibers. These cells are found in a particular region of the brain (cerebellar cortex).

Name _____

Date _____

Section _____

The 🅰 corresponds to the Learning Outcome(s) listed at the beginning of the laboratory exercise.

Nervous Tissue and Nerves

Part A Assessments

Match the terms in column A with the descriptions in column B. Place the letter of your choice in the space provided. 🔺1 🔺2

Column A	Column B
a. Astrocyte	_____ **1.** Sheath of Schwann cell containing cytoplasm and nucleus that encloses myelin
b. Axon	
c. Collateral	_____ **2.** Corresponds to rough endoplasmic reticulum in other cells
d. Dendrite	_____ **3.** Network of fine threads of actin within cell body
e. Myelin	
f. Neurilemma	_____ **4.** Substance of Schwann cell composed of lipoprotein that insulates axons and increases impulse speed
g. Neurofibrils	
h. Nissl bodies (chromatophilic substance)	_____ **5.** Neuron process with many tiny, thornlike spines that conducts an impulse toward the cell body
	_____ **6.** Branch of an axon
	_____ **7.** Star-shaped neuroglial cell between neurons and blood vessels
	_____ **8.** Nerve fiber arising from a slight elevation of the cell body that conducts an impulse away from the cell body

Part B Assessments

Match the terms in column A with the descriptions in column B. Place the letter of your choice in the space provided. 🔺1 🔺2

Column A	Column B
a. Effector	_____ **1.** Transmits impulse from sensory to motor neuron within central nervous system
b. Ependymal cell	
c. Ganglion	_____ **2.** Transmits impulse out of the brain or spinal cord to effectors (muscles and glands)
d. Interneuron (association neuron)	
e. Microglia	_____ **3.** Transmits impulse into brain or spinal cord from receptors
f. Motor (efferent) neuron	_____ **4.** Myelin-forming neuroglial cell in brain and spinal cord
g. Oligodendrocyte	_____ **5.** Phagocytic neuroglial cell
h. Sensory (afferent) neuron	_____ **6.** Structure capable of responding to motor impulse
	_____ **7.** Specialized mass of neuron cell bodies outside the brain or spinal cord
	_____ **8.** Cells that cover the inside spaces of the brain ventricles and secrete cerebrospinal fluid

Part C Assessments

1. Sketch a motor (efferent) neuron. Label the cell body, nucleus, nucleolus, axon hillock, and cell processes (dendrite and axon). **3**

2. Sketch and label a sensory (afferent) neuron cell body. **3**

3. Sketch a neuroglial cell. **2**

Part D Assessments

1. Sketch and label a nerve fiber with Schwann cell (cross section). **3**

2. Sketch and label a nerve fiber with Schwann cell (longitudinal section). **3**

Laboratory Exercise 28

Spinal Cord and Meninges

Pre-Lab

1. Carefully read the introductory material and examine the entire lab content.
2. Be familiar with the structures and functions of the spinal cord and meninges (from lecture or the textbook).
3. Visit www.mhhe.com/martinseries1 for pre-lab questions and Anatomy & Physiology Revealed animations list.

Materials Needed

Compound light microscope
Prepared microscope slide of a spinal cord cross section with spinal nerve roots
Spinal cord model with meninges

For Demonstration Activity:
Preserved spinal cord with meninges intact

The spinal cord is a column of nerve fibers that extends down through the vertebral canal. Together with the brain, it makes up the central nervous system. In the cervical and lumbar regions of the spinal cord, enlargements give rise to spinal nerves to the upper limbs and lower limbs, respectively. The inferior end of the spinal cord, the conus medullaris, is located inferior to lumbar vertebra L1 in the adult.

There are 31 pairs of spinal nerves attached to the spinal cord. At close proximity to the cord, the spinal nerve has two branches: a dorsal (posterior) root containing sensory neurons and dorsal root ganglion, and a ventral (anterior) root containing motor neurons. The spinal nerves emerge through the nearby intervertebral foramina; however, most lumbar and sacral nerves extend inferiorly through the vertebral canal as the cauda equina to emerge in their respective regions of the vertebral column.

The spinal cord has a central region, the gray matter, with paired posterior, lateral, and anterior horns connected by a gray commissure containing the central canal. The gray matter is a processing center for spinal reflexes and synaptic integration. The white matter of the spinal cord, represented by paired posterior, lateral, and anterior funiculi (columns), contains ascending (sensory) and descending (motor) tracts. The specific names of the spinal tracts often reflect their respective origins and the destinations of the fibers. At various levels of the spinal cord, many of the tracts cross over (decussate) to the opposite side of the spinal cord or the brainstem.

The meninges consist of three layers of fibrous connective tissue membranes located between the bones of the skull and vertebral column and the soft tissues of the central nervous system. The superficial dura mater is a tough membrane with an epidural space containing blood vessels, loose connective tissue, and adipose tissue between the membrane and the vertebrae. A weblike arachnoid mater adheres to the inside of the dura mater. The subarachnoid space, located beneath the arachnoid mater, contains cerebrospinal fluid (CSF) serving as a protective cushion of the spinal cord and brain. The delicate innermost membrane, the pia mater, adheres to the surface of the spinal cord and brain. The denticulate ligaments, extending from the pia mater to the dura mater, anchor the spinal cord. An inferior extension of the pia mater, the filum terminale, anchors the spinal cord to the coccyx.

Purpose of the Exercise

To review the characteristics of the spinal cord and the meninges and to observe the major features of these structures.

Learning Outcomes

After completing this exercise, you should be able to

1. Identify the major features and functions of the spinal cord.
2. Arrange the layers of the meninges and describe the structure of each.

Procedure A—Structure of the Spinal Cord

1. Label figures 28.1 and 28.2.
2. Study figures 28.3 and 28.4.
3. Complete Part A of Laboratory Report 28.

FIGURE 28.1 Label the features of the spinal cord cross section and surrounding structures, using the terms provided. ⚠

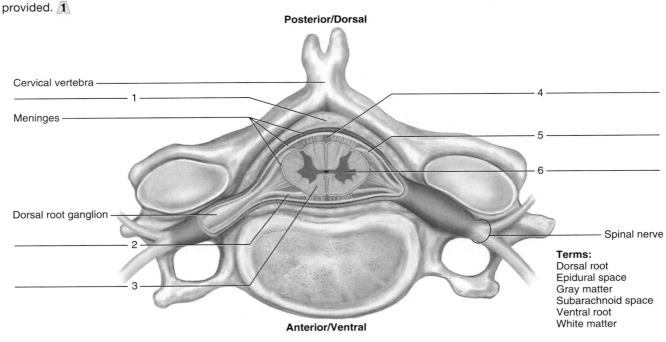

Posterior/Dorsal

Cervical vertebra —————

————————— 1 —————

Meninges ——————

Dorsal root ganglion ——————

————————— 2 —————

————————— 3 —————

Anterior/Ventral

————— 4 —————

————— 5 —————

————— 6 —————

————— Spinal nerve

Terms:
Dorsal root
Epidural space
Gray matter
Subarachnoid space
Ventral root
White matter

FIGURE 28.2 Label this cross section of the spinal cord, including the features of the white and gray matter, using the terms provided. (*Note:* A lateral horn of the gray matter is not present at the level of the spinal cord illustrated.) ⚠

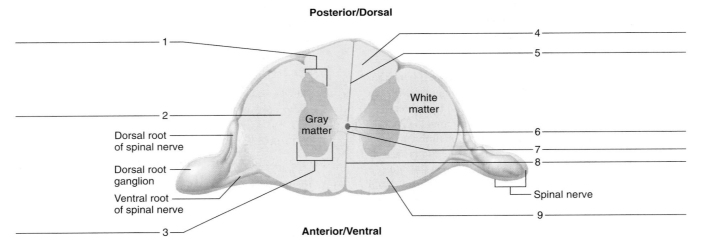

Posterior/Dorsal

————————— 1 —————

————————— 2 —————

Dorsal root of spinal nerve

Dorsal root ganglion

Ventral root of spinal nerve

————————— 3 —————

Gray matter

White matter

Anterior/Ventral

————— 4 —————

————— 5 —————

————— 6 —————

————— 7 —————

————— 8 —————

————— Spinal nerve

————— 9 —————

Terms:
Anterior funiculus   Lateral funiculus
Anterior horn   Posterior funiculus
Anterior median fissure Posterior horn
Central canal   Posterior median sulcus
Gray commissure

FIGURE 28.3 Major ascending (sensory) and descending (motor) tracts within a cross section of the spinal cord. Ascending tracts are in pink, descending tracts in rust, and are shown only on one side. This pattern varies with the level of the spinal cord. This pattern is representative of the midcervical region. (*Note:* These tracts are not visible as individually stained structures on microscope slides.)

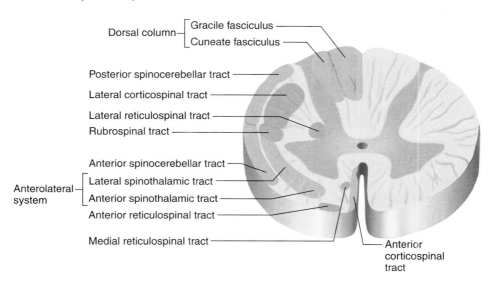

Dorsal column — Gracile fasciculus
Cuneate fasciculus

Posterior spinocerebellar tract

Lateral corticospinal tract

Lateral reticulospinal tract

Rubrospinal tract

Anterolateral system
Anterior spinocerebellar tract
Lateral spinothalamic tract
Anterior spinothalamic tract
Anterior reticulospinal tract

Medial reticulospinal tract

Anterior corticospinal tract

FIGURE 28.4 Posterior view of cervical region of spinal cord and associated nerves and meninges of a cadaver.

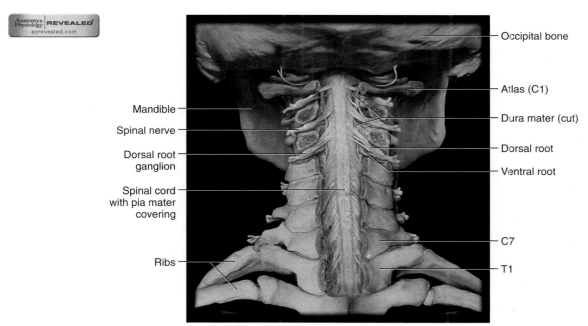

Anatomy & Physiology REVEALED
aprevealed.com

Mandible

Spinal nerve

Dorsal root ganglion

Spinal cord with pia mater covering

Ribs

Occipital bone

Atlas (C1)

Dura mater (cut)

Dorsal root

Ventral root

C7

T1

4. Obtain a prepared microscope slide of a spinal cord cross section. Use the low power of the microscope to locate the following features:

posterior median sulcus

anterior median fissure

central canal

gray matter

    gray commissure

    posterior (dorsal) horn

    lateral horn

    anterior (ventral) horn

white matter

    posterior (dorsal) funiculus (column)

    lateral funiculus (column)

    anterior (ventral) funiculus (column)

roots of spinal nerve

    dorsal roots

    dorsal root ganglia

    ventral roots

5. Observe the model of the spinal cord, and locate the features listed in step 4.
6. Complete Part B of the laboratory report.

Procedure B—Meninges

1. Study figure 28.4 and label figure 28.5.
2. Observe the spinal cord model with meninges and locate the following features:

dura mater (dural sheath)

arachnoid mater (membrane)

subarachnoid space

denticulate ligament

pia mater

3. Complete Part C of the laboratory report.

 Demonstration Activity

Observe the preserved section of spinal cord. Note the heavy covering of dura mater, firmly attached to the cord on each side by a set of ligaments (denticulate ligaments) originating in the pia mater. The intermediate layer of meninges, the arachnoid mater, is devoid of blood vessels, but in a live human being, the space beneath this layer contains cerebrospinal fluid. The pia mater, closely attached to the surface of the spinal cord, contains many blood vessels. What are the functions of these layers?

_____

_____

_____

_____

FIGURE 28.5 Label the meninges, associated near the spinal cord and nerves, using the terms provided. **2**

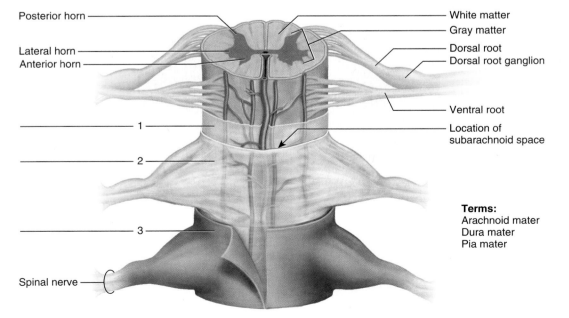

Posterior horn

Lateral horn

Anterior horn

1

2

3

Spinal nerve

White matter

Gray matter

Dorsal root

Dorsal root ganglion

Ventral root

Location of subarachnoid space

Terms:
Arachnoid mater
Dura mater
Pia mater

Name _____

Date _____

Section _____

The ⬛ corresponds to the Learning Outcome(s) listed at the beginning of the laboratory exercise.

Spinal Cord and Meninges

Part A Assessments

Complete the following statements:

1. The spinal cord gives rise to 31 pairs of _____. ⬛

2. The bulge in the spinal cord that gives off nerves to the upper limbs is called the _____ enlargement. ⬛

3. The bulge in the spinal cord that gives off nerves to the lower limbs is called the _____ enlargement. ⬛

4. The _____ is a groove that extends the length of the spinal cord posteriorly. ⬛

5. In a spinal cord cross section, the posterior _____ of the gray matter resemble the upper wings of a butterfly. ⬛

6. The cell bodies of motor neurons are found in the _____ horns of the spinal cord. ⬛

7. The _____ connects the gray matter on the left and right sides of the spinal cord. ⬛

8. The _____ in the gray commissure of the spinal cord contains cerebrospinal fluid and is continuous with the ventricles of the brain. ⬛

9. The white matter of the spinal cord is divided into anterior, lateral, and posterior _____. ⬛

10. Collectively, the dura mater, arachnoid mater, and pia mater are called the _____. ⬛

Part B Assessments

Identify the features indicated in the spinal cord cross section of figure 28.6.

FIGURE 28.6 Micrograph of a spinal cord cross section with spinal nerve roots (35×). Label the features by placing the correct numbers in the spaces provided. ⚠️

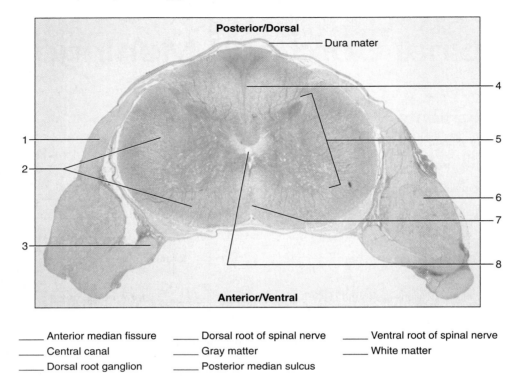

_____ Anterior median fissure   _____ Dorsal root of spinal nerve   _____ Ventral root of spinal nerve
_____ Central canal   _____ Gray matter   _____ White matter
_____ Dorsal root ganglion   _____ Posterior median sulcus

Part C Assessments

Match the terms in column A with the descriptions in column B. Place the letter of your choice in the space provided. 2

Column A	Column B
a. Arachnoid mater	_____ **1.** Connections from pia mater to dura mater that anchor the spinal cord
b. Denticulate ligaments	
c. Dura mater	_____ **2.** Inferior continuation of pia mater to the coccyx
d. Epidural space	_____ **3.** Outermost layer of meninges
e. Filum terminale	
f. Pia mater	_____ **4.** Follows irregular contours of spinal cord surface
g. Subarachnoid space	_____ **5.** Contains cerebrospinal fluid
	_____ **6.** Thin, weblike middle membrane
	_____ **7.** Separates dura mater from bone of vertebra

Reflex Arc and Reflexes

Pre-Lab

1. Carefully read the introductory material and examine the entire lab content.
2. Be familiar with reflexes (from lecture or the textbook).
3. Visit www.mhhe.com/martinseries1 for pre-lab questions and Anatomy & Physiology Revealed animations list.

Materials Needed

Rubber percussion hammer

A reflex arc represents the simplest type of nerve pathway found in the nervous system. This pathway begins with a receptor at the dendrite end of a sensory (afferent) neuron. The sensory neuron leads into the central nervous system and may communicate with one or more interneurons. Some of these interneurons, in turn, communicate with motor (efferent) neurons, whose axons (nerve fibers) lead outward to effectors. Thus, when a sensory receptor is stimulated by a change occurring inside or outside the body, nerve impulses may pass through a reflex arc, and, as a result, effectors may respond. Such an automatic, subconscious response is called a *reflex*.

A *stretch reflex* involves a single synapse (monosynaptic) between a sensory and a motor neuron within the gray matter of the spinal cord. Examples of stretch reflexes include the patellar, calcaneal, biceps, triceps, and plantar reflexes. Other more complex *withdrawal reflexes* involve interneurons (association neurons) in combination with sensory and motor neurons; thus they are polysynaptic. Examples of withdrawal reflexes include responses to touching hot objects or stepping on sharp objects.

Reflexes demonstrated in this lab are stretch reflexes. When a muscle is stretched by a tap over its tendon, stretch receptors (proprioceptors) called *muscle spindles* are stretched within the muscle, which initiates an impulse over a reflex arc. A sensory (afferent) neuron conducts an impulse from the muscle spindle into the gray matter of the spinal cord, where it synapses with a motor (efferent) neuron, which conducts the impulse to the effector muscle. The stretched muscle responds by contracting to resist or reverse further stretching. These stretch reflexes are important to maintaining proper posture, balance, and movements. Observations of many of these reflexes in clinical tests on patients may indicate damage to a level of the spinal cord or peripheral nerves of the particular reflex arc.

Purpose of the Exercise

To review the characteristics of reflex arcs and reflex behavior and to demonstrate some of the reflexes that occur in the human body.

Learning Outcomes

After completing this exercise, you should be able to

1. Describe the components of a reflex arc.
2. Demonstrate and record stretch reflexes that occur in humans.
3. Analyze the components and patterns of stretch reflexes.

Procedure—Reflex Arc and Reflexes

1. Label figure 29.1.
2. Complete Part A of Laboratory Report 29.
3. Work with a laboratory partner to demonstrate each of the reflexes listed. (See fig. 29.2*a–e* also.) *It is important that muscles involved in the reflexes be totally relaxed to observe proper responses.* If a person is trying too hard to experience the reflex or is trying to suppress the reflex, assign a multitasking activity while the stimulus with the rubber percussion hammer occurs. For example, assign a physical task with upper limbs along with a complex mental activity during the patellar reflex. After each demonstration, record your observations in the table provided in Part B of the laboratory report.

   a. *Patellar (knee-jerk) reflex.* Have your laboratory partner sit on a table (or sturdy chair) with legs relaxed and hanging freely over the edge without touching the floor. Gently strike your partner's patellar ligament

FIGURE 29.1 Label this diagram of a withdrawal (polysynaptic) reflex arc by placing the correct numbers in the spaces provided. Reflexes demonstrated in this lab are stretch (monosynaptic) reflex arcs and lack the interneuron. ⚠

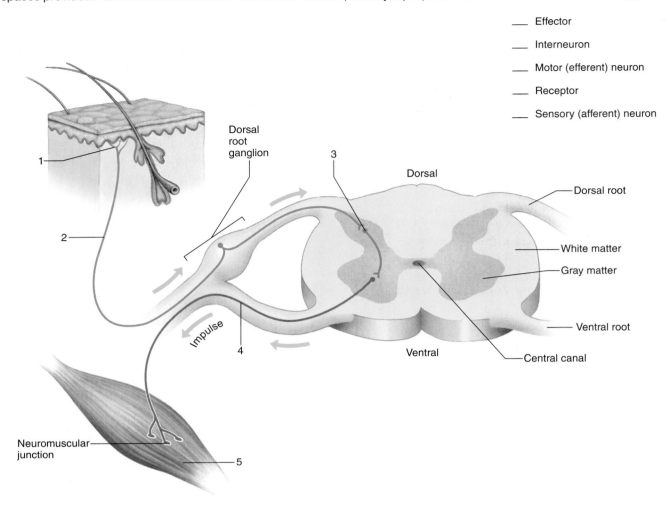

___ Effector

___ Interneuron

___ Motor (efferent) neuron

___ Receptor

___ Sensory (afferent) neuron

(just below the patella) with the blunt side of a rubber percussion hammer (fig. 29.2a). The normal response is a moderate extension of the leg at the knee joint.

b. *Calcaneal (ankle-jerk) reflex.* Have your partner kneel on a chair with back toward you and with feet slightly dorsiflexed over the edge and relaxed. Gently strike the calcaneal tendon (just above its insertion on the calcaneus) with the blunt side of the rubber hammer (fig. 29.2b). The normal response is plantar flexion of the foot.

c. *Biceps (biceps-jerk) reflex.* Have your partner place a bare arm bent about 90° at the elbow on the table. Press your thumb on the inside of the elbow over the tendon of the biceps brachii, and gently strike your thumb with the rubber hammer (fig. 29.2c). Watch the biceps brachii for a response. The response might be a slight twitch of the muscle or flexion of the forearm at the elbow joint.

d. *Triceps (triceps-jerk) reflex.* Have your partner lie supine with an upper limb bent about 90° across the abdomen. Gently strike the tendon of the triceps brachii near its insertion just proximal to the olecranon process at the tip of the elbow (fig. 29.2d). Watch the triceps brachii for a response. The response might be a slight twitch of the muscle or extension of the forearm at the elbow joint.

e. *Plantar reflex.* Have your partner remove a shoe and sock and lie supine with the lateral surface of the foot resting on the table. Draw the metal tip of the rubber hammer, applying firm pressure, over the sole from the heel to the base of the large toe (fig. 29.2e). The normal response is flexion (curling) of the toes and plantar flexion of the foot. If the toes spread apart and dorsiflexion of the great toe occurs, the reflex is the abnormal *Babinski reflex* response (normal in infants until the nerve fibers have complete myelinization). If the Babinski reflex occurs later in life, it may indicate damage to the corticospinal tract of the CNS.

4. Complete Part B of the laboratory report.

FIGURE 29.2 Demonstrate each of the following reflexes: (*a*) patellar reflex; (*b*) calcaneal reflex; (*c*) biceps reflex; (*d*) triceps reflex; and (*e*) plantar reflex.

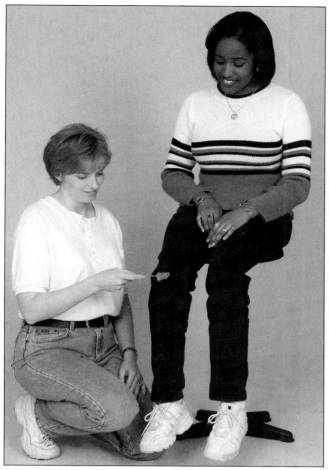

(a) Patellar reflex

(b) Calcaneal reflex

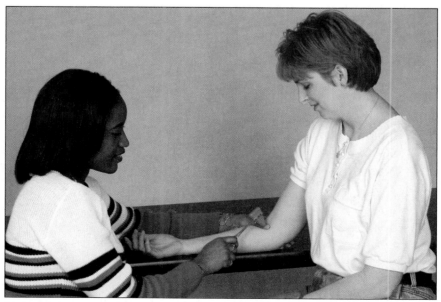

(c) Biceps reflex

FIGURE 29.2 *Continued.*

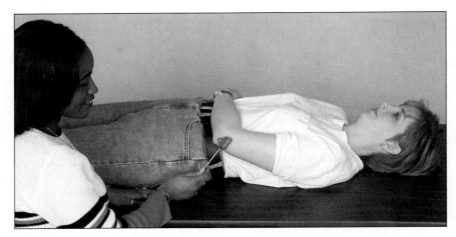

(d) Triceps reflex

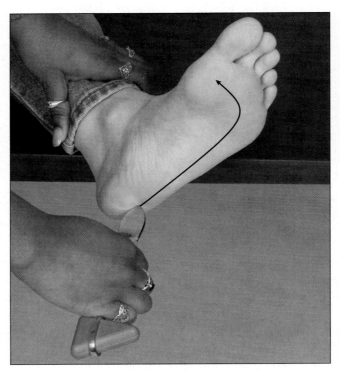

(e) Plantar reflex

Name _____

Date _____

Section _____

The 🄰 corresponds to the Learning Outcome(s) listed at the beginning of the laboratory exercise.

Reflex Arc and Reflexes

Part A Assessments

Complete the following statements:

1. A withdrawal reflex employs _____ neurons in conjunction with sensory and motor neurons. 🄰

2. Interneurons in a withdrawal reflex are located in the _____. 🄰

3. A reflex arc begins with the stimulation of a _____ at the end of a sensory neuron. 🄰

4. Effectors of a reflex arc are glands and _____. 🄰

5. A patellar reflex employs only _____ and motor neurons. 🄰

6. The effector muscle of the patellar reflex is the _____. 🄰

7. The sensory stretch receptors (muscle spindles) of the patellar reflex are located in the _____ muscle. 🄰

8. The dorsal root of a spinal nerve contains the _____ neurons. 🄰

9. The normal plantar reflex results in _____ of toes. 🄰

10. Stroking the sole of the foot in infants results in dorsiflexion and toes that spread apart, called the _____ reflex. 🄰

Part B Assessments

1. Complete the following table: 🄿

Reflex Tested	Response Observed	Effector Muscle Involved
Patellar		
Calcaneal		
Biceps		
Triceps		
Plantar		

2. List the major events that occur in the patellar reflex, from the striking of the patellar ligament to the resulting response. ⚠3

_____

_____

_____

Critical Thinking Activity ⚠3

What characteristics do the reflexes you demonstrated have in common?

_____

_____

_____

Brain and Cranial Nerves

Pre-Lab

1. Carefully read the introductory material and examine the entire lab content.
2. Be familiar with the brain and cranial nerves (from lecture or the textbook).
3. Visit www.mhhe.com/martinseriesl for pre-lab questions and Anatomy & Physiology Revealed animations list.

Materials Needed

Dissectible model of the human brain
Preserved human brain
Anatomical charts of the human brain

The brain, the largest and most complex part of the nervous system, contains nerve centers associated with sensory functions and is responsible for sensations and perceptions. It issues motor commands to skeletal muscles and carries on higher mental activities. It also functions to coordinate muscular movements, and it contains centers and nerve pathways necessary for the regulation of internal organs.

The cerebral cortex, comprised of gray matter and billions of interneurons, represents areas for conscious awareness and decision making processes. Sensory areas receive information from various receptors, association areas interpret sensory input, and motor areas involve planning and controlling muscle movements. All of these functional regions are influenced and integrated together in making complex decisions. Each hemisphere primarily interprets sensory and regulates motor functions on the opposite (contralateral) side of the body.

Twelve pairs of cranial nerves arise from the ventral surface of the brain and are designated by number and name. Although most of these nerves conduct both sensory and motor impulses, some contain only sensory fibers associated with special sense organs. Others are primarily composed of motor fibers and are involved with the activities of muscles and glands.

Purpose of the Exercise

To review the structural and functional characteristics of the human brain and cranial nerves.

Learning Outcomes

After completing this exercise, you should be able to

1. Identify the major external and internal structures in the human brain.
2. Locate the major functional regions of the brain.
3. Identify each of the cranial nerves.
4. Differentiate the functions of each cranial nerve.

Procedure A—Human Brain

1. Study figures 30.1 and 30.2.
2. Label figure 30.3.
3. Complete Part A of Laboratory Report 30.
4. Observe the anatomical charts, dissectible model, and preserved specimen of the human brain. Locate each of the following features:

>   **cerebrum**
>   **cerebral hemispheres**
>   **corpus callosum**
>   **gyri**
>   precentral gyrus
>   postcentral gyrus
>   **sulci**
>   central sulcus
>   lateral sulcus
>   parieto-occipital sulcus
>   **fissures**
>   longitudinal fissure
>   transverse fissure
>   **lobes**
>   frontal lobe
>   parietal lobe
>   temporal lobe
>   occipital lobe
>   insula (insular lobe)

cerebral cortex

basal nuclei (basal ganglia, a widely used clinical term)

  caudate nucleus

  putamen

  globus pallidus

ventricles

  lateral ventricles

  third ventricle

  fourth ventricle

  choroid plexuses

  cerebral aqueduct

diencephalon

  thalamus

  hypothalamus

  optic chiasma (chiasm)

mammillary bodies

pineal gland

cerebellum

  right and left hemispheres

  vermis

  cerebellar cortex

  arbor vitae

  cerebellar peduncles

brainstem

  midbrain

    cerebral aqueduct (aqueduct of midbrain)

    cerebral peduncles

    corpora quadrigemina

  pons

  medulla oblongata

FIGURE 30.1 Superior view of the surface of the brain within the skull of a cadaver.

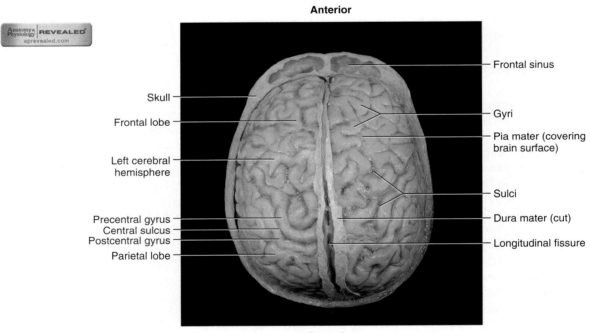

FIGURE 30.2 Lobes of the cerebrum. Retractors are used to expose the deep insula.

Anterior (rostral) | Posterior (caudal)

Precentral gyrus

Central sulcus
Postcentral gyrus

Frontal lobe

Parietal lobe

Insula

Occipital lobe

Parieto-occipital sulcus

Temporal lobe

Lateral sulcus

FIGURE 30.3 Label this diagram of a median section of the brain by placing the correct numbers in the spaces provided.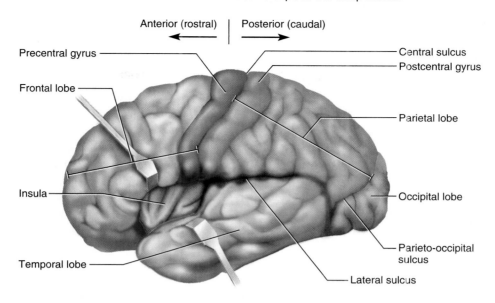

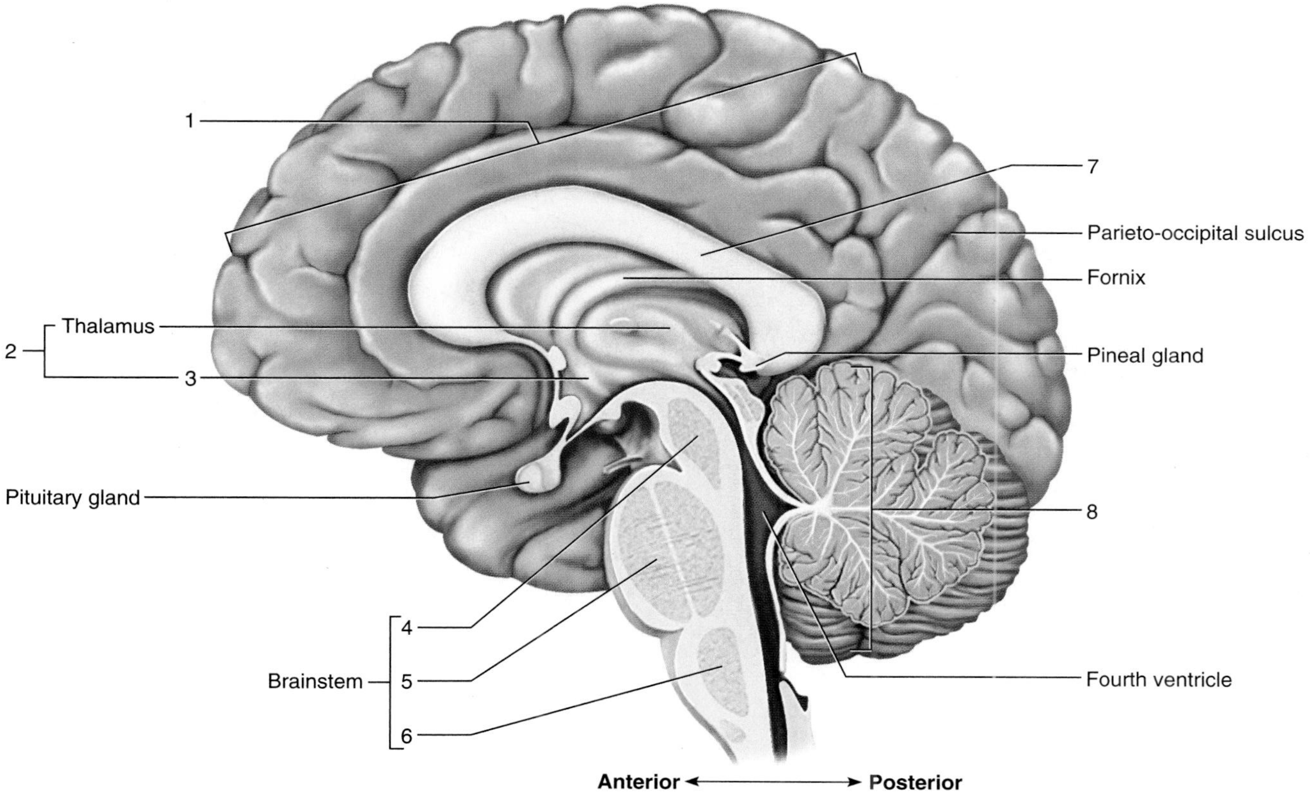

1

7

Parieto-occipital sulcus

Fornix

Thalamus

2

3

Pineal gland

Pituitary gland

8

4

Brainstem 5

6

Fourth ventricle

Anterior ◄──────► Posterior

_____ Cerebellum   _____ Diencephalon   _____ Midbrain
_____ Cerebrum   _____ Hypothalamus   _____ Pons
_____ Corpus callosum   _____ Medulla oblongata

5. Locate the labeled areas in figure 30.4 that represent the following functional regions of the cerebrum:

sensory areas

primary somatosensory cortex—receives information from skin receptors and proprioceptors

somatosensory association cortex—integrates sensory information from primary cortex

Wernicke's area—processes spoken and written language

visual areas—consists of a primary and association (interpretation) area for vision

auditory areas—consists of a primary and association area for hearing

olfactory assocation area—interpretation of foods

gustatory cortex—perceptions of taste

motor areas

primary motor cortex—controls skeletal muscles

motor assocation (premotor) area—planning body movements

Broca's area—planning speech movements

6. Complete Parts B and C of the laboratory report.

Procedure B—Cranial Nerves

1. Study figure 30.5.
2. Observe the model and preserved specimen of the human brain, and locate as many of the following cranial nerves as possible as you differentiate their associated functions:

olfactory nerves (I)—smell

optic nerves (II)—vision

oculomotor nerves (III)—pupil constriction, eye movements, and opens eyelid

trochlear nerves (IV)—stimulate superior oblique muscle

trigeminal nerves (V)—sensory from face and teeth and mastication movements

abducens nerves (VI)—lateral eye movements

facial nerves (VII)—control facial expression muscles and salivation, tear secretions, and taste

vestibulocochlear nerves (VIII)—hearing and balance

FIGURE 30.4 Some structural and functional areas of the left cerebral hemisphere. (*Note:* These areas are not visible as distinct parts of the brain.)

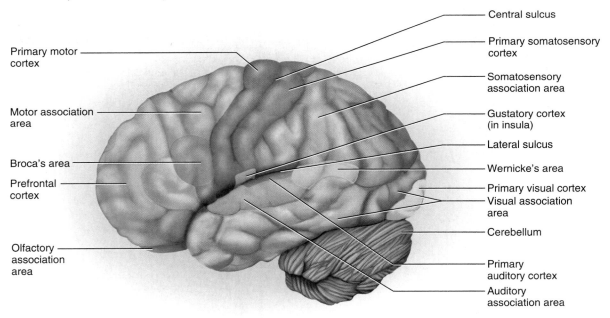

FIGURE 30.5 Photograph of the cranial nerves attached to the base of the human brain.

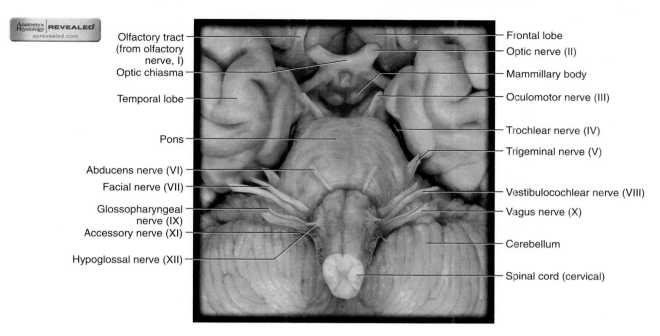

Olfactory tract (from olfactory nerve, I)
Optic chiasma
Temporal lobe
Pons
Abducens nerve (VI)
Facial nerve (VII)
Glossopharyngeal nerve (IX)
Accessory nerve (XI)
Hypoglossal nerve (XII)

Frontal lobe
Optic nerve (II)
Mammillary body
Oculomotor nerve (III)
Trochlear nerve (IV)
Trigeminal nerve (V)
Vestibulocochlear nerve (VIII)
Vagus nerve (X)
Cerebellum
Spinal cord (cervical)

glossopharyngeal nerves (IX)—monitor blood pressure from baroreceptors, salivation, and swallowing

vagus nerves (X)—regulation of many visceral organs, including the heart rate

accessory nerves (XI)—control neck, larynx, and shoulder muscles

hypoglossal nerves (XII)—control tongue movements

The following mnemonic device will help you learn the twelve pairs of cranial nerves in the proper order:

Old **Op**ie **oc**casionally **tr**ies **trig**onometry, and **f**eels **very glo**omy, **vagu**e, and **hypo**active.[1]

3. Complete Parts D and E of the laboratory report.

[1]From *HAPS-Educator,* Winter 2002. An official publication of the Human Anatomy & Physiology Society (HAPS).

Laboratory Report

30

Name _____

Date _____

Section _____

The ⚠ corresponds to the Learning Outcome(s) listed at the beginning of the laboratory exercise.

Brain and Cranial Nerves

Part A Assessments

Match the terms in column A with the descriptions in column B. Place the letter of your choice in the space provided. ⚠

Column A	Column B
a. Central sulcus	_____ **1.** Structure formed by the crossing-over of the optic nerves
b. Cerebral cortex	_____ **2.** Part of diencephalon that forms lower walls and floor of third ventricle
c. Corpus callosum	
d. Gyrus	_____ **3.** Cone-shaped structure in the upper posterior portion of diencephalon
e. Hypothalamus	
f. Insula	_____ **4.** Connects cerebral hemispheres
g. Medulla oblongata	_____ **5.** Ridge on surface of cerebrum
h. Midbrain	_____ **6.** Separates frontal and parietal lobes
i. Optic chiasma	_____ **7.** Part of brainstem between diencephalon and pons
j. Pineal gland	_____ **8.** Rounded bulge on underside of brainstem
k. Pons	_____ **9.** Part of brainstem continuous with the spinal cord
l. Ventricle	_____ **10.** Internal brain chamber filled with CSF
	_____ **11.** Cerebral lobe located deep within lateral sulcus
	_____ **12.** Thin layer of gray matter on surface of cerebrum

Part B Assessments

Complete the following statements:

1. The central _____ separates the frontal and parietal lobes of the cerebrum. ⚠

2. Grooves on the surface of the brain are sulci; ridges on the surface are _____. ⚠

3. The auditory areas of the brain are part of the _____ lobe. ⚠2

4. The vision areas of the brain are part of the _____ lobe. ⚠2

5. The left cerebral hemisphere primarily controls the _____ side of the body. ⚠2

6. The brainstem includes the pons, the midbrain, and the _____. ⚠

7. The delicate _____ membrane is located on the surface of the brain. ⚠

8. The _____ fissure separates the two cerebral hemispheres. ⚠

9. The primary motor cortex is located within the _____ gyrus. ⚠2

10. Arbor vitae and vermis are components of the _____. ⚠

Part C Assessments

Identify the features indicated in the median section of the right half of the human brain in figure 30.6.

FIGURE 30.6 Label the features on this median section of the right half of the human brain by placing the correct numbers in the spaces provided. ⚠1

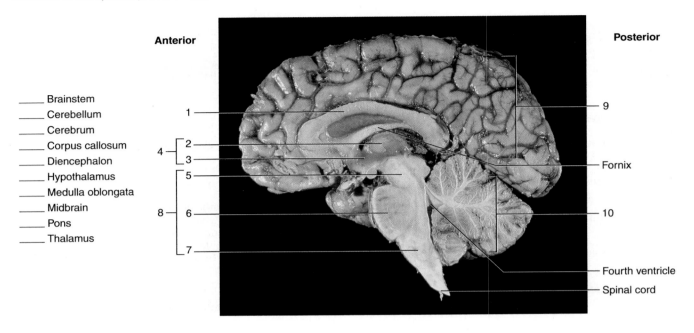

Anterior   Posterior

_____ Brainstem
_____ Cerebellum
_____ Cerebrum
_____ Corpus callosum
_____ Diencephalon
_____ Hypothalamus
_____ Medulla oblongata
_____ Midbrain
_____ Pons
_____ Thalamus

1
4 [2
    3]
  [5
8 [6
    7]

9

Fornix

10

Fourth ventricle

Spinal cord

Part D Assessments

Identify the cranial nerves that arise from the base of the brain in figure 30.7.

FIGURE 30.7 Provide the names of the 12 pairs of cranial nerves as viewed from the base of the brain. The Roman numerals indicated are also often used to reference a cranial nerve. ⚠3

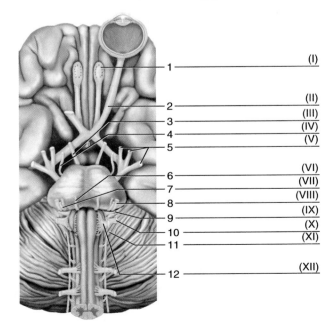

1 ——————— (I)

2 ——————— (II)
3 ——————— (III)
4 ——————— (IV)
5 ——————— (V)

6 ——————— (VI)
7 ——————— (VII)
8 ——————— (VIII)
9 ——————— (IX)
10 ——————— (X)
11 ——————— (XI)

12 ——————— (XII)

Part E Assessments

Match the cranial nerves in column A with the associated functions in column B. Place the letter of your choice in the space provided. 4

Column A	Column B
a. Abducens	_____ **1.** Regulates (slows) heart rate
b. Accessory	_____ **2.** Equilibrium and hearing
c. Facial	_____ **3.** Stimulates superior oblique muscle of eye
d. Glossopharyngeal	_____ **4.** Sensory impulses from teeth
e. Hypoglossal	_____ **5.** Pupil constriction and eyelid opening
f. Oculomotor	_____ **6.** Smell
g. Olfactory	_____ **7.** Controls swallowing and neck and shoulder movements
h. Optic	_____ **8.** Controls tongue movements for speech, swallowing, and food manipulation
i. Trigeminal	
j. Trochlear	_____ **9.** Vision
k. Vagus	_____ **10.** Stimulates lateral rectus muscle of eye
l. Vestibulocochlear	_____ **11.** Taste, salivation, and secretion of tears
	_____ **12.** Sensory from tongue and outer ear, salivation, swallowing, and monitors blood pressure

Electroencephalogram I: BIOPAC© Exercise

Pre-Lab

1. Carefully read introductory material and examine the entire lab content.
2. Be familiar with structures and functions of the brain (from lecture or the textbook).

Materials Needed

Computer system (Mac—minimum 68020; PC running Windows 95/98/NT 4.0; with 4.0 MB RAM)

BIOPAC Student Lab Software ver. 3.0

BIOPAC Acquisition Unit (MP30) with (AC100A) transformer

BIOPAC serial cable (CBLSERA)

BIOPAC electrode lead set (SS2L)

BIOPAC disposable vinyl electrodes (EL503), 3 electrodes per subject

BIOPAC electrode gel (GEL1) and abrasive pad (ELPAD) or alcohol wipes

Suggested: Lycra swim cap or supportive wrap (such as 3M Coban™ self-adhering support wrap) to press electrodes against head for improved contact

Safety

▶ The BIOPAC electrode lead set is safe and easy to use and should be used only as described in the procedures section of the laboratory exercise.

▶ The electrode lead clips are color-coded. Make sure they are connected to the properly placed electrode as demonstrated in figure 31.1.

▶ The vinyl electrodes are meant to be single use. Each subject should use a new set.

FIGURE 31.1 Subject setup.

Source: Courtesy of and © *BIOPAC* Systems Inc.

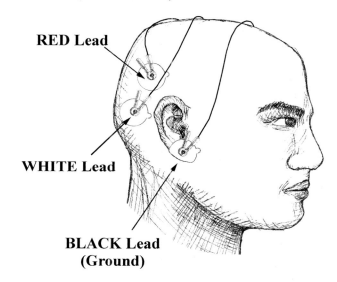

RED Lead

WHITE Lead

BLACK Lead (Ground)

One method of studying brain function is to use the *electroencephalogram (EEG)*. This is basically a noninvasive and painless recording of brain waves produced by the electrical activity of neurons in the brain. Electrodes placed on the surface of the skull can detect the electrical activity of the regions of the brain beneath them. The EEG may then be used to diagnose disorders as well as to indicate normal brain function.

If all the neurons fire (send an impulse) at the same time, the neurons are said to be synchronous and the amplitude (height) of the wave is high. If the neuronal firings are asynchronous, the amplitude is low.

Four basic waves or rhythms are produced: alpha, beta, theta, and delta. These will vary in frequency—that is, the number of times a wave repeats in a given interval of time—as well as amplitude. *Alpha* waves are 8–13 Hz (cycles per second) with amplitudes around 20–200 microvolts (μV), and they are usually found in quiet, awake states with the eyes closed. Alpha waves are suppressed when the eyes are open; this is referred to as desynchronization or alpha block. *Beta* waves are 13–30 Hz with amplitudes of 5–10 μV. They occur in awake, alert individuals responding to external stimuli. *Theta* waves are 4–8 Hz, have an amplitude of 10 µV, and occur in some sleep states. *Delta* waves are less than 5 Hz, have an amplitude of 20–200 µV, and are usually found in deep sleep. These frequencies and amplitudes are in the normal range for a clinical setting and may vary due to stress (such as may exist in the classroom) and also due to the area of the brain recorded (parietal versus occipital).

In this exercise you will be using bipolar recording. In this method, a pair of electrodes is placed over the same cortical region and the difference in electrical potential (voltage) between the two electrodes is detected. A third electrode is placed on the earlobe as a reference or ground of the body's baseline voltage.

Purpose of the Exercise

To record an EEG from an awake, resting subject with eyes open and closed and to identify and examine alpha, beta, delta, and theta components of an EEG.

Learning Outcomes

After completing this exercise, you should be able to:

1 Identify alpha, beta, delta, and theta components of an EEG.

2 Distinguish variations in these components because of differences in stimulation, that is, eyes closed or open.

Procedure A—Setup

1. With your computer turned **ON** and the BIOPAC MP30 unit turned **OFF,** plug the electrode lead set (SSL2) into Channel 1.
2. Turn on the MP30 Data Acquisition Unit.
3. Selection of subject is important; choose someone with easy access to their scalp so that you can be sure you can make good contact between the skin and the electrode. With the subject in a relaxed position, position the electrodes (leads) on the same side of the head as in figure 31.1. The ground electrode (black) is attached to the

earlobe and can be folded under for better adhesion. To help to insure a good contact, make sure the area to be in contact with the electrodes is clean by wiping with the abrasive pad (ELPAD) or alcohol wipe. A small amount of BIOPAC electrode gel (GEL1) may also be used to make better contact between the sensor in the electrode and the skin. For optimal adhesion, hold the electrodes against the skin for about a minute after placement. Wrap the subject's head to secure electrode placement.

4. Start the BIOPAC student Lab Program for Electroencephalogram I, Lesson 3 (LO2-EEG-1).
5. Designate a filename to be used to save the subject's data.

Procedure B—Calibration

1. Click on **Calibrate.** A warning will pop up asking you to check electrode placement.
2. The subject should be in a supine position with the head resting comfortably tilted on one side. After checking the electrode attachments, click **OK.** The calibration will stop automatically after 15 seconds.
3. The calibration recording should look similar to figure 31.2. If not, click **Redo Calibration.**

Procedure C—Recording

1. Read this entire section before proceeding. Before beginning the recording phase, you should designate a **Director** (who will instruct the subject to open or close eyes) and a **Recorder** (who will insert the marker—using Esc for Mac or F9 for PC users—when eyes are open/closed).
2. Click on **Record.** The Director should then instruct the subject to remain relaxed and change eye condition as follows:
   a. 0–10 seconds eyes closed
   b. 10–20 seconds eyes open
   c. 20–30 seconds eyes closed
   The Recorder will insert markers at times: 10 seconds (eyes open) and 20 seconds (eyes closed).

FIGURE 31.2 Calibration recording.

Source: Courtesy of and © *BIOPAC* Systems Inc.

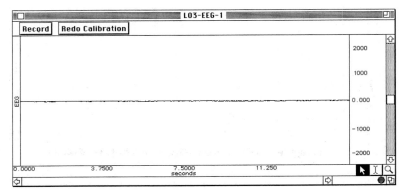

3. Click on **Stop.** If the recording is similar to figure 31.3, go to step 4. If you need to redo, click on **Redo,** this will erase the current data.

4. Click on the frequency buttons in the following sequence: alpha, beta, delta, theta.

5. The data should look like figure 31.4. Look at the alpha frequency band. If there is a decrease in frequency during the "eyes open" segment, click **Done** and move to step 6. If there is not a change, the electrodes may not have been placed properly and you should **redo** the recording from step 2 after checking the electrodes.

6. A pop-up window will appear with four options: **"record from another subject; analyze current data file; analyze another data file; quit."** Make your choice and continue as directed.

Procedure D—Data Analysis

1. You can either analyze the current file or save it and do the analysis later using the **Review Saved Data mode.**

2. Note the channel number designations: CH 1—raw EEG, CH 2—alpha, CH 3—beta, CH 4—delta, CH 5—theta. For optimal viewing, hide CH 1 by holding the Control key (PC) or Option key (Mac) and click on channel box 1 on the left. You may also want to click on "Display menu—Autoscale Waveforms" in order to rescale the bands for better viewing.

3. Across the top of the recordings you will see three boxes for each channel: channel number, measurement type, and the result. Set up the measurement boxes for each channel (CH 2, CH 3, CH 4, CH 5) so that the measurement type is **"stddev."** This will automatically calculate the standard deviation for the area selected.

4. Select using the I-beam tool, in the right bottom corner, the area from time 0–10 seconds (to the first marker). Record the result for each channel in table 31.1 of Part A of Laboratory Report 31.

5. Repeat step 4 for the time period 10–20 seconds (between the first and second markers) and record the results in table 31.1 of the laboratory report.

6. Repeat step 4 for the time period 20–30 seconds (from the second marker to the end) and record the results in table 31.1 of the laboratory report.

7. Now set the measurement boxes for each channel so that the measurement type is **"freq."**

FIGURE 31.3 Raw EEG recording.

Source: Courtesy of and © *BIOPAC* Systems Inc.

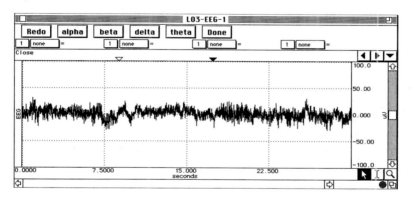

FIGURE 31.4 EEG recording with four different waves.

Source: Courtesy of and © *BIOPAC* Systems Inc.

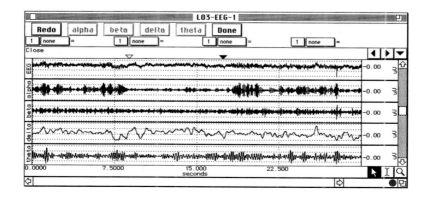

8. Use the I-beam cursor to select an area that represents one cycle in the alpha wave. One cycle is from one peak to the next peak (refer to fig. 31.5). Record the result in table 31.2 of the laboratory report.

9. Repeat for two other alpha cycles and record the results. Calculate the mean and record it in table 31.2 of the laboratory report.

10. Repeat steps 8 and 9 for each of the other waveforms (beta, delta, theta) and record the result in table 31.2 of the laboratory report.

11. When finished, you can save the data on the hard drive, a disk, or a network drive and click EXIT.

12. Complete Part B of the laboratory report.

FIGURE 31.5   Recording with one alpha wave selected.

Source: Courtesy of and © *BIOPAC* Systems Inc.

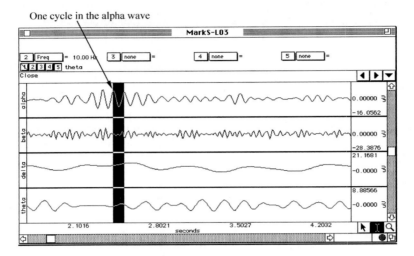

One cycle in the alpha wave

Laboratory Report

31

Name _____

Date _____

Section _____

The ⒜ corresponds to the Learning Outcome(s) listed at the beginning of the
laboratory exercise.

Electroencephalogram I: BIOPAC© Exercise

Subject Profile

Name _____   Height _____

Age _____   Weight _____

Gender: Male / Female

Part A Data and Calculations Assessments

1. EEG Amplitude Measurements ⒜ ⒝

TABLE 31.1 Standard Deviation [stddev]

Rhythm	Channel	Eyes Closed	Eyes Open	Eyes Reclosed
Alpha	CH 2			
Beta	CH 3			
Delta	CH 4			
Theta	CH 5			

2. EEG Frequency Measurements ⒜ ⒝

TABLE 31.2 Frequency [Hz]

Rhythm	Channel	Cycle 1	Cycle 2	Cycle 3	Mean
Alpha					
Beta					
Delta					
Theta					

Part B Assessments

Complete the following:

1. Examine the alpha and beta waveforms for change between the "eyes closed" state and the "eyes open" state. 2

   a. Does desynchronization of the alpha rhythm occur when the eyes are open?

   _____

   b. Does the beta rhythm become more pronounced in the "eyes open" state? 2

   _____

2. Examine the delta and theta rhythms. Is there an increase in delta and theta activity when the eyes are open? Explain your observation. 1

   _____

   _____

   _____

3. Define the following terms: 1

   a. Alpha rhythm

   _____

   _____

   b. Beta rhythm

   _____

   _____

   c. Delta rhythm

   _____

   _____

   d. Theta rhythm

   _____

   _____

Dissection of the Sheep Brain

Pre-Lab

1. Carefully read the introductory material and examine the entire lab content.
2. Be familiar with basic structures and functions (from lecture or the textbook) of the meninges, brain, and the cranial nerves.
3. Visit www.mhhe.com/martinseries1 for pre-lab questions.

Materials Needed

Dissectible model of human brain
Preserved sheep brain
Dissecting tray
Dissection instruments
Long knife

For Demonstration Activity:
Frontal sections of sheep brains

⚠ Safety

▶ Wear disposable gloves when handling the sheep brains.
▶ Save or dispose of the brains as instructed.
▶ Wash your hands before leaving the laboratory.

Mammalian brains have many features in common. Human brains may not be available, so sheep brains often are dissected as an aid to understanding mammalian brain structure. However, the adaptations of the sheep differ from the adaptations of the human, so comparisons of their structural features may not be precise. The sheep is a quadruped, therefore the spinal cord is horizontal, unlike the vertical orientation in a bipedal human. Preserved sheep brains have a different appearance and are firmer than those that are removed directly from the cranial cavity because of the preservatives used.

Purpose of the Exercise

To observe the major features of the sheep brain and to compare these features with those of the human brain.

Learning Outcomes

After completing this exercise, you should be able to

1 Examine the major structures of the sheep brain.
2 Locate the larger cranial nerves of the sheep brain.
3 Summarize several differences and similarities between the sheep brain and the human brain.

Procedure—Dissection of the Sheep Brain

1. Obtain a preserved sheep brain and rinse it thoroughly in water to remove as much of the preserving fluid as possible.
2. Examine the surface of the brain for the presence of meninges. (The outermost layers of these membranes may have been lost during removal of the brain from the cranial cavity.) If meninges are present, locate the following:

    dura mater—the thick, opaque outer layer
    arachnoid mater—the delicate, transparent middle layer attached to the undersurface of the dura mater
    pia mater—the thin, vascular layer that adheres to the surface of the brain (should be present)

3. Remove any remaining dura mater by pulling it gently from the surface of the brain.
4. Position the brain with its ventral surface down in the dissecting tray. Study figure 32.1, and locate the following structures on the specimen:

    longitudinal fissure
    cerebral hemispheres
    gyri
    sulci
    frontal lobe
    parietal lobe
    temporal lobe
    occipital lobe
    cerebellum
    medulla oblongata
    spinal cord

FIGURE 32.1 Dorsal surface of the sheep brain.

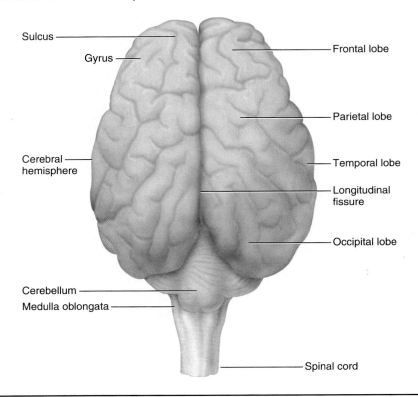

Sulcus
Gyrus
Cerebral hemisphere
Cerebellum
Medulla oblongata

Frontal lobe
Parietal lobe
Temporal lobe
Longitudinal fissure
Occipital lobe
Spinal cord

FIGURE 32.2 Gently bend the cerebellum and medulla oblongata away from the cerebrum to expose the pineal gland and the corpora quadrigemina.

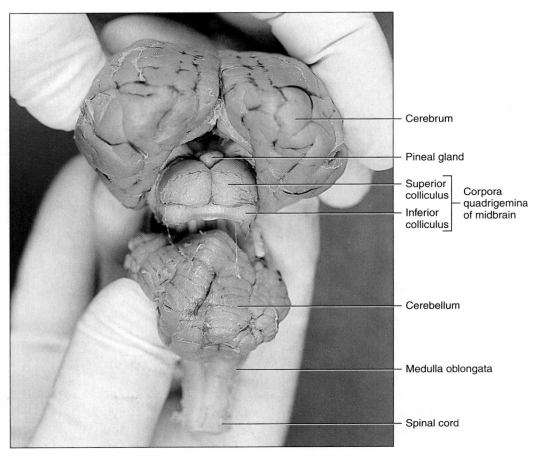

Cerebrum
Pineal gland
Superior colliculus
Inferior colliculus
Corpora quadrigemina of midbrain
Cerebellum
Medulla oblongata
Spinal cord

5. Gently separate the cerebral hemispheres along the longitudinal fissure and expose the transverse band of white fibers within the fissure that connects the hemispheres. This band is the *corpus callosum*.

6. Bend the cerebellum and medulla oblongata slightly downward and away from the cerebrum (fig. 32.2). This will expose the *pineal gland* in the upper midline and the *corpora quadrigemina*, which consists of four rounded structures, called *colliculi*, associated with the midbrain.

7. Position the brain with its ventral surface upward. Study figures 32.3 and 32.4, and locate the following structures on the specimen:

longitudinal fissure
olfactory bulbs
optic nerves
optic chiasma
optic tract
mammillary bodies

FIGURE 32.3 Lateral surface of the sheep brain.

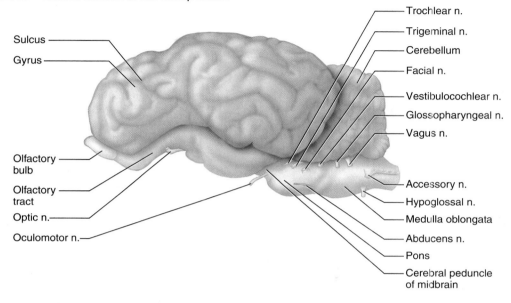

FIGURE 32.4 Ventral surface of the sheep brain.

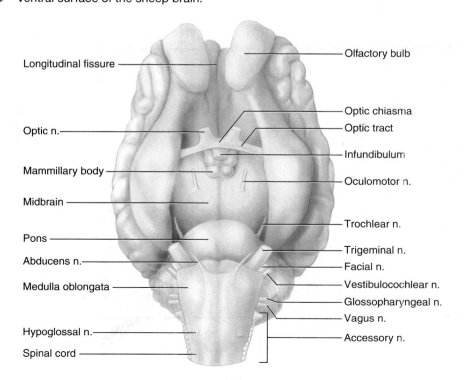

infundibulum (pituitary stalk)

midbrain

pons

8. Although some of the cranial nerves may be missing or are quite small and difficult to find, locate as many of the following as possible, using figures 32.3 and 32.4 as references:

oculomotor nerves

trochlear nerves

trigeminal nerves

abducens nerves

facial nerves

vestibulocochlear nerves

glossopharyngeal nerves

vagus nerves

accessory nerves

hypoglossal nerves

9. Using a long, sharp knife, cut the sheep brain along the midline to produce a median section. Study figures 32.2 and 32.5, and locate the following structures on the specimen:

cerebrum

olfactory bulb

corpus callosum

cerebellum

  white matter

  gray matter

lateral ventricle (one in each cerebral hemisphere)

third ventricle (within diencephalon)

fourth ventricle (between brainstem and cerebellum)

diencephalon

  optic chiasma

  infundibulum

  pituitary gland (this structure may be missing)

  mammillary bodies

  thalamus

  hypothalamus

  pineal gland

midbrain

  corpora quadrigemina

    superior colliculus

    inferior colliculus

pons

medulla oblongata

Demonstration Activity

Observe a frontal section from a sheep brain. Note the longitudinal fissure, gray matter, white matter, corpus callosum, lateral ventricles, third ventricle, and thalamus.

10. Dispose of the sheep brain as directed by the laboratory instructor.

11. Complete Parts A and B of Laboratory Report 32.

FIGURE 32.5 Median section of the right half of the sheep brain dissection.

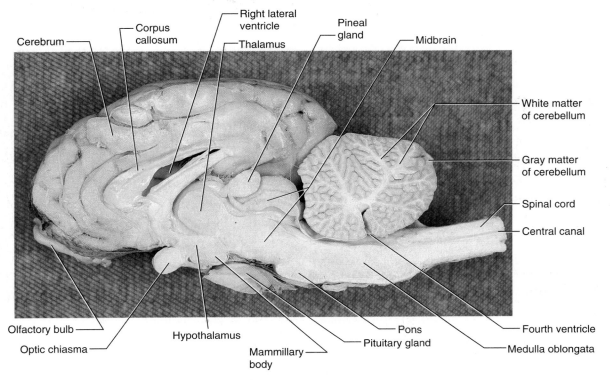

Cerebrum — Corpus callosum — Right lateral ventricle — Thalamus — Pineal gland — Midbrain — White matter of cerebellum — Gray matter of cerebellum — Spinal cord — Central canal — Olfactory bulb — Optic chiasma — Hypothalamus — Mammillary body — Pituitary gland — Pons — Fourth ventricle — Medulla oblongata

Name _____

Date _____

Section _____

The ⒶＡ corresponds to the Learning Outcome(s) listed at the beginning of the laboratory exercise.

Dissection of the Sheep Brain

Part A Assessments

Answer the following questions:

1. Describe the location of any meninges observed to be associated with the sheep brain. ⓵ _____

_____

_____

2. How do the relative sizes of the sheep and human cerebral hemispheres differ? ⓷ _____

_____

_____

3. How do the gyri and sulci of the sheep cerebrum compare with the human cerebrum in numbers? ⓷ _____

_____

_____

_____

4. What is the significance of the differences you noted in your answers for questions 2 and 3? ⓷ _____

_____

_____

5. What difference did you note in the structures of the sheep cerebellum and the human cerebellum? ⓷ _____

_____

_____

6. How do the sizes of the olfactory bulbs of the sheep brain compare with those of the human brain? ⓷ _____

_____

_____

7. Based on their relative sizes, which of the cranial nerves seems to be most highly developed in the sheep brain? ⓶

_____

_____

8. What is the significance of the observations you noted in your answers for questions 6 and 7? ⓷ _____

_____

_____

Part B Assessments

Prepare a list of at least six features to illustrate ways in which the brains of sheep and humans are similar. ⚠1 ⚠3

1. _____

2. _____

3. _____

4. _____

5. _____

6. _____

Interpret the significance of these similarities. ⚠3 _____

_____

_____

Laboratory Exercise 33

General Senses

Materials Needed

Marking pen (washable)

Millimeter ruler

Bristle or sharp pencil

Forceps (fine points)

Blunt metal probes

Three beakers (250 mL)

Warm tap water or 45°C (113°F) water bath

Cold water (ice water)

Thermometer

For Demonstration Activity:

Prepared microscope slides of tactile (Meissner's) and lamellated (Pacinian) corpuscles

Compound light microscope

Sensory receptors are sensitive to changes that occur within the body and its surroundings. Each type of receptor is particularly sensitive to a distinct kind of environmental change and is much less sensitive to other forms of stimulation. When receptors are stimulated, they initiate nerve impulses that travel into the central nervous system. The raw form in which these receptors send information to the brain is called *sensation.* The way our brains interpret this information is called *perception.*

The sensory receptors found widely distributed throughout skin, muscles, joints, and visceral organs are associated with **general senses.** These senses include touch, pressure, temperature, pain, and the senses of muscle movement and body position.

The general senses associated with the body surface can be classified according to the stimulus type. Mechanoreceptors include tactile (Meissner's) corpuscles, which are stimulated by light touch, stretch, or vibration, and lamellated (Pacinian) corpuscles, which are stimulated by deep pressure, stretch, or vibration. Thermoreceptors include those associated with temperature changes. Warm receptors are most sensitive to temperatures between 25°C (77°F) and 45°C (113°F). Cold receptors are most sensitive to temperatures between 10°C (50°F) and 20°C (68°F). Nociceptors are pain receptors that respond to tissue trauma, which may include a cut or pinch, extreme heat or cold, or excessive pressure. Receptors possess structural differences at the dendrite ends of sensory neurons. Those with unencapsulated free nerve endings include pain, cold, and heat receptors; those with encapsulated nerve endings include tactile and lamellated corpuscles.

A sensation may fade away when receptors are continuously stimulated. This is known as *sensory adaptation,* like adjusting to a room temperature or an odor of our environment. However, sensory adaptation to pain is not as prevalent because impulses may continue into the CNS for longer periods of time. The importance of perceiving pain not only alerts us to a change in our environment and potential injury, but it also allows us to detect body abnormalities so that proper adjustments can be taken. Unfortunately, with some spinal cord injuries or diseases such as diabetes mellitus, where the sense of pain is lost or becomes less noticeable, unwarranted tissue damages can occur.

Sensory receptors that are more specialized and confined to the head are associated with **special senses.** Laboratory Exercises 34 through 38 describe the special senses.

Purpose of the Exercise

To review the characteristics of sensory receptors and general senses and to investigate some of the general senses associated with the skin.

Learning Outcomes

After completing this exercise, you should be able to

1. Associate types of sensory receptors with general senses throughout the body.
2. Determine and record the distribution of touch, warm, and cold receptors in various regions of the skin.
3. Measure the two-point threshold of various regions of the skin.

Procedure A—Receptors and General Senses

1. Reexamine the introduction to this laboratory exercise.
2. Complete Part A of Laboratory Report 33.

Demonstration Activity

Observe the tactile (Meissner's) corpuscle with the microscope set up by the laboratory instructor. This type of receptor is abundant in the superficial dermis in outer regions of the body, such as in the fingertips, soles, lips, and external genital organs. It is responsible for the sensation of light touch. (See fig. 33.1.)

Observe the lamellated (Pacinian) corpuscle in the second demonstration microscope. This corpuscle is composed of many layers of connective tissue cells and has a nerve fiber in its central core. Lamellated corpuscles are numerous in the hands, feet, joints, and external genital organs. They are responsible for the sense of deep pressure (fig. 33.2). How are tactile and lamellated corpuscles similar?

_____

_____

How are they different?

_____

_____

_____

Procedure B—Sense of Touch

1. Investigate the distribution of touch receptors in your laboratory partner's skin. To do this, follow these steps:
   a. Use a marking pen and a millimeter ruler to prepare a square with 2.5 cm on each side of the skin on your partner's inner wrist, near the palm.
   b. Divide the square into smaller squares with 0.5 cm on a side, producing a small grid.
   c. Ask your partner to rest the marked wrist on the tabletop and to keep his or her eyes closed throughout the remainder of the experiment.

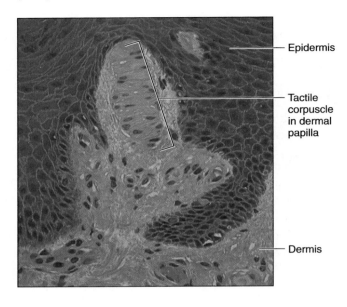

FIGURE 33.1 Tactile (Meissner's) corpuscles, such as this one, are responsible for the sensation of light touch (250×).

- Epidermis
- Tactile corpuscle in dermal papilla
- Dermis

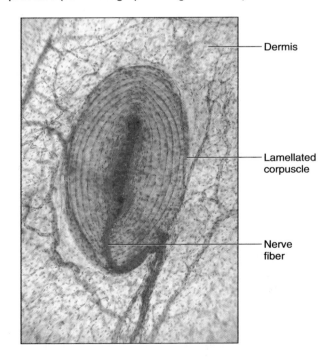

FIGURE 33.2 Lamellated (Pacinian) corpuscles, such as this one, are responsible for the sensation of deep pressure (25× micrograph enlarged to 100×).

- Dermis
- Lamellated corpuscle
- Nerve fiber

   d. Press the end of a bristle on the skin in some part of the grid, using just enough pressure to cause the bristle to bend. A sharp pencil could be used as an alternate device.
   e. Ask your partner to report whenever the touch of the bristle is felt. Record the results in Part B of the laboratory report.

f. Continue this procedure until you have tested twenty-five different locations on the grid. Move randomly through the grid to help prevent anticipation of the next stimulation site.

2. Test two other areas of exposed skin in the same manner, and record the results in Part B of the laboratory report.

3. Answer the questions in Part B of the laboratory report.

Procedure C—Two-Point Threshold

1. Test your partner's ability to recognize the difference between one and two points of skin being stimulated simultaneously. To do this, follow these steps:

   a. Have your partner place a hand with the palm up on the table and close his or her eyes.

   b. Hold the tips of a forceps tightly together and gently touch the skin of your partner's index finger.

   c. Ask your partner to report if it feels like one or two points are touching the finger.

   d. Allow the tips of the forceps to spread so they are 1 mm apart, press both points against the skin simultaneously, and ask your partner to report as before.

   e. Repeat this procedure, allowing the tips of the forceps to spread more each time until your partner can feel both tips being pressed against the skin. The minimum distance between the tips of the forceps when both can be felt is called the *two-point threshold*. As soon as you are able to distinguish two points, two separate receptors are being stimulated instead of only one receptor.

   f. Record the two-point threshold for the skin of the index finger in Part C of the laboratory report.

2. Repeat this procedure to determine the two-point threshold of the palm, the back of the hand, the back of the neck, the leg, and the sole. Record the results in Part C of the laboratory report.

3. Answer the questions in Part C of the laboratory report.

Procedure D—Sense of Temperature

1. Investigate the distribution of *warm (heat) receptors* in your partner's skin. To do this, follow these steps:

   a. Mark a square with 2.5 cm sides on your partner's palm.

   b. Prepare a grid by dividing the square into smaller squares, 0.5 cm on a side.

   c. Have your partner rest the marked palm on the table and close his or her eyes.

   d. Heat a blunt metal probe by placing it in a beaker of warm (hot) water (about 40–45°C/ 104–113°F) for a minute or so. (*Be sure the probe does not get so hot that it burns the skin.*) Use a thermometer to monitor the appropriate warm water from the tap or the water bath.

   e. Wipe the probe dry and touch it to the skin on some part of the grid.

   f. Ask your partner to report if the probe feels warm. Then record the results in Part D of the laboratory report.

   g. Keep the probe warm, and repeat the procedure until you have randomly tested twenty-five different locations on the grid.

2. Investigate the distribution of *cold receptors* by repeating the procedure. Use a blunt metal probe that has been cooled by placing it in ice water for a minute or so. Record the results in Part D of the laboratory report.

3. Answer the questions in Part D of the laboratory report.

Learning Extension Activity

Prepare three beakers of water of different temperatures. One beaker should contain warm water (about 40°C/104°F), one should be room temperature (about 22°C/72°F), and one should contain cold water (about 10°C/50°F). Place the index finger of one hand in the warm water and, at the same time, place the index finger of the other hand in the cold water for about 2 minutes. Then, simultaneously move both index fingers into the water at room temperature. What temperature do you sense with each finger? How do you explain the resulting perceptions?

_____

_____

_____

_____

Name _____

Date _____

Section _____

The ⚠ corresponds to the Learning Outcome(s) listed at the beginning of the laboratory exercise.

General Senses

Part A—Receptors and General Senses Assessments

Complete the following statements:

1. Whenever tissues are damaged, _____ receptors are likely to be stimulated. ⚠

2. Receptors that are sensitive to temperature changes are called _____. ⚠

3. A sensation may seem to fade away when receptors are continuously stimulated as a result of _____ adaptation. ⚠

4. Tactile (Meissner's) corpuscles are responsible for the sense of light _____. ⚠

5. Lamellated (Pacinian) corpuscles are responsible for the sense of deep _____. ⚠

6. _____ receptors are most sensitive to temperatures between 25°C (77°F) and 45°C (113°F). ⚠

7. _____ receptors are most sensitive to temperatures between 10°C (50°F) and 20°C (68°F). ⚠

8. Widely distributed sensory receptors throughout the body are associated with _____ senses in contrast to special senses. ⚠

Part B—Sense of Touch Assessments

1. Record a + to indicate where the bristle was felt and a *0* to indicate where it was not felt. ⚠2

Skin of wrist

2. Show the distribution of touch receptors in two other regions of skin. ⚠2

Region tested _____   Region tested _____

3. Answer the following questions:

   a. How do you describe the pattern of distribution for touch receptors in the regions of the skin you tested? ⚠2

   _____
   _____

   b. How does the concentration of touch receptors seem to vary from region to region? ⚠2 _____

   _____

Part C—Two-Point Threshold Assessments

1. Record the two-point threshold in millimeters for skin in each of the following regions: [3]

   Index finger _____

   Palm _____

   Back of hand _____

   Back of neck _____

   Leg _____

   Sole _____

2. Answer the following questions:

   a. What region of the skin tested has the greatest ability to discriminate two points? [3] _____

   b. What region of the skin has the least sensitivity to this test? [3] _____

   c. What is the significance of these observations in questions *a* and *b*? [3] _____

   _____

   _____

Part D—Sense of Temperature Assessments

1. Record a + to indicate where warm was felt and a *0* to indicate where it was not felt. [2]

   Skin of palm

2. Record a + to indicate where cold was felt and a 0 to indicate where it was not felt. [2]

   Skin of palm

3. Answer the following questions:

   a. How do temperature receptors appear to be distributed in the skin of the palm? [2] _____

   _____

   b. Compare the distribution and concentration of warm and cold receptors in the skin of the palm. [2] _____

   _____

   _____

Smell and Taste

34

Pre-Lab

1. Carefully read the introductory material and examine the entire lab content.
2. Be familiar with the basic structures and functions (from lecture or the textbook) of the receptors associated with smell and taste.
3. Visit www.mhhe.com/martinseriesl for pre-lab questions.

Materials Needed

For Procedure A—Sense of Smell (Olfaction)
Set of substances in stoppered bottles: cinnamon, sage, vanilla, garlic powder, oil of clove, oil of wintergreen, and perfume

For Procedure B—Sense of Taste (Gustation)
Paper cups (small)
Cotton swabs (sterile; disposable)
5% sucrose solution (sweet)
5% NaCl solution (salt)
1% acetic acid or unsweetened lemon juice (sour)
0.5% quinine sulfate solution or 0.1% Epsom salt solution (bitter)
1% monosodium glutamate (MSG) solution (umami)

For Demonstration Activities:
Compound light microscope
Prepared microscope slides of olfactory epithelium and of taste buds

For Learning Extension Activity:
Pieces of apple, potato, carrot, and onion or packages of mixed flavors of LifeSavers

Safety

▸ Be aware of possible food allergies when selecting test solutions and foods to taste.
▸ Prepare fresh solutions for use in Procedure B.
▸ Wash your hands before starting the taste experiment.
▸ Wear disposable gloves when performing taste tests on your laboratory partner.
▸ Use a clean cotton swab for each test. Do not dip a used swab into a test solution.
▸ Dispose of used cotton swabs and paper towels as directed.
▸ Wash your hands before leaving the laboratory.

The senses of smell (olfaction) and taste (gustation) are dependent upon chemoreceptors that are stimulated by various chemicals dissolved in liquids. The receptors of smell are found in the olfactory organs, which are located in the superior parts of the nasal cavity and in a portion of the nasal septum. The receptors of taste occur in the taste buds, which are sensory organs primarily found on the surface of the tongue. Chemicals are considered odorless and tasteless if receptor sites for them are absent.

Olfactory receptor cells are actually neurons, surrounded by columnar epithelial cells, with their apical ends covered with cilia, also called hair cells, embedded in the mucus of the superior nasal cavity. In order to detect an odor, the molecules must first dissolve in the mucus before they bind to the receptor sites of the cilia (olfactory hairs). When receptor cells temporarily bind to an odorant molecule, it results in a nerve impulse (action potential) over the olfactory neurons passing through the foramina of the cribriform plate and nerve fibers in the olfactory bulb. Eventually the impulses arrive at interpreting centers located deep within the temporal lobes and the inferior frontal lobes of the cerebrum. Although humans have the ability to distinguish nearly 10,000 different odors, coded by perhaps less than 1,000 genes, various odors can result from combinations of the receptor cells stimulated. We become sensory adapted to an odor very quickly, but exposure to a different substance can be quickly noticed.

Taste receptor cells are located in taste buds on the tongue, but receptor cells are also distributed in other areas of the mouth cavity and pharynx. A taste bud contains taste cells with terminal microvilli, called taste hairs, projecting through a taste pore on the epithelium of the tongue.

Taste sensations are grouped into five recognized categories: sweet, sour, salt, bitter, and umami. The sweet sensation is produced from sugars, the sour sensation from acids, the salt sensation from ionized inorganic salts, the bitter sensation from alkaloids and spoiled foods, and the umami sensation from aspartic and glutamic acids or a derivative such as monosodium glutamate. When a molecule binds to a receptor cell, a neurotransmitter is secreted and an action potential occurs over a sensory neuron. The nerve impulse continues through the medulla oblongata and the thalamus, and is interpreted in the insula lobe of the cerebrum. Sensory adaptation also occurs rather quickly, which helps explain the reason the first few bites of a particular food have the most vivid flavor.

The senses of smell and taste function closely together, because substances that are tasted often are smelled at the same moment, and they play important roles in the selection of foods. The taste and aroma of foods are also influenced by appearance, texture, temperature, and the person's mood.

Purpose of the Exercise

To review the structures of the organs of smell and taste and to investigate the abilities of smell and taste receptors to discriminate various chemical substances.

Learning Outcomes

After completing this exercise, you should be able to

(1) Compare the characteristics of the smell (olfactory) receptors and the taste (gustatory) receptors.

(2) Explain how the senses of smell and taste function.

(3) Record recognized odors and the time needed for olfactory sensory adaptation to occur.

(4) Locate the distribution of taste receptors on the surface of the tongue and mouth cavity.

Procedure A—Sense of Smell (Olfaction)

1. Reexamine the introduction to this laboratory exercise.
2. Label figure 34.1.
3. Complete Part A of Laboratory Report 34.
4. Test your laboratory partner's ability to recognize the odors of the bottled substances available in the laboratory. To do this, follow these steps:
   a. Have your partner keep his or her eyes closed.
   b. Remove the stopper from one of the bottles, and hold it about 4 cm under your partner's nostrils for about 2 seconds.
   c. Ask your partner to identify the odor, and then replace the stopper.
   d. Record your partner's response in Part B of the laboratory report.
   e. Repeat steps b–d for each of the bottled substances.
5. Repeat the preceding procedure, using the same set of bottled substances, but present them to your partner in a different sequence. Record the results in Part B of the laboratory report.
6. Wait 10 minutes and then determine the time it takes for your partner to experience olfactory sensory adaptation. To do this, follow these steps:
   a. Ask your partner to breathe in through the nostrils and exhale through the mouth.
   b. Remove the stopper from one of the bottles, and hold it about 4 cm under your partner's nostrils.
   c. Keep track of the time that passes until your partner is no longer able to detect the odor of the substance.
   d. Record the result in Part B of the laboratory report.
   e. Wait 5 minutes and repeat this procedure, using a different bottled substance.
   f. Test a third substance in the same manner.
   g. Record the results as before.
7. Complete Part B of the laboratory report.

FIGURE 34.1 Label this diagram of the olfactory organ by placing the correct numbers in the spaces provided. ⚠

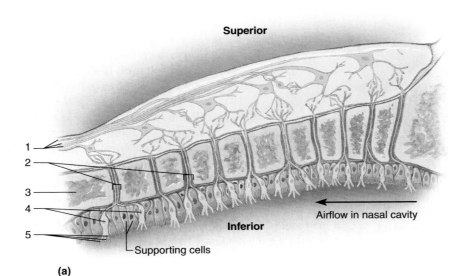

_____ Cilia (olfactory hairs)

_____ Cribriform plate of ethmoid bone

_____ Nerve fibers within olfactory bulb

_____ Olfactory nerves

_____ Olfactory receptor cells

Superior

1
2
3
4
5

Supporting cells

Airflow in nasal cavity

Inferior

(a)

Olfactory tract   Olfactory bulb   Cribriform plate

Olfactory area of nasal cavity

Superior nasal concha

Nasal cavity

(b)

Demonstration Activity

Observe the olfactory epithelium in the microscope set up by the laboratory instructor. The olfactory receptor cells are spindle-shaped, bipolar neurons with spherical nuclei. They also have six to eight cilia at their apical ends. The supporting cells are pseudostratified columnar epithelial cells. However, in this region the tissue lacks goblet cells. (See fig. 34.2.)

Procedure B—Sense of Taste (Gustation)

1. Reexamine the introduction to this laboratory exercise.
2. Label figure 34.3.
3. Complete Part C of the laboratory report.
4. Map the distribution of the receptors for the primary taste sensations on your partner's tongue. To do this, follow these steps:

FIGURE 34.2 Olfactory receptors have cilia at their distal ends (250×).

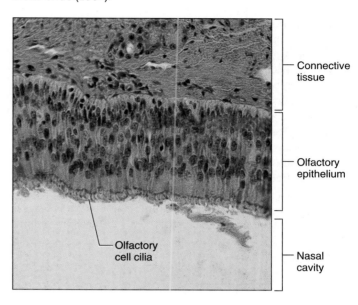

- Connective tissue
- Olfactory epithelium
- Olfactory cell cilia
- Nasal cavity

FIGURE 34.3 Label this diagram by placing the correct numbers in the spaces provided. 🔺

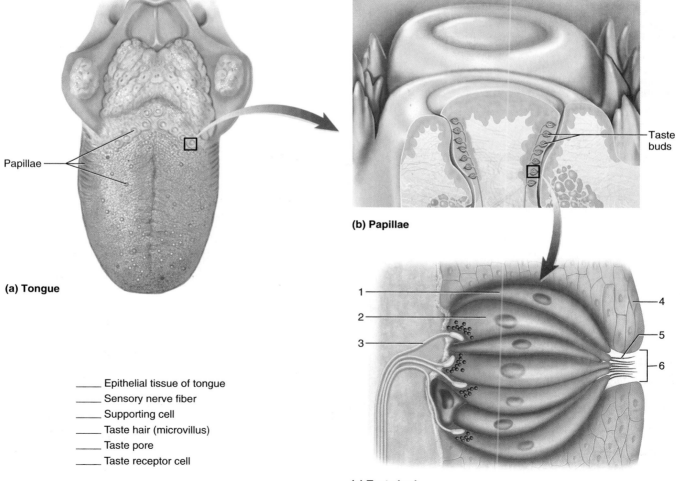

(a) Tongue

Papillae

(b) Papillae

Taste buds

(c) Taste bud

_____ Epithelial tissue of tongue
_____ Sensory nerve fiber
_____ Supporting cell
_____ Taste hair (microvillus)
_____ Taste pore
_____ Taste receptor cell

283

a. Ask your partner to rinse his or her mouth with water and then partially dry the surface of the tongue with a paper towel.

b. Moisten a clean cotton swab with 5% sucrose solution, and touch several regions of your partner's tongue with the swab.

c. Each time you touch the tongue, ask your partner to report if a sweet sensation is experienced.

d. Test the tip, sides, and back of the tongue in this manner.

e. Test some other representative areas inside the mouth cavity such as the cheek, gums, and roof of the mouth (hard and soft palate) for a sweet sensation.

f. Record your partner's responses in Part D of the laboratory report.

g. Have your partner rinse his or her mouth and dry the tongue again, and repeat the preceding procedure, using each of the other four test solutions—NaCl, acetic acid, quinine or Epsom salt, and MSG solution. Be sure to use a fresh swab for each test substance and dispose of used swabs and paper towels as directed.

5. Complete Part D of the laboratory report.

Demonstration Activity

Observe the oval-shaped taste bud in the microscope set up by the laboratory instructor. Note the surrounding epithelial cells. The taste pore, an opening into the taste bud, may be filled with taste hairs (microvilli). Within the taste bud there are supporting cells and thinner taste-receptor cells, which often have lightly stained nuclei (fig. 34.4).

Learning Extension Activity

Test your laboratory partner's ability to recognize the tastes of apple, potato, carrot, and onion. (A package of mixed flavors of LifeSavers is a good alternative.) To do this, follow these steps:

1. Have your partner close his or her eyes and hold the nostrils shut.

2. Place a small piece of one of the test substances on your partner's tongue.

3. Ask your partner to identify the substance without chewing or swallowing it.

4. Repeat the procedure for each of the other substances.

How do you explain the results of this experiment?

_____

_____

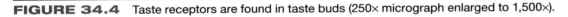

FIGURE 34.4 Taste receptors are found in taste buds (250× micrograph enlarged to 1,500×).

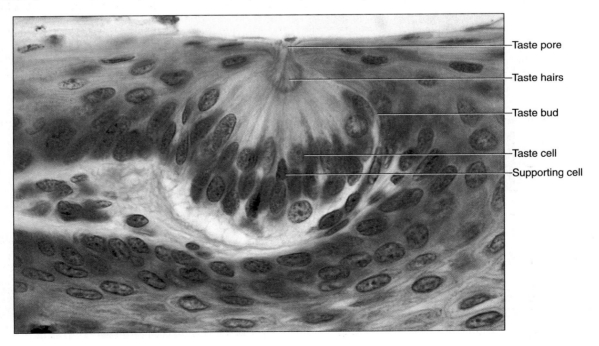

Taste pore

Taste hairs

Taste bud

Taste cell

Supporting cell

Name _____

Date _____

Section _____

The ⚠ corresponds to the Learning Outcome(s) listed at the beginning of the
laboratory exercise.

Smell and Taste

Part A Assessments

Complete the following statements:

1. The distal ends of the olfactory neurons are covered with hairlike _____. ⚠

2. Before gaseous substances can stimulate the olfactory receptors, they must be dissolved in _____ that surrounds the cilia. ⚠

3. The axons of olfactory receptors pass through small openings in the _____ of the ethmoid bone. ⚠

4. The olfactory interpreting centers are located deep within the temporal lobes and at the base of the _____ lobes of the cerebrum. ⚠

5. Olfactory sensations usually fade rapidly as a result of _____. ⚠

6. A chemical would be considered _____ if a person lacks a particular receptor site on the cilia of the olfactory neurons. ⚠

Part B—Sense of Smell Assessments

1. Record the results (as +, if recognized; as 0, if unrecognized) from the tests of odor recognition in the following table: ⚠

Substance Tested	Odor Reported	
	First Trial	Second Trial

2. Record the results of the olfactory sensory adaptation time in the following table: ⚠

Substance Tested	Adaptation Time in Seconds

285

3. Complete the following:

    a. How do you describe your partner's ability to recognize the odors of the substances you tested? **2**

_____

_____

    b. Compare your experimental results with those of others in the class. Did you find any evidence to indicate that individuals may vary in their abilities to recognize odors? Explain your answer. **2**

_____

_____

_____

Critical Thinking Activity

Does the time it takes for sensory adaptation to occur seem to vary with the substances tested? Explain your answer.

_____

_____

_____

Part C Assessments

Complete the following statements:

1. Taste is interpreted in the _____ lobe of the cerebrum. **2**

2. The opening to a taste bud is called a _____. **1**

3. The _____ of a taste cell are its sensitive part. **1**

4. Before the taste of a substance can be detected, the substance must be dissolved in _____. **1**

5. Substances that stimulate taste cells bind with _____ sites on the surfaces of taste hairs. **1**

6. Sour receptors are mainly stimulated by _____. **1**

7. Salt receptors are mainly stimulated by ionized inorganic _____. **1**

8. Alkaloids usually have a _____ taste. **1**

Part D—Sense of Taste Assessments

1. *Taste receptor distribution.* Record a + to indicate where a taste sensation seemed to originate and a *0* if no sensation occurred when the spot was stimulated. **4**

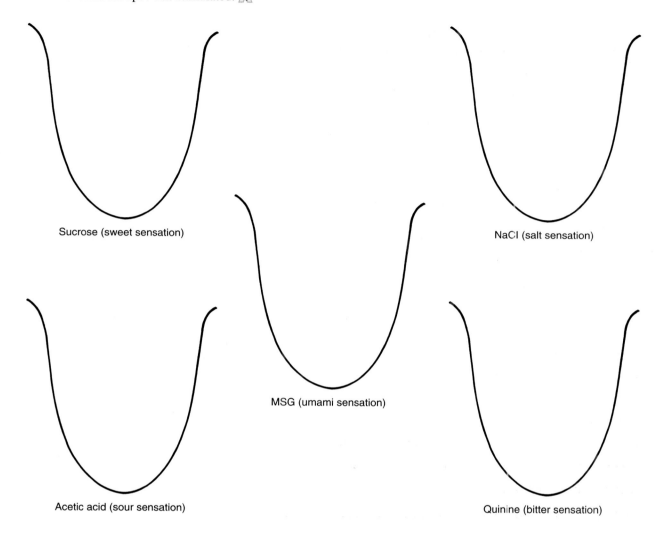

Sucrose (sweet sensation)

MSG (umami sensation)

NaCl (salt sensation)

Acetic acid (sour sensation)

Quinine (bitter sensation)

2. Complete the following:

   a. Describe how each type of taste receptor is distributed on the surface of your partner's tongue. **4**

   _____
   _____
   _____
   _____
   _____

   b. Describe other locations inside the mouth where any sensations of sweet, salt, sour, bitter, or umami were located. **4**

   _____
   _____
   _____
   _____

c. How does your taste distribution map on the tongue compare to those of other students in the class? **4**

_____

_____

_____

_____

Eye Structure

Pre-Lab

1. Carefully read the introductory material and examine the entire lab content.
2. Be familiar with the basic structures and functions (from lecture or the textbook) of parts of the eye.
3. Visit www.mhhe.com/martinseriesl for pre-lab questions and Anatomy & Physiology Revealed animations.

Materials Needed

Dissectible eye model
Compound light microscope
Prepared microscope slide of a mammalian eye (sagittal section)
Sheep or beef eye (fresh or preserved)
Dissecting tray
Dissecting instruments—forceps, sharp scissors, and dissecting needle

For Learning Extension Activity:
Ophthalmoscope

Safety

▶ Wear disposable gloves when working on the eye dissection.
▶ Dispose of tissue remnants and gloves as instructed.
▶ Wash the dissecting tray and instruments as instructed.
▶ Wash your laboratory table.
▶ Wash your hands before leaving the laboratory.

The special sense of vision involves the complex structure of the eye, accessory structures associated with the orbit, the optic nerve, and the brain. The eyebrow, eyelids, conjunctiva, and lacrimal apparatus protect the eyes. The vascular conjunctiva, a mucous membrane, covers all of the anterior surface except the cornea, and it lines the eyelids. Its mucus prevents the eye from drying out, and the lacrimal fluid from the lacrimal (tear) gland cleanses the eye and prevents infections due to the antibacterial agent lysozyme in its secretions. Two voluntary muscles involve control of the eyelids, while six extrinsic muscles allow eye movements in multiple directions.

Three layers (tunics) compose the main structures of the eye. The outer (fibrous) layer includes the protective white sclera and a transparent anterior portion, the cornea, which allows light to enter the eye cavities. The middle (vascular) layer includes the heavily pigmented choroid, the ciliary body that controls the shape of the lens, and the iris with its smooth muscle fibers that control the amount of light passing through its central opening, the pupil. The inner layer includes pigmented and neural components of the retina and the optic nerve, which contains sensory neurons that carry nerve impulses from the rods and cones. A large posterior cavity is filled with a transparent gel, the vitreous humor. A smaller anterior cavity has anterior and posterior chambers separated at the iris, which is filled with a clear aqueous humor.

The photoreceptor cells, the rods and cones, can absorb light to produce visual images. The more numerous rods can be stimulated in dim light and are most abundant in the peripheral retina. The cones are stimulated only in brighter light, but they allow us to perceive sharper images, and they allow us to see them in color because there are three types of cones, with each type most sensitive to wavelengths peaking in the blue, green, or red portion of the visible light spectrum. Located at the posterior center of the retina is a region containing mostly cones, called the macula lutea ("yellow spot"). A tiny pit in the center of the macula lutea, the fovea centralis, contains only cones and represents our best image location. Slightly to the medial side of the macula lutea is the optic disc, where the nerve fibers of the retina leave the eye to become parts of the optic nerve. It is also referred to as the "blind spot" because rods and cones are absent at this location. Sensory impulses from the optic nerves pass through the thalamus and eventually arrive at the primary visual cortex in the occipital lobe of the cerebrum, where the perception of visual images occurs.

Purpose of the Exercise

To review the structure and function of the eye and to dissect a mammalian eye.

Learning Outcomes

After completing this exercise, you should be able to

1. Identify the major structures of an eye.
2. Describe the functions of the structures of an eye.
3. Trace the structures through which light passes as it travels from the cornea to the retina.
4. Dissect a mammalian eye and locate its major features.

Procedure A—Structure and Function of the Eye

1. Reexamine the introduction to this laboratory exercise.
2. Study figures 35.1, 35.2, and 35.3.

FIGURE 35.1 Structures of the lacrimal apparatus.

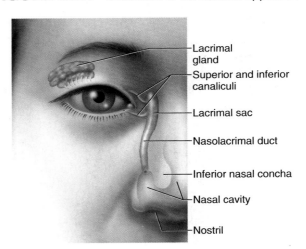

- Lacrimal gland
- Superior and inferior canaliculi
- Lacrimal sac
- Nasolacrimal duct
- Inferior nasal concha
- Nasal cavity
- Nostril

FIGURE 35.2 Extrinsic muscles of the right eye (anterior view). The arrows indicate the direction of the pull by each of the six muscles.

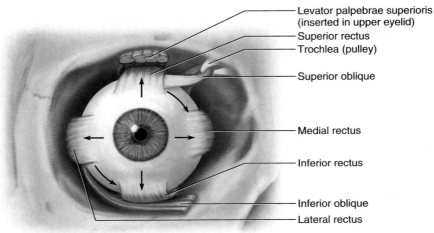

- Levator palpebrae superioris (inserted in upper eyelid)
- Superior rectus
- Trochlea (pulley)
- Superior oblique
- Medial rectus
- Inferior rectus
- Inferior oblique
- Lateral rectus

FIGURE 35.3 Structures of the eye (sagittal section).

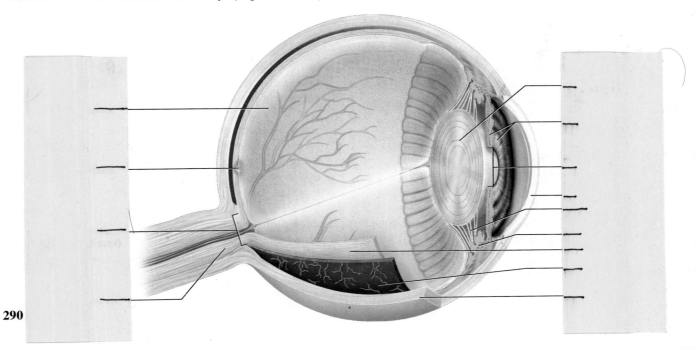

3. Examine the dissectible model of the eye and locate the following features:

eyelid
conjunctiva
orbicularis oculi
levator palpebrae superioris
lacrimal apparatus
  lacrimal gland
  canaliculi
  lacrimal sac
  nasolacrimal duct
extrinsic muscles
  superior rectus
  inferior rectus
  medial rectus
  lateral rectus
  superior oblique
  inferior oblique
trochlea (pulley)
outer (fibrous) layer (tunic)
  sclera
  cornea
middle (vascular) layer (tunic)
  choroid
  ciliary body
    ciliary processes
    ciliary muscles
    suspensory ligaments

  iris
  pupil (opening in center of iris)
inner layer (tunic)
  retina
    macula lutea
    fovea centralis
    optic disc
  optic nerve
lens
anterior cavity
  anterior chamber
  posterior chamber
  aqueous humor
posterior cavity
  vitreous humor

4. Obtain a microscope slide of a mammalian eye section, and locate as many of the features in step 3 as possible.
5. Using high-power magnification, observe the posterior portion of the eye wall, and locate the sclera, choroid, and retina.
6. Using high-power magnification, examine the retina, and note its layered structure (fig. 35.4). Locate the following:

nerve fibers leading to the optic nerve (innermost layer of the retina)
layer of ganglion cells
layer of bipolar neurons
nuclei of rods and cones
receptor ends of rods and cones
pigmented epithelium (outermost layer of the retina)

7. Complete Part A of Laboratory Report 35.

FIGURE 35.4 The cells of the retina are arranged in distinct layers (75×).

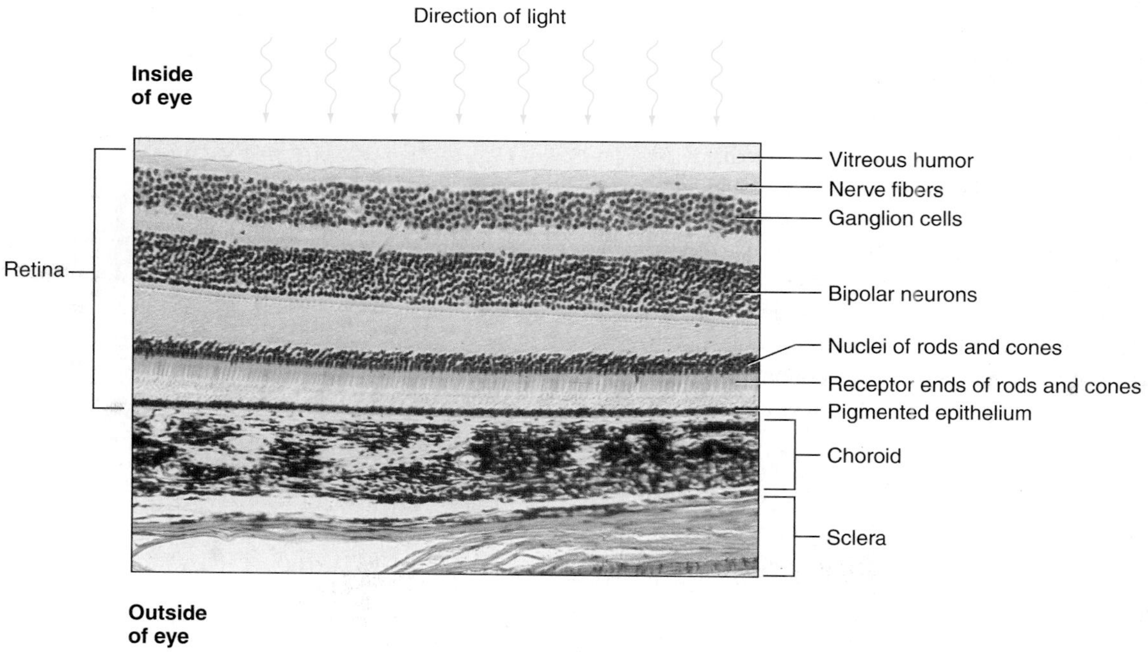

Direction of light

Inside of eye

Retina

Outside of eye

Vitreous humor
Nerve fibers
Ganglion cells
Bipolar neurons
Nuclei of rods and cones
Receptor ends of rods and cones
Pigmented epithelium
Choroid
Sclera

291

Learning Extension Activity

Use an ophthalmoscope to examine the interior of your laboratory partner's eye. This instrument consists of a set of lenses held in a rotating disc, a light source, and some mirrors that reflect the light into the test subject's eye.

The examination should be conducted in a dimly lit room. Have your partner seated and staring straight ahead at eye level. Move the rotating disc of the ophthalmoscope so that the *O* appears in the lens selection window. Hold the instrument in your right hand with the end of your index finger on the rotating disc (fig. 35.5). Direct the light at a slight angle from a distance of about 15 cm into the subject's right eye. The light beam should pass along the inner edge of the pupil. Look through the instrument, and you should see a reddish, circular area—the interior of the eye. Rotate the disc of lenses to higher values until sharp focus is achieved.

Move the ophthalmoscope to within about 5 cm of the eye being examined *being very careful that the instrument does not touch the eye,* and again rotate the lenses to sharpen the focus (fig. 35.6). Locate the optic disc and the blood vessels that pass through it. Also locate the yellowish macula lutea by having your partner stare directly into the light of the instrument (fig. 35.7).

Examine the subject's iris by viewing it from the side and by using a lens with a +15 or +20 value.

Procedure B—Eye Dissection

1. Obtain a mammalian eye, place it in a dissecting tray, and dissect it as follows:
   a. Trim away the fat and other connective tissues but leave the stubs of the *extrinsic muscles* and of the *optic nerve.* This nerve projects outward from the posterior region of the eyeball.
   b. The *conjunctiva* lines the inside of the eyelid and is reflected over the anterior surface of the eye, except for the cornea. Lift some of this thin membrane away from the eye with forceps and examine it.
   c. Locate and observe the *cornea, sclera,* and *iris.* Also note the *pupil* and its shape. The cornea from a fresh eye will be transparent; when preserved, it becomes opaque.
   d. Use sharp scissors to make a frontal section of the eye. To do this, cut through the wall about 1 cm from the margin of the cornea and continue all the way around the eyeball. Try not to damage the internal structures of the eye (fig. 35.8).
   e. Gently separate the eyeball into anterior and posterior portions. Usually the jellylike vitreous humor will remain in the posterior portion, and the lens may adhere to it. Place the parts in the dissecting tray with their contents facing upward.

FIGURE 35.5 The ophthalmoscope is used to examine the interior of the eye.

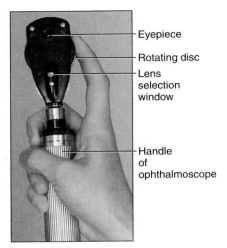

- Eyepiece
- Rotating disc
- Lens selection window
- Handle of ophthalmoscope

FIGURE 35.6 (*a*) Rotate the disc of lenses until sharp focus is achieved. (*b*) Move the ophthalmoscope to within 5 cm of the eye to examine the optic disc.

(a)

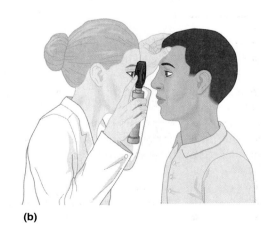

(b)

FIGURE 35.7   Right retina as viewed through the pupil using an ophthalmoscope.

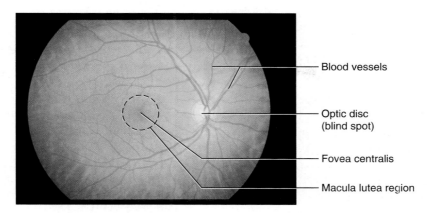

Blood vessels

Optic disc
(blind spot)

Fovea centralis

Macula lutea region

FIGURE 35.8   Prepare a frontal section of the eye.

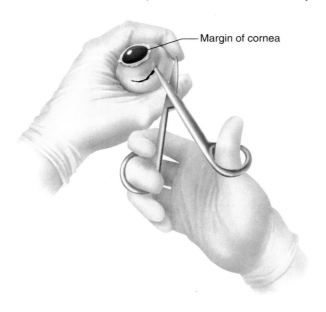

Margin of cornea

FIGURE 35.9   Human lens showing a cataract and one type of lens implant (intraocular lens replacement) on tips of fingers to show their relative size. A normal lens is transparent.

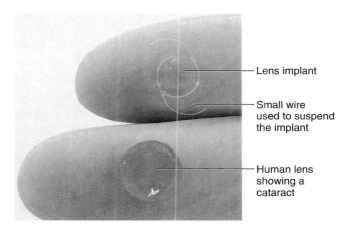

Lens implant

Small wire used to suspend the implant

Human lens showing a cataract

f. Examine the anterior portion of the eye, and locate the *ciliary body,* which appears as a dark, circular structure. Also note the *iris* and the *lens* if it remained in the anterior portion. The lens is normally attached to the ciliary body by many *suspensory ligaments,* which appear as delicate, transparent threads.

g. Use a dissecting needle to gently remove the lens, and examine it. If the lens is still transparent, hold it up and look through it at something in the distance and note that the lens inverts the image. The lens of a preserved eye is usually too opaque for this experience. If the lens of the human eye becomes opaque, the defect is called a cataract (fig. 35.9).

h. Examine the posterior portion of the eye. Note the *vitreous humor.* This jellylike mass helps to hold the lens in place anteriorly and helps to hold the *retina* against the choroid.

i. Carefully remove the vitreous humor and examine the retina. This layer will appear as a thin, nearly colorless to cream-colored membrane that detaches easily from the choroid coat. Compare the structures identified to figure 35.10.

j. Locate the *optic disc*—the point where the retina is attached to the posterior wall of the eyeball and where the optic nerve originates. There are no receptor cells in the optic disc, so this region is also called the "blind spot."

k. Note the iridescent area of the choroid beneath the retina. This colored surface in ungulates (mammals having hoofs) is called the *tapetum fibrosum.* It serves to reflect light back through the retina, an action thought to aid the night vision of some animals. The tapetum fibrosum is lacking in the human eye.

2. Discard the tissues of the eye as directed by the laboratory instructor.

3. Complete Parts B and C of the laboratory report.

FIGURE 35.10   Internal structures of the beef eye dissection.

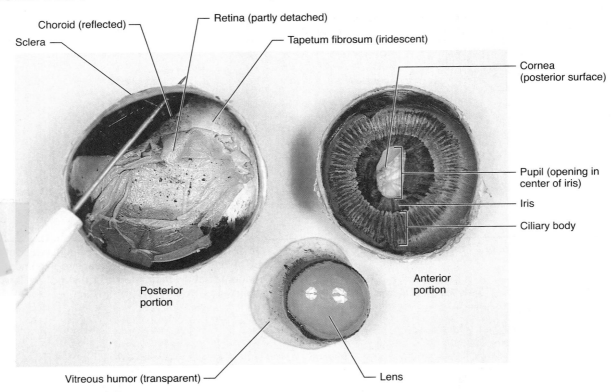

Choroid (reflected)

Retina (partly detached)

Sclera

Tapetum fibrosum (iridescent)

Cornea (posterior surface)

Pupil (opening in center of iris)

Iris

Ciliary body

Posterior portion

Anterior portion

Vitreous humor (transparent)

Lens

35.10 Beef Eye

Critical Thinking Activity

A strong blow to the head might cause the retina to detach. From observations made during the eye dissection, explain why this could happen.

_____

_____

_____

_____

Name _____

Date _____

Section _____

The ⚠ corresponds to the Learning Outcome(s) listed at the beginning of the laboratory exercise.

Eye Structure

Part A Assessments

Identify the features of the eye indicated in figure 35.11. ⚠

FIGURE 35.11 Sections of the eye: (*a*) anterior portion (10×); (*b*) posterior portion (53×). Identify the numbered features, using the terms provided.

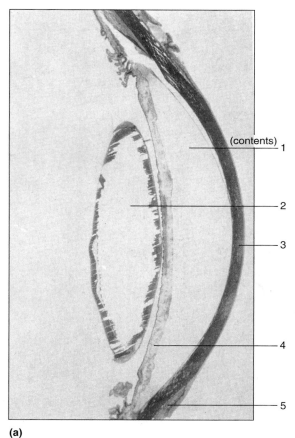

(a)

Terms:

Aqueous humor	Optic disc
Choroid/middle layer	Optic nerve
Conjunctiva	Retina/inner layer
Cornea	Sclera/outer layer
Iris	Vitreous humor
Lens	

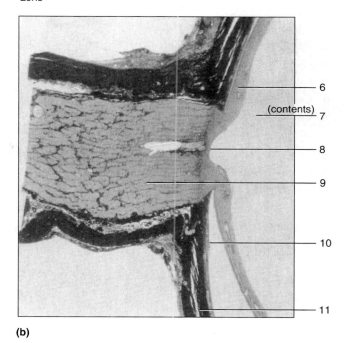

(b)

a. Anterior portion of eye:

1. _____

2. _____

3. _____

4. _____

5. _____

b. Posterior portion of eye:

6. _____

7. _____

8. _____

9. _____

10. _____

11. _____

Part B Assessments

Match the terms in column A with the descriptions in column B. Place the letter of your choice in the space provided. ⓵ ⓶

Column A	Column B
a. Aqueous humor	_____ **1.** Posterior five-sixths of middle (vascular) layer
b. Choroid	_____ **2.** White part of outer (fibrous) layer
c. Ciliary muscles	_____ **3.** Transparent anterior portion of outer layer
d. Conjunctiva	_____ **4.** Inner lining of eyelid
e. Cornea	_____ **5.** Secretes tears
f. Iris	_____ **6.** Fills posterior cavity of eye
g. Lacrimal gland	_____ **7.** Area where optic nerve exits the eye
h. Optic disc	_____ **8.** Smooth muscle that controls light entering the eye
i. Retina	_____ **9.** Fills anterior and posterior chambers of the anterior cavity of the eye
j. Sclera	_____ **10.** Contains visual receptors called rods and cones
k. Suspensory ligament	_____ **11.** Connects lens to ciliary body
l. Vitreous humor	_____ **12.** Cause lens to change shape

Complete the following:

13. List the structures and fluids through which light passes as it travels from the cornea to the retina. ⓷ _____

_____

_____

_____

14. List three ways in which rods and cones differ in structure or function. ◢2◣ _____

_____

_____

_____

_____

Part C Assessments

Complete the following:

1. Which layer/tunic of the eye was the most difficult to cut? ◢4◣ _____

2. What kind of tissue do you think is responsible for this quality of toughness? ◢4◣ _____

3. How do you compare the shape of the pupil in the dissected eye with your own pupil? ◢4◣ _____

_____

_____

4. Where do you find aqueous humor in the dissected eye? ◢4◣ _____

_____

_____

5. What is the function of the dark pigment in the choroid? ◢2◣ _____

_____

_____

6. Describe the lens of the dissected eye. ◢4◣ _____

_____

_____

_____

7. Describe the vitreous humor of the dissected eye. ◢4◣ _____

_____

_____

Visual Tests and Demonstrations

Pre-Lab

1. Carefully read the introductory material and examine the entire lab content.
2. Be familiar with common eye tests and defects (from lecture or the textbook).
3. Visit www.mhhe.com/martinseriesl for pre-lab questions and Anatomy & Physiology Revealed animations.

Materials Needed

Snellen eye chart
3" × 5" card (plain)
3" × 5" card with word typed in center
Astigmatism chart
Meterstick
Metric ruler
Pen flashlight
Ichikawa's or Ishihara's color plates for color blindness test

Normal vision (emmetropia) results when light rays from objects in the external environment are refracted by the cornea and lens of the eye and focused onto the photoreceptors of the retina. If a person has a condition of nearsightedness (myopia), the image focuses in front of the retina because the eyeball is too long; nearby objects are clear, but distant objects are blurred. If a person has a condition of farsightedness (hyperopia), the image focuses behind the retina because the eyeball is too short; distant objects are clear, but nearby objects are blurred. A concave (diverging) lens is required to correct vision for nearsightedness; a convex (converging) lens is required to correct vision for farsightedness. If a defect occurs in the curvature of the cornea or lens, a condition called astigmatism results. Focusing in different planes cannot occur simultaneously. Corrective cylindrical lenses will allow proper refraction of light onto the retina to compensate for the unequal curvatures.

As part of a natural aging process, the lens's elasticity decreases, which degrades our ability to focus on nearby objects. The near point of accommodation test is used to determine the ability to accommodate. Often a person will use "reading glasses" to compensate for this condition. Hereditary defects in color vision result from a lack of certain cones needed to absorb certain wavelengths of light. Special color test plates are used to diagnose a condition of color blindness.

As you conduct this laboratory exercise, it does not matter in what order the visual tests or visual demonstrations are performed. None of the tests or demonstrations depends upon the results of any of the others. If you wear glasses, perform the tests with and without the corrective lenses. If you wear contact lenses, it is not necessary to run the tests under both conditions, but indicate in the laboratory report that all tests were performed with contact lenses in place.

Purpose of the Exercise

To conduct and interpret the tests for visual acuity, astigmatism, accommodation, color vision, the blind spot, and certain reflexes of the eye.

Learning Outcomes

After completing this exercise, you should be able to

1 Conduct and interpret the tests used to evaluate visual acuity, astigmatism, the ability to accommodate for close vision, and color vision.

2 Describe four conditions that can lead to defective vision.

3 Demonstrate and describe the blind spot, photopupillary reflex, accommodation pupillary reflex, and convergence reflex.

Procedure A—Visual Tests

Perform the following visual tests, using your laboratory partner as a test subject. If your partner usually wears glasses, test each eye with and without the glasses.

1. *Visual acuity test.* Visual acuity (sharpness of vision) can be measured by using a Snellen eye chart (fig. 36.1). This chart consists of several sets of letters in different

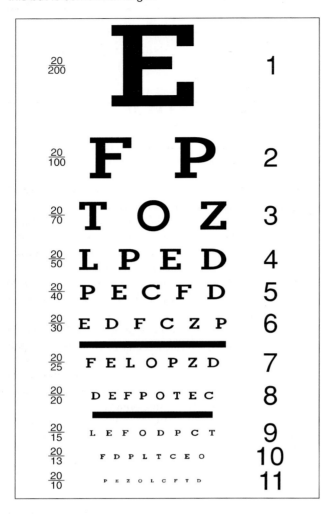

$\frac{20}{200}$ **E**	1
$\frac{20}{100}$ **F P**	2
$\frac{20}{70}$ **T O Z**	3
$\frac{20}{50}$ **L P E D**	4
$\frac{20}{40}$ **P E C F D**	5
$\frac{20}{30}$ **E D F C Z P**	6
$\frac{20}{25}$ **F E L O P Z D**	7
$\frac{20}{20}$ D E F P O T E C	8
$\frac{20}{15}$ L E F O D P C T	9
$\frac{20}{13}$ F D P L T C E O	10
$\frac{20}{10}$ P E Z O L C F T D	11

b. Have your partner stand 20 feet in front of the chart, gently cover the left eye with a 3" × 5" card, and read the smallest set of letters possible using your right eye.

c. Record the visual acuity value for that set of letters in Part A of Laboratory Report 36.

d. Repeat the procedure, using the left eye.

2. *Astigmatism test.* Astigmatism is a condition that results from a defect in the curvature of the cornea or lens. As a consequence, some portions of the image projected on the retina are sharply focused, and other portions are blurred. Astigmatism can be evaluated by using an astigmatism chart (fig. 36.2). This chart consists of sets of black lines radiating from a central spot like the spokes of a wheel. To a normal eye, these lines appear sharply focused and equally dark; however, if the eye has an astigmatism some sets of lines appear sharply focused and dark, while others are blurred and less dark.

To conduct the astigmatism test, follow these steps:

a. Hang the astigmatism chart on a well-illuminated wall at eye level.

b. Stand 20 feet in front of the chart, and gently cover the left eye with a 3" × 5" card. Using your right eye, focus on the spot in the center of the radiating lines, and report which lines, if any, appear more sharply focused and darker.

c. Repeat the procedure, using the left eye.

d. Record the results in Part A of the laboratory report.

3. *Accommodation test.* Accommodation is the changing of the shape of the lens that occurs when the normal eye is focused for close vision. It involves a reflex in which muscles of the ciliary body are stimulated to contract,

sizes printed on a white card. The letters near the top of the chart are relatively large, and those in each lower set become smaller. At one end of each set of letters is an acuity value in the form of a fraction. One of the sets near the bottom of the chart, for example, is marked 20/20. The normal eye can clearly see these letters from the standard distance of 20 feet and thus is said to have 20/20 vision. The letter at the top of the chart is marked 20/200. The normal eye can read letters of this size from a distance of 200 feet. Thus, an eye able to read only the top letter of the chart from a distance of 20 feet is said to have 20/200 vision. This person has less than normal vision. A line of letters near the bottom of the chart is marked 20/15. The normal eye can read letters of this size from a distance of 15 feet, but a person might be able to read it from 20 feet. This person has better than normal vision.

To conduct the visual acuity test, follow these steps:

a. Hang the Snellen eye chart on a well-illuminated wall at eye level.

FIGURE 36.2 Astigmatism is evaluated using a chart such as this one.

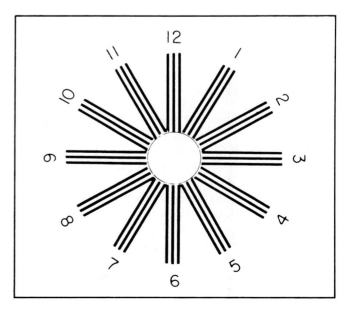

releasing tension on the suspensory ligaments that are fastened to the lens capsule. This allows the capsule to rebound elastically, causing the surface of the lens to become more convex. The ability to accommodate is likely to decrease with age because the tissues involved tend to lose their elasticity.

To evaluate the ability to accommodate, follow these steps:

a. Hold the end of a meterstick against your partner's chin so that the stick extends outward at a right angle to the plane of the face (fig. 36.3).

b. Have your partner close the left eye.

c. Hold a 3" × 5" card with a word typed in the center at the distal end of the meterstick.

d. Slide the card along the stick toward your partner's open right eye, and locate the *point closest to the eye* where your partner can still see the letters of the word sharply focused. This distance is called the *near point of accommodation,* and it tends to increase with age (table 36.1).

e. Repeat the procedure with the right eye closed.

f. Record the results in Part A of the laboratory report.

4. *Color vision test.* Some people exhibit defective color vision because they lack certain cones, usually those sensitive to the reds or greens. This trait is an X-linked (sex-linked) inheritance, so the condition is more prevalent in males (7%) than in females (0.4%). People who lack or possess decreased sensitivity to the red-sensitive cones possess protanopia color blindness; those who lack or possess decreased sensitivity to green-sensitive cones possess deuteranopia color blindness. The color blindness condition is often more of a deficiency or a

TABLE 36.1 Near Point of Accommodation

Age (years)	Average Near Point (cm)
10	7
20	10
30	13
40	20
50	45
60	90

weakness than one of blindness. Laboratory Exercise 61 describes the genetics for color blindness.

To conduct the color vision test, follow these steps:

a. Examine the color test plates in Ichikawa's or Ishihara's book to test for any red-green color vision deficiency. Also examine figure 36.4.

b. Hold the test plates approximately 30 inches from the subject in bright light. All responses should occur within 3 seconds.

c. Compare your responses with the correct answers in Ichikawa's or Ishihara's book. Determine the percentage of males and females in your class who exhibit any color-deficient vision. If an individual exhibits color-deficient vision, determine if the condition is protanopia or deuteranopia.

d. Record the class results in Part A of the laboratory report.

5. Complete Part A of the laboratory report.

FIGURE 36.3 To determine the near point of accommodation, slide the 3" × 5" card along the meterstick toward your partner's open eye until it reaches the closest location where your partner can still see the word sharply focused.

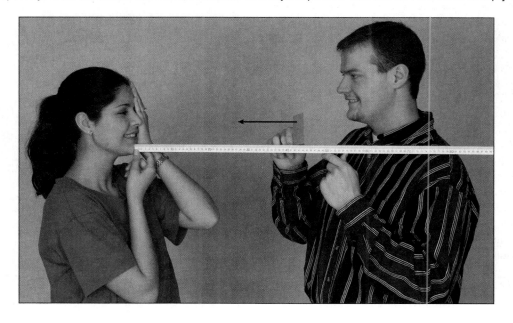

FIGURE 36.4 Samples of Ishihara's color plates. These plates are reproduced from *Ishihara's Tests for Colour Blindness* published by KANEHARA & CO., LTD., Tokyo, Japan, but tests for color blindness cannot be conducted with this material. For accurate testing, the original plates should be used.

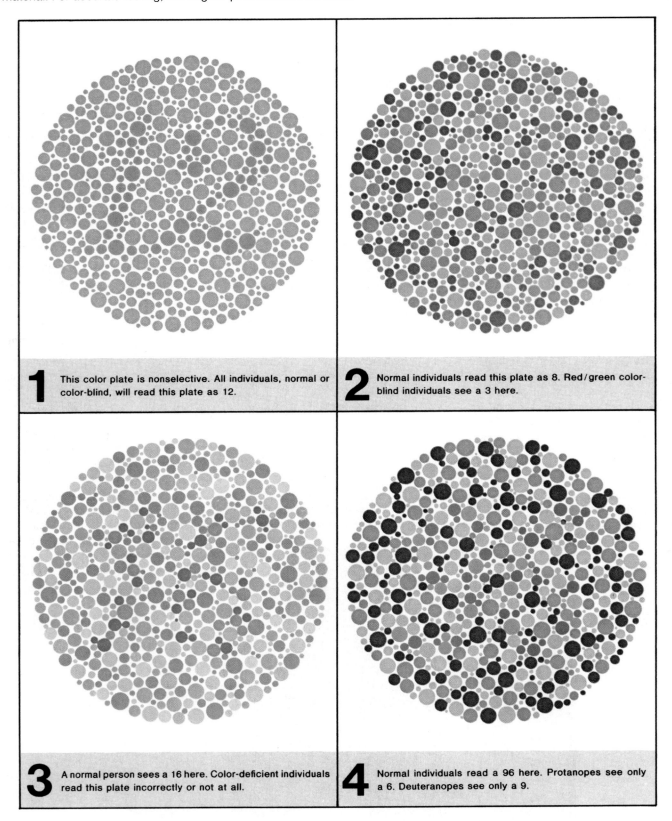

1 This color plate is nonselective. All individuals, normal or color-blind, will read this plate as 12.

2 Normal individuals read this plate as 8. Red/green color-blind individuals see a 3 here.

3 A normal person sees a 16 here. Color-deficient individuals read this plate incorrectly or not at all.

4 Normal individuals read a 96 here. Protanopes see only a 6. Deuteranopes see only a 9.

Procedure B—Visual Demonstrations

Perform the following demonstrations with the help of your laboratory partner.

1. *Blind-spot demonstration.* There are no photoreceptors in the optic disc, located where the nerve fibers of the retina leave the eye and enter the optic nerve. Consequently, this region of the retina is commonly called the *blind spot.*

   To demonstrate the blind spot, follow these steps:

   a. Close your left eye, hold figure 36.5 about 34 cm away from your face, and stare at the + sign in the figure with your right eye.

   b. Move the figure closer to your face as you continue to stare at the + until the dot on the figure suddenly disappears. This happens when the image of the dot is focused on the optic disc. Measure the right eye distance using a metric ruler or a meterstick.

   c. Repeat the procedures with your right eye closed. This time stare at the dot with your left eye, and the + will disappear when the image falls on the optic disc. Measure the distance.

   d. Record the results in Part B of the laboratory report.

🧠 Critical Thinking Activity

Under normal visual circumstances, explain why small objects are not lost from our vision.

_____

_____

_____

_____

---

2. *Photopupillary reflex.* The smooth muscles of the iris control the size of the pupil. For example, when the intensity of light entering the eye increases, a photopupillary reflex is triggered, and the circular muscles of

FIGURE 36.5 Blind-spot demonstration.

the iris are stimulated to contract. As a result, the size of the pupil decreases, and less light enters the eye.

To demonstrate this photopupillary reflex, follow these steps:

   a. Ask your partner to sit with his or her hands thoroughly covering his or her eyes for 2 minutes.

   b. Position a pen flashlight close to one eye with the light shining on the hand that covers the eye.

   c. Ask your partner to remove the hand quickly.

   d. Observe the pupil and note any change in its size.

   e. Have your partner remove the other hand, but keep that uncovered eye shielded from extra light.

   f. Observe both pupils and note any difference in their sizes.

3. *Accommodation pupillary reflex.* The pupil constricts as a normal accommodation reflex response to focusing on close objects. To demonstrate the accommodation pupillary reflex, follow these steps:

   a. Have your partner stare for several seconds at some dimly illuminated object in the room that is more than 20 feet away.

   b. Observe the size of the pupil of one eye. Then hold a pencil about 25 cm in front of your partner's face, and have your partner stare at it.

   c. Note any change in the size of the pupil.

4. *Convergence reflex.* The eyes converge as a normal convergence response to focusing on close objects. To demonstrate the convergence reflex, follow these steps:

   a. Repeat the procedure outlined for the accommodation pupillary reflex.

   b. Note any change in the position of the eyeballs as your partner changes focus from the distant object to the pencil.

5. Complete Part B of the laboratory report.

Name _____

Date _____

Section _____

The △ corresponds to the Learning Outcome(s) listed at the beginning of the laboratory exercise.

Visual Tests and Demonstrations

Part A Assessments

1. Visual acuity test results: △1

Eye Tested	Acuity Values
Right eye	
Right eye with glasses (if applicable)	
Left eye	
Left eye with glasses (if applicable)	

2. Astigmatism test results: △1

Eye Tested	Darker Lines
Right eye	
Right eye with glasses (if applicable)	
Left eye	
Left eye with glasses (if applicable)	

3. Accommodation test results: △1

Eye Tested	Near Point (cm)
Right eye	
Right eye with glasses (if applicable)	
Left eye	
Left eye with glasses (if applicable)	

4. Color vision test results: 🔺1

Condition	Males			Females		
	Class Number	Class Percentage	Expected Percentage	Class Number	Class Percentage	Expected Percentage
Normal color vision			93			99.6
Deficient red-green color vision			7			0.4
Protanopia (lack red-sensitive cones)			less-frequent type			less-frequent type
Deuteranopia (lack green-sensitive cones)			more-frequent type			more-frequent type

5. Complete the following:

    a. What is meant by 20/70 vision? 🔺2 _____

_____

    b. What is meant by 20/10 vision? 🔺2 _____

_____

    c. What visual problem is created by astigmatism? 🔺2 _____

_____

    d. Why does the near point of accommodation often increase with age? 🔺2 _____

_____

    e. Describe the eye defect that causes color-deficient vision. 🔺2 _____

_____

Part B Assessments

1. Blind-spot results: 🔺3

    a. Right eye distance _____

    b. Left eye distance _____

Complete the following:

2. Explain why an eye has a blind spot. 🔺3 _____

_____

3. Describe the photopupillary reflex. 🔺3 _____

_____

4. What difference did you note in the size of the pupils when one eye was exposed to bright light and the other eye was shielded from the light? /3\ _____

_____

5. Describe the accommodation pupillary reflex. /3\ _____

_____

6. Describe the convergence reflex. /3\ _____

_____

Ear and Hearing

Pre-Lab

1. Carefully read the introductory material and examine the entire lab content.
2. Be familiar with the basic structures and functions (from lecture or the textbook) of the parts of the ear associated with hearing.
3. Visit www.mhhe.com/martinseriesl for pre-lab questions and Anatomy & Physiology Revealed animations.

Materials Needed

Dissectible ear model
Watch that ticks
Tuning fork (128 or 256 cps)
Rubber hammer
Cotton
Meterstick

For Demonstration Activities:
Compound light microscope
Prepared microscope slide of cochlea (section)
Audiometer

The ear is composed of outer (external), middle, and inner (internal) parts. The large external auricle gathers sound waves (vibrations) of air and directs them into the external acoustic meatus to the tympanic membrane. The middle ear auditory ossicles conduct and amplify the waves from the tympanic membrane to the oval window. To allow the same pressure in the outer and middle ear, the pharyngo-tympanic (auditory) tube connects the middle ear to the pharynx. The sound waves are further conducted through the fluids of the cochlea in the inner ear. Finally, some hair cells of the spiral organ are stimulated. As certain hair cells (receptor cells) are stimulated, these receptors initiate nerve impulses to pass over the cochlear branch of the vestibulo-cochlear nerve into the brain. When nerve impulses arrive at the auditory cortex of the temporal lobe of the cerebrum, the impulses are interpreted and the sensations of hearing are created.

The human hearing range is from approximately 20 to 20,000 waves per second, known as hertz (Hz). We perceive these different frequencies as pitch. Different frequencies stimulate particular hair cells in specific regions of the spiral organ, allowing us to interpret different pitches. An additional quality of sound is loudness. The intensity of the sound is measured in units called decibels (dB). Permanent damage can happen to the hair cells of the spiral organ if frequent and prolonged exposures occur over 90 decibels.

Although the ear is considered here as a hearing organ, it also has important functions for equilibrium (balance). The inner ear semicircular canals, utricle, and saccule have equilibrium functions. Equilibrium will be studied in Laboratory Exercise 38.

Purpose of the Exercise

To review the structural and functional characteristics of the ear and to conduct some ordinary hearing tests.

Learning Outcomes

After completing this exercise, you should be able to

1. Identify the major structures of the ear.
2. Trace the pathway of sound vibrations from the tympanic membrane to the hearing receptors.
3. Describe the functions of the structures of the ear.
4. Conduct four ordinary hearing tests and summarize the results.

Procedure A—Structure and Function of the Ear

1. Study figures 37.1, 37.2, and 37.3.

FIGURE 37.1 Major structures of the ear. The three divisions of the ear are roughly indicated by the dashed lines.

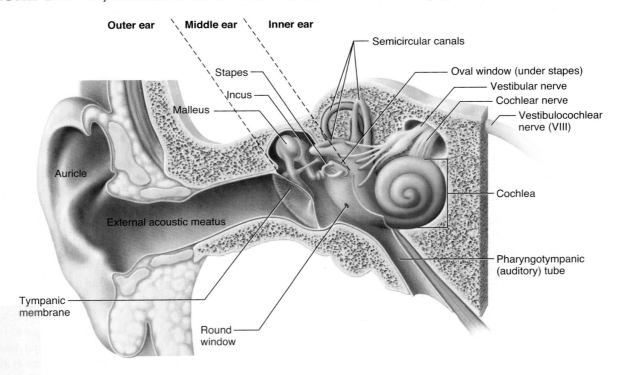

Outer ear   Middle ear   Inner ear

Semicircular canals
Stapes
Incus
Malleus
Oval window (under stapes)
Vestibular nerve
Cochlear nerve
Vestibulocochlear nerve (VIII)
Auricle
Cochlea
External acoustic meatus
Pharyngotympanic (auditory) tube
Tympanic membrane
Round window

FIGURE 37.2 Structures of the middle and inner ear.

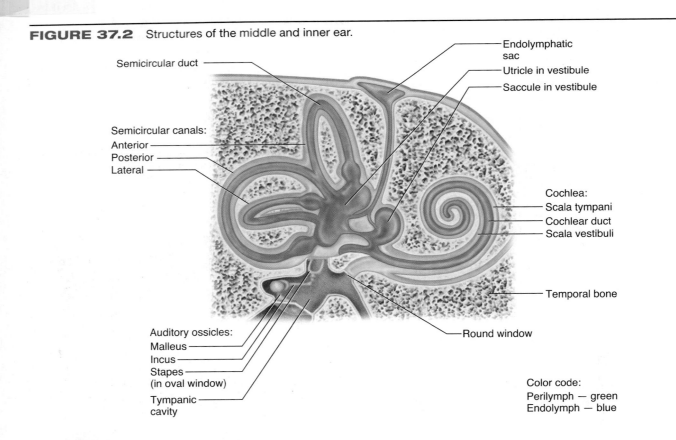

Semicircular duct
Endolymphatic sac
Utricle in vestibule
Saccule in vestibule
Semicircular canals:
Anterior
Posterior
Lateral
Cochlea:
Scala tympani
Cochlear duct
Scala vestibuli
Temporal bone
Auditory ossicles:
Malleus
Incus
Stapes (in oval window)
Round window
Tympanic cavity
Color code:
Perilymph — green
Endolymph — blue

FIGURE 37.3 Anatomy of the cochlea: (*a*) section through cochlea; (*b*) view of one turn of cochlea; (*c*) view of spiral organ (organ of Corti).

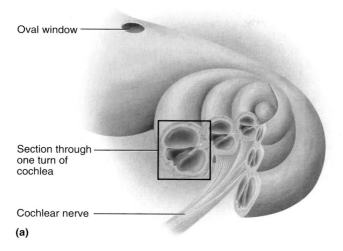

Oval window

Section through one turn of cochlea

Cochlear nerve

(a)

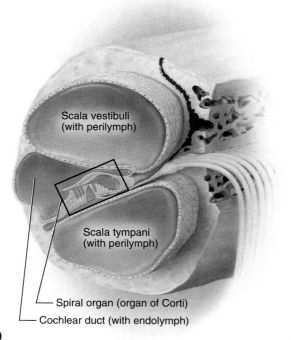

Scala vestibuli (with perilymph)

Scala tympani (with perilymph)

Spiral organ (organ of Corti)

Cochlear duct (with endolymph)

(b)

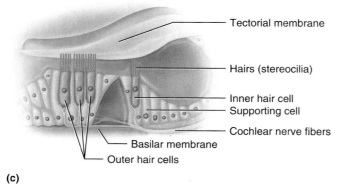

Tectorial membrane

Hairs (stereocilia)

Inner hair cell

Supporting cell

Cochlear nerve fibers

Basilar membrane

Outer hair cells

(c)

2. Examine the dissectible model of the ear and locate the following features:

outer (external) ear
- auricle (pinna)
- external acoustic meatus
- tympanic membrane (eardrum)

middle ear
- tympanic cavity
- auditory ossicles (fig. 37.4)
  - malleus
  - incus
  - stapes
- oval window
- round window
- tensor tympani
- stapedius
- pharyngotympanic tube (auditory tube; eustachian tube)

inner (internal) ear
- bony labyrinth
- membranous labyrinth
- cochlea
- semicircular canals and ducts
- vestibule
  - utricle
  - saccule

vestibulocochlear nerve
- vestibular nerve (balance branch)
- cochlear nerve (hearing branch)

3. Complete Parts A, B, and C of Laboratory Report 37.

FIGURE 37.4 Middle ear bones (auditory ossicles) superimposed on a penny to show their relative size. The arrangement of the malleus, incus, and stapes in the enlarged photograph resembles the articulations in the middle ear.

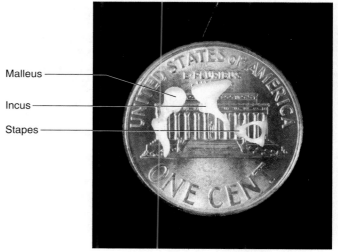

Malleus

Incus

Stapes

Demonstration Activity

Observe the section of the cochlea in the demonstration microscope. Locate one of the turns of the cochlea, and using figures 37.3 and 37.5 as a guide, identify the *scala vestibuli, cochlear duct (scala media), scala tympani, vestibular membrane, basilar membrane,* and the *spiral organ (organ of Corti).*

FIGURE 37.5 A section through the cochlea (22×).

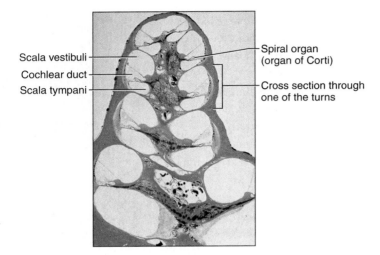

Procedure B—Hearing Tests

Perform the following tests in a quiet room, using your laboratory partner as the test subject.

1. *Auditory acuity test.* To conduct this test, follow these steps:
   a. Have the test subject sit with eyes closed.
   b. Pack one of the subject's ears with cotton.
   c. Hold a ticking watch close to the open ear, and slowly move it straight out and away from the ear.
   d. Have the subject indicate when the sound of the ticking can no longer be heard.
   e. Use a meterstick to measure the distance in centimeters from the ear to the position of the watch.
   f. Repeat this procedure to test the acuity of the other ear.
   g. Record the test results in Part D of the laboratory report.
2. *Sound localization test.* To conduct this test, follow these steps:
   a. Have the subject sit with eyes closed.
   b. Hold the ticking watch somewhere within the audible range of the subject's ears, and ask the subject to point to the watch.
   c. Move the watch to another position and repeat the request. In this manner, determine how accurately

the subject can locate the watch when it is in each of the following positions: in front of the head, behind the head, above the head, on the right side of the head, and on the left side of the head.
   d. Record the test results in Part D of the laboratory report.
3. *Rinne test.* This test is done to assess possible conduction deafness by comparing bone and air conduction. To conduct this test, follow these steps:
   a. Obtain a tuning fork and strike it with a rubber hammer, or on the heel of your hand, causing it to vibrate.
   b. Place the end of the fork's handle against the subject's mastoid process behind one ear. Have the prongs of the fork pointed downward and away from the ear, and be sure nothing is touching them (fig. 37.6a). The sound sensation is that of bone conduction. If no sound is experienced, nerve deafness exists.
   c. Ask the subject to indicate when the sound is no longer heard.
   d. Then quickly remove the fork from the mastoid process and position it in the air close to the opening of the nearby external acoustic meatus (fig. 37.6b).

   If hearing is normal, the sound (from air conduction) will be heard again; if there is conductive impairment, the sound will not be heard. Conductive impairment involves outer or middle ear defects. Hearing aids can improve hearing for conductive deafness because bone conduction transmits the sound into the inner ear. Surgery could possibly correct this type of defect.
   e. Record the test results in Part D of the laboratory report.
4. *Weber test.* This test is used to distinguish possible conduction or sensory deafness. To conduct this test, follow these steps:
   a. Strike the tuning fork with the rubber hammer.
   b. Place the handle of the fork against the subject's forehead in the midline (fig. 37.7).
   c. Ask the subject to indicate if the sound is louder in one ear than in the other or if it is equally loud in both ears.

   If hearing is normal, the sound will be equally loud in both ears. If there is conductive impairment, the sound will appear louder in the affected ear. If some degree of sensory (nerve) deafness exists, the sound will be louder in the normal ear. The impairment involves the spiral organ or the cochlear nerve. Hearing aids will not improve sensory deafness.
   d. Have the subject experience the effects of conductive impairment by packing one ear with cotton and repeating the Weber test. Usually the sound appears louder in the plugged (or impaired) ear because extraneous sounds from the room are blocked out.
   e. Record the test results in Part D of the laboratory report.
5. Complete Part D of the laboratory report.

FIGURE 37.6 Rinne test: (*a*) first placement of vibrating tuning fork until sound is no longer heard; (*b*) second placement of tuning fork to assess air conduction.

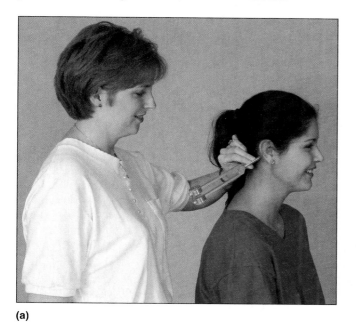

(a)

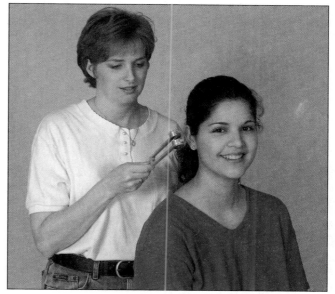

(b)

FIGURE 37.7 Weber test.

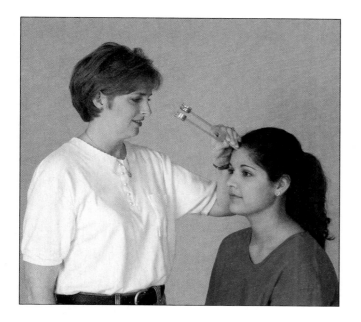

Critical Thinking Activity

Ear structures from the outer ear into the inner ear are progressively smaller. Using results obtained from the hearing tests, explain this advantage.

_____

_____

_____

_____

Demonstration Activity

Ask the laboratory instructor to demonstrate the use of the audiometer. This instrument produces sound vibrations of known frequencies transmitted to one or both ears of a test subject through earphones. The audiometer can be used to determine the threshold of hearing for different sound frequencies, and, in the case of hearing impairment, it can be used to determine the percentage of hearing loss for each frequency.

NOTES

Name _____

Date _____

Section _____

The ⚠ corresponds to the Learning Outcome(s) listed at the beginning of the laboratory exercise.

Ear and Hearing

Part A Assessments

1. Label the structures indicated on the removable part of an ear model in figure 37.8.

FIGURE 37.8 Identify the features indicated on this removable part of an ear model, using the terms provided. ⚠

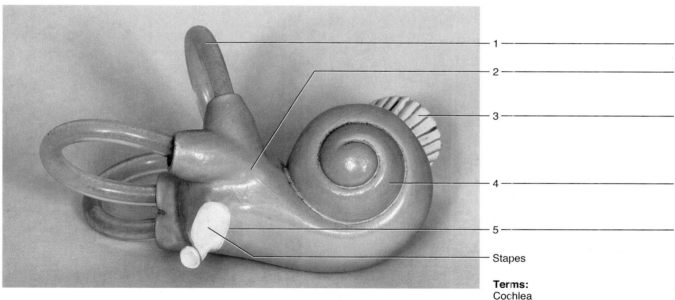

1 — _____

2 — _____

3 — _____

4 — _____

5 — _____

— Stapes

Terms:
Cochlea
Oval window
Semicircular canal
Vestibule
Vestibulocochlear nerve

2. Label the structures indicated in the micrograph of the spiral organ in figure 37.9.

FIGURE 37.9 Label the structures associated with this spiral organ region of a cochlea, using the terms provided. (75× micrograph enlarged to 300×). **1**

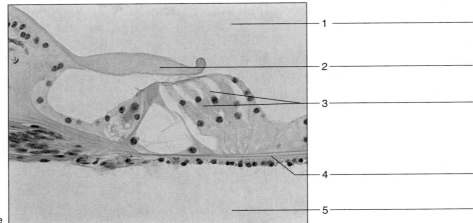

Terms:
Basilar membrane
Cochlear duct
Hair cells
Scala tympani
Tectorial membrane

Part B Assessments

Number the following structures (1–9) to indicate their respective positions in relation to the pathway of the sound vibrations. Assign number 1 to the most superficial portion of the outer ear. **2**

_____ Auricle (air vibrations within)

_____ External acoustic meatus (air vibrations within)

_____ Fluids in cochlear canals

_____ Hair cells of spiral organ

_____ Incus

_____ Malleus

_____ Oval window

_____ Stapes

_____ Tympanic membrane

Part C Assessments

Match the terms in column A with the descriptions in column B. Place the letter of your choice in the space provided. 1 3

Column A	Column B
a. Bony labyrinth	_____ **1.** Auditory ossicle attached to tympanic membrane
b. Ceruminous gland	_____ **2.** Air-filled space containing auditory ossicles
c. External acoustic meatus	_____ **3.** Contacts hairs of hearing receptors
d. Malleus	_____ **4.** Leads from oval window to apex of cochlea
e. Membranous labyrinth	_____ **5.** S-shaped tube leading to tympanic membrane
f. Pharyngotympanic (auditory) tube	_____ **6.** Wax-secreting structure
g. Scala tympani	_____ **7.** Cone-shaped, semitransparent membrane attached to malleus
h. Scala vestibuli	_____ **8.** Auditory ossicle attached to oval window
i. Stapes	_____ **9.** Contains endolymph
j. Tectorial membrane	_____ **10.** Bony canal of inner ear in temporal bone
k. Tympanic cavity	_____ **11.** Connects middle ear and pharynx
l. Tympanic membrane (eardrum)	_____ **12.** Extends from apex of cochlea to round window

Part D Assessments

1. Results of auditory acuity test: 4

Ear Tested	Audible Distance (cm)
Right	
Left	

2. Results of sound localization test: 4

Actual Location	Reported Location
Front of the head	
Behind the head	
Above the head	
Right side of the head	
Left side of the head	

3. Results of experiments using tuning forks: 4

Test	Left Ear (Normal or Impaired)	Right Ear (Normal or Impaired)
Rinne		
Weber		

4. Summarize the results of the hearing tests you conducted on your laboratory partner. ◢

_____

_____

_____

_____

_____

_____

_____

Equilibrium

Pre-Lab

1. Carefully read the introductory material and examine the entire lab content.
2. Be familiar with the basic structures and functions (from lecture or the textbook) of the parts of the ear associated with equilibrium.
3. Visit www.mhhe.com/martinseriesl for pre-lab questions.

Materials Needed

Swivel chair
Bright light

For Demonstration Activity:
Compound light microscope
Prepared microscope slide of semicircular duct (cross section through ampulla)

Safety

▶ Do not pick subjects who have frequent motion sickness.
▶ Have four people surround the subject in the swivel chair in case the person falls from vertigo or loss of balance.
▶ Stop your experiment if the subject becomes nauseated.

The sense of equilibrium involves two sets of sensory organs. One set helps to maintain the stability of the head and body when they are motionless or during linear acceleration and produces a sense of static (gravitational) equilibrium. The other set is concerned with balancing the head and body when angular acceleration produces a sense of dynamic (rotational) equilibrium.

The sense organs associated with the sense of static equilibrium are located within the vestibules of the inner ears. Two chambers, the utricle and saccule, contain receptors called maculae. Each macula is composed of hair cells embedded within a gelatinous otolithic membrane. Embedded within the otolithic membrane are numerous tiny calcium carbonate "ear stones" called otoliths. As we change our head position, gravitational forces allow the shift of the otoliths to bend the stereocilia of the hair cells of the macula, resulting in stimulation of sensory vestibular neurons. Stimulation of the maculae also occurs during linear acceleration. Horizontal acceleration (as occurs when riding in a car) involves the utricle; vertical acceleration (as occurs when riding in an elevator) involves the saccule. As a result of static equilibrium, recognition of movements such as falling are detected and postural adjustments are accomplished.

The sense organs associated with the sense of dynamic equilibrium are located within the ampullae of the three semicircular ducts of the inner ear. Each membranous semicircular duct is located within a semicircular canal of the temporal bone. A small elevation within each ampulla possesses the crista ampullaris. Each crista ampullaris contains hair cells with stereocilia embedded within a gelatinous cap called the cupula. During rotational movements, the cupula is bent, which stimulates hair cells and sensory neurons of the vestibular nerve. Because the three semicircular ducts are in different planes, rotational movements in any direction result in stimulation of the associated hair cells. Impulses from vestibular neurons of the semicircular ducts result in reflex movements of the eye. During rotational movements, characteristic twitching movements of the eyes called nystagmus occur, often accompanied by dizziness (vertigo).

Impulses from inner ear receptors travel over the vestibular neurons of the vestibulocochlear nerve and include destinations in the brainstem and cerebellum. The sense of equilibrium works subconsciously to initiate appropriate corrections to body position and movements. Additional senses work in conjunction with the inner ear and equilibrium. Stretch receptors in muscles and tendons, touch, and vision complement the inner ear for equilibrium and proper body adjustments.

Purpose of the Exercise

To review the structure and function of the organs of equilibrium and to conduct some tests of equilibrium.

FIGURE 38.1 Structures of static equilibrium; (*a*) utricle and saccule each contain a macula; (*b*) macula and otolithic membrane orientation when body is upright; (*c*) macula and otolithic membrane changes when the body is tilted.

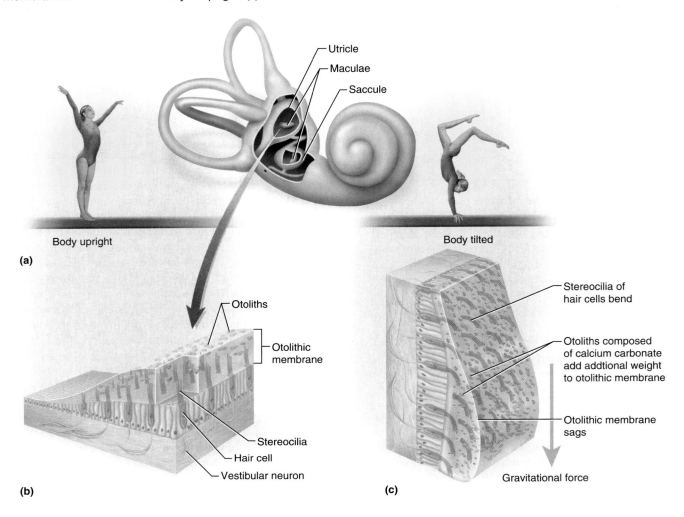

(a) Body upright

Utricle
Maculae
Saccule

(b)

Otoliths
Otolithic membrane
Stereocilia
Hair cell
Vestibular neuron

(c) Body tilted

Stereocilia of hair cells bend
Otoliths composed of calcium carbonate add addtional weight to otolithic membrane
Otolithic membrane sags
Gravitational force

Learning Outcomes

After completing this exercise, you should be able to

1. Locate the organs of static and dynamic equilibrium and describe their functions.
2. Explain the role of vision in the maintenance of equilibrium.
3. Conduct and record the results of the Romberg and Bárány tests of equilibrium.

Procedure A—Organs of Equilibrium

1. Study figures 38.1 and 38.2, which present the structures and functions of static equilibrium.
2. Study figures 38.3 and 38.4, which present the structures and functions of dynamic equilibrium.
3. Complete Part A of Laboratory Report 38.

FIGURE 38.2 Particularly large otoliths are found in a freshwater drum (*Aplodinotus grunniens*). They have been used for lucky charms and jewelry. Human otoliths are microscopic size.

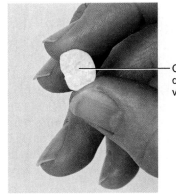

Otolith from a fish (freshwater drum), which helps maintain vertical orientation

FIGURE 38.3 Structures of dynamic equilibrium; (*a*) each semicircular duct has an ampulla containing a crista ampullaris; (*b*) crista ampullaris when the body is stationary; (*c*) changes in crista ampullaris during body rotation.

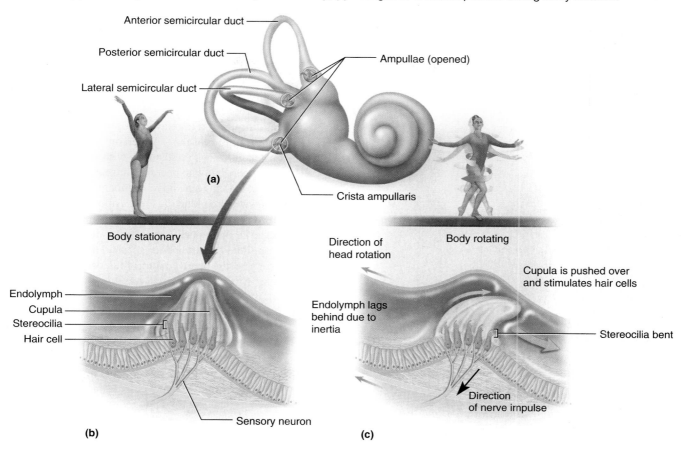

Anterior semicircular duct

Posterior semicircular duct

Lateral semicircular duct

Ampullae (opened)

Crista ampullaris

(a)

Body stationary

Body rotating

Direction of head rotation

Endolymph

Cupula

Stereocilia

Hair cell

Endolymph lags behind due to inertia

Cupula is pushed over and stimulates hair cells

Stereocilia bent

Direction of nerve impulse

Sensory neuron

(b)

(c)

FIGURE 38.4 Semicircular canal superimposed on a penny to show its relative size.

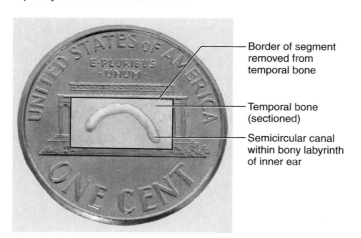

Border of segment removed from temporal bone

Temporal bone (sectioned)

Semicircular canal within bony labyrinth of inner ear

 Demonstration Activity

Observe the cross section of the semicircular duct through the ampulla in the demonstration microscope. Note the crista ampullaris (fig. 38.5) projecting into the lumen of the membranous labyrinth, which in a living person is filled with endolymph. The space between the membranous and bony labyrinths is filled with perilymph.

Procedure B—Tests of Equilibrium

Perform the following tests, using a person as a test subject who is not easily disturbed by dizziness or rotational movement. Also have some other students standing close by to help prevent the test subject from falling during the tests.

FIGURE 38.5 A micrograph of a crista ampullaris (1,400×). The crista ampullaris is located within the ampulla of each semicircular duct.

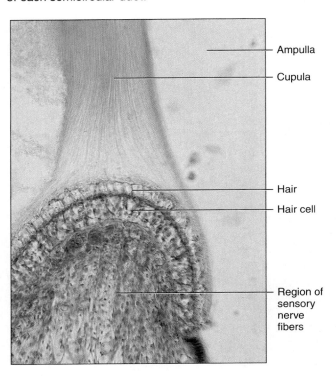

- Ampulla
- Cupula
- Hair
- Hair cell
- Region of sensory nerve fibers

The tests should be stopped immediately if the test subject begins to feel uncomfortable or nauseated.

1. *Vision and equilibrium.* To demonstrate the importance of vision in the maintenance of equilibrium, follow these steps:
   a. Have the test subject stand erect on one foot for 1 minute with his or her eyes open.
   b. Observe the subject's degree of unsteadiness.
   c. Repeat the procedure with the subject's eyes closed. *Be prepared to prevent the subject from falling.*
2. *Romberg test.* To conduct this test, follow these steps:
   a. Position the test subject close to a chalkboard with the back toward the board.
   b. Place a bright light in front of the subject so that a shadow of the body is cast on the board.
   c. Have the subject stand erect with feet close together and eyes staring straight ahead for 3 minutes.
   d. During the test, make marks on the chalkboard along the edge of the shadow of the subject's shoulders to indicate the range of side-to-side swaying.

e. Measure the maximum sway in centimeters and record the results in Part B of the laboratory report.
   f. Repeat the procedure with the subject's eyes closed.
   g. Position the subject so one side is toward the chalkboard.
   h. Repeat the procedure with the eyes open.
   i. Repeat the procedure with the eyes closed.

   The Romberg test is used to evaluate a person's ability to integrate sensory information from proprioceptors and receptors within the organs of static equilibrium and to relay appropriate motor impulses to postural muscles. A person who shows little unsteadiness when standing with feet together and eyes open, but who becomes unsteady when the eyes are closed, has a positive Romberg test.

3. *Bárány test.* To conduct this test, follow these steps:
   a. Have the test subject sit on a swivel chair with his or her eyes open and focused on a distant object, the head tilted forward about 30°, and the hands gripped firmly to the seat. Position four people around the chair for safety. *Be prepared to prevent the subject and the chair from tipping over.*
   b. Rotate the chair ten rotations within 20 seconds.
   c. Abruptly stop the movement of the chair. The subject will still have the sensation of continuous movement and might experience some dizziness (vertigo).
   d. Have the subject look forward and immediately note the nature of the eye movements and their direction. (Such reflex eye twitching movements are called *nystagmus.*) Also note the time it takes for the nystagmus to cease. Nystagmus will continue until the cupula is returned to an original position.
   e. Record your observations in Part B of the laboratory report.
   f. Allow the subject several minutes of rest, then repeat the procedure with the subject's head tilted nearly 90° onto one shoulder.
   g. After another rest period, repeat the procedure with the subject's head bent forward so that the chin is resting on the chest.

   In this test, when the head is tilted about 30°, the lateral semicircular canals receive maximal stimulation, and the nystagmus is normally from side to side. When the head is tilted at 90°, the superior canals are stimulated, and the nystagmus is up and down. When the head is bent forward with the chin on the chest, the posterior canals are stimulated, and the nystagmus is rotary.

4. Complete Part B of the laboratory report.

Name _____

Date _____

Section _____

The ⓐ corresponds to the Learning Outcome(s) listed at the beginning of the laboratory exercise.

Equilibrium

Part A Assessments

Complete the following statements:

1. The organs of static equilibrium are located within two expanded chambers called the _____ and the saccule. ⓐ

2. All of the balance organs are found within the _____ bone of the skull. ⓐ

3. Otoliths are small grains composed of _____. ⓐ

4. Sensory impulses travel from the organs of equilibrium to the brain on vestibular neurons of the _____ nerve. ⓐ

5. The sensory organ of a semicircular duct lies within a swelling called the _____. ⓐ

6. The sensory organ within the ampulla of a semicircular duct is called a _____. ⓐ

7. The _____ of this sensory organ consists of a dome-shaped gelatinous cap. ⓐ

8. Parts of the brainstem and the _____ of the brain interpret impulses from the equilibrium receptors. ⓐ

Part B—Tests of Equilibrium Assessments

1. Vision and equilibrium test results:

   a. When the eyes are open, what sensory organs provide information needed to maintain equilibrium? ②

   _____

   _____

   b. When the eyes are closed, what sensory organs provide such information? ②

   _____

   _____

2. Romberg test results:

   a. Record the test results in the following table: **3**

Conditions	Maximal Movement (cm)
Back toward board, eyes open	
Back toward board, eyes closed	
Side toward board, eyes open	
Side toward board, eyes closed	

   b. Did the test subject's unsteadiness increase when the eyes were closed? _____ What is the significance of

      this observation? **2** _____

      _____

      _____

   c. Why would you expect a person with impairment of the organs of equilibrium to become more unsteady when the

      eyes are closed? **2** _____

      _____

      _____

3. Bárány test results:

   a. Record the test results in the following table: **3**

Position of Head	Description of Eye Movements	Time for Movement to Cease
Tilted 30° forward		
Tilted 90° onto shoulder		
Tilted forward, chin on chest		

   b. Summarize the results of this test. **3**

      _____

      _____

      _____

Critical Thinking Activity

What additional sensory information would you expect a person with impairment of organs of equilibrium to use to supplement their relative lack of some sensory information?

_____

_____

_____

Endocrine Structure and Function

Pre-Lab

1. Carefully read the introductory material and examine the entire lab.
2. Be familiar with basic structures and functions (from lecture or the textbook) of the pituitary, thyroid, parathyroid, adrenal, and pancreas glands.
3. Visit www.mhhe.com/martinseriesl for pre-lab questions and Anatomy & Physiology Revealed animations.

Materials Needed

Human torso model
Compound light microscope
Prepared microscope slides of the following:
    Pituitary gland
    Thyroid gland
    Parathyroid gland
    Adrenal gland
    Pancreas
Ph.I.L.S. 3.0

For Learning Extension Activity:
Water bath equipped with temperature control mechanism set at 37°C (98.6°F)
Laboratory thermometer

The endocrine system consists of ductless glands that act together with parts of the nervous system to help control body activities. The endocrine glands secrete regulatory molecules, called hormones, into the internal environment of interstitial fluid. Hormones are then transported in the blood and influence target cells, usually some distance from the original endocrine gland. At target cells, hormones bind to receptors, resulting in the activation of processes leading to functional changes within the target cells. In this way, hormones influence the rate of metabolic reactions, the transport of substances through cell membranes, the regulation of water and electrolyte balances, and many other functions.

Although some of the glands included in this laboratory exercise have ducts, the endocrine portion of the organ is ductless, and blood still transports the hormones. The pancreas, ovaries, and testes have ducts that function within additional systems other than the endocrine system. Additional secreted substances that are regulatory, although not considered true hormones, include paracrine and autocrine secretions. Paracrine secretions influence only neighboring cells; autocrine secretions influence only the secreting cell itself.

The nervous system and the endocrine system are both regulatory and involve chemical secretions and receptor sites. The neurons of the nervous system secrete neurotransmitters. Receptor sites for neurotransmitters are on the postsynaptic cell, resulting in rapid responses of a brief duration unless additional stimulation occurs. In contrast, the endocrine cells secrete hormones transported by the blood to receptor sites on target cells. The responses from the hormones last longer and might even continue for hours to days even without additional hormonal secretions. By controlling cellular processes, endocrine glands play important roles in the maintenance of homeostasis.

You walk outside on a cold winter day. As your body temperature begins to drop, specific physiological processes will be activated to help maintain your body temperature. Initially, increased amounts of thyroid hormone will be released from the thyroid gland (Temperature receptors in your skin send neural signals to the hypothalamus to trigger the release of thyroid releasing hormone [TRH]; TRH stimulates the anterior pituitary to release thyroid stimulating hormone [TSH]; and TSH stimulates the thyroid gland to release the thyroid hormones, T_3 and T_4). Thyroid hormone is able to help maintain body temperature by increasing the metabolic rate of almost all cells, especially neurons. Neurons are specifically stimulated to increase their number of Na^+/K^+ pumps. Since there are more Na^+/K^+ pumps, more energy is converted (chemical energy of ATP is converted to mechanical energy of pumping Na^+ and K^+), more heat is produced (second law of thermodynamics) and body temperature is maintained. If your body temperature continues to decrease, the hypothalamus will initiate shivering, involuntary contractions of skeletal muscle. Shivering increases body temperature as the muscles contract and generate heat.

However, what occurs when you walk outside on a cold winter day if your thyroid gland is impaired and insufficient thyroid hormone is released? Will your body temperature drop or will you begin to shiver sooner to maintain body temperature?

Purpose of the Exercise

To review the structure and function of major endocrine glands and to examine microscopically the tissues of these glands. To measure and compare the effect of cooling on metabolic rate in animals that differ in their ability to release thyroid hormone.

Learning Outcomes

After completing this exercise, you should be able to

1. Name and locate the major endocrine glands.
2. Sketch and label tissue sections from the pituitary gland, thyroid gland, parathyroid glands, adrenal glands, and pancreas.
3. Associate the principal hormones secreted by each of the major glands.
4. Explain the role of the hypothalamus in controlling the release of thyroid hormone.
5. Explain the relationship of thyroid hormone levels, metabolic rate, body temperature, and oxygen consumption.
6. Interpret changes in oxygen consumption that occur with decreases in environmental temperature.
7. Apply the concepts of thyroid hormone levels, and its affects on metabolic rate, oxygen consumption, and temperature regulation to the human body.

Procedure A—Endocrine Gland Histology

1. Locate the major endocrine glands indicated in figure 39.1.
2. Examine the human torso model and locate the following:

   hypothalamus
   pituitary stalk (infundibulum)
   pituitary gland (hypophysis)
     anterior lobe (adenohypophysis)
     posterior lobe (neurohypophysis)
   thyroid gland
   parathyroid glands
   adrenal glands (suprarenal glands)
     adrenal medulla
     adrenal cortex
   pancreas
   pineal gland
   thymus
   ovaries
   testes

Learning Extension Activity

The secretions of endocrine glands are usually controlled by negative feedback systems. As a result, the concentrations of hormones in body fluids remain relatively stable, although they will fluctuate slightly within a normal range.

Similarly, the mechanism used to maintain the temperature of a laboratory water bath involves negative feedback. In this case, a temperature-sensitive thermostat in the water allows a water heater to operate whenever the water temperature drops below the thermostat's set point. Then, when the water temperature reaches the set point, the thermostat causes the water heater to turn off (a negative effect), and the water bath begins to cool again.

Use a laboratory thermometer to monitor the temperature of the water bath in the laboratory. Measure the temperature at regular intervals until you have recorded ten readings. What was the lowest temperature you recorded?_____ The highest temperature? _____ What was the average temperature of the water bath? _____
How is the water bath temperature control mechanism similar to a hormonal control mechanism in the body?

_____

_____

_____

_____

_____

3. Examine the microscopic tissue sections of the following glands, and identify the features described:

   Pituitary gland. To examine the pituitary tissue, follow these steps:

   a. Observe the tissues using low-power magnification (fig. 39.2).
   b. Locate the *pituitary stalk,* the *anterior lobe* (the largest part of the gland), and the *posterior lobe.* The anterior lobe makes hormones and appears glandular; the posterior lobe stores and releases hormones made by the hypothalamus and is nervous in appearance.
   c. Observe an area of the anterior lobe with high-power magnification (fig. 39.3). Locate a cluster of relatively large cells and identify some *acidophil cells,* which contain pink-stained granules, and some *basophil cells,* which contain blue-stained granules. These acidophil and basophil cells are the hormone-secreting cells. The hormones secreted include TSH, ACTH, prolactin, growth hormone, FSH, and LH.

FIGURE 39.1 Label the major endocrine glands by placing the correct numbers in the spaces provided. 🔺

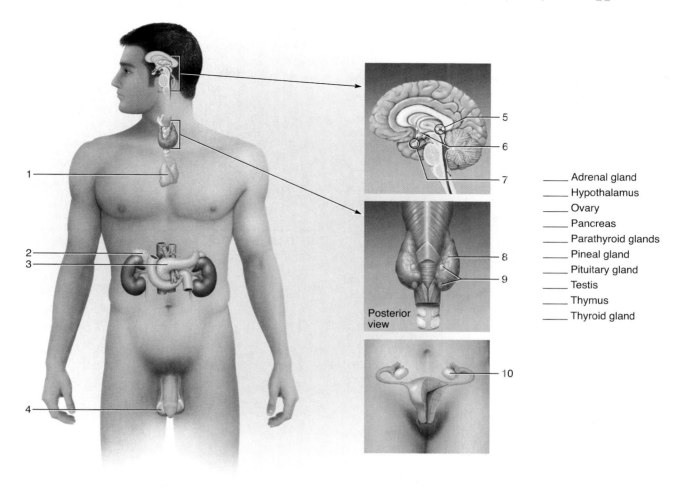

Posterior view

_____ Adrenal gland
_____ Hypothalamus
_____ Ovary
_____ Pancreas
_____ Parathyroid glands
_____ Pineal gland
_____ Pituitary gland
_____ Testis
_____ Thymus
_____ Thyroid gland

FIGURE 39.2 Micrograph of the pituitary gland (6×).

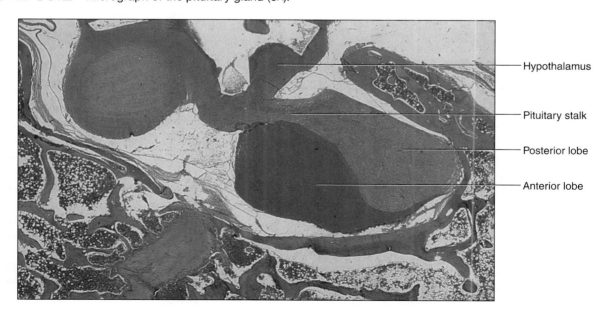

Hypothalamus

Pituitary stalk

Posterior lobe

Anterior lobe

d. Observe an area of the posterior lobe with high-power magnification (fig. 39.4). Note the numerous unmyelinated nerve fibers present in this lobe. Also locate some *pituicytes,* a type of glial cell, scattered among the nerve fibers. The posterior lobe secretes ADH and oxytocin.

e. Prepare labeled sketches of representative portions of the anterior and posterior lobes of the pituitary gland in Part A of Laboratory Report 39.

Thyroid gland. Locate the thyroid gland and associated structures (fig. 39.5). To examine the thyroid tissue, follow these steps:

a. Use low-power magnification to observe the tissue (fig. 39.6). Note the numerous *follicles,* each of which consists of a layer of cells surrounding a colloid-filled cavity.

b. Observe the tissue using high-power magnification. The cells forming the wall of a follicle are simple cuboidal epithelial cells. These cells secrete thyroid hormone (T_3 and T_4). The parafollicular cells secrete calcitonin.

c. Prepare a labeled sketch of a representative portion of the thyroid gland in Part A of the laboratory report.

Parathyroid gland. Locate the parathyroid glands and associated structures (fig. 39.7). To examine the parathyroid tissue, follow these steps:

a. Use low-power magnification to observe the tissue (fig. 39.8). The gland consists of numerous tightly packed secretory cells.

b. Switch to high-power magnification and locate two types of cells—a smaller form (chief cells) arranged in cordlike patterns and a larger form (oxyphil cells) that have distinct cell boundaries and are present in clusters. The *chief cells* secrete parathyroid hormone. The function of the *oxyphil cells* is not well understood.

c. Prepare a labeled sketch of a representative portion of the parathyroid gland in Part A of the laboratory report.

Adrenal gland. Locate an adrenal gland and associated structures (fig. 39.9). To examine the adrenal tissue, follow these steps:

a. Use low-power magnification to observe the tissue (fig. 39.10). Note the thin capsule that covers the gland. Just beneath the capsule, there is a relatively thick *adrenal cortex.* The central portion of the gland is the *adrenal medulla.* The cells of the cortex are in three poorly defined layers. Those of the outer layer (zona glomerulosa) are arranged irregularly; those of the middle layer (zona fasciculata) are in long cords; and those of the inner layer (zona reticularis) are arranged in an interconnected network of cords. The most noted corticosteroids secreted from the cortex include aldosterone from the zona glomerulosa, cortisol from the zona fasciculata, and estro-

FIGURE 39.3 Micrograph of the anterior lobe of the pituitary gland (240×).

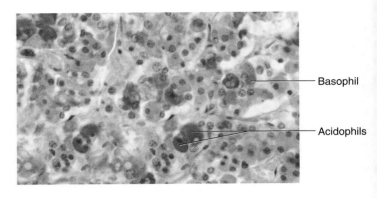

Basophil

Acidophils

FIGURE 39.4 Micrograph of the posterior lobe of the pituitary gland (240×).

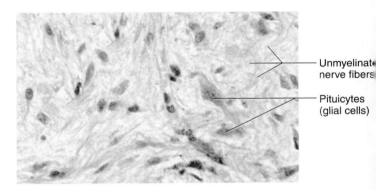

Unmyelinate nerve fibers

Pituicytes (glial cells)

FIGURE 39.5 Anterior view of thyroid gland and associated structures.

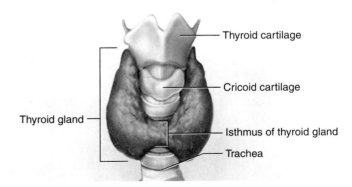

Thyroid cartilage

Cricoid cartilage

Thyroid gland

Isthmus of thyroid gland

Trachea

gens and androgens from the zona reticularis. The cells of the medulla are relatively large, irregularly shaped, and often occur in clusters. The medulla cells secrete epinephrine and norepinephrine.

b. Using high-power magnification, observe each of the layers of the cortex and the cells of the medulla.

c. Prepare labeled sketches of representative portions of the adrenal cortex and medulla in Part A of the laboratory report.

FIGURE 39.6 Micrograph of the thyroid gland (100× micrograph enlarged to 300×).

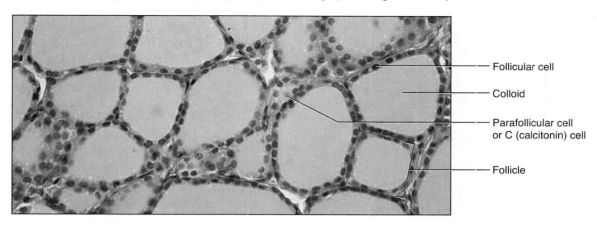

- Follicular cell
- Colloid
- Parafollicular cell or C (calcitonin) cell
- Follicle

FIGURE 39.7 Posterior view of four parathyroid glands and associated structures.

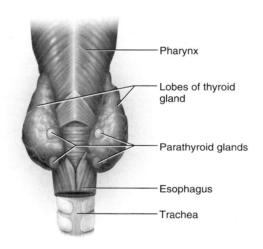

- Pharynx
- Lobes of thyroid gland
- Parathyroid glands
- Esophagus
- Trachea

FIGURE 39.8 Micrograph of the parathyroid gland (65×).

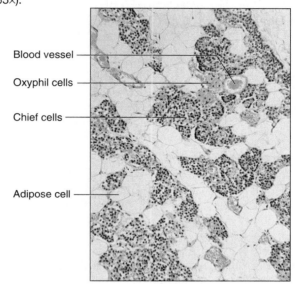

- Blood vessel
- Oxyphil cells
- Chief cells
- Adipose cell

FIGURE 39.9 Adrenal gland: (*a*) frontal section with associated structures; (*b*) diagram of the layers of the adrenal cortex and adrenal medulla.

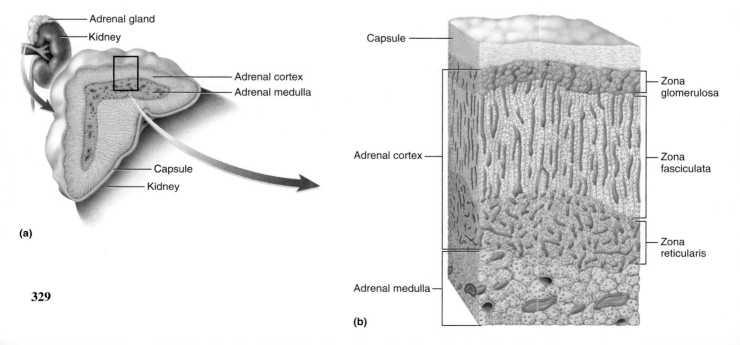

- Adrenal gland
- Kidney
- Adrenal cortex
- Adrenal medulla
- Capsule
- Kidney

(a)

- Capsule
- Adrenal cortex
- Adrenal medulla
- Zona glomerulosa
- Zona fasciculata
- Zona reticularis

(b)

FIGURE 39.10 Micrograph of the adrenal cortex and the adrenal medulla (75×).

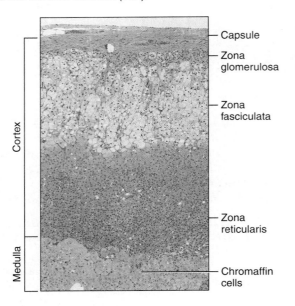

- Capsule
- Zona glomerulosa
- Zona fasciculata
- Zona reticularis
- Chromaffin cells

Cortex

Medulla

Pancreas. Locate the pancreas and associated structures (fig. 39.11). To examine the pancreas tissue, follow these steps:

a. Use low-power magnification to observe the tissue (fig. 39.12). The gland largely consists of deeply stained exocrine cells arranged in clusters around secretory ducts. These exocrine cells (acinar cells) secrete pancreatic juice rich in digestive enzymes. There are circular masses of lightly stained cells scattered throughout the gland. These clumps of cells constitute the *pancreatic islets (islets of Langerhans),* and they represent the endocrine portion of the pancreas. The beta cells secrete insulin, the alpha cells secrete glucagon, and the delta cells secrete somatostatin.

b. Examine an islet, using high-power magnification. Unless special stains are used, it is not possible to distinguish alpha, beta, and delta cells.

c. Prepare a labeled sketch of a representative portion of the pancreas in Part A of the laboratory report.

4. Complete Parts B and C of the laboratory report.

FIGURE 39.11 (*a*) Pancreas and associated structures. (*b*) Diagram of a pancreatic islet.

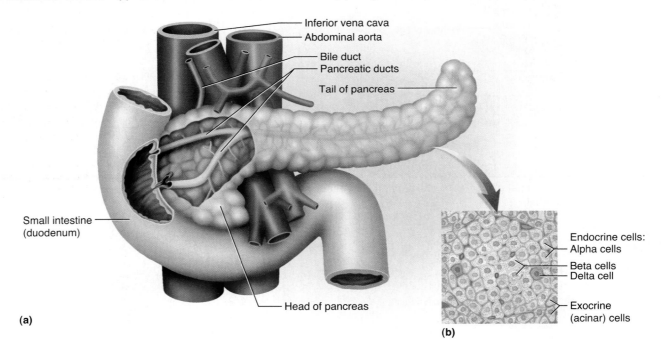

- Inferior vena cava
- Abdominal aorta
- Bile duct
- Pancreatic ducts
- Tail of pancreas
- Small intestine (duodenum)
- Head of pancreas

(a)

Endocrine cells:
- Alpha cells
- Beta cells
- Delta cell
- Exocrine (acinar) cells

(b)

FIGURE 39.12 Micrograph of the pancreas (100× micrograph enlarged to 400×).

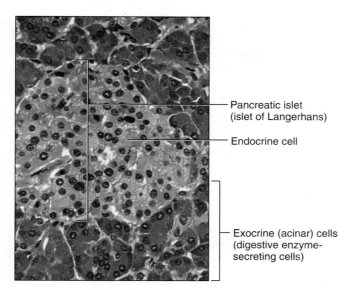

— Pancreatic islet (islet of Langerhans)

— Endocrine cell

— Exocrine (acinar) cells (digestive enzyme-secreting cells)

Procedure B—Ph.I.L.S. Lesson 17 Endocrine Function: Thyroid Gland and Metabolic Rate

Hypothesis

Would you predict an increase or decrease in metabolic rate if thyroid hormone levels decreased? _____

Would you predict an increase or decrease in oxygen consumption if thyroid gland hormone levels decreased?

_____

Would you predict an increase or decrease in temperature if thyroid gland hormone levels decreased?_____

1. Open Exercise 17, Endocrine Function: Thyroid Gland and Metabolic Rate.
2. Read the objectives and introduction and take the pre-lab quiz.
3. After completing the pre-lab quiz, read through the wet lab. Be sure to click open and view the videos that are indicated in red.
4. The lab exercise will open when you click Continue after completing the wet lab (fig. 39.13).

Weighing the Mouse

5. Click the power switch to turn on the scale and then click the power switch to turn on the thermostat.
6. Click Tare to set scale to zero.
7. To weigh a mouse, click on one of the mice and drag it to the scale.

Setting Up the Chamber

8. After weighing the mouse, place it in the chamber by clicking and dragging it to the chamber.
9. Place bubbles on the end of the calibrated tube. Be sure to touch the pipette to the calibration tube.

Measuring the Bubbles

10. Measure the initial position of the bubbles (if at the end of the tube it will be 10 mm) and record at 0:00 time in the data table. (Notice that the data table includes a change in time every 15 seconds between 0:00 and 2:00.)
11. After clicking the Start button, measure the position of the soap bubble within the calibration tube at 15-second intervals (you will hear a beep) and record it in the data table. (The timer will begin when you hit Start.) You may find it more manageable to click Pause after each

FIGURE 39.13 Opening screen for the laboratory exercise on Endocrine Function: Thyroid Gland and Metabolic Rate.

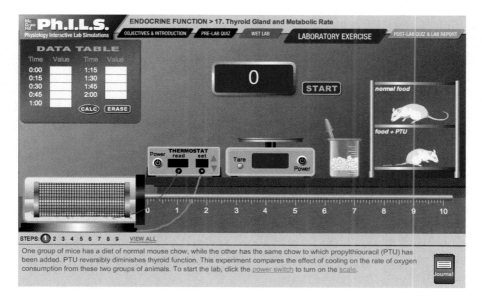

15-second interval, measure, record, and then click Start to begin again.
12. When finished click Pause.
13. Click Calc to see graph (linear regression). Journal will open.

Graph on left: Examine the linear regression graph that relates time and oxygen consumption. Does the amount of oxygen consumed for this mouse increase or decrease over time (i.e., the longer the mouse is in the chamber)? _____ Is that because the mouse is becoming more relaxed the longer it is in the chamber?

_____

Final graph on right: Includes a graph and a table. Notice the table (with temperatures ranging from 8°C to 24°C) will have the first data point entered for the average oxygen consumed per minute (total amount of oxygen divided by 2 minutes). But no points will appear on the graph until all data points are entered.

Repeating the Experiment

14. To complete with the same animal, repeat steps 9–13 but decrease the temperature 2 degrees (by clicking on the down arrow on the thermostat) prior to beginning the steps until data is recorded for all temperatures.
15. After returning the first animal to the cage, repeat steps 5–13 for the second animal.

Interpreting Results

16. *With the graph still on the screen,* answer the questions in Part D of the laboratory report. If you accidentally close the graph, click on the journal panel (red rectangle at bottom right of screen).
17. Complete the post-lab quiz by answering the ten questions on the computer screen.
18. Read the conclusion on the computer screen.
19. You may print the lab report for Endocrine Function: Thyroid Gland and Metabolic Rate.

Name _____

Date _____

Section _____

The ▲ corresponds to the Learning Outcome(s) listed at the beginning of the laboratory exercise.

Endocrine Structure and Function

Part A Assessments

Sketch and label representative portions of the following endocrine glands: ▲2

Pituitary gland (_____×) (anterior lobe)	Pituitary gland (_____×) (posterior lobe)
Thyroid gland (_____×)	Parathyroid gland (_____×)
Adrenal gland (_____×) (cortex and medulla)	Pancreas (_____×)

Part B Assessments

Complete the following:

1. Name six hormones secreted by the anterior lobe of the pituitary gland. **3** _____

_____

2. Name two hormones secreted by the posterior lobe of the pituitary gland. **3** _____

3. Name two thyroid hormones secreted from the follicular cells. **3** _____

4. Name the hormone secreted from the parafollicular cells of the thyroid gland. **3** _____

5. Name two hormones secreted by the adrenal medulla. **3** _____

6. Name the most important corticosteroid secreted by the zona fasciculata cells of the adrenal cortex. **3** _____

Part C Assessments

Match the endocrine gland in column A with a characteristic of the gland in column B. Place the letter of your choice in the space provided. **1** **3**

Column A	Column B
a. Adrenal cortex	_____ 1. Located in sella turcica of sphenoid bone
b. Adrenal medulla	_____ 2. Contains alpha, beta, and delta cells
c. Hypothalamus	_____ 3. Contains colloid-filled cavities
d. Pancreas	_____ 4. Attached to thyroid gland
e. Parathyroid gland	_____ 5. Secretes corticosteroids
f. Pituitary gland	_____ 6. Attached to pituitary gland by a stalk
g. Thymus	_____ 7. Gland inside another gland
h. Thyroid gland	_____ 8. Located in mediastinum

Part D—Ph.I.L.S. Lesson 17, Endocrine Function: Thyroid Gland and Metabolic Rate Assessments

The Hypothalamus and Thyroid Hormone Release

1. Complete the statements about thyroid hormone release in response to a decrease in environmental temperature. **4**

   a. The sensory receptors of the skin detect a decrease in the temperature of the environment; neural signals are sent to

   the _____, causing the release of this hormone: _____

   b. This hormone is released into the hypothalamo-hypophyseal portal vessel and travels to the

   _____, causing the release of this hormone: _____

   c. This hormone is released into the general circulation and travels to the _____, causing the release

   of this hormone: _____

Measuring Oxygen Consumption

2. Complete the statements about the relationship of thyroid hormone levels, metabolic rate, body temperature, and oxygen consumption. **5**

    a. Thyroid hormone _____ (increases or decreases) the metabolic rate, _____ (increases or decreases) body temperature, and _____ (increases or decreases) demand for ATP.

    b. As a result, aerobic cellular respiration _____ (increases or decreases) and oxygen consumption _____ (increases or decreases).

    c. Thus, there is a _____ (direct or inverse) relationship between oxygen consumption and metabolic rate, and the measurement of oxygen consumption _____ (can or cannot) be used to measure the changes in metabolic rate.

Interpreting Results

3. Click the Norm button to display the graph generated for the animal on the normal diet (normal functioning thyroid gland) to answer these questions. **6**

    a. How many lines are on the graph? _____

    b. Examine the change in oxygen consumption as the environmental temperature decreases from 24°C to 18°C. Is oxygen consumption increasing, decreasing, or staying the same? _____

    c. Given the relationship of oxygen consumption and metabolic rate, metabolic rate is _____ (increasing, decreasing, or staying the same).

    d. This part of the graph can be explained by _____ (the release of thyroid hormone or shivering).

    e. Examine the change in oxygen consumption as the environmental temperature decreases from 18°C to 8°C. Is oxygen consumption increasing, decreasing, or staying the same? _____

    f. Given the relationship of oxygen consumption and metabolic rate, metabolic rate is_____ (increasing, decreasing, or staying the same).

    g. This part of the graph can be explained by _____ (the release of thyroid hormone or shivering).

4. Click the PTU button to display the graph generated for the animal on the PTU diet (malfunctioning thyroid gland) to answer these questions. **6**

    a. How many lines are on the graph? _____

    b. Examine the change in oxygen consumption as the environmental temperature decreases from 24°C to 18°C. Is oxygen consumption increasing, decreasing, or staying the same? _____

    c. Given the relationship of oxygen consumption and metabolic rate, metabolic rate is_____ (increasing, decreasing, or staying the same).

    d. This part of the graph can be explained by _____ (the impaired release of thyroid hormone or shivering).

    e. Examine the change in oxygen consumption as the environmental temperature decreases from 18°C to 8°C. Is oxygen consumption increasing, decreasing, or staying the same? _____

    f. Given the relationship of oxygen consumption and metabolic rate, metabolic rate is _____ (increasing, decreasing, or staying the same).

    g. This part of the graph can be explained by _____ (the impaired release of thyroid hormone or shivering).

5. Click the Both button to display the graphs generated for both groups of animals. **6**

  a. For both animals, explain the relationship of temperature and oxygen consumption. _____

  _____

  _____

  _____

  _____

  _____

  When is the greatest amount of oxygen consumed—at cooler or warmer temperatures? _____

  b. Compare the red and black lines between 18°C and 24°C. Which is steeper? _____

  Which animal has a greater oxygen consumption? _____

  Which animal has a higher metabolic rate? _____

  c. The point of intersection of the two black lines (results from the mice with normal functioning thyroid gland)

  represents the point at which the animal begins to shiver. At what temperature does this occur? _____

  d. The point of intersection of the two red lines (results from the mice with malfunctioning thyroid gland) represents

  the point at which the animal begins to shiver. At what temperature does this occur? _____

  e. Which animals begin to shiver first? _____

  Explain _____

  _____

  _____

  _____

  _____

Critical Thinking Activity 7

Would you predict your thyroid hormone level to increase, decrease, or stay the same in the winter? _____
Predict the results (increased, decreased, stay the same) for the following for a person with a hyperthyroid condition (over-active thyroid) and one with a hypothyroid condition (underactive thyroid):

Variable	Change Predicted with Hyperthyroidism	Change Predicted with Hypothyroidism
Thyroid hormone level		
Metabolic rate		
Respiratory rate		
Body temperature		
Body weight		

Which individual, one with a hyperthyroid or a hypothyroid condition, would be more likely to shiver when exposed to

decreases in environmental temperature? _____

Diabetic Physiology

Pre-Lab

1. Carefully read the introductory material and examine the entire lab content.
2. Be familiar with diabetes mellitus (from lecture or the textbook).
3. Visit www.mhhe.com/martinseriesl for pre-lab questions and Anatomy & Physiology Revealed animations.

Materials Needed

500 mL beaker

Live small fish (1" to 1.5" goldfish, guppy, rosy red feeder, or other)

Small fish net

Insulin (regular U-100) (HumulinR in 10 mL vials has 100 units/mL; store in refrigerator—do not freeze)

Syringes for U-100 insulin

10% glucose solution

Clock with second hand or timer

Compound light microscope

Prepared microscope slides of the following:

Normal human pancreas (stained for alpha and beta cells)

Human pancreas of a diabetic

Safety

▶ Review all the safety guidelines inside the front cover.

▶ Wear disposable gloves when handling the fish and syringes.

▶ Dispose of the used syringe and needle in the puncture-resistant container.

▶ Wash your hands before leaving the laboratory.

The primary "fuel" for the mitochondria of cells is sugar (glucose). Insulin produced by the beta cells of the pancreatic islets has a primary regulation role in the transport of blood sugar across the cell membranes. Some individuals do not produce enough insulin; others might not have adequate transport of glucose into the body cells when cells lose insulin receptors. This rather common disease is called *diabetes mellitus*. Diabetes mellitus is a functional disease that might have its onset either during childhood or later in life.

The protein hormone insulin has several functions: it stimulates the liver and skeletal muscles to form glycogen from glucose; it inhibits the conversion of noncarbohydrates into glucose; it promotes the facilitated diffusion of glucose across cell membranes of cells possessing insulin receptors (adipose tissue, cardiac muscle, and resting skeletal muscle); it decreases blood sugar (glucose); it increases protein synthesis; and it promotes fat storage in adipose cells. In a normal person, as blood sugar increases during nutrient absorption after a meal, the rising glucose levels directly stimulate beta cells of the pancreas to secrete insulin. This increase in insulin prevents blood sugar from sudden surges (hyperglycemia) by promoting glycogen production in the liver and an increase in the entry of glucose into muscle and adipose cells.

Between meals and during sleep, blood glucose levels decrease and insulin also decreases. When insulin concentration decreases, more glucose is available to enter cells that lack insulin receptors. Neurons, liver cells, kidney cells, and red blood cells either lack insulin receptors or express limited insulin receptors, and they obtain their glucose by facilitated diffusion. These cells depend upon blood glucose concentrations to enable cellular respiration and ATP production. When insulin is given as a medication for diabetes mellitus, blood glucose levels may drop drastically (hypoglycemia) if the individual does not eat. This can have significant effects on the nervous system, resulting in insulin shock.

Type 1 diabetes mellitus usually has a rapid onset relatively early in life, but it can occur in adults. It occurs when insulin is no longer produced; hence sometimes it is referred to as *insulin-dependent diabetes mellitus (IDDM)*. This autoimmune disorder occurs when antibodies from the immune system destroy beta cells, resulting in a decrease in insulin production. This disorder affects about 10 to 15% of diabetics. The treatment for type 1 diabetes mellitus is to administer insulin, usually by injection.

Type 2 diabetes mellitus has a gradual onset, usually in those over age 40. It is sometimes called *non-insulin-dependent diabetes mellitus (NIDDM)*. This disease results as body cells become less sensitive to insulin or lose insulin receptors and thus cannot respond to insulin even though insulin levels may remain normal. This situation is called *insulin resistance*. Risk factors for the onset of type 2 diabetes mellitus, which afflicts about 85 to 90% of diabetics, include heredity, obesity, and lack of exercise. Obesity is a major indicator for predicting type 2 diabetes; a weight loss of a mere 10 pounds will increase the body's sensitivity to insulin as a means of prevention. The treatments for type 2 diabetes mellitus include a diet that avoids foods that stimulate insulin production, weight control, exercise, and medications. (The causes and treatments for type 2 diabetes closely resemble those for coronary artery disease!)

Purpose of the Exercise

To observe behavior changes that occur during insulin shock and to examine microscopically the pancreatic islets.

Learning Outcomes

After completing this exercise, you should be able to

1 Compare the causes, symptoms, and treatments for type 1 and type 2 diabetes mellitus.

2 Examine and record the behavioral changes caused by insulin shock.

3 Examine and record the recovery from insulin shock when sugar is provided to cells.

4 Distinguish tissue sections of normal pancreatic islets from those indicating diabetes mellitus, and sketch both types.

Procedure A—Insulin Shock

1. Reexamine the introduction to this laboratory exercise.
2. Complete the table in Part A of Laboratory Report 40.
3. A fish can be used to observe normal behavior, induced insulin shock, and recovery from insulin shock. The pancreatic islets are located in the pyloric ceca or other scattered regions in fish. The behavioral changes of fish when they experience insulin shock and recovery mimic those of humans. Observe a fish in 200 mL of aquarium water in a 500 mL beaker. Make your observations of swimming, gill cover (operculum), and mouth movements for 5 minutes. Record your observations in Part B of the laboratory report.
4. Add 200 units of room-temperature insulin slowly, using the syringe, into the water with the fish to induce insulin shock. Record the time or start the timer when the insulin is added. Watch for changes of the swimming, gill cover, and mouth movements as insulin diffuses into the blood at the gills. Signs of insulin shock might include rapid and irregular movements. As hypoglycemia occurs from the increased insulin absorption, the brain cells obtain less of the much-needed glucose, and epinephrine is secreted. Consequently, a rapid heart rate, anxiety, sweating, mental disorientation, impaired vision, dizziness, convulsions, and possible unconsciousness are complications during insulin shock. Record your observations of any behavioral changes and the time of the onset in Part B of the laboratory report.

5. After definite behavioral changes are observed, use the small fish net to move the fish into another 500 mL beaker containing 200 mL of a 10% room-temperature glucose solution. Record the time when the fish is transferred into this container. Make observations of any recovery of normal behaviors, including the amount of time involved. Record your observations in Part B of the laboratory report.

6. When the fish appears to be fully recovered, return the fish to its normal container. Dispose of the syringe according to directions from your laboratory instructor.

7. Complete Part B of the laboratory report.

Procedure B—Pancreatic Islets

1. Examine the normal specially stained pancreas under low-power and high-power magnification (fig. 40.1). Locate the clumps of cells that constitute the *pancreatic islets (islets of Langerhans)* that represent the endocrine portions of the organ. A pancreatic islet contains four distinct cells that secrete different hormones: *alpha cells,* which secrete glucagon, *beta cells,* which secrete insulin, *delta cells,* which secrete somatostatin, and *F cells,* which secrete pancreatic polypeptide. Some of the specially stained pancreatic islets allow for distinguishing the different types of cells (figs. 40.1 and 40.2). The normal pancreatic islet has about 80% beta cells concentrated in the central region of the islet, about 15% alpha cells near the periphery of an islet, about 5% delta cells, and a small number of F cells. There are complex interactions among the four hormones affecting blood sugar, but in general, glucagon will increase blood sugar while insulin will lower blood sugar. Somatostatin acts as a paracrine secretion to inhibit alpha and beta cell secretions, and pancreatic polypeptide inhibits secretions from delta cells and digestive enzyme secretions from the pancreatic acini.

2. Prepare a labeled sketch of a normal pancreatic islet in Part C of the laboratory report. Include labels for alpha cells and beta cells.

3. Examine the pancreas of a diabetic under high-power magnification (fig. 40.3). Locate a pancreatic islet and note the number of beta cells as compared to a normal number of beta cells in the normal pancreatic islet.

4. Prepare a labeled sketch of a pancreatic islet that shows some structural changes due to diabetes mellitus. Place the sketch in Part C of the laboratory report.

5. Describe the differences that you observed between a normal pancreatic islet and a pancreatic islet of a person with diabetes mellitus. **4**

_____
_____
_____
_____
_____
_____

FIGURE 40.1 Micrograph of a specially stained pancreatic islet surrounded by exocrine (acinar) cells (500×). The insulin-containing beta cells are stained dark purple.

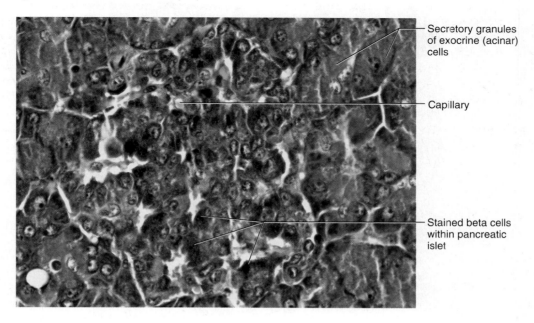

Secretory granules of exocrine (acinar) cells

Capillary

Stained beta cells within pancreatic islet

FIGURE 40.2 Micrograph of a pancreatic islet using immunofluorescence stains to visualize alpha and beta cells. Only the areas of interest are visible in the fluorescence microscope.

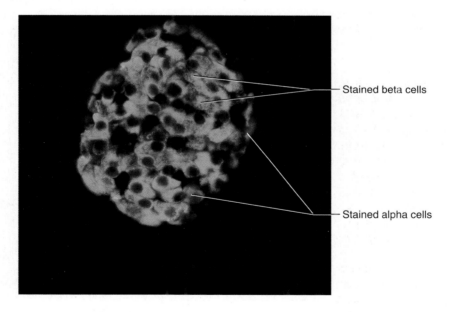

Stained beta cells

Stained alpha cells

FIGURE 40.3 Micrograph of pancreas showing indications of changes from diabetes mellitus (100×).

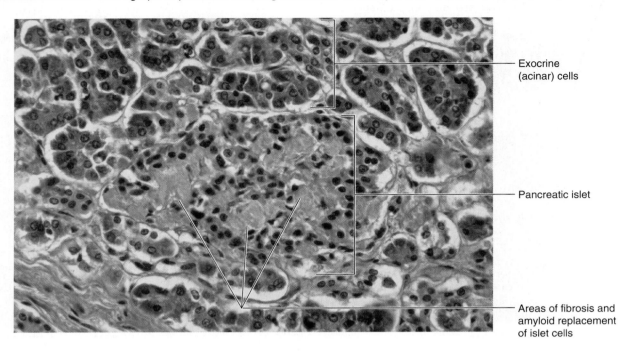

Exocrine (acinar) cells

Pancreatic islet

Areas of fibrosis and amyloid replacement of islet cells

Name _____

Date _____

Section _____

The Ⓐ corresponds to the Learning Outcome(s) listed at the beginning of the laboratory exercise.

Diabetic Physiology

Part A Assessments

Complete the missing parts of the table of diabetes mellitus: Ⓐ1

Characteristic	Type 1 Diabetes	Type 2 Diabetes
Onset age	Early age or adult	
Onset of symptoms		Slow
Percentage of diabetics		85–90%
Natural insulin levels	Below normal	
Beta cells of pancreatic islets		Not destroyed
Pancreatic islet cell antibodies	Present	
Risk factors of having the disease	Heredity	
Typical treatments	Insulin administration	
Untreated blood sugar levels		Hyperglycemia

Part B Assessments

1. Record your observations of normal behavior of the fish.

    a. Swimming movements: _____

    _____

    b. Gill cover movements: _____

    _____

    c. Mouth movements: _____

    _____

2. Record your observations of the fish after insulin was added to the water. Ⓐ2

    a. Recorded time that insulin was added: _____

    b. Swimming movements: _____

    _____

    c. Gill cover movements: _____

    _____

341

    d. Mouth movements: _____

_____

    e. Elapsed time until the insulin shock symptoms occurred: _____

3. Record your observations after the fish was transferred into a glucose solution. **3**

    a. Recorded time that fish was transferred into the glucose solution: _____

    b. Describe any changes in the behavior of the fish that indicates a recovery from insulin shock. _____

_____

_____

_____

    c. Elapsed time until indications of a recovery from insulin shock occurred: _____

Part C Assessments

Sketch and label representative portions of the following pancreatic islets: **4**

Pancreatic islet (normal)	Pancreatic islet (showing changes from diabetes mellitus)

Critical Thinking Activity

Justify the importance for a type 1 diabetic of regulating insulin administration, meals, and exercise. **1**

_____

_____

_____

_____

_____

_____

_____

Blood Cells

Pre-Lab

1. Carefully read the introductory material and examine the entire lab content.
2. Be familiar with RBCs, WBCs, and platelets (from lecture or the textbook).
3. Visit www.mhhe.com/martinseriesl for pre-lab questions and Anatomy & Physiology Revealed animations.

Materials Needed

Compound light microscope
Prepared microscope slides of human blood (Wright's stain)
Colored pencils

For Demonstration Activity:
Mammal blood other than human or contaminant-free human blood is suggested as a substitute for collected blood
Microscope slides (precleaned)
Sterile disposable blood lancets
Alcohol swabs (wipes)
Slide staining rack and tray
Wright's stain
Distilled water

For Learning Extension Activity:
Prepared slides of pathological blood, such as eosinophilia, leukocytosis, leukopenia, and lymphocytosis

Blood is a type of connective tissue whose cells are suspended in a liquid extracellular matrix. These cells are mainly formed in red bone marrow, and they include *red blood cells, white blood cells,* and some cellular fragments called *platelets.*

Red blood cells contain hemoglobin and transport gases between the body cells and the lungs, white blood cells defend the body against infections, and platelets play an important role in stoppage of bleeding (hemostasis).

Clinics and hospitals test blood using a modern hematology blood analyser. The more traditional procedures per-

Safety

▶ It is important that students learn and practice correct procedures for handling body fluids. Consider using either mammal blood other than human or contaminant-free blood that has been tested and is available from various laboratory supply houses. Some of the procedures might be accomplished as demonstrations only. If student blood is used, it is important that students handle only their own blood.

▶ Use an appropriate disinfectant to wash the laboratory tables before and after the procedures.

▶ Wear disposable gloves when handling blood samples.

▶ Clean end of a finger with an alcohol swab before the puncture is performed.

▶ The sterile blood lancet should be used only once.

▶ Dispose of used lancets and blood-contaminated items in an appropriate container (never use the wastebasket).

▶ Wash your hands before leaving the laboratory.

formed in these laboratory exercises will help you better understand each separate blood characteristic. On occasion, a doctor might question a blood test result from the hematology blood analyser and request additional verification using traditional procedures.

Purpose of the Exercise

To review the characteristics of blood cells, to examine them microscopically, and to perform a differential white blood cell count.

Learning Outcomes

After completing this exercise, you should be able to

1. Identify and sketch red blood cells, five types of white blood cells, and platelets.
2. Describe the structure and function of red blood cells, white blood cells, and platelets.
3. Perform and interpret the results of a differential white blood cell count.

Warning

Because of the possibility of blood infections being transmitted from one student to another if blood slides are prepared in the classroom, it is suggested that commercially prepared blood slides be used in this exercise. The instructor, however, may wish to demonstrate the procedure for preparing such a slide. Observe all safety procedures for this lab.

Demonstration Activity— Blood Slide Preparation

To prepare a stained blood slide, follow these steps:

1. Obtain two precleaned microscope slides. Avoid touching their flat surfaces.
2. Thoroughly wash hands with soap and water and dry them with paper towels.
3. Cleanse the end of the middle finger with an alcohol swab and let the finger dry in the air.
4. Remove a sterile disposable blood lancet from its package without touching the sharp end.
5. Puncture the skin on the side near the tip of the middle finger with the lancet and properly discard the lancet.
6. Wipe away the first drop of blood with the alcohol swab. Place a drop of blood about 2 cm from the end of a clean microscope slide. Cover the lanced finger location with a bandage.
7. Use a second slide to spread the blood across the first slide, as illustrated in figure 41.1. Discard the slide used for spreading the blood in the appropriate container.
8. Place the blood slide on a slide staining rack and let it dry in the air.
9. Put enough Wright's stain on the slide to cover the smear but not overflow the slide. Count the number of drops of stain that are used.
10. After 2–3 minutes, add an equal volume of distilled water to the stain and let the slide stand for 4 minutes. From time to time, gently blow on the liquid to mix the water and stain.
11. Flood the slide with distilled water until the blood smear appears light blue.
12. Tilt the slide to pour off the water, and let the slide dry in the air.
13. Examine the blood smear with low-power magnification, and locate an area where the blood cells are well distributed. Observe these cells, using high-power magnification and then an oil immersion objective if one is available.

FIGURE 41.1 To prepare a blood smear: (a) place a drop of blood about 2 cm from the end of a clean slide; (b) hold a second slide at about a 45° angle to the first one, allowing the blood to spread along its edge; (c) push the second slide over the surface of the first so that it pulls the blood with it; (d) observe the completed blood smear. The ideal smear should be 1.5 inches in length, be evenly distributed, and contain a smooth, feathered edge.

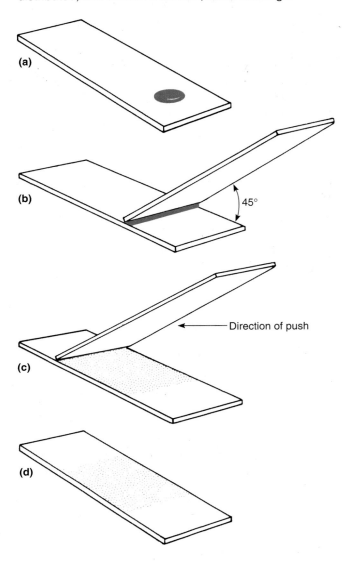

Procedure A—Types of Blood Cells

1. Refer to figure 41.2 as an aid in identifying the various types of blood cells. Study the functions of the blood cells listed in table 41.1. Use the prepared slide of blood and locate each of the following:

   red blood cell (erythrocyte)

   white blood cell (leukocyte)

granulocytes	agranulocytes
neutrophil	lymphocyte
eosinophil	monocyte
basophil	**platelet (thrombocyte)**

FIGURE 41.2 Micrographs of blood cells illustrating some of the numerous variations of each type. Appearance characteristics for each cell pertain to a thin blood film using Wright's stain (1,000 ×).

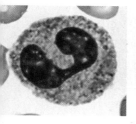 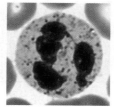 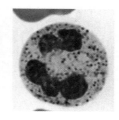

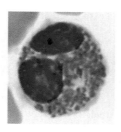 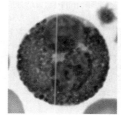 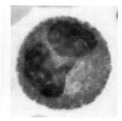

Neutrophils (3 of many variations)
- Fine light-purple granules
- Nucleus single to five lobes (highly variable)
- Immature neutrophils, called bands, have a single C-shaped nucleus
- Mature neutrophils, called segs, have a lobed nucleus
- Often called polymorphonuclear leukocytes when older

Eosinophils (3 of many variations)
- Coarse reddish granules
- Nucleus usually bilobed

 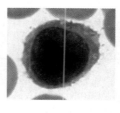 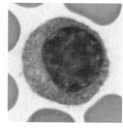

Basophils (3 of many variations)
- Coarse deep blue to almost black granules
- Nucleus often almost hidden by granules

Lymphocytes (3 of many variations)
- Slightly larger than RBCs
- Thin rim of nearly clear cytoplasm
- Nearly round nucleus appears to fill most of cell in smaller lymphocytes
- Larger lymphocytes hard to distinguish from monocytes

Never
Let
Monkeys
Eat
Bananas

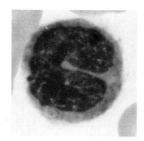 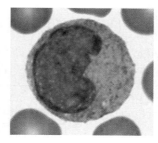 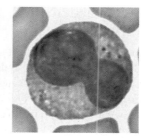

Monocytes (3 of many variations)
- Largest WBC; 2–3x larger than RBCs
- Cytoplasm nearly clear
- Nucleus round, kidney-shaped, oval, or lobed

     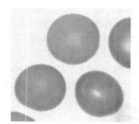

Platelets (several variations)
- Cell fragments
- Single to small clusters

Erythrocytes (several variations)
- Lack nucleus (mature cell)
- Biconcave discs
- Thin centers appear almost hollow

345

TABLE 41.1 Cellular Components of Blood

Component	Function
Red blood cell (erythrocyte)	Transports oxygen and carbon dioxide
White blood cell (leukocyte)	Destroys pathogenic microorganisms and parasites and removes worn cells
Granulocytes	
1. Neutrophil	Phagocytizes small particles
2. Eosinophil	Kills parasites and helps control inflammation and allergic reactions
3. Basophil	Releases heparin and histamine
Agranulocytes	
1. Monocyte	Phagocytizes large particles
2. Lymphocyte	Provides immunity
Platelet (thrombocyte)	Helps control blood loss from broken vessels

2. In Part A of Laboratory Report 41, prepare sketches of single blood cells to illustrate each type. Pay particular attention to relative size, nuclear shape, and color of granules in the cytoplasm (if present). The sketches should be accomplished using either the high-power objective or the oil immersion objective of the compound light microscope.

3. Complete Part B of the laboratory report.

Procedure B—Differential White Blood Cell Count

A differential white blood cell count is performed to determine the percentage of each of the various types of white blood cells present in a blood sample. The test is useful because the relative proportions of white blood cells may change in particular diseases as indicated in table 41.2. Neutrophils, for example, usually increase during bacterial infections, whereas eosinophils may increase during certain parasitic infections and allergic reactions.

1. To make a differential white blood cell count, follow these steps:
   a. Using high-power magnification or an oil immersion objective, focus on the cells at one end of a prepared blood slide where the cells are well distributed.
   b. Slowly move the blood slide back and forth, following a path that avoids passing over the same cells twice (fig. 41.3).

FIGURE 41.3 Move the blood slide back and forth to avoid passing the same cells twice.

   c. Each time you encounter a white blood cell, identify its type and record it in Part C of the laboratory report.
   d. Continue searching for and identifying white blood cells until you have recorded 100 cells in the data table. *Percent* means "parts of 100" for each type of white blood cell, so the total number observed is equal to its percentage in the blood sample.

2. Complete Part C of the laboratory report.

Learning Extension Activity

Obtain a prepared slide of pathological blood that has been stained with Wright's stain. Perform a differential white blood cell count, using this slide, and compare the results with the values for normal blood listed in table 41.2. What differences do you note?

_____

_____

_____

TABLE 41.2 Differential White Blood Cell Count

Cell Type	Normal Value (percent)	Elevated Levels May Indicate
Neutrophil	54–62	Bacterial infections, stress
Lymphocyte	25–33	Mononucleosis, whooping cough, viral infections
Monocyte	3–9	Malaria, tuberculosis, fungal infections
Eosinophil	1–3	Allergic reactions, autoimmune diseases, parasitic worms
Basophil	<1	Cancers, chicken pox, hypothyroidism

Name _____

Date _____

Section _____

The 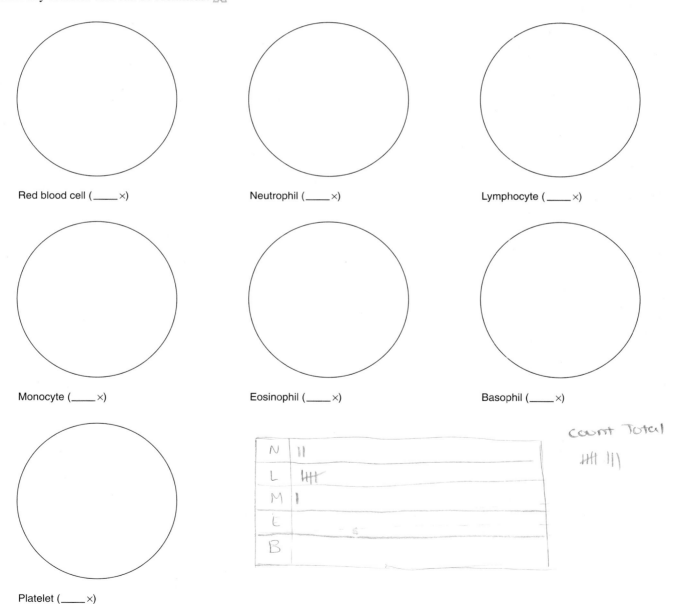 corresponds to the Learning Outcome(s) listed at the beginning of the laboratory exercise.

Blood Cells

Part A Assessments

Sketch a single blood cell of each type in the following spaces. Use colored pencils to represent the stained colors of the cells. Label any features that can be identified. **1**

Red blood cell (_____ ×)

Neutrophil (_____ ×)

Lymphocyte (_____ ×)

Monocyte (_____ ×)

Eosinophil (_____ ×)

Basophil (_____ ×)

Platelet (_____ ×)

Count Total

‖‖ ‖‖‖

N	‖
L	‖‖‖‖
M	‖
E	
B	

Part B Assessments

Complete the following statements:

1. Mature red blood cells are also called _____. 2

2. The shape of a red blood cell can be described as a _____ disc. 2

3. The functions of red blood cells are _____. 2

4. _____ is the oxygen-carrying substance in a red blood cell. 2

5. A mature red blood cell cannot reproduce because it lacks the _____ that was extruded during late development. 2

6. White blood cells are also called _____. 2

7. White blood cells with granular cytoplasm are called _____. 2

8. White blood cells lacking granular cytoplasm are called _____. 2

9. Polymorphonuclear leukocyte is another name for a _____ with a segmented nucleus. 2

10. Normally, the most numerous white blood cells are _____. 2

11. White blood cells whose cytoplasmic granules stain red in acid stain are called _____. 2

12. _____ are normally the least abundant of the white blood cells. 2

13. _____ are the largest of the white blood cells. 2

14. _____ are small agranulocytes that have relatively large, round nuclei with thin rims of cytoplasm. 2

15. Small cell fragments that function to prevent blood loss from an injury site are called _____. 2

Part C Assessments

1. *Differential White Blood Cell Count Data Table.* As you identify white blood cells, record them on the table by using a tally system, such as ⩜ ||. Place tally marks in the "Number Observed" column and total each of the five WBCs when the differential count is completed. Obtain a total of all five WBCs counted to determine the percent of each WBC type. ⑶

Type of WBC	Number Observed	Total	Percent
Neutrophil			
Lymphocyte			
Monocyte			
Eosinophil			
Basophil			
		Total of column	

2. How do the results of your differential white blood cell count compare with the normal values listed in table 41.2? ⑶

_____

_____

_____

_____

_____

Critical Thinking Activity

What is the difference between a differential white blood cell count and a total white blood cell count?

_____

_____

_____

_____

Blood Testing

Pre-Lab

1. Carefully read the introductory material and examine the entire lab content.
2. Be familiar with hematocrit, hemoglobin, and coagulation (from lecture or the textbook).
3. Visit www.mhhe.com/martinseriesl for pre-lab questions and LabCam videos.

Materials Needed

Mammal blood other than human or contaminant-free human blood is suggested as a substitute for collected blood

Sterile disposable blood lancets

Alcohol swabs (wipes)

Ph.I.L.S. 3.0

For Procedure A:

Heparinized microhematocrit capillary tube

Sealing clay (or Critocaps)

Microhematocrit centrifuge

Microhematocrit reader

For Procedure B:

Hemoglobinometer

Lens paper

Hemolysis applicator

For Procedure C:

Small triangular file

Capillary tube (nonheparinized)

Timer

For Alternative Hemoglobin Content Activity:

Tallquist test kit

As an aid in identifying various disease conditions, tests are often performed on blood to determine how its composition compares with normal values. These tests commonly include hematocrit (red blood cell percentage), hemoglobin content, and coagulation.

Safety

▶ It is important that students learn and practice correct procedures for handling body fluids. Consider using either mammal blood other than human or contaminant-free blood that has been tested and is available from various laboratory supply houses. Some of the procedures might be accomplished as demonstrations only. If student blood is used, it is important that students handle only their own blood.

▶ Use an appropriate disinfectant to wash the laboratory tables before and after the procedures.

▶ Wear disposable gloves when handling blood samples.

▶ Clean the end of a finger with alcohol swabs before the puncture is performed.

▶ The sterile blood lancet should be used only once.

▶ Dispose of used lancets and blood-contaminated items in an appropriate container (never use the wastebasket).

▶ Wash your hands before leaving the laboratory.

When a blood sample is collected in a heparinized capillary tube and left standing, the heavier cellular components settle to the bottom of the tube. Spinning the tube in a centrifuge can accelerate this process. The red layer at the bottom portion of the tube represents the compacted red blood cells (RBCs). This RBC portion of the entire volume of blood in the tube is called the *hematocrit,* normally about 45% of the entire volume. A thin, whitish layer on the surface of the RBCs, known as the *buffy coat,* contains the white blood cells and platelets, and it represents less than 1% of the total volume. The remaining straw-colored upper portion of the contents contains the plasma, nearly 55% of the volume.

The contents of a red blood cell are about one-third *hemoglobin,* a protein. Each hemoglobin molecule has four bound heme portions, each containing an atom of iron (Fe). An oxygen molecule can combine with each of the iron atoms, forming bright red *oxyhemoglobin.* When oxygen detaches from the hemoglobin in the capillaries of the tissues, it becomes a darker red and is called *deoxyhemoglobin,* which can appear bluish when viewed through the skin and blood vessel walls. Decreased values of hematocrits or hemoglobin levels could be attributed to hemorrhage, dietary deficiencies, infections, bone marrow cancer, or radiation.

Stoppage of bleeding, called *hemostasis,* involves three defense mechanisms: blood vessel spasm, platelet plug formation, and *coagulation.* The coagulation mechanism involves a series of chain reactions that result in a blood clot. The final stage of the series of reactions occurs when thrombin converts a soluble plasma protein called fibrinogen into insoluble *fibrin.* The fibrin mesh traps cellular components of the blood, stopping the blood loss. The fibrin formation of blood coagulation normally takes 2–10 minutes. Clotting deficiencies could be attributed to low platelet count, leukemia, hemophilia, liver disease, malnutrition, radiation, or anticoagulant drugs.

The clinical significance or conclusive diagnosis of a disease is often difficult to achieve. Most often, several blood and urine characteristics along with other symptoms are assessed collectively in order to make a determination of the abnormality. A self-diagnosis should never be made as a result of a test result conducted in the biology laboratory. Always obtain proper medical exams and treatments from medical personnel.

You take in a deep breath of air into your lungs. The inspired oxygen diffuses from the alveoli of your lungs into your red blood cells, where it binds with a hemoglobin molecule. Oxygen is then transported to systemic capillaries and released from the hemoglobin to diffuse into your body tissues. The amount of oxygen transported and released depends in part on the molecular structure of hemoglobin. Each hemoglobin molecule is composed of four proteins (two alpha proteins and two beta proteins) each containing a heme pigment with an atom of iron. Each iron binds one oxygen molecule (O_2). As a result, each hemoglobin molecule can bind as many as four oxygen molecules. How much oxygen is bound to the hemoglobin molecules is expressed as the percent *(%) saturation of hemoglobin* (i.e., what percentage of the iron-binding sites of hemoglobin is bound with oxygen). The % saturation of hemoglobin is dependent on a number of variables. The most important is the partial pressure of oxygen. This relationship of % saturation of hemoglobin to partial pressure of oxygen can be graphed to produce an *oxygen dissociation curve.* Other variables that influence % saturation of hemoglobin include temperature, carbon dioxide levels, and pH, and the change of any of these variables will trigger a "shift" in the oxygen dissociation curve. A "shift right" indicates that for any given partial pressure of oxygen, there is a decrease in the % saturation of hemoglobin. A "shift left" indicates that for any given partial pressure of oxygen, there is an increase in the % saturation of hemoglobin.

Purpose of the Exercise

To observe the blood tests used to determine hematocrit, hemoglobin content, and coagulation. To measure and compare the percent saturation of hemoglobin when the partial pressure of oxygen and the pH are changed.

Learning Outcomes

After completing this exercise, you should be able to

1 Determine the hematocrit, hemoglobin, and coagulation in a blood sample.

2 Judge the results of the blood tests compared to normal values.

3 Select the blood tests performed in this exercise that could indicate anemia.

4 Graph and explain the relationship of the percent transmittance of light to the partial pressure of oxygen.

5 Explain the relationship of percent transmittance of light to percent saturation of hemoglobin.

6 Explain the relationship of partial pressure of oxygen to percent saturation of hemolgobin.

7 Interpret the influence of pH on the oxygen dissociation curve.

8 Apply the concepts of how pH affects the percent saturation of hemoglobin to the human body.

 Warning

Because of the possibility of blood infections being transmitted from one student to another during blood-testing procedures, it is suggested that the following procedures be performed as demonstrations by the instructor. Observe all safety procedures listed for this lab.

Procedure A—Hematocrit

To determine the hematocrit (percentage of red blood cells) in a whole blood sample, the cells must be separated from the liquid plasma. This separation can be rapidly accomplished by placing a tube of blood in a centrifuge. The force created by the spinning motion of the centrifuge causes the cells to be packed into the lower end of the tube. The quantities of cells and plasma can then be measured, and the percentage of cells (hematocrit or packed cell volume) can be calculated.

1. To determine the hematocrit in a blood sample, follow these steps:
   a. Thoroughly wash hands with soap and water and dry them with paper towels.
   b. Cleanse the end of the middle finger with an alcohol swab and let the finger dry in the air.
   c. Remove a sterile disposable blood lancet from its package without touching the sharp end.
   d. Puncture the skin on the side near the tip of the middle finger with the lancet and properly discard the lancet. Wipe away the first drop of blood with the alcohol swab.
   e. Touch a drop of blood with the colored end of a heparinized capillary tube. Hold the tube tilted

slightly upward so that the blood will easily move into it by capillary action (fig. 42.1a). To prevent an air bubble, keep the tip in the blood until filled.

f. Allow the blood to fill about two-thirds of the length of the tube. Cover the lanced finger location with a bandage.

g. Hold a finger over the tip of the dry end so that blood will not drain out while you seal the blood end. Plug the blood end of the tube by pushing it with a rotating motion into sealing clay or by adding a plastic Critocap (fig. 42.1b).

FIGURE 42.1 Steps of the hematocrit (red blood cell percentage) procedure: (a) load a heparinized capillary tube with blood; (b) plug the blood end of the tube with sealing clay; (c) place the tube in a microhematocrit centrifuge.

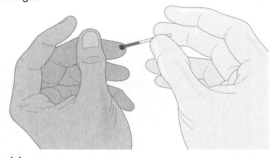

(a)

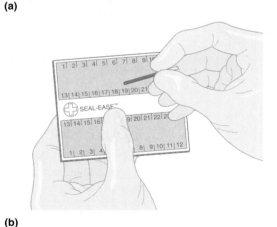

(b)

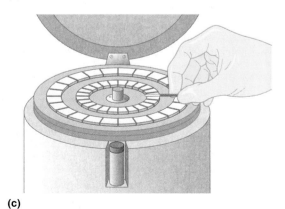

(c)

h. Place the sealed tube into one of the numbered grooves of a microhematocrit centrifuge. The tube's sealed end should point outward from the center and should touch the rubber lining on the rim of the centrifuge (fig. 42.1c).

i. The centrifuge should be balanced by placing specimen tubes on opposite sides of the moving head, the inside cover should be tightened with the lock wrench, and the outside cover should be securely fastened.

j. Run the centrifuge for 3–5 minutes.

k. After the centrifuge has stopped, remove the specimen tube. The red blood cells have been packed into the bottom of the tube. The clear liquid on top of the cells is plasma.

l. Use a microhematocrit reader to determine the percentage of red blood cells in the tube. If a microhematocrit reader is not available, measure the total length of the blood column in millimeters (red cells plus plasma) and the length of the red blood cell column alone in millimeters. Divide the red blood cell length by the total blood column length and multiply the answer by 100 to calculate the percentage of red blood cells.

2. Record the test result in Part A of Laboratory Report 42.

Procedure B—Hemoglobin Content

Although the hemoglobin content of a blood sample can be measured in several ways, a common method uses a hemoglobinometer. This instrument is designed to compare the color of light passing through a hemolyzed blood sample with a standard color. The results of the test are expressed in grams of hemoglobin per 100 mL of blood or in percentage of normal values.

1. To measure the hemoglobin content of a blood sample, follow these steps:

a. Obtain a hemoglobinometer and remove the blood chamber from the slot in its side.

b. Separate the pieces of glass from the metal clip and clean them with alcohol swabs and lens paper. One of the pieces of glass has two broad, U-shaped areas surrounded by depressions. The other piece is flat on both sides.

c. Obtain a large drop of blood from a finger, as before.

d. Place the drop of blood on one of the U-shaped areas of the blood chamber glass (fig. 42.2a).

e. Stir the blood with the tip of a hemolysis applicator until the blood appears clear rather than cloudy. This usually takes about 45 seconds (fig. 42.2b).

f. Place the flat piece of glass on top of the blood plate and slide both into the metal clip of the blood chamber.

g. Push the blood chamber into the slot on the side of the hemoglobinometer, making sure that it is in all the way (fig. 42.2c).

FIGURE 42.2 Steps of the hemoglobin content procedure: (*a*) load the blood chamber with blood; (*b*) stir the blood with a hemolysis applicator; (*c*) place the blood chamber in the slot of the hemoglobinometer; (*d*) match the colors in the green area by moving the slide on the side of the instrument.

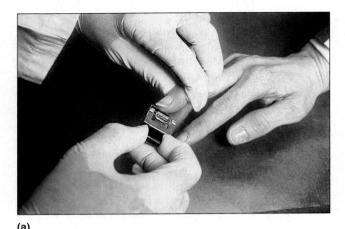

(a)

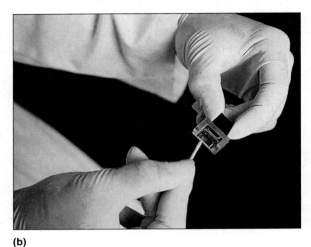

(b)

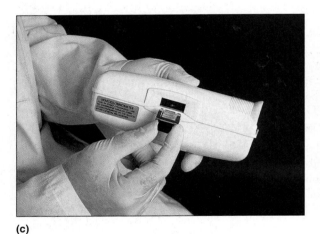

(c)

(d)

h. Hold the hemoglobinometer in the left hand with the thumb on the light switch on the underside (fig. 42.2*d*).

i. Look into the eyepiece and note the green area split in half.

j. Slowly move the slide on the side of the instrument back and forth with the right hand until the two halves of the green area look the same.

k. Note the value in the upper scale (grams of hemoglobin per 100 mL of blood), indicated by the mark in the center of the movable slide.

2. Record the test result in Part A of the laboratory report.

Alternative Activity

The *Tallquist method* can also be used to determine reasonably accurate hemoglobin content. The Tallquist test kit contains a special test paper, color scale, and directions.

Procedure C—Coagulation

Coagulation time, often called clotting time, is the time from the onset of bleeding until the insoluble protein fibrin is formed. This happens when thrombin converts soluble fibrinogen into insoluble fibrin. This clotting time normally ranges from 2 to 10 minutes. This process is prolonged if the person has clotting deficiencies or is being treated with anticoagulants such as heparin (some is produced by basophils and mast cells), warfarin (Coumadin), or aspirin. In this laboratory exercise we will try to determine the time to the nearest minute.

1. To determine coagulation time, follow these steps:

a. Prepare the finger to be lanced by following the directions in Procedure A. Lance the end of a finger to obtain a drop of blood. Wipe away the first drop of blood with the alcohol swab.

b. Touch the next drop of blood with one end of a nonheparinized capillary tube. Hold the tube tilted slightly upward so that the blood will easily move into it by capillary action (fig. 42.1*a*). Keep the tip in the blood until the tube is nearly filled. If the tube is nearly filled, it will allow enough tube length for breaking it several times.

c. Place the capillary tube on a paper towel. Cover the lanced finger location with a bandage. Record the time: _____

d. At 1-minute intervals, use the small triangular file and make a scratch on the capillary tube starting near one end of the tube. Hold the tube with fingers on each side of the scratch, the weakened location of the tube, and break the tube away from you, being careful to keep the two pieces close together after the break. Gently pull the two ends of the tube apart while observing carefully to see if it breaks cleanly apart. If it breaks cleanly, fibrin has not formed yet (fig. 42.3a).

e. Continue breaking the capillary tube each minute until fibrin is noted spanning the two parts of the capillary tube (fig. 42.3b). Note the time for coagulation.

FIGURE 42.3 Steps of the coagulation procedure: (a) clean break of capillary tube before any fibrin formed; (b) fibrin spans between the broken capillary tube segments when coagulation occurs.

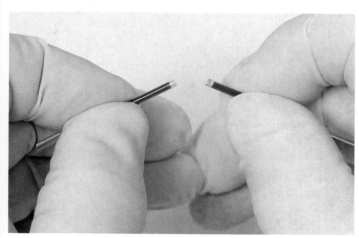
(a)

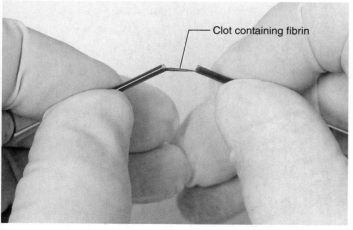
(b)

2. Record the test results in Part A of the laboratory report.

3. Complete Part B of the laboratory report.

Alternative Activity

Laboratories in modern hospitals and clinics use updated hematology analyzers (fig. 42.4) for evaluations of the blood factors that are described in Laboratory Exercise 42. The more traditional procedures performed in these laboratory activities will help you to better understand each blood characteristic.

Procedure D—Ph.I.L.S. Lesson 31 Blood: pH & Hb-Oxygen Binding

Hypothesis

Would you predict the amount of oxygen bound to hemoglobin to increase, decrease, or stay the same if the partial pressure of oxygen increased? _____

If hydrogen ions interfere with the ability of hemoglobin to bind oxygen, would you predict the amount of oxygen bound to hemoglobin to increase, decrease, or stay the same if the H^+ concentration increased (pH decreased)? _____ So, would you predict that the hemoglobin is better able or less able to bind oxygen in conditions of low pH? _____

1. Open Exercise 31, Blood: pH & Hb-Oxygen Binding
2. Read the objectives and introduction and take the prelab quiz.

FIGURE 42.4 Modern hematology analyzer being used in the laboratory of a clinic. (*Note:* A tour of a laboratory in a hospital or clinic might be arranged.)

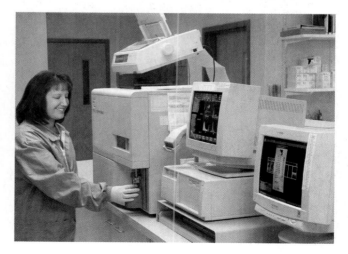

3. After completing the pre-lab quiz, read through the wet lab. Be sure to click open and view the videos that are indicated in red in the wet lab.
4. The lab exercise will open when you click Continue after completing the wet lab (fig. 42.5).

Setup

5. Click the power switch to turn on the spectrophotometer.
6. Set the wavelength at 620 nm (wavelength at which hemoglobin absorbs light if intact).

Calibrate Spectrophotometer

7. Set the transmittance value at 0 using the arrows at zero.
8. Open the lid of the "holder" of the spectrophotometer (if unsure, click on the word *spectrophotometer* in the instructions at the bottom of the screen).
9. Click on the globe and drag it to the top of one of the tubes (it will snap in place) to form the "tonometer." (To "click and drag," move the mouse to position the arrow on the globe; left-click on mouse and hold; drag the globe to the top of one of the tubes; and when positioned, release the left click).
10. Place the tonometer into the spectrophotometer. (The journal will open a table with the partial pressures of oxygen ranging from 160 mm Hg to 0 mm Hg for three different pH values—6.8, 7.4, and 8.0). Set the transmittance at 100% using the arrows at Calibrate.

Measuring Transmittance

11. With the tonometer remaining in the spectrophotometer, record the percent transmittance by clicking open the journal (red rectangle at bottom right of screen).
12. Click on the tonometer to remove it. (The tonometer will come up out of the spectrophotometer and the journal will close automatically.) Then drag it back to the rack.

Changing the Partial Pressure of O_2 (PO_2)

13. Position the end of the vacuum tube on the top of the "tonometer."
14. Click the down arrow once to reduce the barometric pressure (which simulates a decrease in partial pressure of oxygen).
15. Click on the vacuum tube to return it to its original position.
16. Place the tonometer in the spectrophotometer.
17. Click on the journal to record.
18. Repeat steps 9–17 until all values for the different partial pressures of oxygen have been recorded in the table on the computer.

Changing the pH

19. When all values have been recorded for a given sample, drag the tonometer to place the sample into the recycling can (three arrows in a circle in bottom right of screen).
20. Repeat steps 9–19 but use one of the two other samples with a different pH.
21. Click on the journal to see the complete table.
22. To view graph, click on "graph" in lower right of journal window.

Interpreting Results

23. *With the graph still on the screen,* answer the questions in Part C of the laboratory report. If you accidentally closed the graph, click on the journal panel (red rectangle at bottom right of screen).
24. Complete the post-lab quiz (click open post laboratory quiz and lab report) by answering the ten questions on the computer screen.
25. Read the conclusion on the computer screen.
26. You may print the lab report for Blood: pH & Hb-Oxygen Binding.

FIGURE 42.5 Opening screen for the laboratory exercise on Blood: pH & Hb-Oxygen Binding.

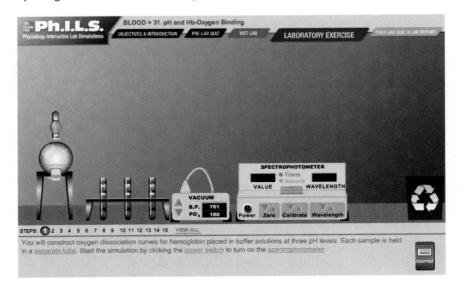

Name _____

Date _____

Section _____

The ⒶA corresponds to the Learning Outcome(s) listed at the beginning of the laboratory exercise.

Blood Testing

Part A Assessments

Blood test data: ⒶA

Blood Test	Test Results	Normal Values
Hematocrit (mL per 100 mL blood)		Men: 40–54% Women: 37–47%
Hemoglobin content (g per 100 mL blood)		Men: 14–18 g/100mL Women: 12–16 g/100mL
Coagulation		2–10 minutes

Part B Assessments

Complete the following:

1. How does the hematocrit from the blood test compare with the normal value? ⒶA

_____

_____

2. How does the hemoglobin content from the blood test compare with the normal value? ⒶA _____

_____

_____

3. How does the coagulation time from the blood test compare with the normal value? ⒶA _____

_____

_____

Which blood tests performed in this lab could be used to determine possible anemia? /3\

_____

_____

Part C Ph.I.L.S. Lesson 31, Blood: pH & Hb-Oxygen Binding Assessments

1. Use the data from the computer simulation collected at a pH of 7.4; graph the partial pressure of oxygen and the percent transmittance of light. (Include the title. Label each axis with units. Plot the points and connect). /4\

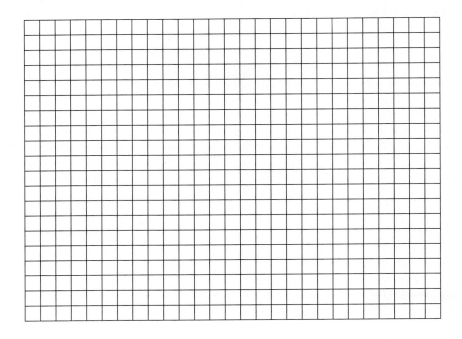

2. Examine the graph you produced in question 1. As the partial pressure of oxygen increases, the % transmittance of light _____. /4\

3. There is a direct relationship between percent transmittance of light and percent saturation of hemoglobin. So, as the percent transmittance of light increases, the percent saturation of hemoglobin _____ (increases or decreases). Thus, as the partial pressure of oxygen increases, the % saturation of hemoglobin _____ (increases, decreases, or stays the same). /5\ /6\

4. Compare the effect of pH on percent transmittance of light by examining the three dissociation curves produced at a pH of 6.8, 7.4, and 8.0 and answer the following questions: **7**

    a. When comparing the curve produced at 6.8 to the curve produced at 7.4, the curve produced when the pH is 6.8 would be described as _____ (a "shift right" or a "shift left") and at any given partial pressure of oxygen, the % transmittance of light is _____ (higher or lower) and thus, the % saturation of hemoglobin would be _____ (higher or lower).

    b. At a given PO_2, hemoglobin has the highest affinity for oxygen (ability to bind oxygen) at this pH _____ and the lowest affinity for oxygen at this pH _____.

Critical Thinking Activity 8

Hydrogen ion (H^+) levels increase within the body from increased production of lactic acid (during anaerobic cellular respiration) and from increased production of carbon dioxide ($CO_2 + H_2O \longleftrightarrow H^+ + HCO_3^-$). The increased amounts of H^+ bind with the amino acids that compose the protein of hemoglobin, slightly altering hemoglobin's shape. This change in shape interferes with the ability of iron within the hemoglobin molecule to bind the oxygen (decreased affinity or attraction). (This is referred to as the Bohr effect).

If the tissues of the body are metabolically active, _____ (more or less) H^+ are produced and the pH is _____ (decreased or increased). As a result of the Bohr effect, when hemoglobin reaches the systemic capillaries, the hemoglobin will bind the oxygen with _____ (greater or less) affinity; _____ (more or less) oxygen is released to the cells; the saturation of hemoglobin will be _____ (greater or less), and hemoglobin returns to the lungs with _____ (more or less) oxygen than would occur at a normal pH.

Blood Typing

Pre-Lab

1. Carefully read the introductory material and examine the entire lab content.
2. Be familiar with the ABO blood group and the Rh blood group (from lecture or the textbook).
3. Visit www.mhhe.com/martinseriesl for pre-lab questions.

Materials Needed

For Procedure A
ABO blood-typing kit
Simulated blood-typing kits are suggested as a substitute for collected blood

For Procedure B
Microscope slide
Alcohol swabs (wipes)
Sterile blood lancet
Toothpicks
Anti-D serum
Slide warming box (Rh blood-typing box or Rh view box)

Safety

▶ It is important that students learn and practice correct procedures for handling body fluids. Consider using simulated blood-typing kits or contaminant-free blood that has been tested and is available from various laboratory supply houses. Some of the procedures might be accomplished as demonstrations only. If student blood is used, it is important that students handle only their own blood.

▶ Use an appropriate disinfectant to wash the laboratory tables before and after the procedures.

▶ Wear disposable gloves when handling blood samples.

▶ Clean the end of a finger with an alcohol swab before the puncture is performed.

▶ The sterile blood lancet should be used only once.

▶ Dispose of used lancets and blood-contaminated items in an appropriate container (never use the wastebasket).

▶ Wash your hands before leaving the laboratory.

Blood typing involves identifying protein substances called *antigens* present in red blood cell membranes. Although there are many different antigens associated with human red blood cells, only a few of them are of clinical importance. These include the antigens of the ABO group and those of the Rh group.

To determine which antigens are present, a blood sample is mixed with blood-typing sera that contain known types of antibodies. If a particular antibody contacts a corresponding antigen, a reaction occurs, and the red blood cells clump together (agglutination). Thus, if blood cells are mixed with serum containing antibodies that react with antigen A and the cells clump together, antigen A must be present in those cells.

Although there are many antigens on the red blood cell membranes other than those of the ABO group and the Rh group, these antigens are of the greatest concern during transfusions. Most other antigens cause little if any transfusion reactions. As a final check on blood compatibility before transfusions, an important cross-matching test occurs. In this test, small samples of the blood from the donor and the recipient are mixed to be sure clumping (agglutination) does not create a transfusion reaction. If surgery is elective, autologous transfusions are promoted, allowing some of the person's own predonated blood from storage to be used during the surgery.

Another antigen commonly found on the human RBC membrane was first identified in rhesus monkeys, and it became known as the Rh factor. Although many different types of antigens are related to the Rh factor, only the D antigen is checked using an anti-D reagent. About 85% of Americans possess the D antigen and are therefore considered to be Rh-positive (Rh^+). Those lacking the D antigen are considered Rh-negative (Rh⁻). If an Rh-negative woman is pregnant carrying an Rh-positive fetus, the mother might obtain some RBCs of the fetus during the birth process or during a miscarriage. As a result, the mother would begin producing anti-D antibodies, creating complications for the second and future pregnancies. This condition is known as hemolytic disease of the fetus and newborn (erythroblastosis fetalis). This condition can be prevented with the proper administration of RhoGAM, which prevents the mother from producing anti-D antibodies.

Purpose of the Exercise

To determine the ABO blood type of a blood sample and to observe an Rh blood-typing test.

Learning Outcomes

After completing this exercise, you should be able to

1. Analyze the basis of ABO blood typing.
2. Match the ABO type of a blood sample.
3. Explain the basis of Rh blood typing.
4. Interpret how the Rh type of a blood sample is determined.

Warning

Because of the possibility of blood infections being transmitted from one student to another if blood testing is performed in the classroom, it is suggested that commercially prepared blood-typing kits containing virus-free human blood be used for ABO blood typing. The instructor may wish to demonstrate Rh blood typing. Observe all of the safety procedures listed for this lab.

Procedure A—ABO Blood Typing

1. Study table 43.1.
2. Compete Part A of Laboratory Report 43.

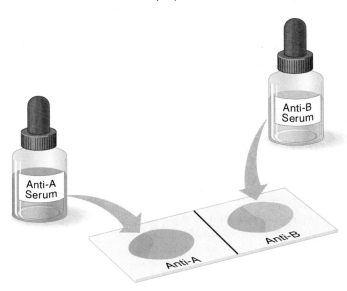

FIGURE 43.1 Slide prepared for ABO blood typing.

3. Perform the ABO blood type test using the blood-typing kit. To do this, follow these steps:
   a. Obtain a clean microscope slide and mark across its center with a wax pencil to divide it into right and left halves. Also write "Anti-A" near the edge of the left half and "Anti-B" near the edge of the right half (fig. 43.1).
   b. Place a small drop of blood on each half of the microscope slide. Work quickly so that the blood will not have time to clot.
   c. Add a drop of anti-A serum to the blood on the left half and a drop of anti-B serum to the blood on the right half. Note the color coding of the anti-A and anti-B typing sera. To avoid contaminating the serum, avoid touching the blood with the serum while it is in the dropper; instead allow the serum to fall from the dropper onto the blood.
   d. Use separate toothpicks to stir each sample of serum and blood together, and spread each over an area about as large as a quarter. Dispose of toothpicks in an appropriate container.

TABLE 43.1 Antigens and Antibodies of the ABO Blood Group and Preferred, Permissible, and Incompatible Donors

Blood Type	RBC Antigens (Agglutinogens)	Plasma Antibodies (Agglutinins)	Preferred Donor Type	Permissible Donor Type in Limited Amounts	Incompatible Donor
A	A	Anti-B	A	O	B, AB
B	B	Anti-A	B	O	A, AB
AB (universal recipient)	A and B	Neither anti-A nor anti-B	AB	A, B, O	None
O (universal donor)	Neither A nor B	Both anti-A and anti-B	O	No alternative types	A, B, AB

TABLE 43.2 Possible Reactions of ABO Blood-Typing Sera

Reactions		Blood Type	U.S. Frequency
Anti-A Serum	**Anti-B Serum**		
Clumping (agglutination)	No clumping	Type A	41%
No clumping	Clumping	Type B	9%
Clumping	Clumping	Type AB	3%
No clumping	No clumping	Type O	47%

Note: The inheritance of the ABO blood groups is described in Laboratory Exercise 61.

e. Examine the samples for clumping of blood cells (agglutination) after 2 minutes.

f. See table 43.2 and figure 43.2 for aid in interpreting the test results.

g. Discard contaminated materials as directed by the laboratory instructor.

4. Complete Part B of the laboratory report.

Critical Thinking Activity

Judging from the observations of the blood-typing results, predict the components in the anti-A and anti-B sera that caused clumping (agglutination).

_____

_____

_____

FIGURE 43.2 Four possible results of the ABO test.

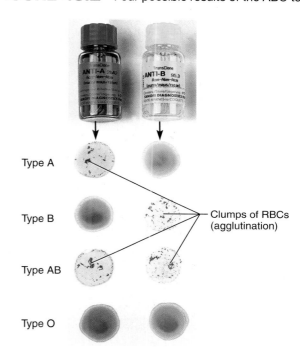

Type A

Type B

Type AB

Type O

Clumps of RBCs (agglutination)

Procedure B—Rh Blood Typing

1. Complete Part C of the laboratory report.
2. To determine the Rh blood type of a blood sample, follow these steps:
   a. Lance the tip of a finger. (See the demonstration procedures in Laboratory Exercise 41 for directions.) Place a small drop of blood in the center of a clean microscope slide.
   b. Add a drop of anti-D serum to the blood and mix them together with a clean toothpick.
   c. Place the slide on the plate of a warming box (Rh blood-typing box or Rh view box) prewarmed to 45°C (113°F) (fig. 43.3).
   d. Slowly rock the box back and forth to keep the mixture moving, and watch for clumping (agglutination) of the blood cells. When clumping occurs in anti-D serum, the clumps usually are smaller than those that appear in anti-A or anti-B sera, so they may be less obvious. However, if clumping occurs, the blood is called Rh positive; if no clumping occurs *within 2 minutes,* the blood is called Rh negative.
   e. Discard all contaminated materials in appropriate containers.
3. Complete Part D of the laboratory report.

FIGURE 43.3 Slide warming box used for Rh blood typing.

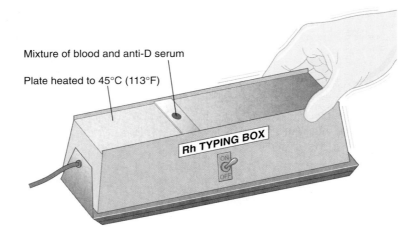

Mixture of blood and anti-D serum

Plate heated to 45°C (113°F)

Rh TYPING BOX

Name _____

Date _____

Section _____

The ⚠ corresponds to the Learning Outcome(s) listed at the beginning of the laboratory exercise.

Blood Typing

Part A Assessments

Complete the following statements:

1. The antigens of the ABO blood group are located in the _____. ⚠

2. The blood of every person contains one of (how many possible?) _____ combinations of antigens. ⚠

3. Type A blood contains antigen _____. ⚠

4. Type B blood contains antigen _____. ⚠

5. Type A blood contains _____ antibody in the plasma. ⚠

6. Type B blood contains _____ antibody in the plasma. ⚠

7. Persons with ABO blood type _____ are often called universal recipients. ⚠

8. Persons with ABO blood type _____ are often called universal donors. ⚠

Part B Assessments

Complete the following:

1. What was the ABO type of the blood tested? ⚠ _____

2. What ABO antigens are present in the red blood cells of this type of blood? ⚠ _____

3. What ABO antibodies are present in the plasma of this type of blood? ⚠ _____

4. If a person with this blood type needed a blood transfusion, what ABO type(s) of blood could be received safely? ⚠

_____

5. If a person with this blood type was serving as a blood donor, what ABO blood type(s) could receive the blood safely? ⚠

_____

Part C Assessments

Complete the following statements:

1. The Rh blood group was named after the _____. ⚠

2. Of the antigens in the Rh group, the most important is _____. ⚠

3. If red blood cells lack Rh antigens, the blood is called _____. ⚠

4. Rh antibodies form only in persons with _____ type blood in response to special stimulation. ⚠

5. If an Rh-negative person who is sensitive to Rh-positive blood receives a transfusion of Rh-positive blood, the donor's cells are likely to _____. **3**

6. An Rh-negative woman who might be carrying an _____ fetus is given an injection of RhoGAM to prevent hemolytic disease of the fetus and newborn. **3**

Part D Assessments

Complete the following:

1. What was the Rh type of the blood tested in the demonstration? **4** _____

2. What Rh antigen is present in the red blood cells of this type of blood? **4** _____

3. What Rh antibody is normally present in the plasma of this type of blood? **4** _____

4. If a person with this blood type needed a blood transfusion, what type of blood could be received safely? **3**

   _____

5. If a person with this blood type was serving as a blood donor, a person with what type of blood could receive the blood safely? **3** _____

Heart Structure

The heart is a muscular pump located within the mediastinum and resting upon the diaphragm. It is enclosed by the lungs, thoracic vertebrae, and sternum, and attached at its superior end (the base) are several large blood vessels. Its inferior end extends downward to the left and terminates as a bluntly pointed apex.

The heart and the proximal ends of the attached blood vessels are enclosed by a double-layered pericardium. The innermost layer of this membrane (visceral layer of serous pericardium) consists of a thin covering closely applied to the surface of the heart, whereas the outer layer (parietal layer of serous pericardium with fibrous pericardium) forms a tough, protective sac surrounding the heart. Between the parietal and visceral layers of the pericardium is a space, the pericardial cavity, that contains a small volume of serous (pericardial) fluid.

Functioning as a muscular pump of blood, the heart consists of four chambers: two superior atria and two inferior ventricles. The interatrial septum and the interventricular septum prevent oxygenated blood in the left-side chambers from mixing with deoxygenated blood in the right chambers. Atrioventricular (AV) valves (mitral and tricuspid) are located between each atrium and ventricle, while semilunar valves (aortic and pulmonary) are at the exits of the ventricles into the two large arteries. The heart valves only allow blood flow in one direction.

Purpose of the Exercise

To review the structural characteristics of the human heart and to examine the major features of a mammalian heart.

Learning Outcomes

After completing this exercise, you should be able to

1 Identify and label the major structural features of the human heart and closely associated blood vessels.
2 Match heart structures with appropriate locations and functions.
3 Compare the features of the human heart with those of another mammal.

Procedure A—The Human Heart

1. Study figures 44.1, 44.2, 44.3, and 44.4.
2. Label figure 44.5.
3. Complete Part A of Laboratory Report 44.

FIGURE 44.1 Anterior view of the human heart.

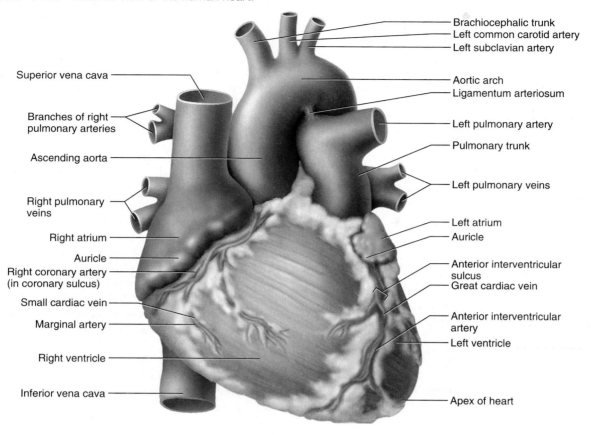

Brachiocephalic trunk
Left common carotid artery
Left subclavian artery

Superior vena cava

Aortic arch
Ligamentum arteriosum

Branches of right pulmonary arteries

Left pulmonary artery

Pulmonary trunk

Ascending aorta

Left pulmonary veins

Right pulmonary veins

Left atrium
Auricle

Right atrium

Auricle

Anterior interventricular sulcus
Great cardiac vein

Right coronary artery (in coronary sulcus)

Small cardiac vein

Anterior interventricular artery

Marginal artery

Left ventricle

Right ventricle

Inferior vena cava

Apex of heart

FIGURE 44.2 Posterior view of the human heart.

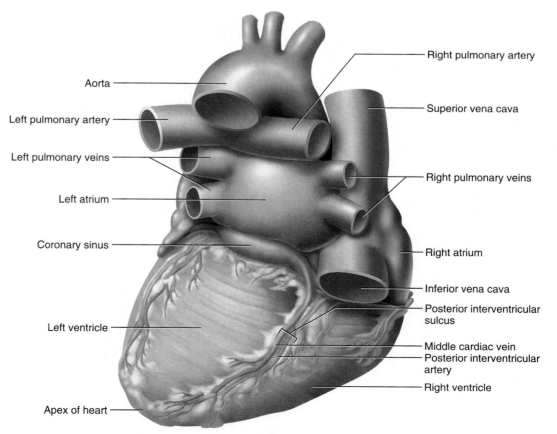

Right pulmonary artery

Aorta

Left pulmonary artery

Superior vena cava

Left pulmonary veins

Right pulmonary veins

Left atrium

Coronary sinus

Right atrium

Inferior vena cava

Posterior interventricular sulcus

Left ventricle

Middle cardiac vein
Posterior interventricular artery

Right ventricle

Apex of heart

368

FIGURE 44.3 Frontal section of the heart wall and pericardium.

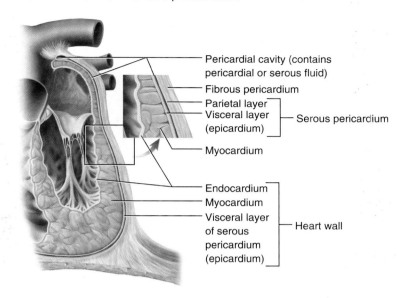

Pericardial cavity (contains pericardial or serous fluid)

Fibrous pericardium

Parietal layer
Visceral layer (epicardium) — Serous pericardium

Myocardium

Endocardium
Myocardium
Visceral layer of serous pericardium (epicardium) — Heart wall

FIGURE 44.4 Superior view of the four heart valves (atria removed).

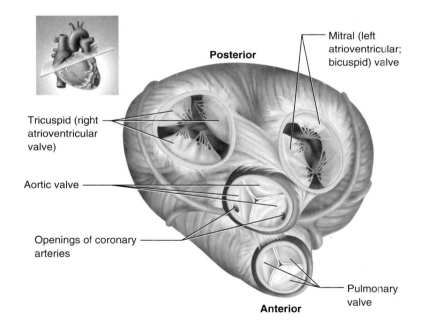

Posterior

Mitral (left atrioventricular; bicuspid) valve

Tricuspid (right atrioventricular valve)

Aortic valve

Openings of coronary arteries

Pulmonary valve

Anterior

4. Examine the human heart model, and locate the following features:

heart
    base of heart (superior region where blood vessels emerge)
    apex of heart (inferior, rounded end)
pericardium (pericardial sac)
    fibrous pericardium (outer layer)

    serous pericardium
        parietal layer (inner fused lining of fibrous pericardium)
        visceral layer (epicardium)
pericardial cavity (between parietal and visceral layers of serous pericardial membranes)
myocardium (cardiac muscle)
endocardium (lines heart chambers)

369

FIGURE 44.5 Label this frontal section of the human heart. The arrows indicate the direction of blood flow. ⚠

Brachiocephalic artery
Left common carotid artery
Left subclavian artery

Brachiocephalic veins

aorta — 1

Right pulmonary artery

7 — *pulmonary artery*
8 — *pulmonary trunk*
9 — *pulmonary vein*
10 — *left atria* (chamber)
11 — *semilunar* (valve)
12 — *bicuspid* (valve)

Superior vena cava — 2

Right pulmonary veins

aortic (semilunar) — 3
(valve)
right atria — 4
(chamber)
Opening of coronary sinus
tricuspid — 5
(valve)
Chordae tendineae

Papillary muscle

Interventricular septum
13 — *left ventricle* (chamber)

Inferior vena cava — 6

14 — *right ventricle* (chamber)

Epicardium
Myocardium
Endocardium

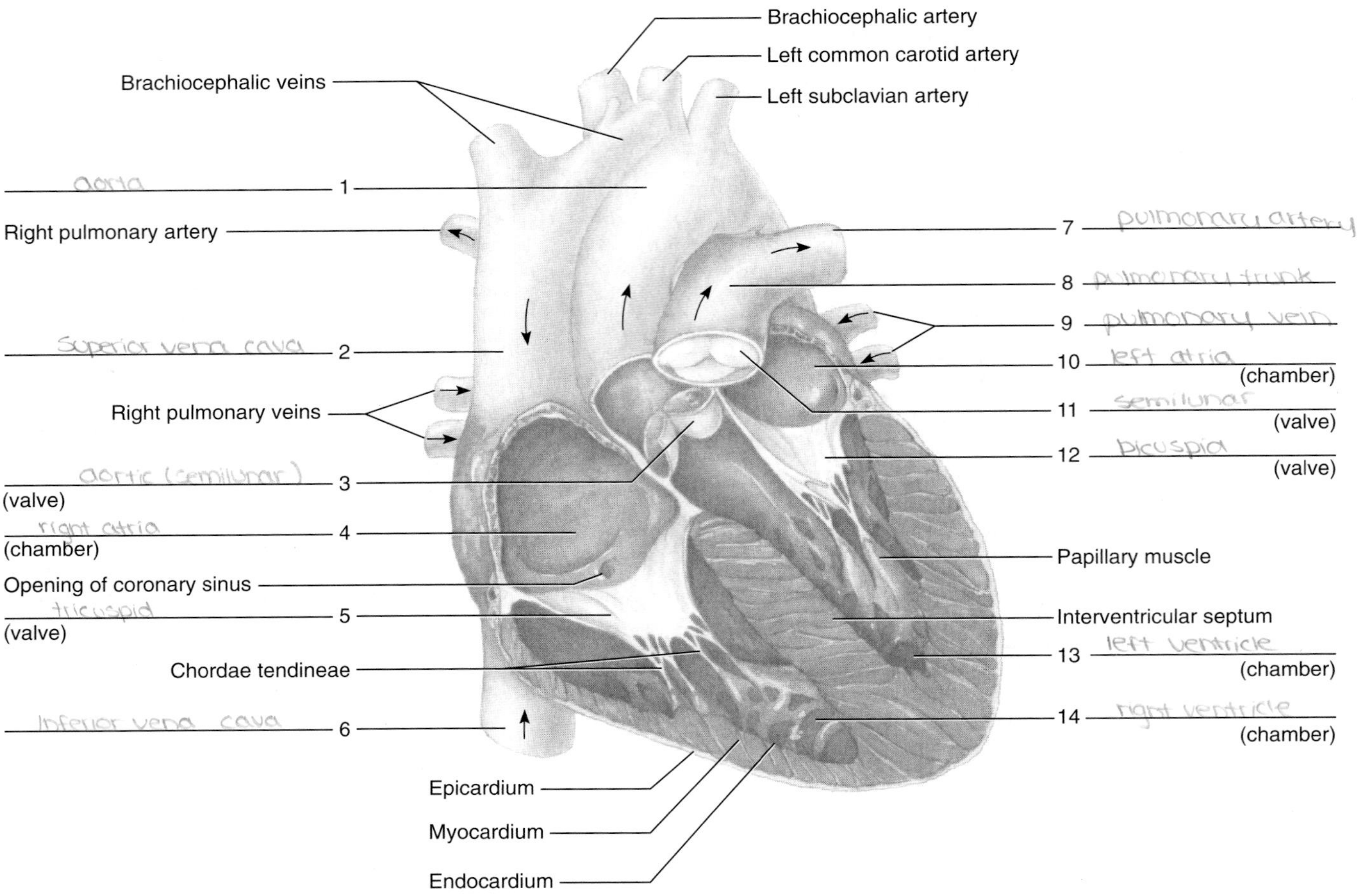

atria
  right atrium
  left atrium
  auricles
ventricles
  right ventricle
  left ventricle
atrioventricular orifices
atrioventricular valves (AV valves)
  tricuspid valve (right atrioventricular valve)
  mitral valve (left atrioventricular valve; bicuspid valve)
semilunar valves
  pulmonary valve
  aortic valve
chordae tendineae (tendinous cords)
papillary muscles
coronary sulcus (atrioventricular sulcus or groove)
interventricular sulci
  anterior sulcus
  posterior sulcus

superior vena cava
inferior vena cava
pulmonary trunk
pulmonary arteries
pulmonary veins
aorta
left coronary artery
  circumflex artery
  anterior interventricular (descending) artery
right coronary artery
  posterior interventricular artery
  marginal artery
cardiac veins
  great cardiac vein
  middle cardiac vein
  small cardiac vein
coronary sinus (for return of blood from cardiac veins into right atrium)

5. Complete Parts B and C of the laboratory report.

Learning Extension Activity

Use red and blue colored pencils to color the blood vessels in figure 44.5. Use red to illustrate a blood vessel high in oxygen, and use blue to illustrate a blood vessel low in oxygen.

Procedure B—Dissection of a Sheep Heart

1. Obtain a preserved sheep heart. Rinse it in water thoroughly to remove as much of the preservative as possible. Also run water into the large blood vessels to force any blood clots out of the heart chambers.
2. Place the heart in a dissecting tray with its ventral surface up (fig. 44.6), and proceed as follows:
   a. Although the relatively thick *pericardial sac* probably is missing, look for traces of this membrane around the origins of the large blood vessels.
   b. Locate the *visceral pericardium,* which appears as a thin, transparent layer on the surface of the heart. Use a scalpel to remove a portion of this layer and expose the *myocardium* beneath. Also note the abundance of fat along the paths of various blood vessels. This adipose tissue occurs in the loose connective tissue that underlies the visceral pericardium.
   c. Identify the following:
      right atrium
      right ventricle
      left atrium
      left ventricle
      atrioventricular sulcus
      anterior interventricular sulcus
   d. Carefully remove the fat from the anterior interventricular sulcus, and expose the blood vessels that pass along this groove. They include a branch of the *left coronary artery* (anterior interventricular artery) and a *cardiac vein.*
3. Examine the dorsal surface of the heart (fig. 44.7), and proceed as follows:
   a. Identify the *atrioventricular sulcus* and the *posterior interventricular sulcus.*
   b. Locate the stumps of two relatively thin-walled veins that enter the right atrium. Demonstrate this connection by passing a slender probe through them. The upper vessel is the *superior vena cava,* and the lower one is the *inferior vena cava.*

FIGURE 44.6 Ventral surface of sheep heart.

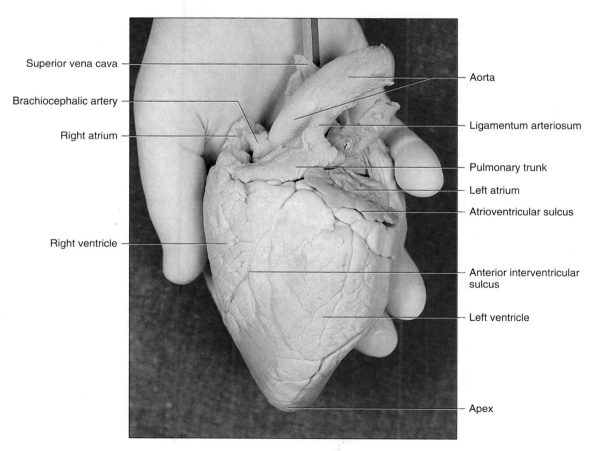

FIGURE 44.7 Dorsal surface of sheep heart. To open the right atrium, insert a blade of the scissors into the superior (anterior in sheep) vena cava and cut downward.

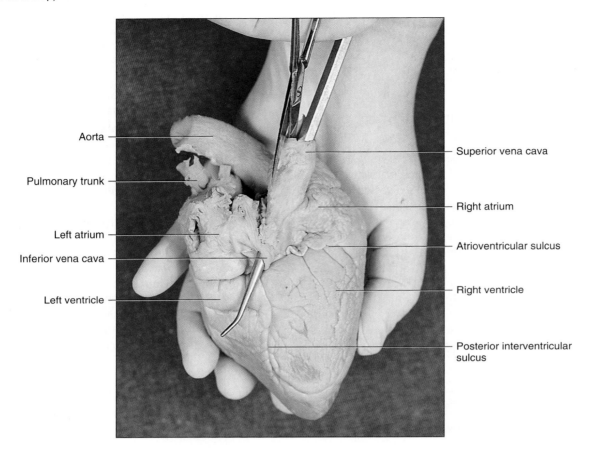

4. Open the right atrium. To do this, follow these steps:
   a. Insert a blade of the scissors into the superior vena cava and cut downward through the atrial wall (fig. 44.7).
   b. Open the chamber, locate the *right atrioventricular valve,* and examine its cusps.
   c. Also locate the opening to the *coronary sinus* between the valve and the inferior vena cava.
   d. Run some water through the right atrioventricular valve to fill the chamber of the right ventricle.
   e. Gently squeeze the ventricles, and watch the cusps of the valve as the water moves up against them.
5. Open the right ventricle as follows:
   a. Continue cutting downward through the right atrioventricular valve and the right ventricular wall until you reach the apex of the heart.
   b. Locate the *chordae tendineae* and the *papillary muscles.*
   c. Find the opening to the *pulmonary trunk,* and use the scissors to cut upward through the wall of the right ventricle. Follow the pulmonary trunk until you have exposed the *pulmonary valve.*
   d. Examine the valve and its cusps.

6. Open the left side of the heart. To do this, follow these steps:
   a. Insert the blade of the scissors through the wall of the left atrium and cut downward to the apex of the heart.
   b. Open the left atrium, and locate the four openings of the *pulmonary veins.* Pass a slender probe through each opening, and locate the stump of its vessel.
   c. Examine the *left atrioventricular valve* and its cusps.
   d. Also examine the left ventricle, and compare the thickness of its wall with that of the right ventricle.
7. Locate the aorta, which leads away from the left ventricle, and proceed as follows:
   a. Compare the thickness of the aortic wall with that of a pulmonary artery.
   b. Use scissors to cut along the length of the aorta to expose the *aortic valve* at its base.
   c. Examine the cusps of the valve, and locate the openings of the *coronary arteries* just distal to them.
8. As a review, locate and identify the stumps of each of the major blood vessels associated with the heart.
9. Discard or save the specimen as directed by the laboratory instructor.
10. Complete Part D of the laboratory report.

Laboratory Report

44

Name _____

Date _____

Section _____

The ⒜ corresponds to the Learning Outcome(s) listed at the beginning of the laboratory exercise.

Heart Structure

Part A Assessments

Examine the labeled features and identify the numbered features in figure 44.8 of the heart of a human cadaver (anterior view).

FIGURE 44.8 Identify the features on this anterior view of the heart region of a cadaver, using the terms provided. ⒜

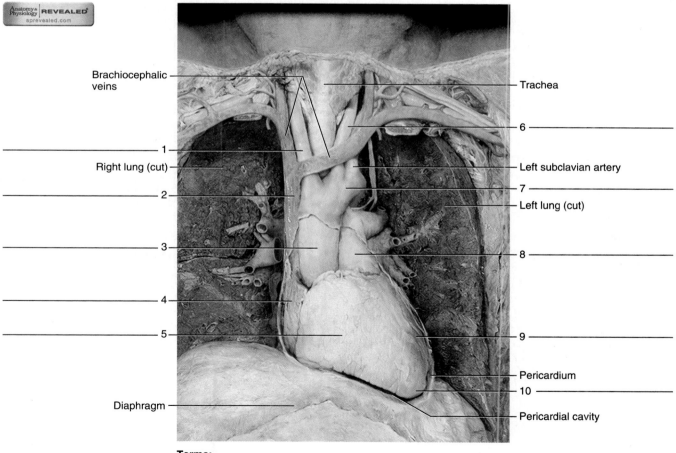

Terms:

Aortic arch	Left ventricle
Apex of heart	Pulmonary trunk
Ascending aorta	Right atrium
Brachiocephalic trunk	Right ventricle
Left common carotid artery	Superior vena cava

Part B Assessments

Locate the labeled features and identify the numbered features in figure 44.9 of a dissectible human heart model (frontal section).

FIGURE 44.9 Identify the features indicated on this anterior view of a frontal section of a human heart model, using the terms provided. (*Note:* The pulmonary valve is not shown on the portion of the model photographed.) /1\

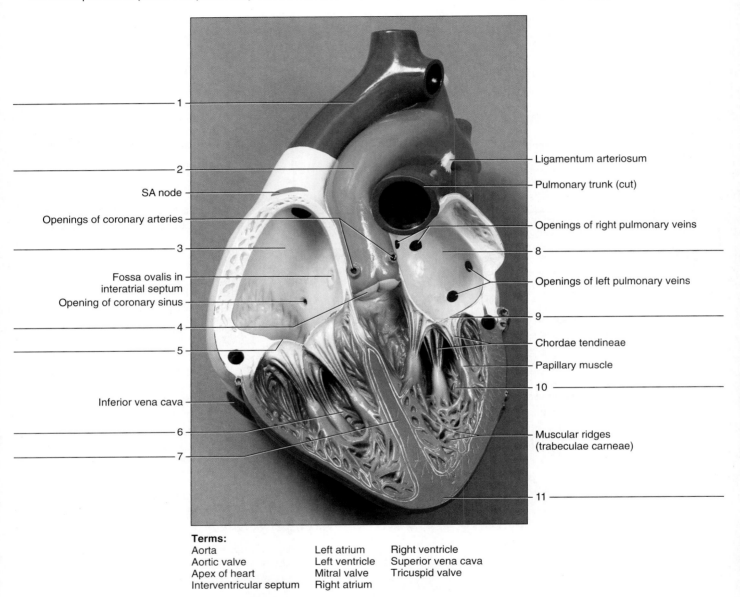

Terms:

Aorta	Left atrium	Right ventricle
Aortic valve	Left ventricle	Superior vena cava
Apex of heart	Mitral valve	Tricuspid valve
Interventricular septum	Right atrium	

Part C Assessments

Match the terms in column A with the descriptions in column B. Place the letter of your choice in the space provided. **2**

Column A	**Column B**
a. Aorta	_____ **1.** Structure from which chordae tendineae originate
b. Cardiac vein	_____ **2.** Prevents blood movement from right ventricle to right atrium
c. Coronary artery	_____ **3.** Membranes around heart
d. Coronary sinus	_____ **4.** Prevents blood movement from left ventricle to left atrium
e. Endocardium	_____ **5.** Gives rise to left and right pulmonary arteries
f. Mitral valve	_____ **6.** Drains blood from myocardium into right atrium
g. Myocardium	_____ **7.** Inner lining of heart chamber
h. Papillary muscle	_____ **8.** Layer largely composed of cardiac muscle tissue
i. Pericardial cavity	_____ **9.** Space containing serous fluid to reduce friction during heartbeats
j. Pericardial sac	_____ **10.** Drains blood from myocardial capillaries
k. Pulmonary trunk	_____ **11.** Supplies blood to heart muscle
l. Tricuspid valve	_____ **12.** Distributes blood to body organs (systemic circuit) except lungs

Part D Assessments

Complete the following:

1. Compare the structure of the right atrioventricular valve with that of the pulmonary valve. **3** _____

_____

_____

_____

_____

2. Describe the action of the right atrioventricular valve when you squeezed the water-filled right ventricle. **3** _____

_____

_____

_____

_____

3. Describe the function of the chordae tendineae and the papillary muscles. **3** _____

_____

_____

_____

_____

4. What is the significance of the difference in thickness between the wall of the aorta and the wall of the pulmonary trunk? **3**

_____

_____

_____

_____

5. List in order the major blood vessels, chambers, and valves through which blood must pass in traveling from a vena cava to the aorta. **3** _____

_____

_____

_____

_____

_____

Critical Thinking Activity

What is the significance of the difference in thickness of the ventricular walls? **3**

_____

_____

_____

_____

Cardiac Cycle

Pre-Lab

1. Carefully read the introductory material and examine the entire lab content.
2. Be familiar with the cardiac cycle, heart sounds, cardiac conduction system, and electrocardiogram (from lecture or the textbook).
3. Visit www.mhhe.com/martinseriesl for pre-lab questions and Anatomy & Physiology Revealed animations.

Materials Needed

Ph.I.L.S. 3.0

For Procedure A—Heart Sounds:
Stethoscope
Alcohol swabs (wipes)

For Procedure B—Electrocardiogram:
Electrocardiograph (or other instrument for recording an ECG)
Cot or table
Alcohol swabs (wipes)
Electrode cream (paste)
Plate electrodes and cables
Self-sticking leads (optional)
Lead selector switch

A set of atrial contractions while the ventricular walls relax, followed by ventricular contractions while the atrial walls relax, constitutes a cardiac cycle. Such a cycle is accompanied by blood pressure changes within the heart chambers, movement of blood in and out of the chambers, and opening and closing of heart valves. These valves closing produce vibrations in the tissues and thus create the sounds associated with the heartbeat. If backflow of blood occurs when a heart valve is closed, creating a turbulence noise, the condition is called a murmur.

The regulation and coordination of the cardiac cycle involves the cardiac conduction system. The pathway of electrical signals originates from the sinoatrial (SA) node located in the right atrium near the entrance of the superior vena cava. The stimulation of the heartbeat and the heart rate originate from the SA node, so it is often called the pacemaker. As the signals pass through the atrial walls toward the atrioventricular (AV) node, contractions of the atria followed by relaxations take place. Once the AV node has been signaled, the rapid continuation of the electrical signals occurs through the AV bundle and the right and left bundle branches within the interventricular septum and terminates via the Purkinje fibers throughout the ventricular walls. Once the myocardium of the ventricles has been stimulated, ventricular contractions followed by relaxations occur, completing one cardiac cycle. The sympathetic and parasympathetic subdivisions of the autonomic nervous system influence the activity of the pacemaker under various conditions. An increased rate occurs from sympathetic responses; a decreased rate occurs from parasympathetic responses.

A number of electrical changes also occur in the myocardium as it contracts and relaxes. These changes can be detected by using metal electrodes and an instrument called an *electrocardiograph*. The recording produced by the instrument is an *electrocardiogram,* or *ECG (EKG)*. Depolarization and repolarization electrical events of the cardiac cycle can be observed and interpreted from the ECG graphic recording. The heart rate can be determined by counting the number of QRS complexes on the ECG during one minute.

Purpose of the Exercise

To review the events of a cardiac cycle, to become acquainted with normal heart sounds, and to record an electrocardiogram.

Learning Outcomes

After completing this exercise, you should be able to

1. Interpret the major events of a cardiac cycle.
2. Associate the sounds produced during a cardiac cycle with the valves closing.
3. Correlate the components of a normal ECG pattern with the phases of a cardiac cycle.

④ Record and interpret an electrocardiogram.

⑤ Diagram and label the waves of an electrocardiogram (ECG) and correlate with the heart sounds.

⑥ Integrate the electrical changes and the heart sounds with the cardiac cycle.

Procedure A—Heart Sounds

1. Listen to your heart sounds. To do this, follow these steps:

   a. Obtain a stethoscope, and clean its earpieces and the diaphragm by using alcohol swabs.

   b. Fit the earpieces into your ear canals so that the angles are positioned in the forward direction.

   c. Firmly place the diaphragm (or bell) of the stethoscope on the chest over the fifth intercostal space near the apex of the heart (fig. 45.1) and listen to the sounds. This is a good location to hear the first sound (*lubb*) of a cardiac cycle when the AV valves are closing, which occurs during ventricular systole (contraction).

   d. Move the diaphragm to the second intercostal space, just to the left of the sternum, and listen to the sounds from this region. You should be able to hear the sec-ond sound (*dupp*) of the cardiac cycle clearly when the semilunar valves are closing, which occurs during ventricular diastole (relaxation).

   e. It is possible to hear sounds associated with the aortic and pulmonary valves by listening from the second intercostal space on either side of the sternum. The aortic valve sound comes from the right, and the pulmonary valve sound from the left. The sound associated with the mitral valve can be heard from the fifth intercostal space at the nipple line on the left. The sound of the tricuspid valve can be heard at the fifth intercostal space just to the right of the sternum (fig 45.1).

2. Inhale slowly and deeply, and exhale slowly while you listen to the heart sounds from each of the locations as before. Note any changes that have occurred in the sounds.

3. Exercise moderately outside the laboratory for a few minutes so that other students listening to heart sounds will not be disturbed. After the exercise period, listen to the heart sounds and note any changes that have occurred in them. *This exercise should be avoided by anyone with health risks.*

4. Complete Parts A and B of Laboratory Report 45.

FIGURE 45.1   The first sound (lubb) of a cardiac cycle can be heard by placing the diaphragm of a stethoscope over the fifth intercostal space near the apex of the heart. The second sound (dupp) can be heard over the second intercostal space, just left of the sternum. The thoracic regions circled indicate where the sounds of each heart valve are most easily heard.

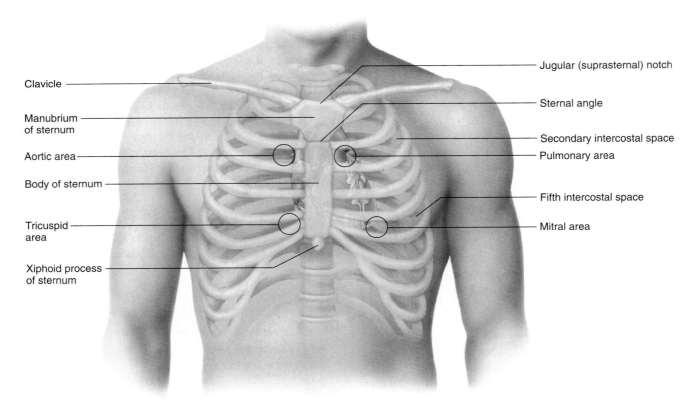

Procedure B—Electrocardiogram

1. Study figures 45.2 and 45.3, and table 45.1.
2. Complete Part C of the laboratory report.
3. The laboratory instructor will demonstrate the proper adjustment and use of the instrument available to record an electrocardiogram.

4. Record your laboratory partner's ECG. To do this, follow these steps:
   a. Have your partner lie on a cot or table close to the electrocardiograph, remaining as relaxed and still as possible.
   b. Scrub the electrode placement locations with alcohol swabs (fig. 45.4). Apply a small quantity of electrode

FIGURE 45.2 The cardiac conduction system pathway, indicated by the arrows, from the SA node to the Purkinje fibers (frontal section of heart).

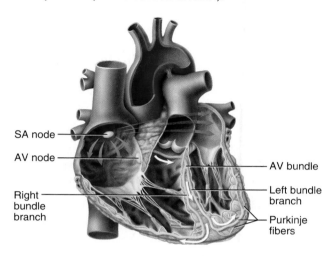

TABLE 45.1 ECG Components, Durations, and Significance

ECG Component	Duration in Seconds	Corresponding Significance
P wave	0.06–0.11	Depolarization of atrial fibers
P-Q (P-R) interval	0.12–0.20	Time for cardiac impulse from SA node through AV node
QRS complex	< 0.12	Depolarization of ventricular fibers
S-T segment	0.12	Time for ventricles to contract
Q-T interval	0.32–0.42	Time from ventricular depolarization to end of ventricular repolarization
T wave	0.16	Repolarization of ventricular fibers (ends pattern)

Note: Atrial repolarization is not observed in the ECG becaue it occurs at the same time that ventricular fibers depolarize. Therefore it is obscured by the QRS complex.

FIGURE 45.3 Components of a normal ECG pattern with a time scale.

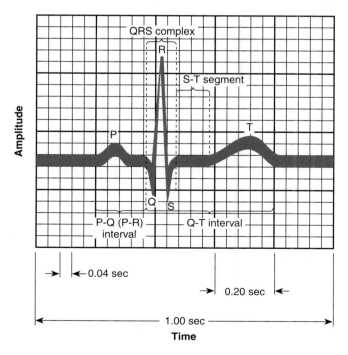

FIGURE 45.4 To record an ECG, attach electrodes to the wrists and ankles.

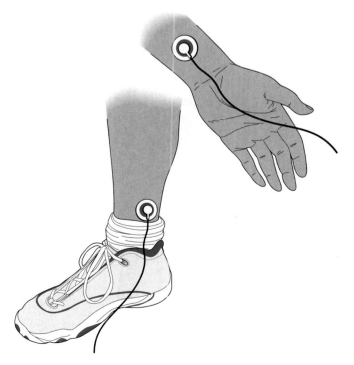

cream to the skin on the insides of the wrists and ankles. (Any jewelry on the wrists or ankles should be removed.)

c. Spread some electrode cream over the inner surfaces of four plate electrodes and attach one to each of the prepared skin areas (fig. 45.4). Make sure there is good contact between the skin and the metal of the electrodes. The electrode plate on the right ankle is the grounding system.

d. Attach the plate electrodes to the corresponding cables of a lead selector switch. When an ECG recording is made, only two electrodes are used at a time, and the selector switch allows various combinations of electrodes (leads) to be activated. Three standard limb leads placed on the two wrists and the left ankle are used for an ECG. This arrangement has become known as *Einthoven's triangle,*[1] which permits the recording of the potential difference between any two of the electrodes (fig. 45.5).

The standard leads I, II, and III are called bipolar leads because they are the potential difference between two electrodes (a positive and a negative). Lead I measures the potential difference between the right wrist (negative) and the left wrist (positive). Lead II measures the potential difference between the right wrist and the left ankle, and lead III measures the

potential difference between the left wrist and the left ankle. The right ankle is always the ground.

e. Turn on the recording instrument and adjust it as previously demonstrated by the laboratory instructor. The paper speed should be set at 2.5 cm/second. This is the standard speed for ECG recordings.

f. Set the lead selector switch to lead I (right wrist, left wrist electrodes), and record the ECG for 1 minute.

g. Set the lead selector switch to lead II (right wrist, left ankle electrodes), and record the ECG for 1 minute.

h. Set the lead selector switch to lead III (left wrist, left ankle electrodes), and record the ECG for 1 minute.

i. Remove the electrodes and clean the cream from the metal and skin.

j. Use figure 45.3 to label the ECG components of the results from leads I, II, and III. The P-Q interval is often called the P-R interval because the Q wave is often small or absent. The normal P-Q interval is 0.12–0.20 seconds. The normal QRS complex duration is less than 0.10 seconds.

5. Complete Part D of the laboratory report.

Procedure C—Ph.I.L.S. Lesson 23 ECG and Heart Function: The Meaning of Heart Sounds

Hypothesis

Would you predict the electrical stimulation of the heart to occur before the heart sounds are heard? _____

Would you predict the QRS wave to be correlated with the first heart sound, "lubb," or the second heart sound "dupp"? _____

Would you predict the T wave to be correlated with the first heart sound, "lubb" or the second heart sound "dupp"? _____

_____

1. Open Exercise 23, ECG and Heart Function: The Meaning of Heart Sounds.
2. Read the objectives and introduction and take the pre-lab quiz.
3. After completing the pre-lab quiz, read through the wet lab. Be sure to click open and view the videos that are indicated in red in the wet lab.
4. The lab exercise will open when you click Continue after completing the wet lab (fig. 45.6).

Setup

5. Check that you can hear heart sounds.
6. View animation on recording when the heart sounds are produced. (Notice that the button is pushed down with the first heart sound (lubb) and released with the second heart sound (dupp). Complete the practice test.
7. Click the power switch to turn on the data acquisition unit.
8. **Electrodes** detect electrical activity of the heart that is conducted through the body and transmit **to** the data

---

[1]Willem Einthoven (1860–1927), a Dutch physiologist, received the Nobel prize for physiology or medicine for his work with electrocardiograms.

FIGURE 45.5 Standard limb leads for electrocardiograms.

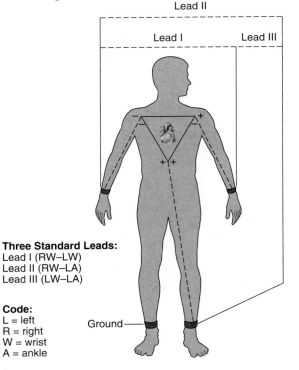

Lead II

Lead I   Lead III

Three Standard Leads:
Lead I (RW–LW)
Lead II (RW–LA)
Lead III (LW–LA)

Code:
L = left
R = right
W = wrist
A = ankle

Ground

FIGURE 45.6 Opening screen for the laboratory exercise on ECG and Heart Function: The Meaning of Heart Sounds.

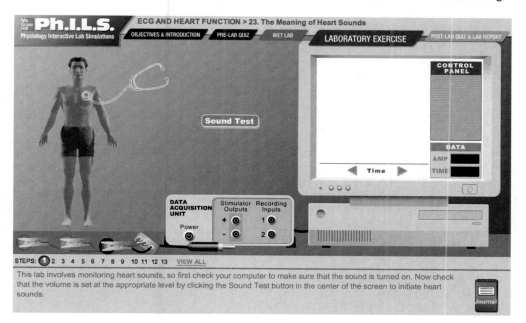

This lab involves monitoring heart sounds, so first check your computer to make sure that the sound is turned on. Now check that the volume is set at the appropriate level by clicking the Sound Test button in the center of the screen to initiate heart sounds.

acquisition unit (input). Connect the electrodes to the data acquisition unit by inserting the black plug into input 1 of the data acquisition unit and attach electrodes to the volunteer (red electrode to left wrist; black electrode [ground] to right wrist; and green electrode to left ankle).

Producing ECG and Recording Heart Sounds

9. Click Start in control panel (red line is ECG and blue line is heart sounds).
10. As the line tracing scrolls across the control panel screen, use the mouse to mark the heart sounds. Record four to six cycles.

11. Print the graph of the ECG and heart sounds by clicking the P in the bottom left of the monitor screen.

Interpreting Results

12. *With the graph still on the screen,* answer the questions in Part E of the laboratory report. If you accidentally closed the graph, click on the journal panel (red rectangle at bottom right of screen).
13. Complete the post-lab quiz (click open post laboratory quiz and lab report) by answering the ten questions on the computer screen.
14. Read the conclusion on the computer screen.
15. You may print the lab report for ECG and Heart Function: The Meaning of Heart Sounds.

Name _____

Date _____

Section _____

The ⚠ corresponds to the Learning Outcome(s) listed at the beginning of the laboratory exercise.

Cardiac Cycle

Part A Assessments

Complete the following statements:

1. The period during which a heart chamber is contracting is called _____. ⚠

2. The period during which a heart chamber is relaxing is called _____. ⚠

3. During ventricular contraction, the AV valves (tricuspid and mitral valves) are _____. ⚠

4. During ventricular relaxation, the AV valves are _____. ⚠

5. The pulmonary and aortic valves open when the pressure in the _____ exceeds the pressure in the pulmonary trunk and aorta. ⚠

6. The first sound of a cardiac cycle occurs when the _____ are closing. 2

7. The second sound of a cardiac cycle occurs when the _____ are closing. 2

8. The sound created when blood leaks back through an incompletely closed valve is called a _____. 2

Part B Assessments

Complete the following:

1. What changes did you note in the heart sounds when you inhaled deeply? 2

_____

_____

2. What changes did you note in the heart sounds following the exercise period? 2

_____

_____

Part C Assessments

Complete the following statements:

1. Normally, the _____ node serves as the pacemaker of the heart. 3

2. The _____ node is located in the inferior portion of the interatrial septum. 3

3. The large fibers on the distal side of the AV node make up the _____. 3

4. The fibers that carry cardiac impulses from the interventricular septum into the myocardium are called _____. 3

5. A(n) _____ is a recording of electrical changes occurring in the myocardium during a cardiac cycle. **3**

6. The P wave corresponds to depolarization of the muscle fibers of the _____. **3**

7. The QRS complex corresponds to depolarization of the muscle fibers of the _____. **3**

8. The T wave corresponds to repolarization of the muscle fibers of the _____. **3**

9. Why is atrial repolarization not observed in the ECG? **3** _____

_____

_____

Part D Assessments

1. Attach a short segment of the ECG recording from each of the three leads you used, and label the waves of each. **4**

Lead I

Lead II

Lead III

2. What differences do you find in the ECG patterns of these leads? ▲ _____

_____

_____

_____

_____

_____

3. How much time passed from the beginning of the P wave to the beginning of the QRS complex (P-Q interval, or P-R interval) in the ECG from lead I? ▲ _____

4. What is the significance of this P-Q (P-R) interval? ▲ _____

_____

_____

_____

_____

5. How can you determine the heart rate from an electrocardiogram? ▲ _____

_____

_____

_____

6. What was your heart rate as determined from the ECG? ▲ _____

Critical Thinking Activity

If a person's heart rate is 72 beats per minute, determine the number of QRS complexes that would have appeared on an ECG during the first 30 seconds. ▲ _____

_____

_____

_____

Part E Ph.I.L.S. Lesson 23, ECG and Heart Function: The Meaning of Heart Sounds Assessments

1. Diagram the graph of the ECG recording from the computer simulation and indicate the heart sounds. 5⃥

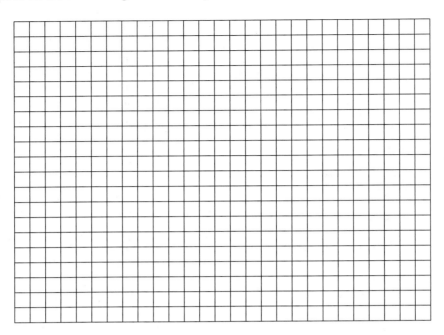

2. Answer the following questions about the graph in question 1. 5⃥

   a. The QRS wave is correlated with the _____ (first heart sound, lubb, or the second heart sound, dupp)?

   b. The T wave is correlated with the _____ (first heart sound, lubb, or the second heart sound, dupp)?

3. Complete the following: 1⃥ 2⃥ 4⃥ 5⃥ 6⃥

   a. Atrial depolarization, causing the _____ wave of the ECG, has triggered contraction of the atria, and the blood is pumped from the atria into the ventricles. The position of the valves (AV valves open and semilunar valves closed, remains unchanged).

   b. Ventricular depolarization, causing the _____ wave of the ECG, has triggered contraction of the ventricles,

   and the blood is pumped from the ventricles into the arteries (aorta and pulmonary trunk). The

   _____ valves close, producing the _____ heart sound and the

   _____ valves remain open.

   c. Ventricular repolarization, causing the _____ wave of the ECG, occurs as the ventricles are relaxing. To prevent

   the backflow of blood from the arteries into the ventricles, the _____ valves close, producing

   the _____ heart sounds.

   d. At what point do the AV valves reopen? _____

Critical Thinking Activity

If a person has a heart murmur (caused by the improper opening or closing of a heart valve), the _____

_____ (ECG or the heart sounds) will be unusual as the blood is forced through a more narrow opening (stenotic valve), or the blood will backflow through a valve that does not close properly (incompetent valve). 2⃥

Electrocardiography: BIOPAC© Exercise

Pre-Lab

1. Carefully read the introductory material and examine the entire lab content.
2. Be familiar with the cardiac cycle, cardiac conduction system, and electrocardiogram (from lecture or the textbook).

Materials Needed

Computer system (Mac—minimum 68020; PC running Windows 95/98/NT 4.0; with 4.0 MB RAM)

BIOPAC Student Lab software ver. 3.0

BIOPAC Acquisition Unit (MP30) with (AC100A) transformer

BIOPAC serial cable (CBLSERA)

BIOPAC electrode lead set (SS2L)

BIOPAC disposable vinyl electrodes (EL503), 6 electrodes per subject

BIOPAC electrode gel (GEL1) and abrasive pad (ELPAD) or alcohol wipes

Cot or lab table with pillow

Chair or stool

Safety

▶ The BIOPAC electrode lead set is safe and easy to use and should be used only as described in the procedures section of the laboratory exercise.

▶ The electrode lead clips are color-coded. Make sure they are connected to the properly placed electrode as demonstrated in figure 46.2.

▶ The vinyl electrodes are meant to be single use. Each subject should use a new set.

To enable the heart to pump blood through the body, electrical impulses are conducted through the cardiac muscle. This electrical activity can be detected on the surface of the body using electrodes and can be recorded as an electrocardiogram or ECG. The ECG detects the electrical activity that occurs just prior to cardiac muscle depolarization and just prior to repolarization.

The arrangement of the electrodes (positive, negative, and ground) with respect to each other on the body is called a lead. The leads have been standardized, and the one used in this exercise is lead II: positive electrode on the left ankle, negative electrode on the right wrist, and ground electrode on the right ankle.

A typical ECG tracing appears in figure 46.1. It consists of the following components:

- The *baseline* (isoelectric line) is a straight line used as a point of departure for the electrical activity of depolarization and repolarization in the cardiac cycle.
- The P wave results from the activity just before atrial depolarization.
- The QRS wave results from activity just prior to ventricular depolarization.
- The T wave results from activity just prior to ventricular repolarization.
- The electrical signal just prior to atrial repolarization is masked by the larger QRS wave.

In addition to the wave components, *intervals* and *segments* can be identified. An interval contains at least one wave and a straight line. For example, the P-R interval includes the P wave and the baseline before the QRS wave. A segment is the period from the end of one wave to the beginning of the next wave. For example, the P-R segment represents the time delay as the signal travels from the AV node to the ventricles.

Typical lead II ECG values are given in table 46.1.

In this lesson you will record the ECG under four conditions: lying quietly, immediately after sitting up, breathing deeply, and after moderate exercise. Using these recordings you will determine changes in rate and rhythm of the cardiac cycle.

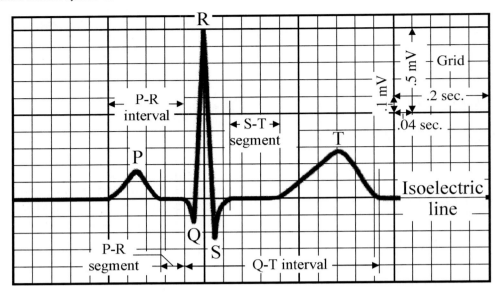

TABLE 46.1 Normal Lead II ECG Values

Phase	Duration (second)	Amplitude (millivolt)
P wave	0.06–0.11	<0.25
P-R interval	0.12–0.20	
P-R segment	0.08	
QRS complex (R)	<0.12	0.8–1.2
S-T segment	0.12	
Q-T interval	0.36–0.44	
T wave	0.16	<0.5

Purpose of the Exercise

To record an ECG and calculate heart rate and rhythm changes that occur with changes in body position and breathing.

Learning Outcomes

After completing this exercise, you should be able to

1. Record the electrical components of an electrocardiogram.
2. Correlate electrical events as displayed on the ECG with the mechanical events that occur during the cardiac cycle.
3. Interpret rate and rhythm changes in the ECG associated with changes in body position and breathing.

Procedure A—Setup

1. With your computer turned **ON** and the BIOPAC MP30 unit turned **OFF,** plug the electrode lead set (SSL2) into Channel 2. Plug the headphones into the back of the unit if desired to listen to the heartbeat.
2. Turn on the MP30 Data Acquisition Unit.

3. Attach the electrodes to the subject as shown in figure 46.2. To help to insure a good contact, make sure the area to be in contact with the electrodes is clean by wiping with the abrasive pad (ELPAD) or alcohol wipe. A small amount of BIOPAC electrode gel (GEL1) may also be used to make better contact between the sensor in the electrode and the skin. For optimal adhesion, the electrodes should be placed on the skin at least 5 minutes before the calibration procedure.
4. Attach the electrode lead set with the pinch connectors as shown in figure 46.2. Make sure the proper color electrode cable is attached to the appropriate electrode: *white lead* on medial surface of right forearm at the wrist; *red lead* on medial surface of left leg; and *black lead* on medial surface of right leg (ground).
5. Have the subject lie down on the cot or table and relax. Make sure to position the electrode cables so that they are not pulling on the electrodes. The cable clips may be attached to the subject's clothing for convenience.
6. Start the BIOPAC student lab program for Electrocardiography I, lesson L05-ECG-1.
7. Designate a unique filename to be used to save the subject's data.

Procedure B—Calibration

1. Make sure the subject is relaxed throughout the calibration procedure. *For all procedures, the subject must be still—no laughing or talking.* Click on **Calibrate.** A dialog box will direct you through the calibration procedure. After reading the box, click **OK.**
2. The calibration will stop automatically (this will take about 8 seconds).
3. The calibration recording should look similar to figure 46.3. If it does not, click **Redo Calibration.** If the baseline is not flat or there are any large spikes or jitters, **Redo Calibration.**

FIGURE 46.2 Electrode lead attachments.

Source: Courtesy of and © BIOPAC Systems Inc.

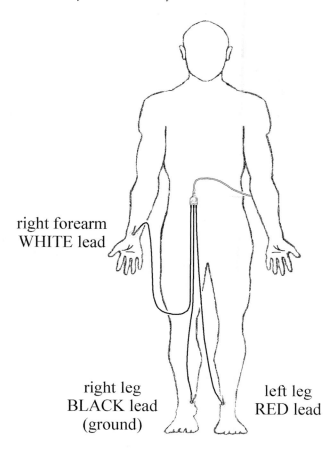

right forearm
WHITE lead

right leg
BLACK lead
(ground)

left leg
RED lead

Segment 1—Lying Down

1. With the subject lying down and relaxed, click **Record.** Record for 20 seconds.
2. Click **Suspend.** The recording should be similar to figure 46.4. If it is not, click **Redo** and repeat step 1.
3. If correct, move on to step 4 (Segment 2).

Segment 2—After Sitting Up

4. Have the subject quickly sit in a chair, with forearms resting comfortably on the chair armrests or thighs. Click on **Resume.**
5. Record for 20 seconds (seconds 21–40), and have the subject take in 5 deep breaths during recording.
6. Click **Suspend.** The recording should be similar to figure 46.5. If it is not click **Redo** and repeat step 4.
7. If correct, move on to step 8 (Segment 3). Ignore the instructions at the bottom of the screen.

Segment 3—After Exercise

8. Have the subject perform moderate exercise to elevate his/her heart rate, such as jumping jacks or running in place. The electrode cable pinch connectors may be removed for this, but not the electrodes. These need to be reattached quickly; then move on to step 9.
9. Click on **Resume.** A marker labeled "Deep Breathing" will come up automatically.
10. Record for 60 seconds (seconds 41–100).
11. Click **Suspend.** The recording should be similar to figure 46.6. If it is not, click **Redo** and repeat step 8.
12. If correct, click **Done.** Remove the electrodes.

FIGURE 46.3 Calibration recording.

Source: Courtesy of and © BIOPAC Systems Inc.

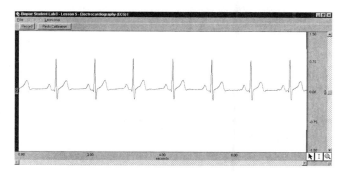

FIGURE 46.4 Segment 1 ECG—lying down.

Source: Courtesy of and © BIOPAC Systems Inc.

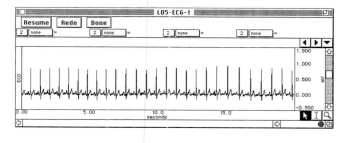

FIGURE 46.5 Segment 2 ECG—after sitting up.

Source: Courtesy of and © BIOPAC Systems Inc.

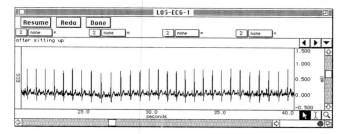

Procedure C—Recording

Read this entire section before proceeding. You will record four segments with the subject: lying down, immediately after sitting up, breathing deeply, and after exercise. You will also need to designate a *Recorder* for Segment 2 who will insert the marker—using Esc for Mac or F9 for PC users—at the beginning of an inhale and another marker at the exhale.

FIGURE 46.6 Segment 3 ECG—after exercise.

Source: Courtesy of and © BIOPAC Systems Inc.

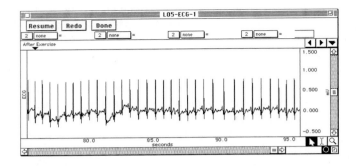

FIGURE 46.7 Selection of ECG from R wave to R wave.

Source: Courtesy of and © BIOPAC Systems Inc.

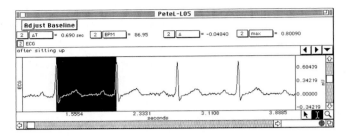

Procedure D—Data Analysis

1. You can either analyze the current file or save it and do the analysis later using the **Review Saved Data mode** and selecting the correct file. Note the channel number designations: **CH 2 ECG Lead II.**

Segment 1—Lying Down

2. Set up your display window for optimal viewing of four successive beats from Segment 1 (0–20 seconds) using the Zoom tool in the lower right corner.
3. Set up the first four pairs of channel/measurement boxes across the top as follows:
   - CH 2/Δ T
   - CH 2/BPM
   - CH 2/Δ (delta amplitude)
   - CH 2/max (maximum amplitude).

   This will give you these measurements:

   Δ **T**—the difference in time between the end and the beginning of the selected area

   BPM—the beats per minute between the end and the beginning of the selected area

   Δ—computes the difference in amplitude between the last point and the first point of the selected area

   max—the maximum amplitude within the selected area

4. After clicking the I-beam box (this will activate the select function), select the area between two successive R waves as shown in figure 46.7.
5. The values for ΔT and BPM in the measurement boxes should be recorded in table 46.2 in Part A of Laboratory Report 46.
6. Repeat for two other intervals in the current waveform display and record the data in table 46.2 in Part A of the laboratory report.
7. Zoom in on a single cardiac cycle from Segment 1. Using the I-beam tool, select each of the following (as

indicated on fig. 46.1) and record the duration (ΔT) and the amplitude (Δ) in table 46.3 in Part A of the laboratory report: P wave, P-R interval, P-R segment, QRS complex, Q-T interval, S-T segment, and T wave. Calculate the means for the three cycles as necessary to complete table 46.3.

Segment 2—After Sitting Up

8. Scroll along the horizontal axis (time indicator) to the data for Segment 2 (21–40 seconds).
9. Use the I-beam tool as in the Segment 1 analysis to select an area between two R waves. Record the data in table 46.4 in Part A of the laboratory report.
10. Repeat for two other intervals in the current waveform display and record the data in table 46.4. Calculate the means for ΔT and BPM.
11. Use the I-beam tool to determine the Q-T interval and the end of the T wave to the subsequent R wave for three areas of this waveform and record the data in table 46.5. Calculate the means of each.

Segment 3—After Exercise

12. Scroll along the horizontal axis (time indicator) to the data for Segment 3 (41–100 seconds) which is marked "Deep Breathing."
13. Use the I-beam tool as in the Segment 1 and 2 analyses to select an area between two R waves. Record the data in table 46.6 in Part A of the laboratory report.
14. Repeat for two other intervals in the current waveform display and record the data in table 46.6. Calculate the means for ΔT and BPM.
15. Use the I-beam tool to determine the Q-T interval and the end of the T wave to the subsequent R wave for three areas of this waveform and record the data in table 46.7. Calculate the means of each.
16. Exit the program.
17. Complete Part B of the laboratory report.

Name _____

Date _____

Section _____

The ⚠ corresponds to the Learning Outcome(s) listed at the beginning of the laboratory exercise.

Electrocardiography: BIOPAC© Exercise

Part A Data and Calculations Assessments

Subject Profile

Name _____   Height _____

Age_____   Weight _____

Gender: Male / Female

1. Supine, resting, regular breathing (using *Segment I* data)

Complete the following tables with the lesson data indicated, and calculate the mean and range as appropriate. ⚠

TABLE 46.2

Measurement	From Channel	Cardiac Cycle			Mean	Range
		1	**2**	**3**		
Δ T	CH 2					
BPM	CH 2					

TABLE 46.3

ECG Component	Duration Δ T				Amplitude (mV) Δ [CH 2]			
	Cycle 1	**Cycle 2**	**Cycle 3**	**Cycle 4**	**Cycle 1**	**Cycle 2**	**Cycle 3**	**Mean**
P wave								
P-R interval								
P-R segment								
QRS complex								
Q-T interval								
S-T segment								
T wave								

2. Sitting **2 3**

TABLE 46.4

Heart Rate	CH #	Cycle 1	Cycle 2	Cycle 3	Mean
Δ T	CH 2				
BPM	CH 2				

TABLE 46.5

| Ventricular Readings | CH 2 Δ T | | | |
	Cycle 1	Cycle 2	Cycle 3	Mean
Q-T interval (corresponds to ventricular systole)				
End of T wave to subsequent R wave (corresponds to ventricular diastole)				

3. After exercise **2 3**

TABLE 46.6

Heart Rate	CH #	Cycle 1	Cycle 2	Cycle 3	Mean
Δ T	CH 2				
BPM	CH 2				

TABLE 46.7

| Ventricular Readings | CH 2 Δ T | | | |
	Cycle 1	Cycle 2	Cycle 3	Mean
Q-T interval (corresponds to ventricular systole)				
End of T wave to subsequent R wave (corresponds to ventricular diastole)				

Part B Assessments

Complete the following:

1. Was there a change in heart rate when the subject went from reclining to sitting? _____

Describe the physiological mechanisms causing this change. **3**

_____

_____

_____

_____

_____

2. Is there a difference between the heart rates when the subject was sitting at rest versus after exercise? _____
Explain any differences. 3

_____

_____

_____

_____

_____

3. What changes occurred in the duration of the Q-T interval (ventricular systole) between resting and post-exercise?
Explain any differences. 3

_____

_____

_____

_____

_____

4. What changes occurred in the duration of end of T wave to R wave (ventricular diastole) between resting and post-exercise? Explain any differences. 3

_____

_____

_____

_____

_____

Blood Vessel Structure, Arteries, and Veins

Pre-Lab

1. Carefully read the introductory material and examine the entire lab content.
2. Be familiar with the structure of blood vessels and the major arteries and veins of the systemic and pulmonary circuits (from lecture or the textbook).
3. Visit www.mhhe.com/martinseriesl for pre-lab questions, Anatomy & Physiology Revealed animations, and LabCam videos.

Materials Needed

Compound light microscope
Prepared microscope slides:
   Artery cross section
   Vein cross section
Live frog or goldfish
Frog Ringer's solution
Paper towel
Rubber bands
Frog board or heavy cardboard (with a 1-inch hole cut in one corner)
Dissecting pins
Thread
Masking tape
Human torso model
Anatomical charts of the cardiovascular system
For Learning Extension Activity:
Ice
Hot plate
Thermometer

The blood vessels form a closed system of tubes that carry blood to and from the heart, lungs, and body cells. These tubes include arteries and arterioles that conduct blood

Safety

▶ Wear disposable gloves when handling the live frogs.
▶ Return the frogs to the location indicated after the experiment.
▶ Wash your hands before leaving the laboratory.

away from the heart; capillaries in which exchanges of substances occur between the blood and surrounding tissues; and venules and veins that return blood to the heart.

The blood vessels of the cardiovascular system can be divided into two major pathways—the pulmonary circuit and the systemic circuit. Within each circuit, arteries transport blood away from the heart. After exchanges of gases, nutrients, and wastes have occurred between the blood and the surrounding tissues, veins return the blood to the heart.

Frequent variations exist in anatomical structures among humans, especially in arteries and veins. The illustrations in this laboratory manual represent normal (normal means the most common variation) anatomy.

Purpose of the Exercise

To review the structure and functions of blood vessels and the circulatory pathways, and to locate major arteries and veins.

Learning Outcomes

After completing this exercise, you should be able to

1. Distinguish and sketch the three major layers in the wall of an artery and a vein.
2. Interpret the types of blood vessels in the web of a frog's foot.
3. Distinguish and trace the pulmonary and systemic circuits.
4. Locate the major arteries in these circuits on a diagram, chart, or model.
5. Locate the major veins in these circuits on a diagram, chart, or model.

Procedure A—Blood Vessel Structure

1. Obtain a microscope slide of an artery cross section, and examine it using low-power and high-power magnification. Identify the three distinct layers (tunics) of the arterial wall. The inner layer (*tunica interna*), is composed of an endothelium (simple squamous epithelium) and appears as a wavy line due to an abundance of elastic fibers that have recoiled just beneath it. The middle layer (*tunica media*) consists of numerous concentrically arranged smooth muscle cells with elastic fibers scattered among them. The outer layer (*tunica externa*) contains connective tissue rich in collagenous fibers (fig. 47.1).

2. Prepare a labeled sketch of the arterial wall in Part A of Laboratory Report 47.

3. Obtain a slide of a vein cross section and examine it as you did the artery cross section. Note the thinner wall and larger lumen relative to an artery of comparable locations. Identify the three layers of the wall, and prepare a labeled sketch in Part A of the laboratory report.

4. Complete Part A of the laboratory report.

5. Observe the blood vessels in the webbing of a frog's foot. (As a substitute for a frog, a live goldfish could be used to observe circulation in the tail.) To do this, follow these steps:

   a. Obtain a live frog. Wrap its body in a moist paper towel, leaving one foot extending outward. Secure the towel with rubber bands, but be careful not to wrap the animal so tightly that it could be injured. Try to keep the nostrils exposed.

   b. Place the frog on a frog board or on a piece of heavy cardboard with the foot near the hole in one corner.

   c. Fasten the wrapped body to the board with masking tape.

   d. Carefully spread the web of the foot over the hole and secure it to the board with dissecting pins and thread (fig. 47.2). Keep the web moist with frog Ringer's solution.

   e. Secure the board on the stage of a microscope with heavy rubber bands, and position it so that the web is beneath the objective lens.

   f. Focus on the web, using low-power magnification, and locate some blood vessels. Note the movement of the blood cells and the direction of the blood flow. You might notice that red blood cells of frogs are nucleated. Identify an arteriole, a capillary, and a venule.

   g. Examine each of these vessels with high-power magnification.

Learning Extension Activity

Investigate the effect of temperature change on the blood vessels of the frog's foot by flooding the web with a small quantity of ice water. Observe the blood vessels with low-power magnification and note any changes in their diameters or the rate of blood flow. Remove the ice water and replace it with water heated to about 35°C (95°F). Repeat your observations. What do you conclude from this experiment?

_____

_____

_____

FIGURE 47.1 Cross section of an artery and a vein (5×).

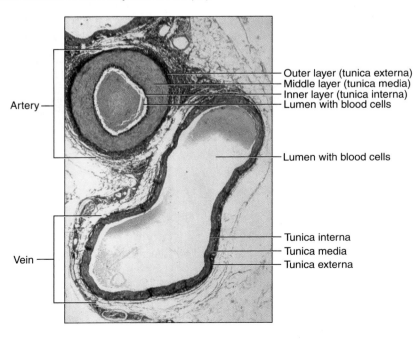

Artery

- Outer layer (tunica externa)
- Middle layer (tunica media)
- Inner layer (tunica interna)
- Lumen with blood cells

- Lumen with blood cells

Vein

- Tunica interna
- Tunica media
- Tunica externa

FIGURE 47.2 Spread the web of the foot over the hole and secure it to the board with pins and thread.

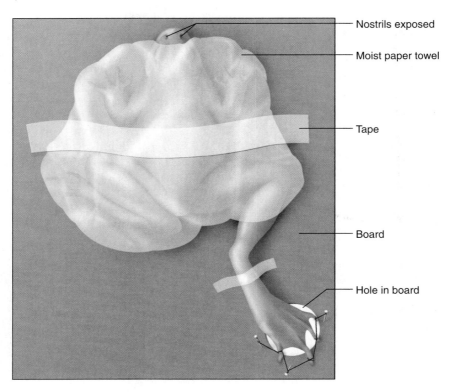

Nostrils exposed

Moist paper towel

Tape

Board

Hole in board

h. When finished, return the frog to the location indicated by your instructor. The microscope lenses and stage will likely need cleaning after the experiment.

6. Complete Part B of the laboratory report.

Procedure B—Paths of Circulation

1. Label figure 47.3.
2. Locate the following blood vessels on the available anatomical charts and the human torso model:

pulmonary trunk	1
pulmonary arteries	2
pulmonary veins	4
aorta	1
superior vena cava	1
inferior vena cava	1

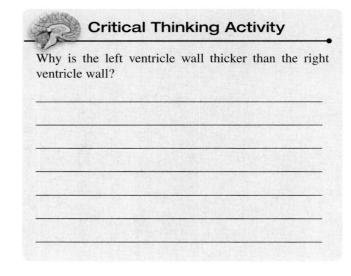

Critical Thinking Activity

Why is the left ventricle wall thicker than the right ventricle wall?

_____

_____

_____

_____

_____

_____

_____

FIGURE 47.3

Label the major blood vessels associated with the pulmonary and systemic circuits, using the terms provided. **3**

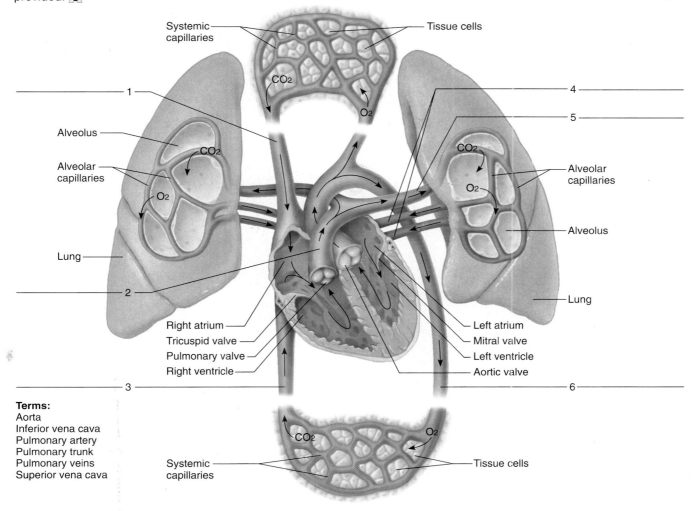

Terms:
Aorta
Inferior vena cava
Pulmonary artery
Pulmonary trunk
Pulmonary veins
Superior vena cava

Procedure C—The Arterial System

1. Study figures 47.4, 47.5, 47.6, and 47.7.
2. Locate the following arteries of the systemic circuit on the charts and human torso model:

   aorta
   ascending aorta
   aortic arch (arch of aorta)
   thoracic aorta
   abdominal aorta
   branches of the aorta
   coronary artery
   brachiocephalic trunk (artery)
   left common carotid artery
   left subclavian artery
   celiac trunk (artery)
   superior mesenteric artery
   renal artery
   inferior mesenteric artery
   median sacral artery
   common iliac artery
   arteries to neck, head, and brain
   vertebral artery
   thyrocervical trunk
   common carotid artery
   external carotid artery
   superior thyroid artery
   superficial temporal artery
   facial artery
   internal carotid artery
   arteries to shoulder and upper limb
   subclavian artery
   axillary artery
   brachial artery
   deep brachial artery
   ulnar artery
   radial artery

arteries to pelvis and lower limb
    common iliac artery
    internal iliac artery
    external iliac artery
    femoral artery
    deep artery of thigh (deep femoral artery)

    popliteal artery
    anterior tibial artery
    dorsalis pedis artery (dorsal artery of foot)
    posterior tibial artery

3. Complete Part C of the laboratory report.

FIGURE 47.4   Arteries supplying the right side of the neck and head. (The clavicle has been removed.)

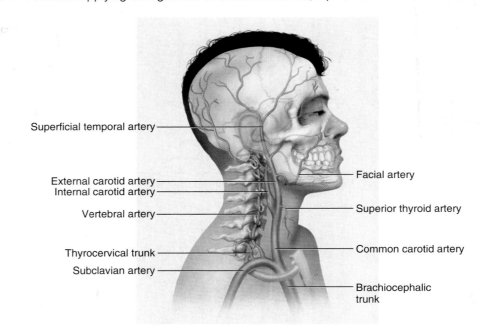

FIGURE 47.5   Major arteries of the right shoulder and upper limb.

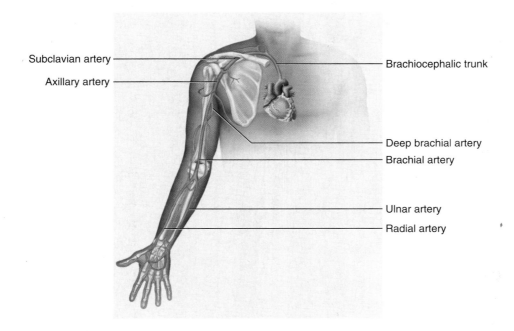

FIGURE 47.6 Major abdominal, mesenteric, and pelvic arteries (anterior view).

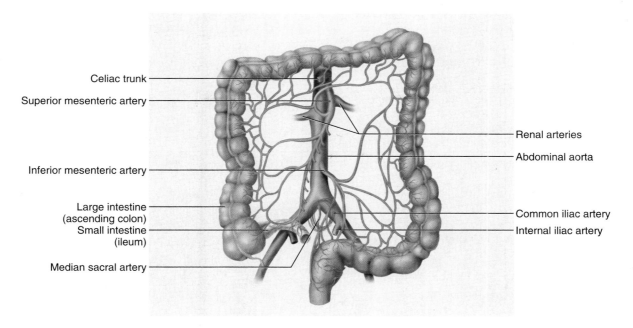

Celiac trunk
Superior mesenteric artery
Inferior mesenteric artery
Large intestine
(ascending colon)
Small intestine
(ileum)
Median sacral artery
Renal arteries
Abdominal aorta
Common iliac artery
Internal iliac artery

FIGURE 47.7 Major arteries supplying the right lower limb (anterior view).

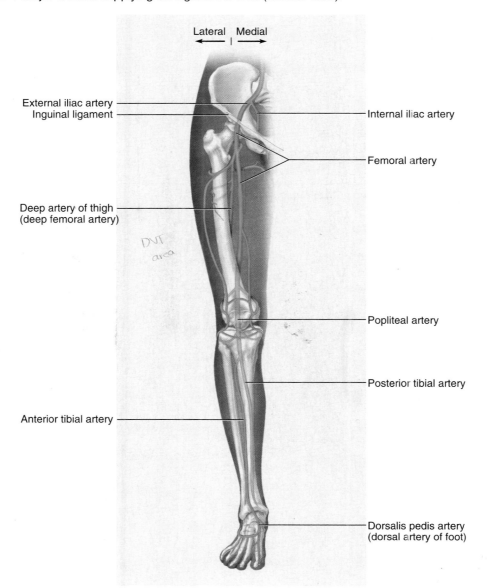

Lateral Medial

External iliac artery
Inguinal ligament
Deep artery of thigh
(deep femoral artery)
Internal iliac artery
Femoral artery
Popliteal artery
Posterior tibial artery
Anterior tibial artery
Dorsalis pedis artery
(dorsal artery of foot)

Procedure D—The Venous System

1. Study figures 47.8, 47.9, 47.10, and 47.11.
2. Locate the following veins of the systemic circuit on the charts and the human torso model:

veins from brain, head, and neck
    dural venous sinus
    external jugular vein
    internal jugular vein
    vertebral vein
    subclavian vein
    brachiocephalic vein
    superior vena cava

veins from upper limb and shoulder
    radial vein
    ulnar vein
    brachial vein
    basilic vein
    cephalic vein
    median cubital vein (antecubital vein)
    axillary vein
    subclavian vein

veins of the abdominal viscera
    hepatic portal vein
    gastric vein
    superior mesenteric vein
    splenic vein
    inferior mesenteric vein
    hepatic vein
    renal vein

veins from lower limb and pelvis
    anterior tibial vein
    posterior tibial vein
    popliteal vein
    femoral vein
    great (long) saphenous vein
    small (short) saphenous vein
    external iliac vein
    internal iliac vein
    common iliac vein
    inferior vena cava

3. Complete Parts D and E of the laboratory report.

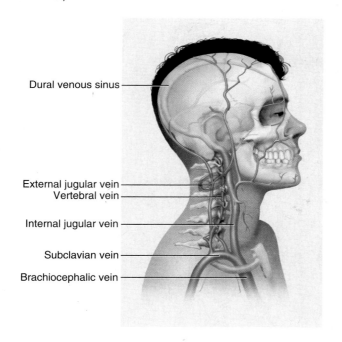

FIGURE 47.8 Major veins associated with the right side of the head and neck. (The clavicle has been removed.)

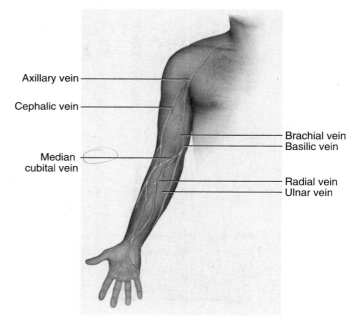

FIGURE 47.9 Anterior view of the veins of the right upper limb. The superficial veins are dark blue and the deep veins are light blue.

FIGURE 47.10 Major abdominal, hepatic portal, and pelvic veins (anterior view).

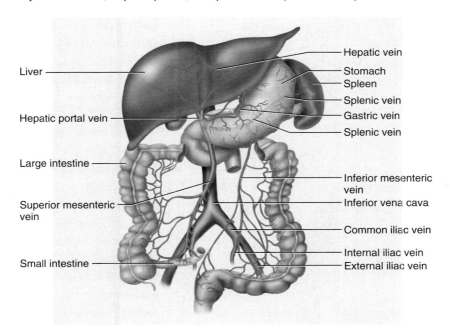

Liver

Hepatic portal vein

Large intestine

Superior mesenteric vein

Small intestine

Hepatic vein
Stomach
Spleen
Splenic vein
Gastric vein
Splenic vein
Inferior mesenteric vein
Inferior vena cava
Common iliac vein
Internal iliac vein
External iliac vein

FIGURE 47.11 Major veins of the right pelvis and lower limb (anterior view).

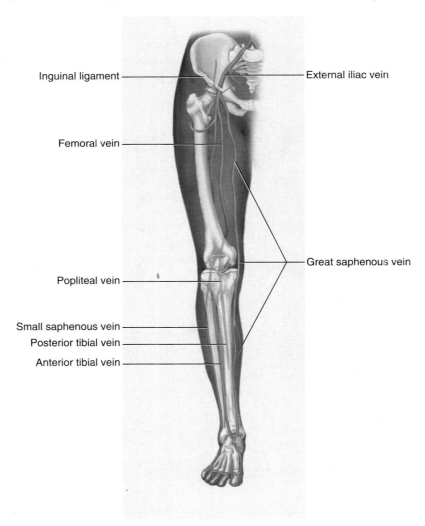

Inguinal ligament

Femoral vein

Popliteal vein

Small saphenous vein
Posterior tibial vein
Anterior tibial vein

External iliac vein

Great saphenous vein

Blood Vessel Structure, Arteries, and Veins

Part A Assessments

1. Sketch and label a section of an arterial wall next to a section of a venous wall. ⚠

   Artery wall   Vein wall

2. Describe the differences you noted in the structures of the arterial and venous walls. Mention each of the three layers of
   the wall. ⚠ _____

   _____

   _____

 Critical Thinking Activity

Explain the functional significance of the differences you noted in the structures of the arterial and venous walls. ⚠

_____

_____

Part B Assessments

Complete the following:

1. How did you distinguish between arterioles and venules when you observed the vessels in the web of the frog's foot? **2**

_____

_____

2. How did you recognize capillaries in the web? **2** _____

_____

3. What differences did you note in the rate of blood flow through the arterioles, capillaries, and venules? **2** _____

_____

_____

Part C Assessments

Provide the name of the missing artery in each of the following sequences: **3**

1. Brachiocephalic trunk, _____, right axillary artery

2. Ascending aorta, _____, descending thoracic aorta

3. Abdominal aorta, _____, ascending colon (right side of large intestine)

4. Brachiocephalic trunk, _____, right external carotid artery

5. Axillary artery, _____, radial artery

6. Common iliac artery, _____, femoral artery

7. Pulmonary trunk, _____, lungs

Part D Assessments

Provide the name of the missing vein or veins in each of the following sequences: **3**

1. Right subclavian vein, _____, superior vena cava

2. Anterior tibial vein, _____, femoral vein

3. Internal iliac vein, _____, inferior vena cava

4. Radial vein, _____, axillary vein

5. Great saphenous vein, _____, external iliac vein

6. Lungs, _____, left atrium

7. Kidney, _____, inferior vena cava

Part E Assessments

Label the major arteries and veins indicated in figures 47.12 and 47.13.

FIGURE 47.12   Label the major systemic arteries. **A**

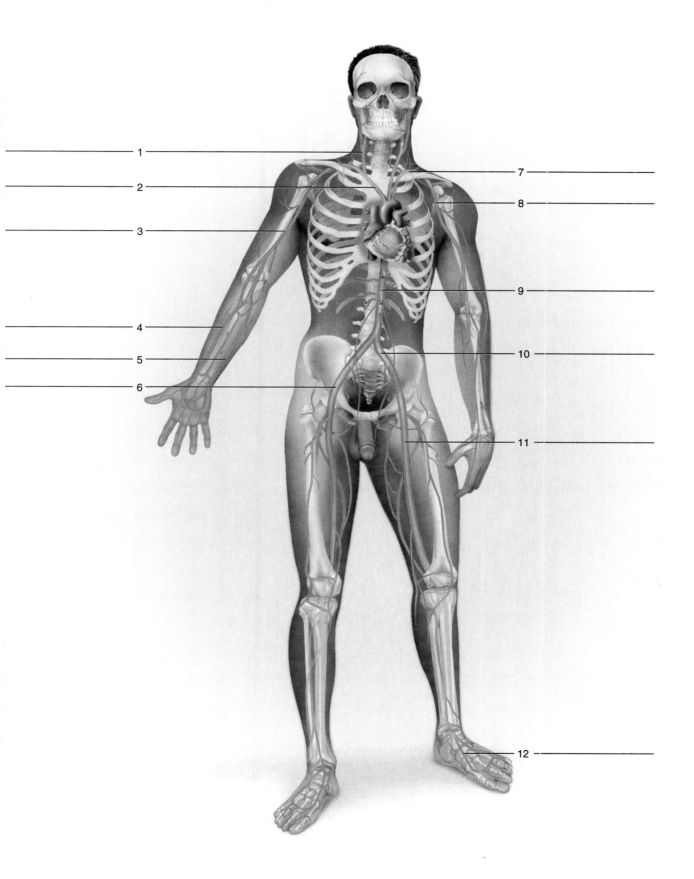

1
2
3
4
5
6
7
8
9
10
11
12

FIGURE 47.13 Label the major systemic veins. 5

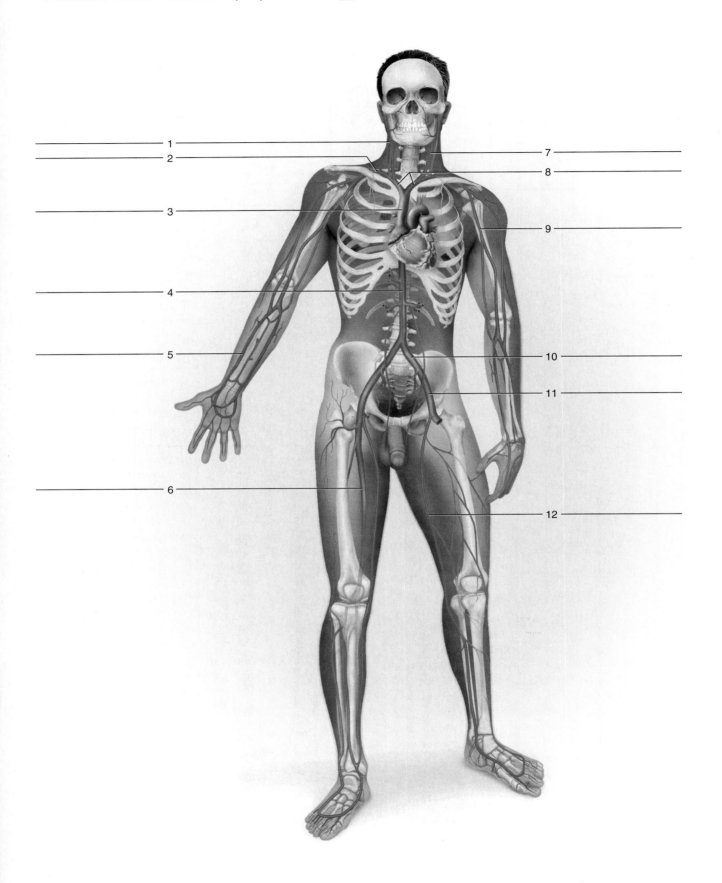

1
2
3
4
5
6
7
8
9
10
11
12

Pulse Rate and Blood Pressure

The sudden surge of blood that enters the arteries each time the ventricles of the heart contract (systole) causes the elastic walls of these vessels to expand. Then, as the ventricles relax (diastole), the arterial walls recoil. This alternate expanding and recoiling of an arterial wall can be palpated (felt) as a *pulse* in any vessel that is near the surface of the body. Because most of the major arteries are deep in the body, there are a limited number of possible locations to palpate the pulse. If any artery is superficial and firm tissue is just beneath the artery, a pulse can be palpated at that location. The number of pulse expansions per minute correlates with the heart rate or cardiac cycles. The left ventricle is the systemic pumping chamber, and it is therefore responsible for the pulse wave and for the blood pressure in the arteries selected for these assessments.

The force exerted by the blood pressing against the inner walls of arteries creates blood pressure. This systemic arterial pressure reaches its maximum, called the *systolic pressure,* during contraction of the left ventricle. Because the left ventricle relaxes when it fills with blood again, the blood pressure then drops to its lowest level, called the *diastolic pressure.* Blood pressure is measured in millimeters of mercury (mm Hg) and is expressed as a systolic pressure over diastolic pressure. A normal resting pressure is considered to be 120/80 mm Hg or slightly lower. If pressures during resting conditions are too high, the condition is called hypertension; if pressures are too low, the condition is called hypotension.

After blood has passed through the capillaries, the blood pressure established by the heart is no longer sufficient to return blood to the heart via the veins. The skeletal muscle pump of the limbs and the respiratory pump of the torso assist in venous return. When the skeletal muscles of the limbs contract, they squeeze in on the blood in the veins of the limbs, forcing the blood into the torso (valves in the veins prevent the backflow of blood). In the torso there are two cavities, the thoracic cavity and abdominopelvic cavity, divided by the diaphragm. When the diaphragm contracts (along with the external intercostal muscles), the thoracic cavity volume increases and the pressure decreases. Simultaneously, the volume of the abdominopelvic cavity decreases and the pressure increases. This difference in pressure between the abdominopelvic cavity and the thoracic cavity assists the movement of blood into the veins of the thoracic cavity and back into the heart. As a result, more blood enters the heart, and the heart rate increases to pump the additional blood. Less blood enters the heart during expiration, since the thoracic cavity volume decreases and its pressure increases.

Purpose of the Exercise

To examine the pulse, determine the pulse rate, measure blood pressure, and investigate the effects of body position and exercise on pulse rate and blood pressure. To investigate the effects of breathing on heart rate and stroke volume.

Learning Outcomes

After completing this exercise, you should be able to

1 Determine pulse rate and pulse characteristics.

2 Test the effects of various factors on pulse rate.

3 Measure and record blood pressure using a sphygmomanometer.

4 Test the effects of various factors on blood pressure.

5 Distinguish components of pulse and blood pressure measurements.

6 Explain the relationship of the heart rate (and stroke volume) to the breathing cycle.

Procedure A—Pulse Rate

1. Identify some of the pulse locations indicated in figure 48.1.

2. Examine your laboratory partner's radial pulse. To do this, follow these steps:

   a. Have your partner sit quietly, remaining as relaxed as possible.

   b. Locate the pulse by placing your index and middle fingers over the radial artery on the anterior surface of the wrist. Do not use your thumb for sensing the pulse because you may feel a pulse coming from an artery in the thumb. The radial artery is most often used because it is easy and convenient to locate.

   c. Note the characteristics of the pulse. That is, could it be described as regular or irregular, strong or weak, hard or soft? The pulse should be regular and the amplitude (magnitude) will decrease as the distance

from the left ventricle increases. The strength of the pulse gives some indications of blood pressure. Under high blood pressure, the pulse feels very hard and strong; under low blood pressure, the pulse feels weak and can be easily compressed.

   d. To determine the pulse rate, count the number of pulses that occur in 1 minute. This can be accomplished by counting pulses in 30 seconds and multiplying that number by 2. (A pulse count for a full minute, although taking a little longer, would give a better opportunity to detect irregularities.)

   e. Record the pulse rate in Part A of Laboratory Report 48.

3. Repeat the procedure and determine the pulse rate in each of the following conditions:

   a. immediately after lying down;

   b. 3–5 minutes after lying down;

   c. immediately after standing;

   d. 3–5 minutes after standing quietly;

   e. immediately after 3 minutes of moderate exercise (omit if the person has health problems);

   f. 3–5 minutes after exercise has ended.

   g. Record the pulse rate in Part A of the laboratory report.

4. Complete Part A of the laboratory report.

Learning Extension Activity

Determine the pulse rate and pulse characteristics in two additional locations on your laboratory partner. Locate and record the pulse rate from the common carotid artery and the dorsalis pedis artery (fig. 48.1).

Common carotid artery pulse rate _____

Dorsalis pedis artery pulse rate _____

Compare the amplitude of the pulse characteristic between these two pulse locations, and interpret any of

the variations noted. _____

_____

Demonstration Activity

If the equipment is available, the laboratory instructor will demonstrate how a photoelectric pulse pickup transducer or plethysmograph can be used together with a physiological recording apparatus to record the pulse. Such a recording allows an investigator to analyze certain characteristics of the pulse more precisely than is possible using a finger to examine the pulse. For example, the pulse rate can be determined very accurately from a recording, and the heights of the pulse waves provide information about the blood pressure.

FIGURE 48.1 Locations where an arterial pulse can be palpated.

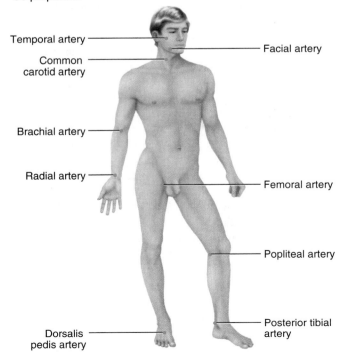

Temporal artery

Facial artery

Common carotid artery

Brachial artery

Radial artery

Femoral artery

Popliteal artery

Posterior tibial artery

Dorsalis pedis artery

FIGURE 48.2 Pressure changes in blood vessels of the systemic circuit.

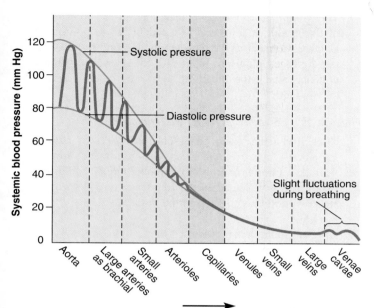

Increasing distance from left ventricle to right atrium

Procedure B—Blood Pressure

1. Although blood pressure is present in all blood vessels, the standard location to record blood pressure is the brachial artery. Examine figure 48.2, which indicates various pressure relationships throughout the systemic circuit.

2. Measure your laboratory partner's arterial blood pressure. To do this, follow these steps:
   a. Obtain a sphygmomanometer and a stethoscope.
   b. Clean the earpieces and the diaphragm of the stethoscope with alcohol swabs.
   c. Have your partner sit quietly with bare upper limb resting on a table at heart level. Have the person remain as relaxed as possible.
   d. Locate the brachial artery at the antecubital space. Wrap the cuff of the sphygmomanometer around the arm so that its lower border is about 2.5 cm above the bend of the elbow. Center the bladder of the cuff in line with the *brachial pulse* (fig. 48.3).
   e. Palpate the *radial pulse.* Close the valve on the neck of the rubber bulb connected to the cuff, and pump air from the bulb into the cuff. Inflate the cuff while watching the sphygmomanometer, and note the pressure when the pulse disappears. (This is a rough estimate of the systolic pressure.) Immediately deflate the cuff.
   f. Position the stethoscope over the brachial artery. Reinflate the cuff to a level 30 mm Hg higher than the point where the pulse disappeared during palpation.
   g. Slowly open the valve of the bulb until the pressure in the cuff drops at a rate of about 2 or 3 mm Hg per second.

FIGURE 48.3 Blood pressure is commonly measured by using a sphygmomanometer (blood pressure cuff). The use of a column of mercury is the most accurate measurement, but due to environmental concerns, it has been replaced by alternative gauges and digital readouts.

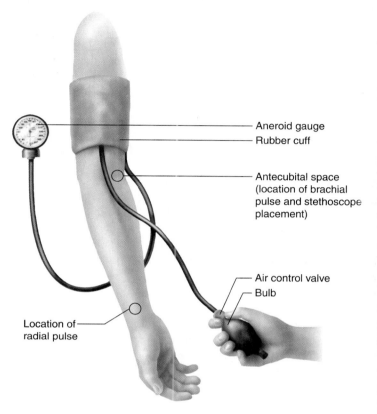

   h. Listen for sounds (Korotkoff sounds) from the brachial artery. When the first loud tapping sound is heard, record the reading as the systolic pressure. This indicates the pressure exerted against the arterial wall during systole.
   i. Continue to listen to the sounds as the pressure drops, and note the level when the last sound is heard. Record this reading as the diastolic pressure, which measures the constant arterial resistance.
   j. Release all of the pressure from the cuff.
   k. Repeat the procedure until you have two blood pressure measurements from each arm, allowing 2–3 minutes of rest between readings.
   l. Average your readings and enter them in the table in Part B of the laboratory report.

3. Measure your partner's blood pressure in each of the following conditions:
   a. 3–5 minutes after lying down;
   b. 3–5 minutes after standing quietly;
   c. immediately after 3 minutes of moderate exercise *(omit if the person has health problems);*
   d. 3–5 minutes after exercise has ended.
   e. Record the blood pressures in Part B of the laboratory report.

4. Complete Parts B and C of the laboratory report.

Procedure C—Ph.I.L.S. Lesson 36 Respiration: Deep Breathing and Cardiac Function

Hypothesis

During inspiration, would you predict the heart rate would increase or decrease? _____ Would you predict the stroke volume would increase or decrease? _____

During expiration; would you predict the heart rate would increase or decrease? _____ Would you predict the stroke volume would increase or decrease? _____

_____

1. Open Exercise 36; Respiration: Deep Breathing and Cardiac Function.
2. Read the objectives and introduction and take the pre-lab quiz.
3. After completing the pre-lab quiz, read through the wet lab. Be sure to click open and view the videos that are indicated in red in the wet lab.
4. The lab exercise will open when you click Continue after completing the wet lab (fig. 48.4).

Setup

5. Click the power switch to turn on the data acquisition unit.
6. The *finger pulse unit* indirectly measures heart rate by detecting pulses of blood that pass through the arteries of the finger, and it transmits *to* the data acquisition unit (input). Connect to the data acquisition unit by insert-ing the blue plug into input 2 of the data acquisition unit and attach the finger pulse unit to the left hand of the volunteer.
7. The *breathing apparatus* detects airflow into and out of the mouth; it transmits *to* the data acquisition unit (input). Connect to the data acquisition unit by inserting the black plug into input 1 of the data acquisition unit and attach the breathing apparatus to the volunteer's mouth.

Recording Pulse and Breathing Pattern

8. Click Start in the control panel. The red line is the breathing pattern (upward deflection is inspiration and downward deflection is expiration); the blue line is the pulse (heart rate is reflected in the frequency of the waves; stroke volume is reflected in the amplitude [height] of the wave).
9. After two complete breathing cycles, click Stop.
10. Print the graph of the breathing pattern and pulse by clicking the P in the bottom left of the monitor screen.

Interpreting Results

11. *With the graph still on the screen* answer the questions in Part D of the laboratory report. If you accidentally closed the graph, click on the journal panel (red rectangle at bottom right of screen).
12. Complete the post-lab quiz (click open post laboratory quiz and lab report) by answering the ten questions on the computer screen.
13. Read the conclusion on the computer screen.
14. You may print the lab report for Respiration: Deep Breathing and Cardiac Function.

FIGURE 48.4 Opening screen for the laboratory exercise on Respiration: Deep Breathing and Cardiac Function.

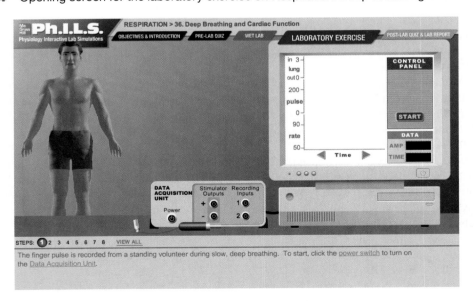

Laboratory Report

48

Name _____

Date _____

Section _____

The 🅐 corresponds to the Learning Outcome(s) listed at the beginning of the laboratory exercise.

Pulse Rate and Blood Pressure

Part A Assessments

1. Enter your observations of pulse characteristics and pulse rates in the table. 🅐1 🅐2

Test Subject	Pulse Characteristics	Pulse Rate (beats/min)
Sitting		
Lying down		
3–5 minutes later		
Standing		
3–5 minutes later		
After exercise		
3–5 minutes later		

2. Summarize the effects of body position and exercise on the characteristics and rates of the pulse. 🅐2 _____

_____

_____

_____

411

Part B Assessments

1. Enter the initial measurements of blood pressure in the table. ⎯3⎯

Reading	Blood Pressure in Right Arm	Blood Pressure in Left Arm
First		
Second		
Average		

2. Enter your test results in the table. ⎯4⎯

Test Subject	Blood Pressure
3–5 minutes after lying down	
3–5 minutes after standing	
After 3 minutes of moderate exercise	
3–5 minutes later	

3. Summarize the effects of body position and exercise on blood pressure. ⎯4⎯ _____

_____

_____

4. Summarize any correlations between pulse rate and blood pressure from any of the experimental conditions. ⎯5⎯

_____

_____

_____

Critical Thinking Activity

When a pulse is palpated and counted, which pressure (systolic or diastolic) would be characteristic at that moment? Explain your answer. ⎯1⎯

_____

_____

_____

Part C Assessments

Complete the following statements:

1. The maximum pressure achieved during ventricular contraction is called _____ pressure. /5\

2. The lowest pressure that remains in the arterial system during ventricular relaxation is called _____ pressure. /5\

3. The pulse rate is equal to the _____ rate. /5\

4. A pulse that feels full and is not easily compressed is produced by an elevated _____. /5\

5. The instrument commonly used to measure systemic arterial blood pressure is called a _____. /5\

6. Blood pressure is expressed in units of _____. /5\

7. The upper number of the fraction used to record blood pressure indicates the _____ pressure. /5\

8. The _____ artery in the arm is the standard systemic artery in which blood pressure is measured. /5\

Part D Ph.I.L.S. Lesson 36, Respiration: Deep Breathing and Cardiac Function Assessments

Use the printed graph to help you answer the following: /6\

1. During inspiration, the breathing curve is _____ (an upward, a downward) deflection, and during expiration the breathing curve is _____ (an upward, a downward) deflection.

2. The heart rate is determined by the _____ (frequency or amplitude) of the pulses.

3. The stroke volume is determined by the _____ (frequency or amplitude) of the pulses.

4. During inspiration:

   a. The heart rate _____ (increases, decreases, or stays the same).

   b. The stroke volume _____ (increases, decreases, or stays the same).

5. During expiration:

   a. The heart rate _____ (increases, decreases, or stays the same).

   b. The stroke volume _____ (increases, decreases, or stays the same).

6. Cardiac output (CO) is determined by heart rate (HR) and stroke volume (SV): HR x SV = CO. To maintain a consistent cardiac output at rest, as heart rate increases, stroke volume _____ and as heart rate decreases, stroke volume _____. Did you observe this relationship on the graph? _____ Is there a direct or inverse relationship between heart rate and stroke volume when maintaining cardiac output at a set value? _____

Critical Thinking Activity

Explain the relationship between inspiration, pressure changes in the thoracic cavity, venous return (movement of blood in veins), heart rate, stroke volume, and cardiac output at rest. /2\ /6\

_____

_____

_____

_____

_____

Lymphatic System

Pre-Lab

1. Carefully read the introductory material and examine the entire lab content.
2. Be familiar with the basic structures and functions (from lecture or the textbook) of lymphatic pathways, lymph nodes, thymus, spleen, and tonsils.
3. Visit www.mhhe.com/martinseriesl for pre-lab questions and Anatomy & Physiology Revealed animations.

Materials Needed

Human torso model
Anatomical chart of the lymphatic system
Compound light microscope
Prepared microscope slides:
    Lymph node section
    Human thymus section
    Human spleen section
    Human tonsil section

The lymphatic system is closely associated with the cardiovascular system and includes a network of vessels that assist in the circulation of body fluids. These vessels provide pathways through which excess fluid can be transported away from the interstitial spaces within most of the tissues and returned to the bloodstream. Without this system, this fluid would accumulate in tissue spaces, producing edema. The lymphatic capillaries are locations where the fluids first enter the lymphatic system. These capillaries drain into lymphatic vessels, then into larger lymphatic trunks, and finally into one of two collecting ducts just before the connections with the two subclavian veins. As a result of the orientation of valves, only one-way flow occurs. This happens during skeletal muscle contraction and breathing, causing compression upon the vessels that forces fluid through the lymphatic vessels. The larger thoracic duct provides drainage for the entire body except for the right upper quadrant. The thoracic duct connects to the left subclavian vein, while a right lymphatic duct drains only the right upper quadrant into the right subclavian vein.

Along the pathway of the lymphatic vessels lymph nodes are positioned; they are especially concentrated in areas of major joints. The lymph nodes contain concentrations of lymphocytes and macrophages that cleanse the lymph and activate an immense response. Swollen and sore lymph nodes represent areas of infection that occur as our bodies fight off the microorganisms.

Other lymphatic organs are part of the lymphatic system. The red bone marrow is a major production site of white blood cells. The thymus, located within the mediastinum, is a location of T lymphocyte maturation. The thymus is proportionately very large in a fetus and in a child, but it begins to atrophy around puberty, and this continues as we age. The spleen, the largest lymphatic organ, is located in the left upper quadrant of the abdominopelvic cavity. The spleen is highly vascular and contains large numbers of red blood cells, lymphocytes, and macrophages. Palatine, pharyngeal, and lingual tonsils are located in regions of the entrance into the pharynx (throat). Numerous nodules within the tonsils contain concentrations of lymphocytes. The organs of the lymphatic system help to defend the tissues against infections by filtering particles from the lymph and by supporting the activities of lymphocytes that furnish immunity against specific disease-causing agents or pathogens.

Purpose of the Exercise

To review the structure of the lymphatic system and to observe the microscopic structure of a lymph node, thymus, spleen, and tonsil.

Learning Outcomes

After completing this exercise, you should be able to

1. Trace the major lymphatic pathways in an anatomical chart or model.
2. Locate and identify the major clusters of lymph nodes in an anatomical chart or model.
3. Locate and sketch the major microscopic structures of a lymph node, the thymus, the spleen, and a tonsil.
4. Describe the structure and function of a lymph node, the thymus, the spleen, and a tonsil.

FIGURE 49.1 Lymphatic vessels, lymph nodes, and lymphatic organs. Label the major regions of lymph nodes by placing the correct numbers in the spaces provided. **2**

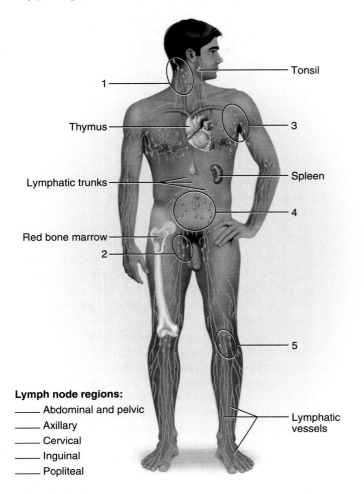

Tonsil

1

Thymus

3

Lymphatic trunks

Spleen

4

Red bone marrow

2

5

Lymph node regions:
_____ Abdominal and pelvic
_____ Axillary
_____ Cervical
_____ Inguinal
_____ Popliteal

Lymphatic vessels

Procedure A—Lymphatic Pathways

1. Study figures 49.1 and 49.2.
2. Observe the human torso model and the anatomical chart of the lymphatic system, and locate the following features:

   lymphatic vessels
   lymph nodes
   lymphatic trunks
     lumbar trunk
     intestinal trunk
     bronchomediastinal trunk
     subclavian trunk
     jugular trunk
   cisterna chyli
   collecting ducts
     thoracic duct
     right lymphatic duct
   internal jugular veins
   subclavian veins

3. Complete Part A of Laboratory Report 49.

Procedure B—Lymph Nodes

1. Observe the anatomical chart of the lymphatic system and the human torso model, and locate the clusters of lymph nodes in the following regions: **2**

   cervical region
   axillary region
   popliteal region
   inguinal region

FIGURE 49.2 Lymphatic pathways into the subclavian veins. The arrows indicate the direction of lymph drainage.

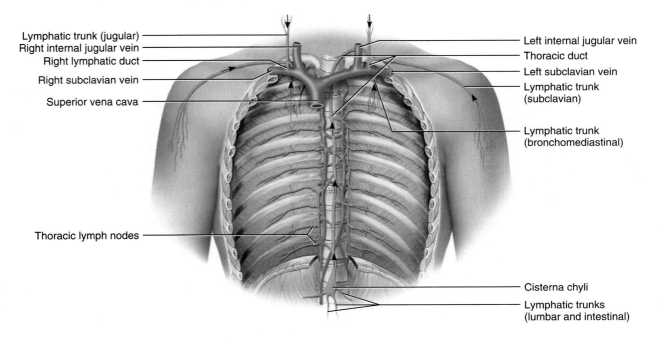

Lymphatic trunk (jugular)
Right internal jugular vein
Right lymphatic duct
Right subclavian vein
Superior vena cava

Left internal jugular vein
Thoracic duct
Left subclavian vein
Lymphatic trunk (subclavian)
Lymphatic trunk (bronchomediastinal)

Thoracic lymph nodes

Cisterna chyli
Lymphatic trunks (lumbar and intestinal)

pelvic cavity

abdominal cavity

thoracic cavity

2. Palpate the lymph nodes in your cervical region. They are located along the lower border of the mandible and between the ramus of the mandible and the sternocleidomastoid muscle. They feel like small, firm lumps.

3. As a review activity, label figure 49.1.

4. Study figures 49.3 and 49.4.

5. Obtain a prepared microscope slide of a lymph node and observe it, using low-power magnification (fig. 49.4b). Identify the *capsule* that surrounds the node and is mainly composed of collagenous fibers, the *lymph nodules* that appear as dense masses near the surface of the node, and the *lymph sinus* that appears as narrow space between the nodules and the capsule.

6. Using high-power magnification, examine a nodule within the lymph node. The nodule contains densely packed *lymphocytes*.

7. Prepare a labeled sketch of a representative section of a lymph node in Part B of the laboratory report.

8. Complete Part C of the laboratory report.

Procedure C—Thymus, Spleen, and Tonsil

1. Locate the thymus, spleen, and tonsils in the anatomical chart of the lymphatic system and on the human torso model.

2. Obtain a prepared microscope slide of human thymus and observe it, using low-power magnification (fig. 49.5). Note how the thymus is subdivided into *lobules* by *septa* of connective tissue that contain blood vessels. Identify the capsule of loose connective tissue that surrounds the thymus, the outer *cortex* of a lobule composed of densely packed cells and deeply stained, and the inner *medulla* of a lobule composed of loosely packed lymphocytes and epithelial cells and lightly stained.

3. Examine the cortex tissue of a lobule using high-power magnification. The cells of the cortex are composed of densely packed *lymphocytes* among some epithelial cells and *macrophages*. Some of these cortical cells may be undergoing mitosis, so their chromosomes may be visible.

FIGURE 49.3 Diagram of a lymph node showing internal structures, attached lymphatic vessels, and arrows indicating lymph flow.

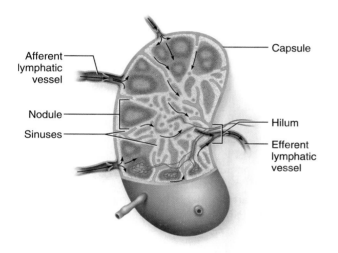

Afferent lymphatic vessel

Nodule

Sinuses

Capsule

Hilum

Efferent lymphatic vessel

FIGURE 49.4 Lymph nodes: (*a*) photograph of a human lymph node on tip of finger; (*b*) micrograph of a lymph node (5×).

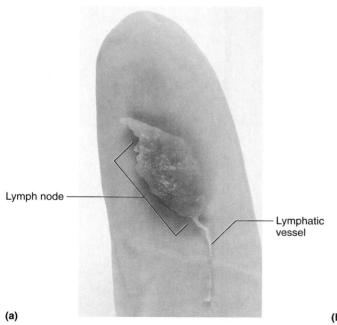

Lymph node

Lymphatic vessel

(a)

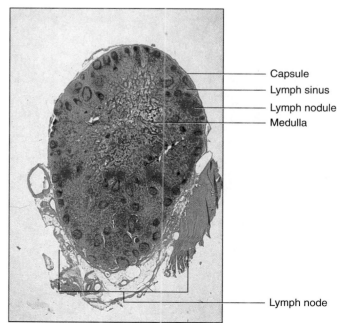

Capsule
Lymph sinus
Lymph nodule
Medulla

Lymph node

(b)

FIGURE 49.5 A section of the thymus (10× micrograph enlarged to 20×).

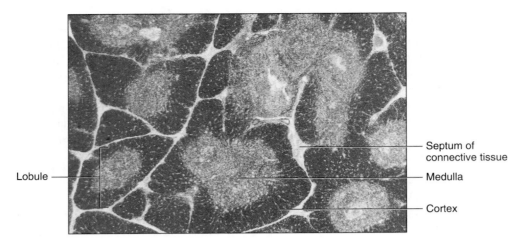

Lobule

Septum of connective tissue

Medulla

Cortex

4. Prepare a labeled sketch of a representative section of the thymus in Part B of the laboratory report.

5. Obtain a prepared slide of the human spleen and observe it, using low-power magnification (fig. 49.6). Identify the *capsule* of dense connective tissue that surrounds the spleen. The tissues of the spleen include circular *nodules of white* (in unstained tissue) *pulp* enclosed in a matrix of *red pulp.*

6. Using high-power magnification, examine a nodule of white pulp and red pulp. The cells of the white pulp are mainly *lymphocytes.* Also, there may be an arteriole centrally located in the nodule. The cells of the red pulp are mostly red blood cells with many lymphocytes and macrophages.

7. Prepare a labeled sketch of a representative section of the spleen in Part B of the laboratory report.

8. Obtain a prepared microscope slide of a tonsil and observe it, using low-power magnification (fig. 49.7). Identify the surface epithelium, deep invaginated pits called *tonsillar crypts,* and numerous lymph nodules. Although the tonsillar crypts often harbor bacteria and food debris, infections are usually prevented within these lymphoid organs.

9. Prepare a labeled sketch of a representative section of a tonsil in Part B of the laboratory report.

10. Complete Part D of the laboratory report.

FIGURE 49.7 Micrograph of a section of the pharyngeal tonsil (5×).

FIGURE 49.6 Micrograph of a section of the spleen (15×).

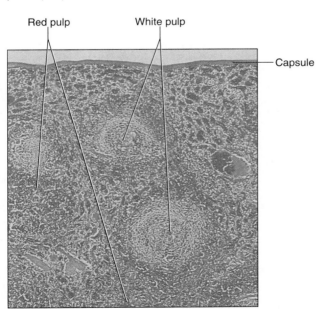

Red pulp

White pulp

Capsule

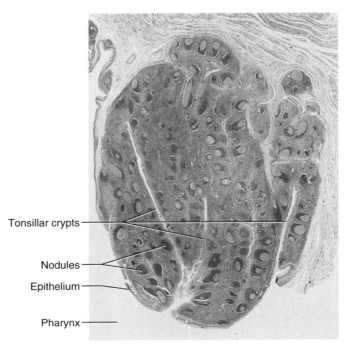

Tonsillar crypts

Nodules

Epithelium

Pharynx

Name _____

Date _____

Section _____

The Ⓐ corresponds to the Learning Outcome(s) listed at the beginning of the laboratory exercise.

Lymphatic System

Part A Assessments

Complete the following statements:

1. Lymphatic pathways begin as lymphatic _____ that merge to form lymphatic vessels. Ⓐ

2. Lymph drainage from collecting ducts enters the _____ veins. Ⓐ

3. Once tissue (interstitial) fluid is inside a lymph capillary, the fluid is called _____.

4. Lymphatic vessels contain _____ that help prevent the backflow of lymph.

5. Lymphatic vessels usually lead to _____ that filter the fluid being transported. Ⓐ

6. The _____ is the larger and longer of the two lymphatic collecting ducts. Ⓐ

Part B Assessments

Sketch the microscopic structure of the following organs: ③

Lymph node sketch (_____×)	Thymus sketch (_____×)
Spleen sketch (_____×)	Tonsil sketch (_____×)

Part C Assessments

Complete the following statements:

1. Lymph nodes contain large numbers of white blood cells called _____ and macrophages that fight invading microorganisms. **4**

2. The indented region where blood vessels and nerves join a lymph node is called the _____. **4**

3. Lymph _____ that contain germinal centers are the structural units of a lymph node. **4**

4. The spaces within a lymph node are called lymph _____ through which lymph circulates. **4**

5. Lymph enters a node through a(an) _____ lymphatic vessel. **4**

6. The partially encapsulated lymph nodes in the pharynx are called _____. **4**

7. The lymph nodes associated with the lymphatic vessels that drain the lower limbs are located in the _____ region. **2**

Part D Assessments

Complete the following statements:

1. The thymus is located in the _____, anterior to the aortic arch. **3**

2. The _____ is very large in a child and will atrophy in advanced age. **4**

3. The _____ is the largest of the lymphatic organs.

4. A _____ of connective tissue surrounds the spleen. **4**

5. The tiny islands (nodules) of tissue within the spleen that contain many lymphocytes comprise the _____ pulp. **4**

6. The _____ pulp of the spleen contains large numbers of red blood cells, lymphocytes, and macrophages. **4**

7. _____ within the spleen engulf and destroy foreign particles and cellular debris. **4**

8. The lymphoid organs in the pharynx (throat) that often are infection sites in children are collectively called _____. **4**

Laboratory Exercise 50

Respiratory Organs

Pre-Lab

1. Carefully read the introductory material and examine the entire lab content.
2. Be familiar with the basic structures and functions (from lecture or the textbook) of the major respiratory organs.
3. Visit www.mhhe.com/martinseriesl for pre-lab questions, Anatomy & Physiology Revealed animations, and LabCam Videos.

Materials Needed

Human skull (sagittal section)
Human torso model
Larynx model
Thoracic organs model
Compound light microscope
Prepared microscope slides of the following:
    Trachea (cross section)
    Lung, human (normal)

For Demonstration Activities:
Animal lung with trachea (fresh or preserved)
Bicycle pump
Prepared microscope slides of the following:
    Lung tissue (smoker)
    Lung tissue (emphysema)

Safety

▶ Wear disposable gloves when working on the fresh or preserved animal lung demonstration.
▶ Wash your hands before leaving the laboratory.

The respiratory system involves movements of oxygen from our external environment eventually to our cells, and the movements of carbon dioxide produced by our cells until it exits our body. Breathing (pulmonary ventilation) involves the nose, nasal cavity, paranasal sinuses, pharynx,

larynx, trachea, and bronchial tree, all serving as passageways for gases into and out of the lungs. The exchange of oxygen and carbon dioxide in the lungs is called external respiration; the exchange of oxygen and carbon dioxide in the tissues is called internal respiration. The blood transports gases to and from the alveolar sacs and the body cells.

Larger bronchial tubes possess supportive cartilaginous rings and plates so that they do not collapse during breathing. Smooth muscle is part of the walls of these bronchial tubes. If the smooth muscle of these tubes relaxes, the air passages dilate, which allows a greater volume of air movement. The epithelial lining of the respiratory tubes includes occasional goblet cells, which secrete protective mucus. Numerous epithelial cells possess cilia that extend into the mucus. The action of the cilia creates a current of mucus that moves any entrapped debris toward the pharynx. The mucus with any entrapped particles is normally swallowed.

The right lung has three lobes; the left lung has two lobes. A right main bronchus branches into three lobar bronchi, each extending into one of the three right lobes. A left main bronchus branches into two lobar bronchi, each extending into a left lobe. The hilum of the lung represents an area where a bronchus and major blood vessels are located. A visceral pleura adheres to the surface of each lung, including the fissures between the lobes. A parietal pleura covers the internal surface of the thoracic wall and the superior surface of the diaphragm. A narrow potential space, called the pleural cavity, is located between the two pleurae and contains pleural fluid secreted by these membranes. This pleural fluid serves as lubrication during breathing movements, and due to the surface tension, it resists separations of the pleurae.

Purpose of the Exercise

To review the structure and function of the respiratory organs and to examine the tissues of some of these organs.

Learning Outcomes

After completing this exercise, you should be able to

1. Locate the major organs and structural features of the respiratory system.
2. Describe the functions of these organs.
3. Sketch and label features of tissue sections of the trachea and lung.

Procedure A—Respiratory Organs

1. Label figures 50.1 and 50.2.
2. Examine the sagittal section of the human skull and the human torso model. Locate the following features:

nasal cavity

    naris (nostril)

    nasal septum

    nasal conchae

    meatuses

paranasal sinuses

    maxillary sinus

    frontal sinus

    ethmoidal sinus

    sphenoidal sinus

FIGURE 50.1 Label the major features of the respiratory system, using the terms provided. ⚠

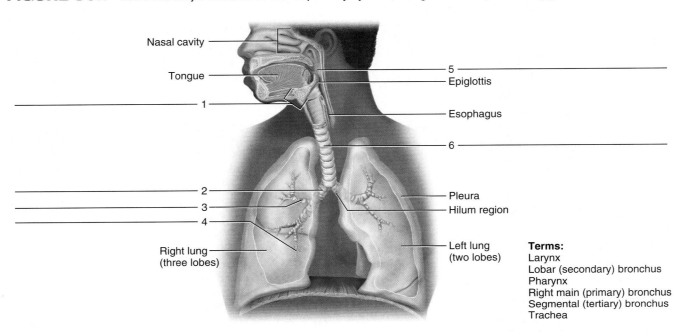

Nasal cavity
Tongue
1
5
Epiglottis
Esophagus
6
2
3
4
Pleura
Hilum region
Right lung (three lobes)
Left lung (two lobes)

Terms:
Larynx
Lobar (secondary) bronchus
Pharynx
Right main (primary) bronchus
Segmental (tertiary) bronchus
Trachea

FIGURE 50.2 Label the features of this sagittal section of the upper respiratory tract, using the terms provided. ⚠

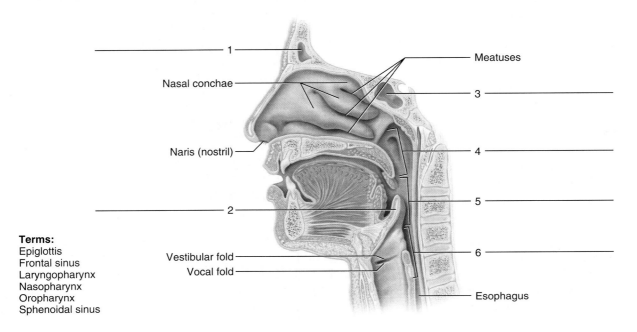

1
Meatuses
Nasal conchae
3
Naris (nostril)
4
2
5
Vestibular fold
Vocal fold
6
Esophagus

Terms:
Epiglottis
Frontal sinus
Laryngopharynx
Nasopharynx
Oropharynx
Sphenoidal sinus

FIGURE 50.3 Major features of the larynx: (*a*) anterior view; (*b*) posterior view.

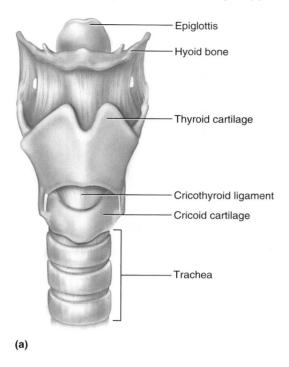

- Epiglottis
- Hyoid bone
- Thyroid cartilage
- Cricothyroid ligament
- Cricoid cartilage
- Trachea

(a)

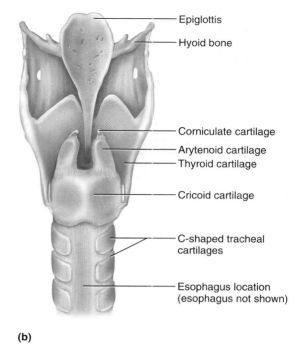

- Epiglottis
- Hyoid bone
- Corniculate cartilage
- Arytenoid cartilage
- Thyroid cartilage
- Cricoid cartilage
- C-shaped tracheal cartilages
- Esophagus location (esophagus not shown)

(b)

FIGURE 50.4 Superior view of larynx with the glottis open as seen using a laryngoscope.

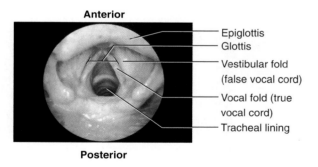

Anterior

- Epiglottis
- Glottis
- Vestibular fold (false vocal cord)
- Vocal fold (true vocal cord)
- Tracheal lining

Posterior

3. Study figures 50.3 and 50.4.
4. Observe the larynx model, the thoracic organs model, and the human torso model. Locate the following features:

   pharynx
     nasopharynx
     oropharynx
     laryngopharynx
   larynx (palpate your larynx)
     vocal cords
       vestibular folds (false vocal cords)
       vocal folds (true vocal cords)
     thyroid cartilage ("Adam's apple")

   cricoid cartilage
   cricothyroid ligament
   epiglottis
   epiglottic cartilage
   arytenoid cartilages
   corniculate cartilages
   cuneiform cartilages
   glottis
   trachea (palpate your trachea)
   bronchi
     right and left main (primary) bronchi
     lobar (secondary) bronchi
     segmental (tertiary) bronchi
   bronchioles
   lung
     hilum of lung
     lobes
       superior (upper) lobe
       middle lobe (right only)
       inferior (lower) lobe
     lobules (smallest visible subdivisions)
   visceral pleura
   parietal pleura
   pleural cavity

5. Complete Part A of Laboratory Report 50.

Demonstration Activity

Observe the animal lung and the attached trachea. Identify the larynx, major laryngeal cartilages, trachea, and the incomplete cartilaginous rings of the trachea. Open the larynx and locate the vocal folds. Examine the visceral pleura on the surface of a lung. A bicycle pump could be used to demonstrate lung inflation. Section the lung and locate some bronchioles and alveoli. Squeeze a portion of a lung between your fingers. How do you describe the texture of the lung?

_____

_____

_____

Procedure B—Respiratory Tissues

1. Obtain a prepared microscope slide of a trachea, and use low-power magnification to examine it. Notice the inner lining of ciliated pseudostratified columnar epithelium and the deep layer of hyaline cartilage, which represents a portion of an incomplete (C-shaped) tracheal ring (fig. 50.5).

FIGURE 50.5 Micrograph of a section of the tracheal wall (63×).

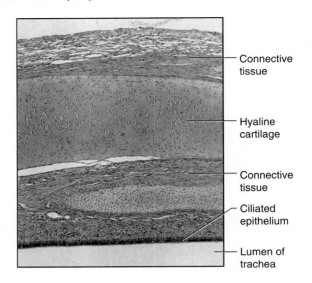

- Connective tissue
- Hyaline cartilage
- Connective tissue
- Ciliated epithelium
- Lumen of trachea

2. Use high-power magnification to observe the cilia on the free surface of the epithelial lining. Locate the wine-glass-shaped goblet cells, which secrete a protective mucus, in the epithelium (fig. 50.6).

FIGURE 50.6 Micrograph of ciliated epithelium in the respiratory tract (275×).

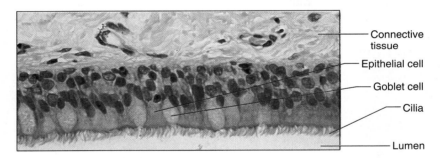

- Connective tissue
- Epithelial cell
- Goblet cell
- Cilia
- Lumen

FIGURE 50.7 Bronchioles, alveoli, and blood vessel network in a lung.

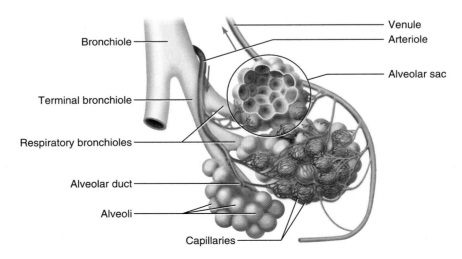

- Bronchiole
- Terminal bronchiole
- Respiratory bronchioles
- Alveolar duct
- Alveoli
- Capillaries
- Venule
- Arteriole
- Alveolar sac

FIGURE 50.8 Micrograph of human lung tissue (35×).

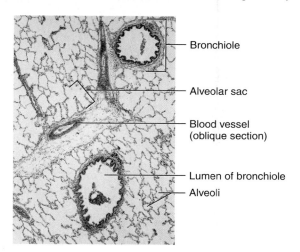

- Bronchiole
- Alveolar sac
- Blood vessel (oblique section)
- Lumen of bronchiole
- Alveoli

FIGURE 50.9 Micrograph of human lung tissue (250×).

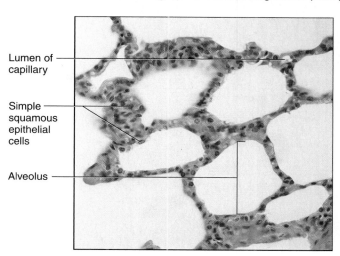

- Lumen of capillary
- Simple squamous epithelial cells
- Alveolus

3. Prepare a labeled sketch of a representative portion of the tracheal wall in Part B of the laboratory report.
4. Study the internal structures of a lung (fig. 50.7).
5. Obtain a prepared microscope slide of human lung tissue. Examine it, using low-power magnification, and note the many open spaces of the air sacs (alveoli). Look for a bronchiole—a tube with a relatively thick wall and a wavy inner lining. Locate the smooth muscle tissue in the wall of this tube (fig. 50.8). You also may see a section of cartilage as part of the bronchiole wall.
6. Use high-power magnification to examine the alveoli (fig. 50.9). Their walls are composed of simple squamous epithelium. You also may see sections of blood vessels containing blood cells.
7. Prepare a labeled sketch of a representative portion of the lung in Part B of the laboratory report.
8. Complete Part C of the laboratory report.

 Demonstration Activity

Examine the prepared microscope slides of the lung tissue of a smoker and a person with emphysema, using low-power magnification. How does the smoker's lung tissue compare with that of the normal lung tissue that you examined previously?

_____

_____

How does the emphysema patient's lung tissue compare with the normal lung tissue?

_____

_____

Name _____

Date _____

Section _____

The ⒶU corresponds to the Learning Outcome(s) listed at the beginning of the laboratory exercise.

Respiratory Organs

Part A Assessments

1. Match the terms in column A with the descriptions in column B. Place the letter of your choice in the space provided. Ⓐ1 Ⓐ2

Column A		Column B
a. Alveolus	_____	**1.** Potential space between visceral and parietal pleurae
b. Cricoid cartilage	_____	**2.** Most inferior portion of larynx
c. Epiglottis	_____	**3.** Serves as resonant chamber and reduces weight of skull
d. Glottis	_____	**4.** Microscopic air sac for gas exchange
e. Lung	_____	**5.** Consists of large lobes
f. Nasal concha	_____	**6.** Vocal folds including the opening between them
g. Pharynx	_____	**7.** Fold of mucous membrane containing elastic fibers responsible for sounds
h. Pleural cavity		
i. Sinus (paranasal sinus)	_____	**8.** Increases surface area of nasal mucous membrane
j. Vocal fold (true vocal cord)	_____	**9.** Passageway for air and food
	_____	**10.** Partially covers opening of larynx during swallowing

2. Label the structures indicated in figure 50.10.

Part B Assessments

Sketch and label a portion of the tracheal wall and a portion of lung tissue. Ⓐ3

Tracheal wall (_____×)	Lung tissue (_____×)

FIGURE 50.10 Label the features of the larynx region of a cadaver (lateral view). 1

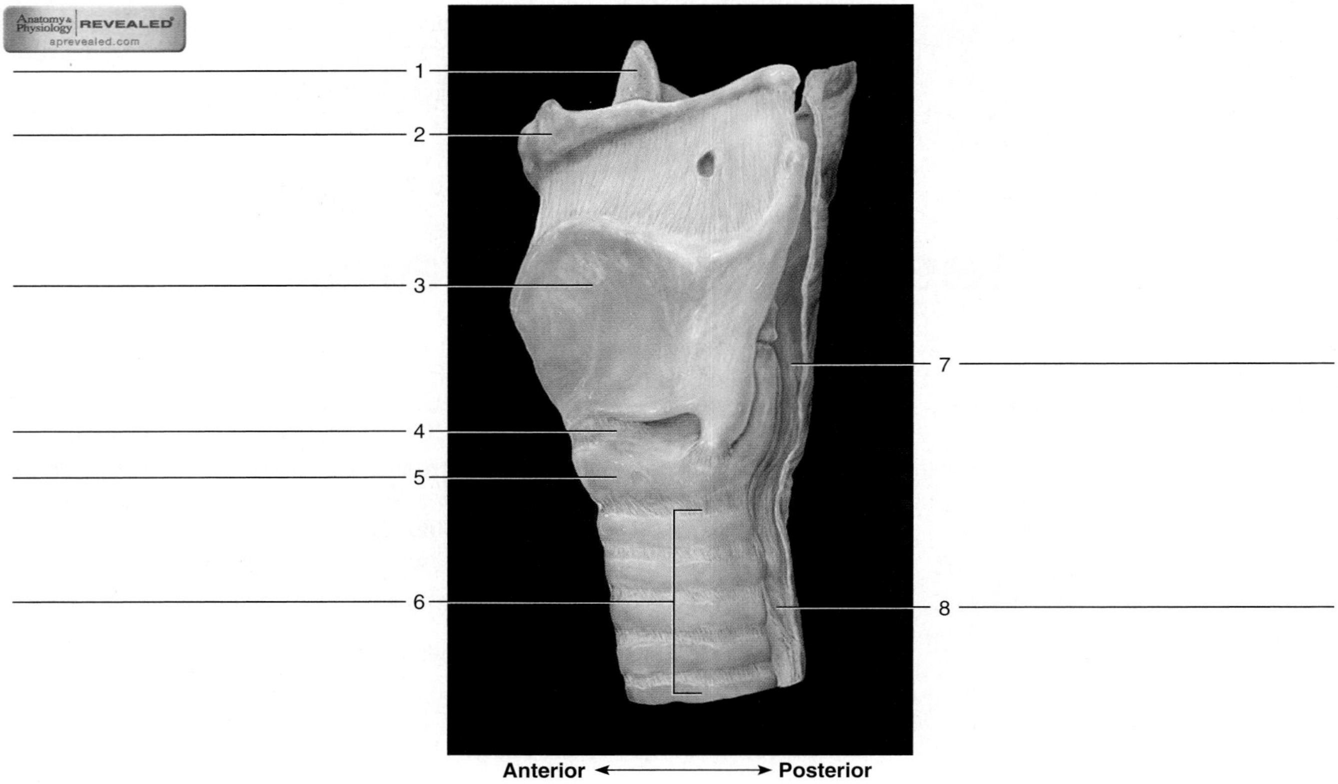

Anterior ← → Posterior

Part C Assessments

Complete the following:

1. What is the function of the mucus secreted by the goblet cells? 2 _____

_____

_____

2. Describe the function of the cilia in the respiratory tubes. 2 _____

_____

_____

3. How is breathing affected if the smooth muscle of the bronchial tree relaxes? 2 _____

_____

_____

Critical Thinking Activity

Why are the alveolar walls so thin? 2

_____

_____

_____

Laboratory Exercise 51

Breathing and Respiratory Volumes

Pre-Lab

1. Carefully read the introductory material and examine the entire lab content.
2. Be familiar with inspiration, expiration, and respiratory volumes (from lecture or the textbook).
3. Visit www.mhhe.com/martinseriesl for pre-lab questions, Anatomy & Physiology Revealed animations, and LabCam videos.

Materials Needed

Lung function model
Spirometer, handheld (dry portable)
Alcohol swabs (wipes)
Disposable mouthpieces
Nose clips
Meterstick
Ph.I.L.S. 3.0

For Learning Extension Activity:
Clock with second hand

Safety

▶ Clean the spirometer with an alcohol swab (wipe) before each use.
▶ Place a new disposable mouthpiece on the stem of the spirometer before each use.
▶ Dispose of the alcohol swabs and mouthpieces according to directions from your laboratory instructor.

Breathing, or pulmonary ventilation, involves the movement of air from outside the body through the bronchial tree and into the alveoli and the reversal of this air movement to allow gas (oxygen and carbon dioxide) exchange between air and blood. These movements are caused by changes in the size of the thoracic cavity that result from skeletal muscle contractions. The size of the thoracic cavity during inspiration is increased by contractions of the diaphragm, external intercostals, internal intercostals (interchondral part), pectoralis minor, sternocleidomastoid, and the scalenes. Expiration is aided by contractions of the internal intercostals (costal part), rectus abdominis, external oblique, and from the elastic recoil of stretched tissues.

The volumes of air that move in and out of the lungs during various phases of breathing are called *respiratory (pulmonary) volumes* and *capacities.* These volumes can be measured by using an instrument called a *spirometer.* Respiratory capacities can be determined by using various combinations of respiratory volumes. However, the values obtained vary with a person's age, sex, height, weight, stress and physical fitness.

With each breath you take, what volume of air reaches the alveoli of your lungs? You take a normal breath of air through your nose or mouth. The volume of air you breathe in (or out) is called the *tidal volume.* The average tidal volume is 500 mL (imagine the equivalent of a 1/2 liter bottle). This volume of air is moved in through your nose and mouth and through your respiratory passageway, including the nasal cavity, pharynx, larynx, trachea, and bronchial tree, to the alveoli. A certain amount of that air never reaches your alveoli—it is within the respiratory passageway. This air is said to be in the "*anatomic dead space.*" The average volume of anatomic dead space is 150 mL. If the tidal volume is 500 mL and the volume of the anatomic dead space is 150 mL, what volume of air reaches your alveoli? Understanding the difference between tidal volume and the volume of air that reaches the lungs is important because *only* the air that reaches the alveoli provides oxygen for gas exchange with the blood.

Purpose of the Exercise

To review the mechanisms of breathing and to measure or calculate certain respiratory air volumes and respiratory capacities.

Learning Outcomes

After completing this exercise, you should be able to

① Differentiate between the mechanisms responsible for inspiration and expiration.

② Measure respiratory volumes using a spirometer.

③ Calculate respiratory capacities using the data obtained from respiratory volumes.

④ Match the respiratory volumes and respiratory capacities with their definitions.

⑤ Explain the changes in tidal volume in response to changes in volume of the anatomic dead space.

⑥ Apply concepts covered in this laboratory exercise to the changes that occur when the bronchioles are dilated or constricted by the autonomic nervous system.

Procedure A—Breathing Mechanisms

1. Observe the mechanical lung function model (fig. 51.1). It consists of a heavy plastic bell jar with a rubber sheeting clamped over its wide open end. Its narrow upper opening is plugged with a rubber stopper through which a Y tube is passed. Small rubber balloons are fastened to the arms of the Y. What happens to the balloons when the rubber sheeting is pulled downward?

_____

_____

What happens when the sheeting is pushed upward?

_____

_____

2. Complete Part A of Laboratory Report 51.

FIGURE 51.1   A lung function model.

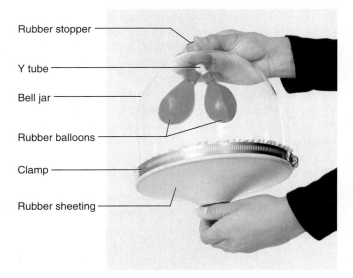

- Rubber stopper
- Y tube
- Bell jar
- Rubber balloons
- Clamp
- Rubber sheeting

Procedure B—Respiratory Volumes and Capacities (This procedure can be in conjunction with Laboratory Exercise 52.)

 Warning

If the subject begins to feel dizzy or light-headed while performing Procedure B, stop the exercise and breathe normally.

1. Get a handheld spirometer (fig. 51.2). Point the needle to zero by rotating the adjustable dial. Before using the instrument, clean it with an alcohol swab and place a new disposable mouthpiece over its stem. The instrument should be held with the dial upward, and *air should be blown only into the disposable mouthpiece* (fig. 51.3). If air tends to exit the nostrils during exhalation, use a nose clip or pinch your nose as a prevention when exhaling into the spirometer. Movement of the needle indicates the air volume that leaves the lungs. The exhalation should be slowly and forcefully performed. Too rapid an exhalation could result in erroneous data or damage to the spirometer.

2. *Tidal volume* (TV) (about 500 mL) is the volume of air that enters (or leaves) the lungs during a *respiratory cycle* (one inspiration plus the following expiration). *Resting tidal volume* is the volume of air that enters (or leaves) the lungs during normal, quiet breathing (fig. 51.4). To measure this volume, follow these steps:

   a. Sit quietly for a few moments.

   b. Position the spirometer dial so that the needle points to zero.

FIGURE 51.2   A handheld spirometer can be used to measure respiratory volumes.

FIGURE 51.3 Demonstration of use of a handheld spirometer. Air should only be blown slowly and forcefully into a disposable mouthpiece. Use a nose clip or pinch your nose when measuring expiratory reserve volume and vital capacity volume if air exits the nostrils.

c. Place the mouthpiece between your lips and exhale three ordinary expirations into it after inhaling through the nose each time. *Do not force air out of your lungs; exhale normally.*

d. Divide the total value indicated by the needle by 3, and record this amount as your resting tidal volume on the table in Part B of the laboratory report.

3. *Expiratory reserve volume* (ERV) (about 1,200 mL) is the volume of air in addition to the tidal volume that leaves the lungs during forced expiration. To measure this volume, follow these steps:

a. Breathe normally for a few moments. Set the needle to zero.

b. At the end of an ordinary expiration, place the mouthpiece between your lips and exhale all of the air you can force from your lungs through the spirometer. Use a nose clip or pinch your nose to prevent any air from exiting your nostrils.

c. Record the results as your expiratory reserve volume in Part B of the laboratory report.

4. *Vital capacity* (VC) (about 4,800 mL) is the maximum volume of air that can be exhaled after taking the deepest breath possible. To measure this volume, follow these steps:

a. Breathe normally for a few moments. Set the needle at zero.

b. Breathe in and out deeply a couple of times, then take the deepest breath possible.

c. Place the mouthpiece between your lips and exhale all the air out of your lungs, slowly and forcefully. Use a nose clip or pinch your nose to prevent any air from exiting your nostrils.

d. Record the value as your vital capacity in Part B of the laboratory report. Compare your result with that expected for a person of your sex, age, and height listed in tables 51.1 and 51.2. Use the meterstick to determine your height in centimeters if necessary or multiply your height in inches times 2.54 to calculate your height in centimeters. Considerable individual variations from the expected will be noted due to parameters other than sex, age, and height, which could include physical shape, health, medications, and others.

FIGURE 51.4 Graphic representation of respiratory volumes and capacities.

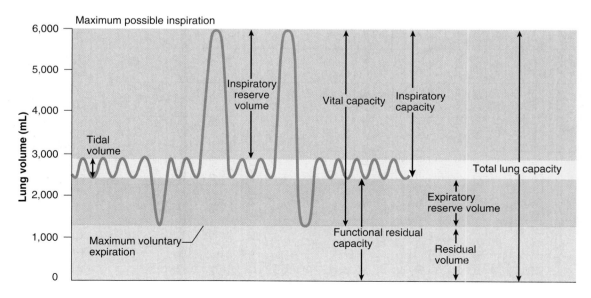

TABLE 51.1 Predicted Vital Capacities (in Milliliters) for Females

| | **Height in Centimeters** | | | | | | | | | | | | | | | | | | | | | | | | |
Age	146	148	150	152	154	156	158	160	162	164	166	168	170	172	174	176	178	180	182	184	186	188	190	192	194
16	2950	2990	3030	3070	3110	3150	3190	3230	3270	3310	3350	3390	3430	3470	3510	3550	3590	3630	3670	3715	3755	3800	3840	3880	3920
18	2920	2960	3000	3040	3080	3120	3160	3200	3240	3280	3320	3360	3400	3440	3480	3520	3560	3600	3640	3680	3720	3760	3800	3840	3880
20	2890	2930	2970	3010	3050	3090	3130	3170	3210	3250	3290	3330	3370	3410	3450	3490	3525	3565	3605	3645	3695	3720	3760	3800	3840
22	2860	2900	2940	2980	3020	3060	3095	3135	3175	3215	3255	3290	3330	3370	3410	3450	3490	3530	3570	3610	3650	3685	3725	3765	3800
24	2830	2870	2910	2950	2985	3025	3065	3100	3140	3180	3220	3260	3300	3335	3375	3415	3455	3490	3530	3570	3610	3650	3685	3725	3765
26	2800	2840	2880	2920	2960	3000	3035	3070	3110	3150	3190	3230	3265	3300	3340	3380	3420	3455	3495	3530	3570	3610	3650	3685	3725
28	2775	2810	2850	2890	2930	2965	3000	3040	3070	3115	3155	3190	3230	3270	3305	3345	3380	3420	3460	3495	3535	3570	3610	3650	3685
30	2745	2780	2820	2860	2895	2935	2970	3010	3045	3085	3120	3160	3195	3235	3270	3310	3345	3385	3420	3460	3495	3535	3570	3610	3645
32	2715	2750	2790	2825	2865	2900	2940	2975	3015	3050	3090	3125	3160	3200	3235	3275	3310	3350	3385	3425	3460	3495	3535	3570	3610
34	2685	2725	2760	2795	2835	2870	2910	2945	2980	3020	3055	3090	3130	3165	3200	3240	3275	3310	3350	3385	3425	3460	3495	3535	3570
36	2655	2695	2730	2765	2805	2840	2875	2910	2950	2985	3020	3060	3095	3130	3165	3205	3240	3275	3310	3350	3385	3420	3460	3495	3530
38	2630	2665	2700	2735	2770	2810	2845	2880	2915	2950	2990	3025	3060	3095	3130	3170	3205	3240	3275	3310	3350	3385	3420	3455	3490
40	2600	2635	2670	2705	2740	2775	2810	2850	2885	2920	2955	2990	3025	3060	3095	3135	3170	3205	3240	3275	3310	3345	3380	3420	3455
42	2570	2605	2640	2675	2710	2745	2780	2815	2850	2885	2920	2955	2990	3025	3060	3100	3135	3170	3205	3240	3275	3310	3345	3380	3415
44	2540	2575	2610	2645	2680	2715	2750	2785	2820	2855	2890	2925	2960	2995	3030	3060	3095	3130	3165	3200	3235	3270	3305	3340	3375
46	2510	2545	2580	2615	2650	2685	2715	2750	2785	2820	2855	2890	2925	2960	2995	3030	3060	3095	3130	3165	3200	3235	3270	3305	3340
48	2480	2515	2550	2585	2620	2650	2685	2715	2750	2785	2820	2855	2890	2925	2960	2995	3030	3060	3095	3130	3160	3195	3230	3265	3300
50	2455	2485	2520	2555	2590	2625	2655	2690	2720	2755	2785	2820	2855	2890	2925	2955	2990	3025	3060	3090	3125	3155	3190	3225	3260
52	2425	2455	2490	2525	2555	2590	2625	2655	2690	2720	2755	2790	2820	2855	2890	2925	2955	2990	3020	3055	3090	3125	3155	3190	3220
54	2395	2425	2460	2495	2530	2560	2590	2625	2655	2690	2720	2755	2790	2820	2855	2885	2920	2950	2985	3020	3050	3085	3115	3150	3180
56	2365	2400	2430	2460	2495	2525	2560	2590	2625	2655	2690	2720	2755	2790	2820	2855	2885	2920	2950	2980	3015	3045	3080	3110	3145
58	2335	2370	2400	2430	2460	2495	2525	2560	2590	2625	2655	2690	2720	2750	2785	2815	2850	2880	2920	2945	2975	3010	3040	3075	3105
60	2305	2340	2370	2400	2430	2460	2495	2525	2560	2590	2625	2655	2685	2720	2750	2780	2810	2845	2875	2915	2940	2970	3000	3035	3065
62	2280	2310	2340	2370	2405	2435	2465	2495	2525	2560	2590	2620	2655	2685	2715	2745	2775	2810	2840	2870	2900	2935	2965	2995	3025
64	2250	2280	2310	2340	2370	2400	2430	2465	2495	2525	2555	2585	2620	2650	2680	2710	2740	2770	2805	2835	2865	2895	2920	2955	2990
66	2220	2250	2280	2310	2340	2370	2400	2430	2460	2495	2525	2555	2585	2615	2645	2675	2705	2735	2765	2800	2825	2860	2890	2920	2950
68	2190	2220	2250	2280	2310	2340	2370	2400	2430	2460	2490	2520	2550	2580	2610	2640	2670	2700	2730	2760	2795	2820	2850	2880	2910
70	2160	2190	2220	2250	2280	2310	2340	2370	2400	2425	2455	2485	2515	2545	2575	2605	2635	2665	2695	2725	2755	2780	2810	2840	2870
72	2130	2160	2190	2220	2250	2280	2310	2335	2365	2395	2425	2455	2480	2510	2540	2570	2600	2630	2660	2685	2715	2745	2775	2805	2830
74	2100	2130	2160	2190	2220	2245	2275	2305	2335	2360	2390	2420	2450	2475	2505	2535	2565	2590	2620	2650	2680	2710	2740	2765	2795

From E. DeF. Baldwin and E. W. Richards, Jr., "Pulmonary Insufficiency 1, Physiologic Classification, Clinical Methods of Analysis, Standard Values in Normal Subjects" in *Medicine* 27:243, © by William & Wilkins. Used by permission.

5.7

432

TABLE 52.2   Predicted Vital Capacities (in Milliliters) for Males

Age	Height in Centimeters																								
	146	148	150	152	154	156	158	160	162	164	166	168	170	172	174	176	178	180	182	184	186	188	190	192	194
16	3765	3820	3870	3920	3975	4025	4075	4130	4180	4230	4285	4335	4385	4440	4490	4540	4590	4645	4695	4745	4800	4850	4900	4955	5005
18	3740	3790	3840	3890	3940	3995	4045	4095	4145	4200	4250	4300	4350	4405	4455	4505	4555	4610	4660	4710	4760	4815	4865	4915	4965
20	3710	3760	3810	3860	3910	3960	4015	4065	4115	4165	4215	4265	4320	4370	4420	4470	4520	4570	4625	4675	4725	4775	4825	4875	4930
22	3680	3730	3780	3830	3880	3930	3980	4030	4080	4135	4185	4235	4285	4335	4385	4435	4485	4535	4585	4635	4685	4735	4790	4840	4890
24	3635	3685	3735	3785	3835	3885	3935	3985	4035	4085	4135	4185	4235	4285	4330	4380	4430	4480	4530	4580	4630	4680	4730	4780	4830
26	3605	3655	3705	3755	3805	3855	3905	3955	4000	4050	4100	4150	4200	4250	4300	4350	4395	4445	4495	4545	4595	4645	4695	4740	4790
28	3575	3625	3675	3725	3775	3820	3870	3920	3970	4020	4070	4115	4165	4215	4265	4310	4360	4410	4460	4510	4555	4605	4655	4705	4755
30	3550	3595	3645	3695	3740	3790	3840	3890	3935	3985	4035	4080	4130	4180	4230	4275	4325	4375	4425	4470	4520	4570	4615	4665	4715
32	3520	3565	3615	3665	3710	3760	3810	3855	3905	3950	4000	4050	4095	4145	4195	4240	4290	4340	4385	4435	4485	4530	4580	4625	4675
34	3475	3525	3570	3620	3665	3715	3760	3810	3855	3905	3950	4000	4045	4095	4140	4190	4225	4285	4330	4380	4425	4475	4520	4570	4615
36	3445	3495	3540	3585	3635	3680	3730	3775	3825	3870	3920	3965	4010	4060	4105	4155	4200	4250	4295	4340	4390	4435	4485	4530	4580
38	3415	3465	3510	3555	3605	3650	3695	3745	3790	3840	3885	3930	3980	4025	4070	4120	4165	4210	4260	4305	4350	4400	4445	4495	4540
40	3385	3435	3480	3525	3575	3620	3665	3710	3760	3805	3850	3900	3945	3990	4035	4085	4130	4175	4220	4270	4315	4360	4410	4455	4500
42	3360	3405	3450	3495	3540	3590	3635	3680	3725	3770	3820	3865	3910	3955	4000	4050	4095	4140	4185	4230	4280	4325	4370	4415	4460
44	3315	3360	3405	3450	3495	3540	3585	3630	3675	3725	3770	3815	3860	3905	3950	3995	4040	4085	4130	4175	4220	4270	4315	4360	4405
46	3285	3330	3375	3420	3465	3510	3555	3600	3645	3690	3735	3780	3825	3870	3915	3960	4005	4050	4095	4140	4185	4230	4275	4320	4365
48	3255	3300	3345	3390	3435	3480	3525	3570	3615	3655	3700	3745	3790	3835	3880	3925	3970	4015	4060	4105	4150	4190	4235	4280	4325
50	3210	3255	3300	3345	3390	3430	3475	3520	3565	3610	3650	3695	3740	3785	3830	3870	3915	3960	4005	4050	4090	4135	4180	4225	4270
52	3185	3225	3270	3315	3355	3400	3445	3490	3530	3575	3620	3660	3705	3750	3795	3835	3880	3925	3970	4010	4055	4100	4140	4185	4230
54	3155	3195	3240	3285	3325	3370	3415	3455	3500	3540	3585	3630	3670	3715	3760	3800	3845	3890	3930	3975	4020	4060	4105	4145	4190
56	3125	3165	3210	3255	3295	3340	3380	3425	3465	3510	3550	3595	3640	3680	3725	3765	3810	3850	3895	3940	3980	4025	4065	4110	4150
58	3080	3125	3165	3210	3250	3290	3335	3375	3420	3460	3500	3545	3585	3630	3670	3715	3755	3800	3840	3880	3925	3965	4010	4050	4095
60	3050	3095	3135	3175	3220	3260	3300	3345	3385	3430	3470	3500	3555	3595	3635	3680	3720	3760	3805	3845	3885	3930	3970	4015	4055
62	3020	3060	3110	3150	3190	3230	3270	3310	3350	3390	3440	3480	3520	3560	3600	3640	3680	3730	3770	3810	3850	3890	3930	3970	4020
64	2990	3030	3080	3120	3160	3200	3240	3280	3320	3360	3400	3440	3490	3530	3570	3610	3650	3690	3730	3770	3810	3850	3900	3940	3980
66	2950	2990	3030	3070	3110	3150	3190	3230	3270	3310	3350	3390	3430	3470	3510	3550	3600	3640	3680	3720	3760	3800	3840	3880	3920
68	2920	2960	3000	3040	3080	3120	3160	3200	3240	3280	3320	3360	3400	3440	3480	3520	3560	3600	3640	3680	3720	3760	3800	3840	3880
70	2890	2930	2970	3010	3050	3090	3130	3170	3210	3250	3290	3330	3370	3410	3450	3480	3520	3560	3600	3640	3680	3720	3760	3800	3840
72	2860	2900	2940	2980	3020	3060	3100	3140	3180	3210	3250	3290	3330	3370	3410	3450	3490	3530	3570	3610	3650	3680	3720	3760	3800
74	2820	2860	2900	2930	2970	3010	3050	3090	3130	3170	3200	3240	3280	3320	3360	3400	3440	3470	3510	3550	3590	3630	3670	3710	3740

From E. DeF. Baldwin and E. W. Richards, Jr. "Pulmonary Classification, Clinical Methods of Analysis, Standard Values in Normal Subjects" in *Medicine* 27:243, © by William & Wilkins. Used by permission.

Critical Thinking Activity

It can be noted from the data in tables 51.1 and 51.2 that vital capacities gradually decrease with age. Propose an explanation for this normal correlation.

_____

_____

_____

_____

5. *Inspiratory reserve volume* (IRV) (about 3,100 mL) is the volume of air in addition to the tidal volume that enters the lungs during forced inspiration. Calculate your inspiratory reserve volume by subtracting your tidal volume (TV) and your expiratory reserve volume (ERV) from your vital capacity (VC):

$$IRV = VC - (TV + ERV)$$

8. *Inspiratory capacity* (IC) (about 3,600 mL) is the maximum volume of air a person can inhale following exhalation of the tidal volume. Calculate your inspiratory capacity by adding your tidal volume (TV) and your inspiratory reserve volume (IRV):

$$IC = TV + IRV$$

7. *Functional residual capacity* (FRC) (about 2,400 mL) is the volume of air that remains in the lungs following exhalation of the tidal volume. Calculate your func-

Learning Extension Activity

Determine your *minute respiratory volume*. To do this, follow these steps:

1. Sit quietly for a while, and then to establish your breathing rate, count the number of times you breathe in 1 minute. This might be inaccurate because conscious awareness of breathing rate can alter the results. You might ask a laboratory partner to record your breathing rate at some time when you are not expecting it to be recorded. A normal resting breathing rate is about 12–15 breaths per minute.
2. Calculate your minute respiratory volume by multiplying your breathing rate by your tidal volume:

_____ × _____ = _____
(breathing   (tidal   (minute respiratory
rate)   volume)   volume)

3. This value indicates the total volume of air that moves into your respiratory passages during each minute of ordinary breathing.

tional residual capacity (FRC) by adding your expiratory reserve volume (ERV) and your residual volume (RV), which you can assume is 1,200 mL:

$$FRC = ERV + 1,200$$

8. *Residual volume* (RV) (about 1,200 mL) is the volume of air that always remains in the lungs after the most forceful expiration. Although it is part of *total lung capacity* (about 6,000 mL), it cannot be measured with a spirometer. The residual air allows gas exchange and the alveoli to remain open during the respiratory cycle.
9. Complete Parts B and C of the laboratory report.

Procedure C—Ph.I.L.S. Lesson 34 Respiration: Altering Airway Volume

Hypothesis

If the anatomic dead space were larger, would you predict the tidal volume would increase, decrease, or stay the same so that the same amount of air reaches the alveoli? _____

_____

1. Open Exercise 34, Respiration: Altering Airway Volume
2. Read the objectives and introduction and take the pre-lab quiz.
3. After completing the pre-lab quiz, read through the wet lab. Be sure to click open and view the videos that are indicated in red in the wet lab.
4. The lab exercise will open when you click Continue after completing the wet lab (fig. 51.5).

Setup

5. Click the power switch to turn on the data acquisition unit.
6. Connect the breathing apparatus by inserting the black plug into input 1 of the data acquisition unit.
7. Attach the breathing apparatus to the subject's mouth by dragging the other end to the mouth.

Breathing Apparatus Alone

8. Click the Start button. A trace will appear on the control panel. You will see the rhythmic upward deflections of inspiration and the downward deflections of expiration. (This volume of air is tidal volume.) Record three breathing cycles and click Stop.
9. Measure the tidal volume for the first wave on the left. Position the crosshairs (using the mouse) at the top of the wave and click. A black arrow will appear. If not in the correct location, reposition by dragging to a new position. Now position the crosshairs at the bottom of the deflection (after the wave or 0 volts) and click. The result will display in the (yellow) data panel.
10. Click the journal panel (red rectangle at bottom right of screen) to enter your value into the journal. A table for

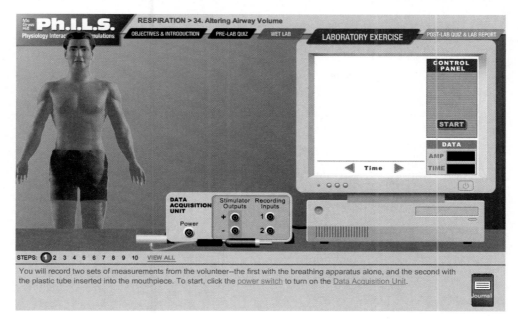

tidal volumes (measured without tube and with tube) will appear. There are three trials for each setup.

11. Repeat steps 9 and 10 for the second and third waves.
12. Print the graph of the tidal volumes by clicking the P in the bottom left of the monitor screen.
13. Close the journal window.

Breathing Apparatus with Plastic Tube

14. Insert the plastic tube into the breathing apparatus (insert into end of plastic tube on the right side).
15. Repeat steps 8 through 13.

Interpreting Results

16. *With the journal still on the screen,* answer the questions in Part D of the laboratory report. If you accidentally closed the graph, click on the journal panel (red rectangle at bottom right of screen).
17. Complete the post-lab quiz (click open post laboratory quiz and lab report) by answering the ten questions on the computer screen.
18. Read the conclusion on the computer screen.
19. You may print the lab report for Respiration: Altering Airway Volume.

Laboratory Report

51

Name _____

Date _____

Section _____

The Ⓐ corresponds to the Learning Outcome(s) listed at the beginning of the laboratory exercise.

Breathing and Respiratory Volumes

Part A Assessments

Complete the following statements:

1. When using the lung function model, what part of the respiratory system is represented by: Ⓐ**1**

   a. The rubber sheeting? _____

   b. The bell jar? _____

   c. The Y tube? _____

   d. The balloons? _____

2. When the diaphragm contracts, the size of the thoracic cavity _____. Ⓐ**1**

3. The ribs are raised by contraction of the _____
   muscles, which increases the size of the thoracic cavity. Ⓐ**1**

4. Muscles that help to force out more than the normal volume of air by pulling the ribs downward and inward include the
   _____. Ⓐ**1**

5. We inhale when the diaphragm _____. Ⓐ**1**

Part B Assessments

1. Test results for respiratory air volumes and capacities: Ⓐ**2** Ⓐ**3**

Respiratory Volume or Capacity	Expected Value* (approximate)	Test Result	Percent of Expected Value (test result/expected value × 100)
Tidal volume (resting) (TV)	500 mL		
Expiratory reserve volume (ERV)	1,200 mL		
Vital capacity (VC)	(enter yours from table 51.1 or 51.2)		
Inspiratory reserve volume (IRV)	3,100 mL		
Inspiratory capacity (IC)	3,600 mL		
Functional residual capacity (FRC)	2,400 mL		

*The values listed are most characteristic for a healthy, tall, young adult male. If your expected value for vital capacity is considerably different than 4,800 mL, your other values would vary accordingly.

2. Complete the following:

    a. How do your test results compare with the expected values? _____

_____

_____

_____

_____

    b. How does your vital capacity compare with the average value for a person of your sex, age, and height?

_____

_____

    c. What measurement in addition to vital capacity is needed before you can calculate your total lung capacity?

_____

3. If your experimental results are considerably different than the predicted vital capacities, propose reasons for the differences. As you write this paragraph, consider factors such as smoking, physical fitness, respiratory disorders, stress, and medications. (Your instructor might have you make some class correlations from class data.)

_____

_____

_____

_____

_____

_____

Part C Assessments

Match the air volumes in column A with the definitions in column B. Place the letter of your choice in the space provided. 4

Column A	Column B
a. Expiratory reserve volume	_____ **1.** Volume in addition to tidal volume that leaves the lungs during forced expiration
b. Functional residual capacity	
c. Inspiratory capacity	_____ **2.** Vital capacity plus residual volume
d. Inspiratory reserve volume	_____ **3.** Volume that remains in lungs after the most forceful expiration
e. Residual volume	
f. Tidal volume	_____ **4.** Volume that enters or leaves lungs during a respiratory cycle
g. Total lung capacity	_____ **5.** Volume in addition to tidal volume that enters lungs during forced inspiration
h. Vital capacity	
	_____ **6.** Maximum volume a person can exhale after taking the deepest possible breath
	_____ **7.** Maximum volume a person can inhale following exhalation of the tidal volume
	_____ **8.** Volume of air remaining in the lungs following exhalation of the tidal volume

Part D Ph.I.L.S. Lesson 34, Respiration: Altering Airway Volume Assessments

1. From the experimental results, complete the following table for the measured tidal volumes comparing breathing apparatus alone and breathing apparatus with a plastic tube. /2\

Trial	Without Tube	With Tube
Trial # 1		
Trial # 2		
Trial # 3		
Average (mean)		

2. Which setup had the larger average tidal volume? _____ /5\

3. To maintain normal alveolar ventilation, if the anatomic dead space is increased, the tidal volume will _____ (increase, decrease, or remain the same). /5\

4. In the setup with the tube attached to the breathing apparatus, did additional air reach the alveoli? _____ /5\

5. Complete the following table for the effect on each variable if a person breathes with the tube. /5\

Volume of Air	Change (increase, decrease, or remain the same)
Anatomic dead space	
Tidal volume	
Alveolar ventilation	

Critical Thinking Activity /6\

If the bronchioles are dilated (when the sympathetic nervous system is activated, for example), the anatomic dead space _____ (increases, decreases, or remains the same), and tidal volume will _____ _____ (increase, decrease, or remain the same). In patients with emphysema, the anatomic dead space increases (since there is a loss of elasticity of lung tissue and the thoracic cage can not effectively "be pulled back in"). Predict the changes in breathing of a patient with emphysema as he/she tries to compensate for the increase in anatomic dead space. _____

_____

_____

_____

Spirometry: BIOPAC© Exercise

The volume of air an individual can inhale and exhale can be measured using a spirometer. Two types have primarily been used in the past, a bell spirometer and a handheld spirometer. In this exercise you will measure pulmonary volumes using the BIOPAC airflow transducer. Your BIOPAC student laboratory manual will contain complete instructions if you need to troubleshoot. These are the basic steps that should take you through the exercise fairly easily.

The volumes you will measure are:

Tidal volume (TV): The amount of air a person inhales or exhales while resting (about 500 mL).

Inspiratory reserve volume (IRV): The amount of air a person can forcefully inhale above tidal inhalation (about 3,100 mL).

Expiratory reserve (ERV): The amount of air a person can forcefully exhale beyond tidal exhalation (about 1,200 mL).

Residual volume (RV): The volume of air left in the lungs after forceful exhalation. This cannot be measured and is preset at 1 L on the BIOPAC software.

Vital capacity (VC): The amount of air a person can forcefully exhale after forceful inhalation (about 4,800 mL). This can also be determined by the following equation:

$$VC = TV + IRV + ERV$$

Figure 52.1 demonstrates what these would look like on a computer-generated spirogram.

Vital capacity can also be predicted based on an individual's sex, height in centimeters (H), and age in years (A) using the following equations:

Male: $\quad VC = 0.052\,H - 0.022\,A - 3.60$

Female: $\quad VC = 0.041\,H - 0.018\,A - 2.69$

The pulmonary volumes described previously can also be used to determine other lung capacities.

Inspiratory capacity (IC): $IC = TV + IRV$

Expiratory capacity (EC): $EC = TV + RV$

Functional residual capacity (FRC): $FRC = ERV + RV$

Total lung capacity (TLC):
$$TLC = IRV + TV + ERV + RV$$

The measures of these volumes and capacities can be used when evaluating pulmonary health.

FIGURE 52.1 Example of computer-generated respiratory volumes and capacities.

Source: Courtesy of and © BIOPAC Systems Inc.

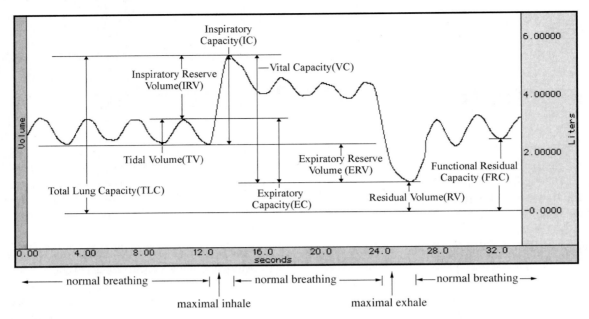

Purpose of the Exercise

To measure and calculate pulmonary volumes and capacities using the BIOPAC airflow transducer.

Learning Outcomes

After completing this exercise, you should be able to

1. Examine experimentally and record and/or calculate selected pulmonary volumes and capacities.
2. Compare the observed volumes and capacities with average values.
3. Compare the volumes and capacities of subjects differing in sex, age, height, and weight.

Procedure A—Setup

1. With your computer **ON** and the BIOPAC MP30 unit turned **OFF,** plug the airflow transducer (SS11L or SSL11LA) into Channel 1.
2. Turn on the MP30 Data Acquisition Unit.
3. Connect a bacteriological filter and calibration syringe/filter assembly into the airflow transducer as shown in figure 52.2. (Be sure it is inserted on the side labeled "Inlet.")
4. Start the BIOPAC student lab program for Pulmonary Function I, lesson L12-Lung-1.
5. Designate a filename to be used to save the subject's data.

FIGURE 52.2 Setup for calibration.

Source: Courtesy of and © BIOPAC Systems Inc.

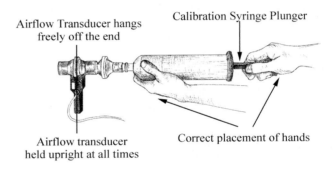

Procedure B—Calibration

1. Pull the calibration syringe plunger all the way out and hold the calibration syringe/filter assembly parallel.
2. The calibration occurs in two stages. During the first stage, simply hold the assembly still and click on the **Calibrate** button.
3. For the second calibration stage, click **Yes** after reading the alert box. Then push the syringe plunger in and out completely 5 times (10 strokes). This is to mimic restful breathing and should be at a rate of: push in 1 second, wait 2 seconds, pull out 1 second, wait 2 seconds, and repeat.
4. Click on **End Calibration.** If your calibration data resembles Figure 52.3, proceed to the **Recording** section. If it does not, **Redo** calibration.

FIGURE 52.3 Sample calibration recording.

Source: Courtesy of and © BIOPAC Systems Inc.

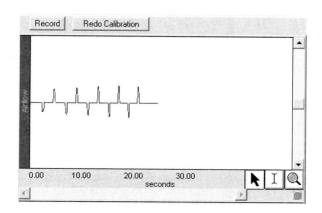

FIGURE 52.5 Sample subject recording.

Source: Courtesy of and © BIOPAC Systems Inc.

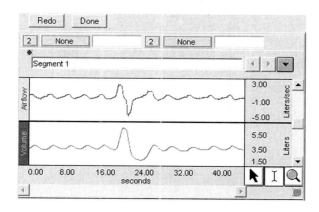

Procedure C—Recording

1. Insert a clean mouthpiece (and filter if applicable) into the airflow transducer. With nose clip on, begin breathing into the transducer (fig. 52.4).
2. Have the subject breathe normally for about 20 seconds, then click on **Record** and have the subject
   a. breathe normally for 5 breaths
      - inhale as deeply as possible
      - exhale as deeply as possible
   b. breathe normally for 5 breaths
   c. click on **Stop**
3. The recording should look similar to figure 52.5. If you need to redo, click on **Redo,** otherwise, click **Done.**

FIGURE 52.4 Subject setup.

Source: Courtesy of and © BIOPAC Systems Inc.

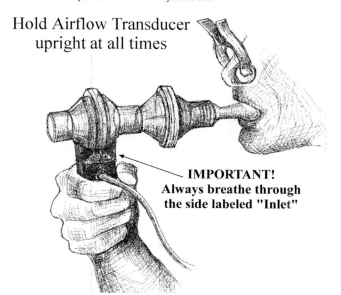

Hold Airflow Transducer upright at all times

IMPORTANT!
Always breathe through the side labeled "Inlet"

Procedure D—Data Analysis

1. You can either analyze the current file or save it and do the analysis later using the **Review Saved Data mode.** Since you are interested in volumes, you need to first turn off the Airflow channel. With the data file open, click the number 1 (for Channel 1) in the upper left corner. To turn it off: with a PC, hold down the Ctrl key, or with a Mac the Option key, then click on Channel 1. Only one recording channel, Volumes, should be on the screen.
2. Prepare the first measurement box after the number 2 (for Channel 2) with p-p (point to point). This will automatically give you your subject's value for the volume selected by the I-beam cursor. To help you with determining where to set the cursor, refer to figure 52.1.
3. With the I-beam cursor, select the entire region during the first 5 normal breaths. The p-p indicates the tidal volume (TV). Record this value on Part A of Laboratory Report 52.
4. Select the area from the peak of the last normal breath of the first 5 breaths to the peak of the deep inspiration. The p-p represents inspiratory reserve volume (IRV). Record this value on Part A of the laboratory report.
5. Select the area from the lowest point of deep expiration to the end of the normal breath following it. The p-p represents expiratory reserve volume (ERV). Record this value on Part A of the laboratory report.
6. Select the area from the beginning of deep inspiration to the end of deep expiration. The p-p represents vital capacity (VC). Record this value on Part A of the laboratory report.
7. When finished, you can save the data on the hard drive, a disk, or a network drive and click **EXIT.**
8. The values obtained can be used to calculate the various pulmonary capacities in order to complete Part A of the laboratory report.
9. Complete Part B of the laboratory report.

Name _____

Date _____

Section _____

The ⚠ corresponds to the Learning Outcome(s) listed at the beginning of the laboratory exercise.

Spirometry: BIOPAC© Exercise

Part A Assessments

1. Test results for respiratory air volumes: ⚠

Volume	Expected Value	Observed
Tidal volume (TV)		
Inspiratory reserve volume (IRV)		
Expiratory reserve volume (ERV)		

Capacity	Expected Value	Observed	Predicted (from H & A)
Vital capacity			

Residual volume (RV) used: _____ (default is 1 liter)

2. Use the preceding data to calculate the following capacities: ⚠

Capacity	Formula	Calculation
Inspiratory capacity (IC)	IC = TV + IRV	
Expiratory capacity (EC)	EC = TV + RV	
Functional residual capacity (FRC)	FRC = ERV + RV	
Total lung capacity (TLC)	TLC = IRV + TV + ERV + RV	

Part B Assessments

Complete the following:

1. How do the experimental results compare to the expected or normal values presented in the introduction? **2**

_____

_____

_____

2. Why is the predicted vital capacity dependent upon the height of the subject? **3**

_____

_____

_____

3. Why might vital capacity and expiratory reserve decrease with age? **3**

_____

_____

_____

Laboratory Exercise 53

Control of Breathing

Pre-Lab

1. Carefully read the introductory material and examine the entire lab content.
2. Be familiar with control of breathing (from lecture or the textbook).
3. Visit www.mhhe.com/martinseries1 for pre-lab questions and Anatomy & Physiology Revealed animations.

Materials Needed

Clock with second hand
Paper bags, small

For Demonstration Activity:
Flasks
Glass tubing
Rubber stoppers, two-hole
Calcium hydroxide solution (limewater)

For Learning Extension Activity:
Pneumograph
Physiological recording apparatus

Breathing is controlled from regions of the brainstem called the *respiratory areas,* which control both inspiration and expiration. These areas initiate and regulate nerve impulses that travel to various breathing muscles, causing rhythmic breathing movements and adjustments to the rate and depth of breathing to meet various cellular needs. The respiratory area includes the *medullary respiratory area,* which is composed of two bilateral groups of neurons in the medulla oblongata. They are called the *dorsal respiratory group* and the *ventral respiratory group.* Neurons in another part of the brainstem, the pons, compose the *pontine respiratory group* (formerly the pneumotaxic center). These neurons make connections with the medullary rhythmicity center, and they may contribute to the basic rhythm of breathing.

Various factors can influence the respiratory areas and thus affect the rate and depth of breathing. These factors include stretch of the lung tissues, emotional state, and the presence in the blood of certain chemicals, such as carbon dioxide, hydrogen ions, and oxygen. For example, the breathing rate increases as the blood concentration of carbon dioxide or hydrogen ions increases or as the concentration of oxygen decreases.

Purpose of the Exercise

To review the muscles and the mechanisms that control breathing and to investigate some of the factors that affect the rate and depth of breathing.

Learning Outcomes

After completing this exercise, you should be able to

1. Locate the breathing centers in the brainstem.
2. Describe the mechanisms that control and influence breathing.
3. Select the respiratory muscles involved in inspiration and forced expiration.
4. Test and record the effect of various factors on the rate and depth of breathing.

Procedure A—Control of Breathing

1. Several skeletal muscles contract during breathing. The principal muscles involved during inspiration are the diaphragm and external intercostals. The sternocleidomastoid, scalenes, and pectoralis minor are synergistic during more forceful inhalation. During quiet respiration, there is minimal involvement of the expiratory muscles. During forced expiration, the principal muscles are the internal intercostals, but the rectus abdominis and the external oblique muscles can provide extra force (fig. 53.1).
2. Respiratory areas in the brainstem control the cycle of breathing. The dorsal respiratory group of neurons in the medulla oblongata receives input from the pontine respiratory group (pneumotaxic center), various regions

FIGURE 53.1 Respiratory muscles involved in inspiration and forced expiration. The blue arrows indicate the direction of muscle contraction during inspiration; the green arrows indicate the direction of muscle contraction during forced expiration.

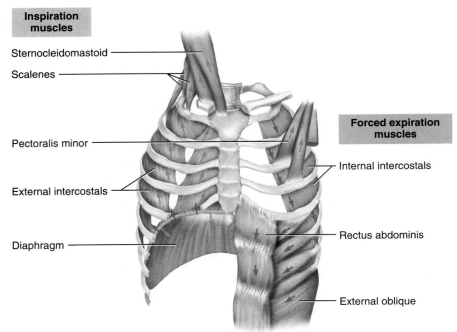

of the brain, cranial nerves (vagus and glossopharyngeal), and chemosensitive areas of the medulla oblongata (fig. 53.2). Impulses from the dorsal respiratory group stimulate the muscles of inspiration and lead to the basic rhythm of breathing. This results in a normal respiratory rate of 12 to 15 breaths per minute. The ventral respiratory group of neurons will fire during inspiration and expiration to stimulate the appropriate muscles of inspiration and expiration. The ventral respiratory group is more active during forceful breathing (fig. 53.3).

3. Complete Part A of Laboratory Report 53.

Procedure B—Factors Affecting Breathing

Perform each of the following tests, using your laboratory partner as a test subject.

1. *Normal breathing.* To determine the subject's normal breathing rate and depth, follow these steps:
   a. Have the subject sit quietly for a few minutes.
   b. After the rest period, ask the subject to count backwards mentally, beginning with five hundred.
   c. While the subject is distracted by counting, watch the subject's chest movements, and count the breaths taken in a minute. Use this value as the normal breathing rate (breaths per minute).
   d. Note the relative depth of the breathing movements.
   e. Record your observations in the table in Part B of the laboratory report.

 Demonstration Activity

When a solution of calcium hydroxide is exposed to carbon dioxide, a chemical reaction occurs, and a white precipitate of calcium carbonate is formed as indicated by the following reaction:

$$Ca(OH)_2 + CO_2 \longrightarrow CaCO_3 + H_2O$$

Thus, a clear water solution of calcium hydroxide (limewater) can be used to detect the presence of carbon dioxide because the solution becomes cloudy if this gas is bubbled through it.

The laboratory instructor will demonstrate this test for carbon dioxide by drawing some air through limewater in an apparatus such as that shown in figure 53.4. Then the instructor will blow an equal volume of expired air through a similar apparatus. (*Note:* A new sterile mouthpiece should be used each time the apparatus is demonstrated.) Watch for the appearance of a precipitate that causes the limewater to become cloudy. Was there any carbon dioxide in the atmospheric air drawn through the limewater?

_____

If so, how did the amount of carbon dioxide in the atmospheric air compare with the amount in the expired air?

_____

_____

FIGURE 53.2 The respiratory areas are located in the pons and the medulla oblongata.

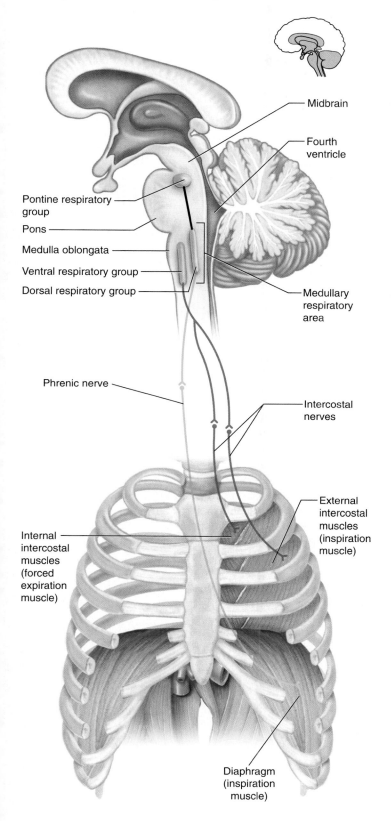

- Midbrain
- Fourth ventricle
- Pontine respiratory group
- Pons
- Medulla oblongata
- Ventral respiratory group
- Dorsal respiratory group
- Medullary respiratory area
- Phrenic nerve
- Intercostal nerves
- External intercostal muscles (inspiration muscle)
- Internal intercostal muscles (forced expiration muscle)
- Diaphragm (inspiration muscle)

FIGURE 53.3 The medullary rhythmicity area and the pontine respiratory group control breathing.

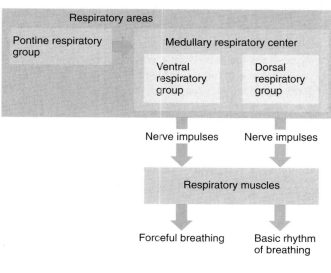

FIGURE 53.4 Apparatus used to demonstrate the presence of carbon dioxide in air: (*a*) atmospheric air is drawn through limewater; (*b*) expired air is blown through limewater.

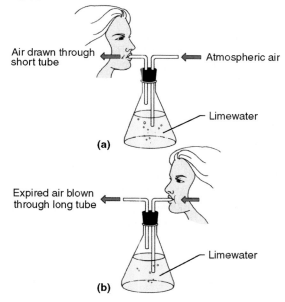

Air drawn through short tube — Atmospheric air

Limewater

(a)

Expired air blown through long tube

Limewater

(b)

2. *Effect of hyperventilation.* To test the effect of hyperventilation on breathing, follow these steps:
   a. Seat the subject and *guard to prevent the possibility of the subject falling over.*
   b. Have the subject breathe rapidly and deeply for a maximum of 1 minute. *If the subject begins to feel dizzy, the hyperventilation should be halted immediately to prevent the subject from fainting from complications of alkalosis. The increased blood pH causes vasoconstriction of cerebral arterioles, which decreases circulation and oxygen to the brain.*

c. After the period of hyperventilation, determine the subject's breathing rate, and judge the breathing depth as before.

d. Record the results in Part B of the laboratory report.

3. *Effect of rebreathing air.* To test the effect of rebreathing air on breathing, follow these steps:

  a. Have the subject sit quietly (approximately 5 minutes) until the breathing rate returns to normal.

  b. Have the subject breathe deeply into a small paper bag that is held tightly over the nose and mouth. *If the subject begins to feel light-headed or like fainting, the rebreathing air should be halted immediately to prevent further acidosis and fainting.*

  c. After 2 minutes of rebreathing air, determine the subject's breathing rate, and judge the depth of breathing.

  d. Record the results in Part B of the laboratory report.

4. *Effect of breath holding.* To test the effect of breath holding on breathing, follow these steps:

  a. Have the subject sit quietly (approximately 5 minutes) until the breathing rate returns to normal.

  b. Have the subject hold his or her breath as long as possible. *If the subject begins to feel light-headed or like fainting, breath holding should be halted immediately to prevent further acidosis and fainting.*

  c. As the subject begins to breathe again, determine the rate of breathing, and judge the depth of breathing.

  d. Record the results in Part B of the laboratory report.

5. *Effect of exercise.* To test the effect of exercise on breathing, follow these steps:

  a. Have the subject sit quietly (approximately 5 minutes) until breathing rate returns to normal.

  b. Have the subject exercise by moderately running in place for 3-5 minutes. *This exercise should be avoided by anyone with health risks.*

  c. After the exercise, determine the breathing rate, and judge the depth of breathing.

  d. Record the results in Part B.

6. The importance of the respiratory rate and blood pH relationship is illustrated in figure 53.5. A condition of respiratory acidosis or alkalosis, leading to possible death, can occur from changes in the depth or rate of breathing.

7. Complete Part B of the laboratory report.

FIGURE 53.5 The relationship of respiratory rate and blood pH. The normal breathing rate of 12–15 breaths per minute relates to a blood pH within the normal range.

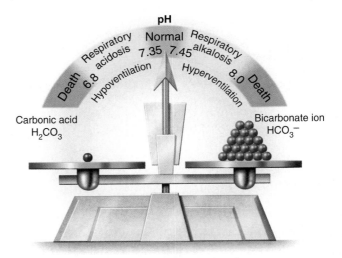

pH

Death Respiratory acidosis 6.8 Hypoventilation 7.35 Normal 7.45 Respiratory alkalosis Hyperventilation 8.0 Death

Carbonic acid H_2CO_3

Bicarbonate ion HCO_3^-

Learning Extension Activity

A *pneumograph* is a device that can be used together with some type of recording apparatus to record breathing movements. The laboratory instructor will demonstrate the use of this equipment to record various movements, such as those that accompany coughing, laughing, yawning, and speaking.

Devise an experiment to test the effect of some factor, such as hyperventilation, rebreathing air, or exercise, on the length of time a person can hold the breath. *After the laboratory instructor has approved your plan,* carry out the experiment, using the pneumograph and recording equipment. What conclusion can you draw from the results of your experiment?

_____

_____

_____

_____

Laboratory Report

53

Name _____

Date _____

Section _____

The △ corresponds to the Learning Outcome(s) listed at the beginning of the laboratory exercise.

Control of Breathing

Part A Assessments

Complete the following statements:

1. The respiratory areas are widely scattered throughout the _____ and medulla oblongata of the brainstem. △1

2. The _____ respiratory group within the medulla oblongata primarily stimulates the diaphragm. △1

3. The _____ respiratory group within the medulla oblongata primarily stimulates the intercostals and abdominal muscles. △1

4. Chemosensitive areas are stimulated by changes in the blood concentrations of hydrogen ions and _____. △2

5. As the blood concentration of carbon dioxide increases, the breathing rate _____. △2

6. As a result of increased breathing, the blood concentration of carbon dioxide is _____. △2

7. As a result of hyperventilation, breath-holding time is _____. △2

8. The principal muscles of forced expiration are the _____. △3

9. The principal muscles of inspiration are the _____ and the external intercostal muscles. △3

Part B Assessments

1. Record the results of your breathing tests in the table. △4

Factor Tested	Breathing Rate (breaths/minute)	Breathing Depth (+, + +, + + +)
Normal		
Hyperventilation		
Rebreathing air		
Breath holding		
Exercise		

2. Briefly explain the reason for the changes in breathing that occurred in each of the following cases: /2\

   a. Hyperventilation

   _____

   _____

   _____

   _____

   b. Rebreathing air

   _____

   _____

   _____

   _____

   c. Breath holding

   _____

   _____

   _____

   _____

   d. Exercise

   _____

   _____

   _____

   _____

3. Complete the following:

   a. Why is it important to distract a person when you are determining the normal rate of breathing?

   _____

   _____

   b. How can the depth of breathing be measured accurately?

   _____

   _____

Critical Thinking Activity

Why is it dangerous for a swimmer to hyperventilate in order to hold the breath for a longer period of time? /2\

_____

_____

_____

_____

Laboratory Exercise 54

Digestive Organs

Pre-Lab

1. Carefully read the introductory material and examine the entire lab content.
2. Be familiar with the basic structures and functions (from lecture or the textbook) of the oral cavity, esophagus, stomach, small intestine, and large intestine.
3. Visit www.mhhe.com/martinseriesl for pre-lab questions and Anatomy & Physiology Revealed animations.

Materials Needed

Human torso model
Head model, sagittal section
Skull with teeth
Teeth, sectioned
Tooth model, sectioned
Paper cup
Compound light microscope
Prepared microscope slides of the following:
    Sublingual gland (salivary gland)
    Esophagus
    Stomach (fundus)
    Pancreas (exocrine portion)
    Small intestine (jejunum)
    Large intestine

The digestive system includes the organs associated with the alimentary canal and several accessory structures. The alimentary canal, a muscular tube, passes through the body from the opening of the mouth to the anus. It includes the mouth, pharynx, esophagus, stomach, small intestine, and large intestine. The canal is adapted to move substances throughout its length. It is specialized in various regions to store, digest, and absorb food materials and to eliminate the residues. The accessory organs, which include the salivary glands, liver, gallbladder, and pancreas, secrete products into the alimentary canal that aid digestive functions.

The oral cavity is a primary area for mechanical digestion where we masticate the food using the teeth and jaw muscles. Three pairs of salivary glands secrete mucus and salivary amylase into the oral cavity, where chemical digestion begins. The pharynx and esophagus secrete mucus and serve as passageways for the food and liquids to reach the stomach. The contents are forced along by peristaltic waves.

Within the stomach, swallowed contents mix with gastric juice, which contains hydrochloric acid and the enzyme pepsin. The hydrochloric acid creates a pH near 2, serving to destroy most ingested bacteria, and it activates pepsin to begin the chemical digestion of proteins. The gastric folds (rugae) allow the stomach to hold large quantities of food. Only a limited absorption occurs in the stomach. The partially digested contents, called chyme, enter the small intestine periodically through the muscular pyloric sphincter.

Within the small intestine, chyme is mixed with bile and pancreatic juice. Bile is produced in the liver and then stored in the gallbladder. Bile will emulsify the fat, while pancreatic juice contains multiple enzymes and bicarbonate ions for the chemical digestion of carbohydrates, fats, and proteins. The small intestine is nearly 21 feet long and possesses villi that increase its surface area, so chemical digestion is completed and most nutrient absorption occurs before the contents reach the large intestine.

Contents not absorbed enter the large intestine through the ileocecal valve. Within the large intestine, mucus is secreted and water and electrolytes are absorbed. Bacteria that inhabit the large intestine can break down some remaining residues, producing some vitamins that are absorbed. Feces composed of water, undigested substances, mucus, and bacteria are formed and stored until elimination.

Purpose of the Exercise

To review the structure and function of the digestive organs and to examine the tissues of these organs.

FIGURE 54.1 Label the features of the oral cavity by placing the correct numbers in the spaces provided (anterior view). ⓐ

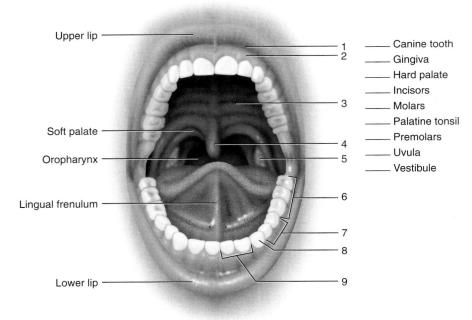

Upper lip

Soft palate
Oropharynx

Lingual frenulum

Lower lip

1
2
3
4
5
6
7
8
9

_____ Canine tooth
_____ Gingiva
_____ Hard palate
_____ Incisors
_____ Molars
_____ Palatine tonsil
_____ Premolars
_____ Uvula
_____ Vestibule

Learning Outcomes

After completing this exercise, you should be able to

① Locate and label the digestive organs and their major structures.

② Examine and sketch the structures of tissue sections of these organs.

③ Match digestive organs with their descriptions.

④ Describe the functions of these organs.

Procedure A—Oral Cavity and Salivary Glands

1. Study and label figure 54.1.
2. Examine the head model (sagittal section) and a skull. Locate the following structures:

   oral cavity (mouth)
   vestibule
   tongue
      lingual frenulum (frenulum of tongue)
      papillae
   palate
      hard palate
      soft palate
      uvula
   palatine tonsils
   gingivae (gums)
   teeth
      incisors
      canines (cuspids)

   premolars (bicuspids)
   molars

3. Study figure 54.2
4. Examine a sectioned tooth and a tooth model. Locate the following features:

   crown
      enamel
      dentin
   neck

FIGURE 54.2 Longitudinal section of a molar.

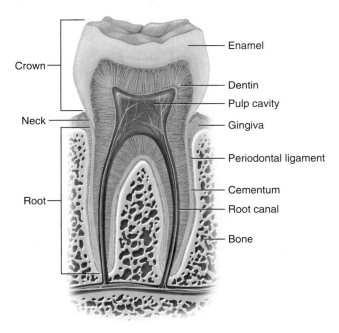

Crown

Neck

Root

Enamel
Dentin
Pulp cavity
Gingiva
Periodontal ligament
Cementum
Root canal
Bone

Terms:
Masseter muscle
Parotid duct
Parotid gland
Sublingual ducts
Sublingual gland
Submandibular duct
Submandibular gland
Tongue

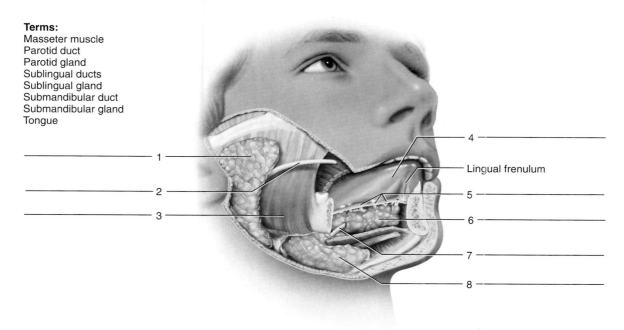

1 ——————
2 ——————
3 ——————

4 ——————
—— Lingual frenulum
5 ——————
6 ——————
7 ——————
8 ——————

root

pulp cavity

cementum (cement)

root canal

5. Study and label figure 54.3.
6. Observe the head of the human torso model, and locate the following:

    parotid salivary gland

    parotid duct (Stensen's duct)

    submandibular salivary gland

    submandibular duct (Wharton's duct)

    sublingual salivary gland

    sublingual ducts (10–12 ducts)

7. Examine a microscopic section of a sublingual gland, using low- and high-power magnification. Note the

mucous cells that produce mucus and serous cells that produce enzymes. Serous cells sometimes form caps (demilunes) around mucous cells. Also note any larger secretory duct surrounded by cuboidal epithelial cells (fig. 54.4).

8. Prepare a labeled sketch of a representative section of a salivary gland in Part A of Laboratory Report 54.

Procedure B—Pharynx and Esophagus

1. Reexamine figure 50.2 in Laboratory Exercise 50.
2. Observe the human torso model, and locate the following features:

    pharynx

    nasopharynx

    oropharynx

    laryngopharynx

    epiglottis

    esophagus

    lower esophageal sphincter (cardiac sphincter; gastroesophageal sphincter)

3. Have your partner take a swallow from a cup of water. Carefully watch the movements in the anterior region of the neck. What steps in the swallowing process did you observe?

————————————————

————————————————

————————————————

————————————————

FIGURE 54.4 Micrograph of the sublingual salivary gland (300×).

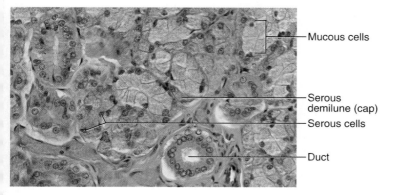

—— Mucous cells

—— Serous demilune (cap)

—— Serous cells

—— Duct

4. Examine a microscopic section of esophagus wall, using low-power magnification (fig. 54.5). The inner lining is composed of stratified squamous epithelium, and there are layers of muscle tissue in the wall. Locate some mucous glands in the submucosa. They appear as clusters of lightly stained cells.

5. Prepare a labeled sketch of the esophagus wall in Part A of the laboratory report.

Procedure C—The Stomach

1. Study and label figure 54.6.

2. Observe the human torso model, and locate the following features of the stomach:

    gastric folds (rugae)

    cardia (cardiac region)

    fundus of stomach (fundic region)

    body of stomach (body region)

    pyloric part (pyloric region)

        pyloric antrum

        pyloric canal

        pylorus

    pyloric sphincter

    lesser curvature

    greater curvature

3. Examine a microscopic section of stomach wall, using low-power magnification. Note how the inner lining of simple columnar epithelium dips inward to form gastric pits. The gastric glands are tubular structures that open

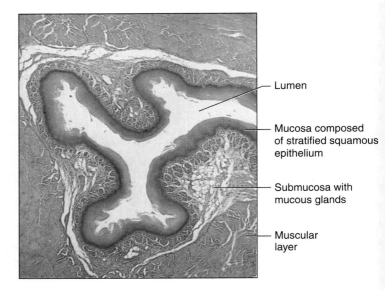

FIGURE 54.5 Micrograph of a cross section of the esophagus (10×).

- Lumen
- Mucosa composed of stratified squamous epithelium
- Submucosa with mucous glands
- Muscular layer

into the gastric pits. Near the deep ends of these glands, you should be able to locate some intensely stained (bluish) chief cells and some lightly stained (pinkish) parietal cells (fig. 54.7). The parietal cells secrete hydrochloric acid; the chief cells produce pepsinogen, the inactive form of pepsin.

4. Prepare a labeled sketch of a representative section of the stomach wall in Part A of the laboratory report.

FIGURE 54.6 Label the major features of the stomach and associated structures by placing the correct numbers in the spaces provided (frontal section of internal structures). ⚠

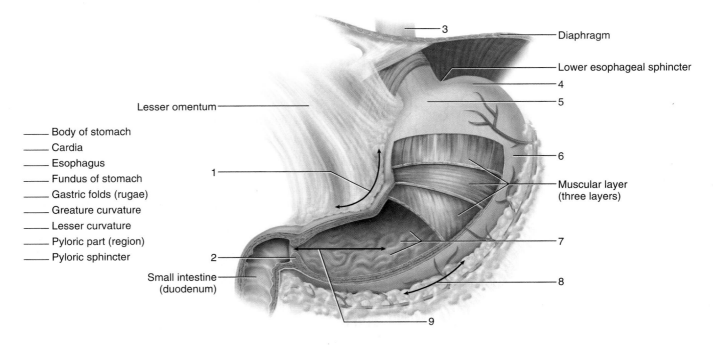

- Lesser omentum
- _____ Body of stomach
- _____ Cardia
- _____ Esophagus
- _____ Fundus of stomach
- _____ Gastric folds (rugae)
- _____ Greature curvature
- _____ Lesser curvature
- _____ Pyloric part (region)
- _____ Pyloric sphincter
- Small intestine (duodenum)

- 3
- Diaphragm
- Lower esophageal sphincter
- 4
- 5
- 1
- 6
- Muscular layer (three layers)
- 7
- 2
- 8
- 9

FIGURE 54.7 Micrograph of the mucosa of the stomach wall (50× micrograph enlarged to 100×).

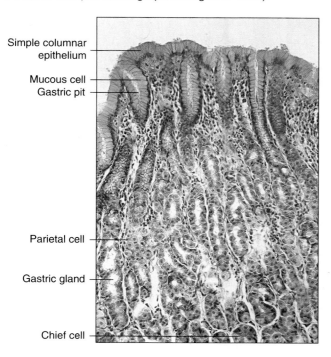

Simple columnar epithelium

Mucous cell
Gastric pit

Parietal cell

Gastric gland

Chief cell

Procedure D—Pancreas and Liver

1. Study and label figure 54.8.
2. Observe the human torso model, and locate the following structures:

   pancreas
   - tail of pancreas
   - head of pancreas
   - pancreatic duct
   - accessory pancreatic duct

   liver
   - right lobe
   - quadrate lobe
   - caudate lobe
   - left lobe

   gallbladder
   hepatic ducts
   common hepatic duct
   cystic duct
   bile duct (common)
   hepatopancreatic sphincter (sphincter of Oddi)

3. Examine the pancreas slide, using low-power magnification. Observe the exocrine (acinar) cells that secrete

FIGURE 54.8 Label the features associated with the liver and pancreas, using the terms provided. /1\

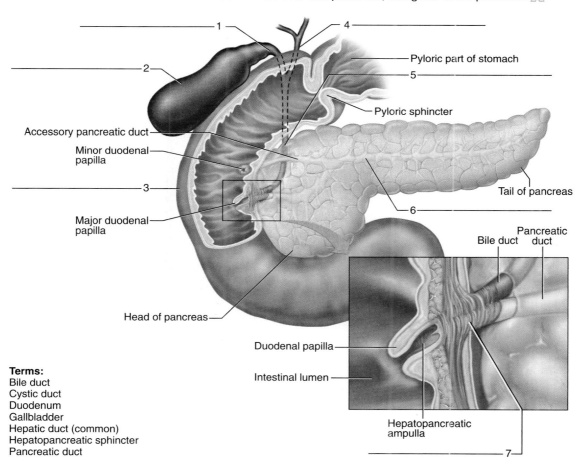

Pyloric part of stomach

Pyloric sphincter

Accessory pancreatic duct

Minor duodenal papilla

Tail of pancreas

Major duodenal papilla

Head of pancreas

Duodenal papilla

Intestinal lumen

Bile duct

Pancreatic duct

Hepatopancreatic ampulla

Terms:
Bile duct
Cystic duct
Duodenum
Gallbladder
Hepatic duct (common)
Hepatopancreatic sphincter
Pancreatic duct

pancreatic juice. See figure 39.12 in Laboratory Exercise 39 for a micrograph of the pancreas.

4. Prepare a labeled sketch of a representative section of the pancreas in Part A of the laboratory report.

Procedure E—Small and Large Intestines

1. Study and label figures 54.9, 54.10, and 54.11.

FIGURE 54.9 Label the features (anterior view) of the small intestine and associated structures, using the terms provided. (*Note:* The small intestine is pulled aside to expose the ileocecal junction.)

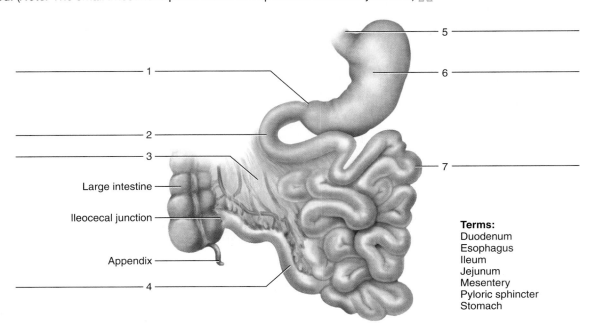

Large intestine
Ileocecal junction
Appendix

Terms:
Duodenum
Esophagus
Ileum
Jejunum
Mesentery
Pyloric sphincter
Stomach

FIGURE 54.10 Label the features of the large intestine, using the terms provided (anterior view).

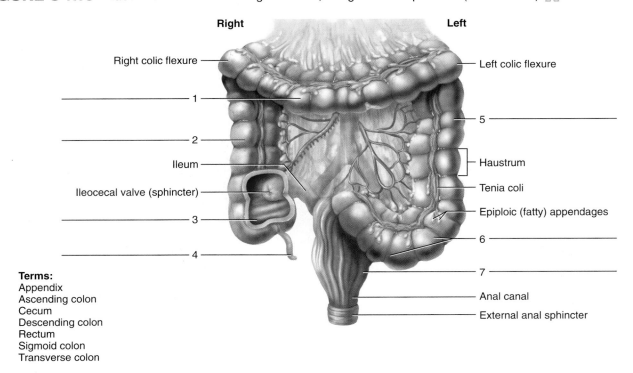

Right   Left

Right colic flexure   Left colic flexure

Ileum

Ileocecal valve (sphincter)

Haustrum
Tenia coli
Epiploic (fatty) appendages

Anal canal
External anal sphincter

Terms:
Appendix
Ascending colon
Cecum
Descending colon
Rectum
Sigmoid colon
Transverse colon

FIGURE 54.11 Normal appendix.

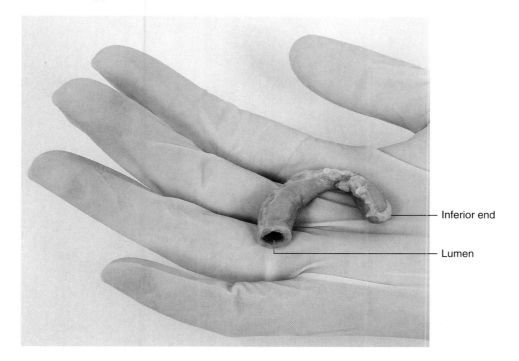

— Inferior end

— Lumen

2. Observe the human torso model, and locate each of the following features:

small intestine
  duodenum
  jejunum
  ileum
mesentery
ileocecal valve (sphincter)
large intestine
  large intestinal wall
    haustrum
    tenia coli
    epiploic (fatty) appendage
  cecum
  appendix (vermiform appendix)
  ascending colon
  right colic (hepatic) flexure
  transverse colon
  left colic (splenic) flexure
  descending colon
  sigmoid colon
  rectum
  anal canal
  anal columns

anal sphincter muscles
    external anal sphincter
    internal anal sphincter
  anus

3. Using low-power magnification, examine a microscopic section of small intestine wall. Identify the mucosa, submucosa, muscular layer, and serosa. Note the villi that extend into the lumen of the tube. Study a single villus, using high-power magnification. Note the core of connective tissue and the covering of simple columnar epithelium that contains some lightly stained goblet cells (fig. 54.12). The villi greatly increase the surface area for absorption of digestive products.

4. Prepare a labeled sketch of the wall of the small intestine in Part A of the laboratory report.

5. Examine a microscopic section of large intestine wall. Note the lack of villi. Also note the tubular mucous glands that open on the surface of the inner lining and the numerous lightly stained goblet cells. Locate the four layers of the wall (fig. 54.13). The mucus functions as a lubrication and holds the particles of fecal matter together.

6. Prepare a labeled sketch of the wall of the large intestine in Part A of the laboratory report.

7. Complete Parts B, C, and D of the laboratory report.

FIGURE 54.12   Micrograph of the small intestine wall (40×).

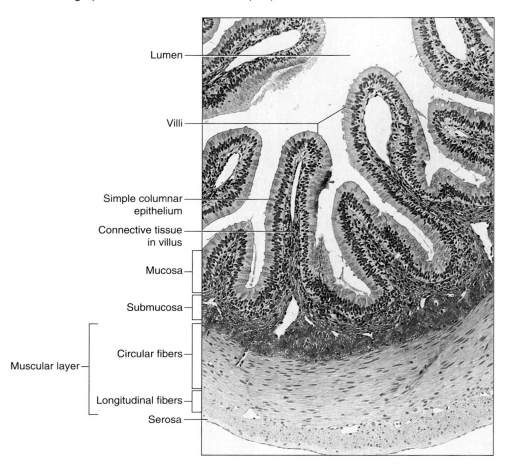

Lumen

Villi

Simple columnar epithelium

Connective tissue in villus

Mucosa

Submucosa

Muscular layer

Circular fibers

Longitudinal fibers

Serosa

FIGURE 54.13   Micrograph of the large intestine wall (64×).

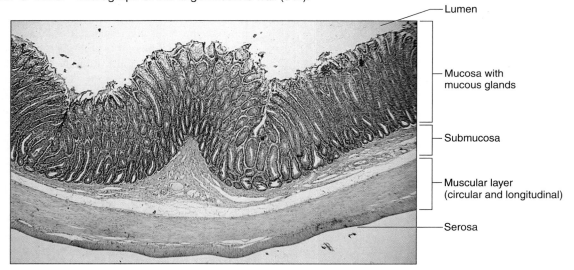

Lumen

Mucosa with mucous glands

Submucosa

Muscular layer (circular and longitudinal)

Serosa

   Critical Thinking Activity

How is the structure of the small intestine better adapted for absorption than the large intestine?

_____

_____

Name _____

Date _____

Section _____

The Ⓐ corresponds to the Learning Outcome(s) listed at the beginning of the laboratory exercise.

Digestive Organs

Part A Assessments

In the space that follows, sketch a representative area of the organ indicated. Label any of the structures observed, and indicate the magnification used for each sketch. ②

Salivary gland (_____×)	Esophagus (_____×)
Stomach wall (_____×)	Pancreas (_____×)
Small intestine wall (_____×)	Large intestine wall (_____×)

Part B Assessments

Identify the numbered features in figures 54.14 and 54.15 of a cadaver.

FIGURE 54.14 Label the features of the stomach and nearby regions in this frontal section of a cadaver (anterior view). 🄰

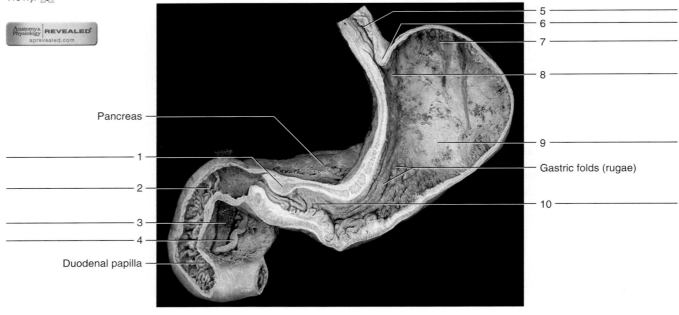

FIGURE 54.15 Label the digestive structures of this abdominopelvic cavity of a cadaver (anterior view). 🄰

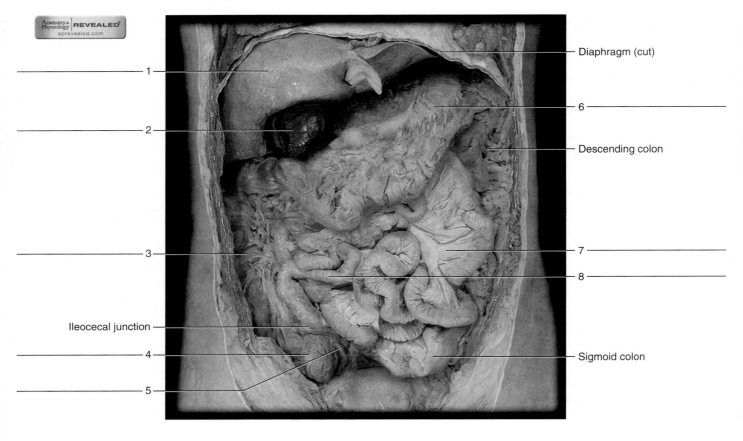

Part C Assessments

Match the terms in column A with the descriptions in column B. Place the letter of your choice in the space provided. ⓷

<table>
<tr><td>Column A</td><td colspan="2">Column B</td></tr>
<tr><td>a. Cardia</td><td>_____</td><td>1. Smallest of major salivary glands</td></tr>
<tr><td>b. Crown</td><td>_____</td><td>2. Secrete hydrochloric acid into stomach</td></tr>
<tr><td>c. Cystic duct</td><td>_____</td><td>3. Last section of small intestine</td></tr>
<tr><td>d. Gastric folds</td><td>_____</td><td>4. Region of stomach near lower esophageal sphincter</td></tr>
<tr><td>e. Ileum</td><td>_____</td><td>5. Attaches tooth to jaw</td></tr>
<tr><td>f. Major duodenal papilla</td><td>_____</td><td>6. Responsible for peristaltic waves</td></tr>
<tr><td>g. Mucosa</td><td>_____</td><td>7. Allow stomach to expand</td></tr>
<tr><td>h. Muscular layer</td><td>_____</td><td>8. Space between teeth, cheeks, and lips</td></tr>
<tr><td>i. Parietal cells</td><td>_____</td><td>9. Attached to gallbladder</td></tr>
<tr><td>j. Periodontal ligament</td><td>_____</td><td>10. Portion of tooth projecting beyond gingivae</td></tr>
<tr><td>k. Sublingual gland</td><td>_____</td><td>11. Layer nearest lumen of alimentary canal</td></tr>
<tr><td>l. Vestibule</td><td>_____</td><td>12. Common opening region for bile and pancreatic secretions</td></tr>
</table>

Part D Assessments

Complete the following:

1. Summarize the functions of the oral cavity. ⓸ _____

_____

_____

2. Summarize the functions of the esophagus. ⓸ _____

_____

_____

3. Name the four regions of the stomach. ⓵ _____

4. Name the valve that prevents regurgitation of food from the small intestine back into the stomach. ⓵ _____

5. Name the gastric cells that secrete digestive enzymes. ⓵ _____

6. Summarize the functions of the stomach. ⓸ _____

_____

_____

_____

7. Name the three portions of the small intestine. ⓵ _____

8. Name the four portions of the colon. ⓵ _____

9. Name the valve that controls the movement of material between the small and large intestines. ⓵ _____

10. Name the small projection that contains lymphatic tissue attached to the cecum. ⓵ _____

11. Summarize the functions of the small intestine. 4 _____

_____

_____

_____

12. Summarize the functions of the large intestine. 4 _____

_____

_____

Action of a Digestive Enzyme

Pre-Lab

1. Carefully read the introductory material and examine the entire lab content.
2. Be familiar with enzyme structure and activity (from lecture or the textbook).
3. Visit www.mhhe.com/martinseriesl for pre-lab questions.

Materials Needed

0.5% amylase solution*
Beakers (50 and 500 mL)
Distilled water
Funnel
Pipettes (1 and 10 mL)
Pipette rubber bulbs
0.5% starch solution
Graduated cylinder (10 mL)
Test tubes
Test-tube clamps
Wax marker
Iodine-potassium-iodide solution
Medicine dropper
Ice
Water bath, 37°C (98.6°F)
Porcelain test plate
Benedict's solution
Hot plates
Test-tube rack
Thermometer

For Alternative Procedure Activity:
Small disposable cups
Distilled water

*The amylase should be free of sugar for best results; a low-maltose solution of amylase yields good results. See Appendix 1.

 Safety

▶ Use only a mechanical pipetting device (never your mouth). Use pipettes with rubber bulbs or dropping pipettes.
▶ Wear safety glasses when working with acids and when heating test tubes.
▶ Use test-tube clamps when handling hot test tubes.
▶ If an open flame is used for heating the test solutions, keep clothes and hair away from the flame.
▶ Review all the safety guidelines inside the front cover.
▶ If student saliva is used as a source of amylase, it is important that students wear disposable gloves and only handle their own materials.
▶ Use an appropriate disinfectant to wash the laboratory tables before and after the procedures.
▶ Dispose of chemicals according to appropriate directions.
▶ Wash your hands before leaving the laboratory.

The digestive enzyme in salivary secretions is called *salivary amylase*. Pancreatic amylase is secreted among several other pancreatic enzymes. A bacterial extraction of amylase is available for laboratory experiments. This enzyme splits starch molecules into sugar (disaccharide) molecules, which is the first step in the digestion of complex carbohydrates.

As in the case of other enzymes, amylase is a protein catalyst. Its activity is affected by exposure to certain environmental factors, including various temperatures, pH, radiation, and electricity. As temperatures increase, faster chemical reactions occur as the collisions of molecules happen at a greater frequency. Eventually, temperatures increase to a point that the enzyme is denatured, and the rate of the enzyme activity rapidly declines. As temperatures decrease, enzyme activity also decreases due to fewer collisions of the molecules; however, the colder temperatures do not denature the enzyme. Normal body temperature provides an environment for enzyme activity near the optimum for enzymatic reactions.

The pH of the environment where enzymes are secreted also has a major influence on the enzyme reactions. The optimum pH for amylase activity is between 6.8 and 7.0, typical of salivary secretions. When salivary amylase arrives in the stomach, hydrochloric acid deactivates the enzyme, diminishing any further chemical digestion of remaining starch in the stomach. However, pancreatic amylase is secreted into the small intestine, where an optimum pH for amylase is once more provided. Other enzymes in the digestive system have different optimum pH ranges of activity compared to amylase. For example, pepsin from stomach secretions has an optimum activity around pH 2, while trypsin from pancreatic secretions operates best around pH 7–8.

Purpose of the Exercise

To investigate the action of amylase and the effect of heat on its enzymatic activity.

Learning Outcomes

After completing this exercise, you should be able to

① Explain the action of amylase.

② Test a solution for the presence of starch or the presence of sugar.

③ Test the effects of varying temperatures on the activity of amylase.

FIGURE 55.1 Lock-and-key model of enzyme action of amylase on starch digestion.

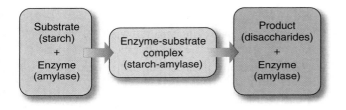

Procedure A—Amylase Activity

1. Study the lock-and-key model of amylase action on starch digestion (fig. 55.1).
2. Examine the locations where starch is digested into glucose (fig. 55.2).
3. Mark three clean test tubes as *tubes 1, 2,* and *3,* and prepare the tubes as follows (fig. 55.3):

   Tube 1: **Add 6 mL of amylase solution.**

   Tube 2: **Add 6 mL of starch solution.**

   Tube 3: **Add 5 mL of starch solution and 1 mL of amylase solution.**

Alternative Procedure Activity

Human saliva could be used as a source of amylase solutions instead of bacterial amylase preparations. Collect about 5 mL of saliva into a small disposable cup. Add an equal amount of distilled water and mix them together for the amylase solutions during the laboratory procedures. Be sure to follow all of the safety guidelines.

---

4. Shake the tubes well to mix the contents, and place them in a warm water bath, 37°C (98.6°F), for 10 minutes.
5. At the end of the 10 minutes, test the contents of each tube for the presence of starch. To do this, follow these steps:

   a. Place 1 mL of the solution to be tested in a depression of a porcelain test plate.

FIGURE 55.2 Flowchart of starch digestion.

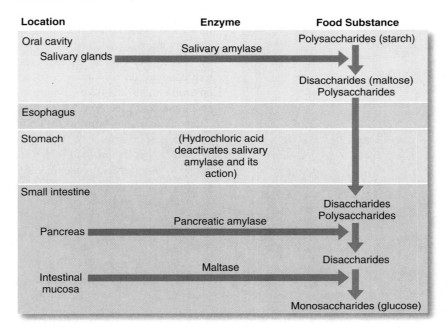

466

FIGURE 55.3 Test tubes prepared for testing amylase activity.

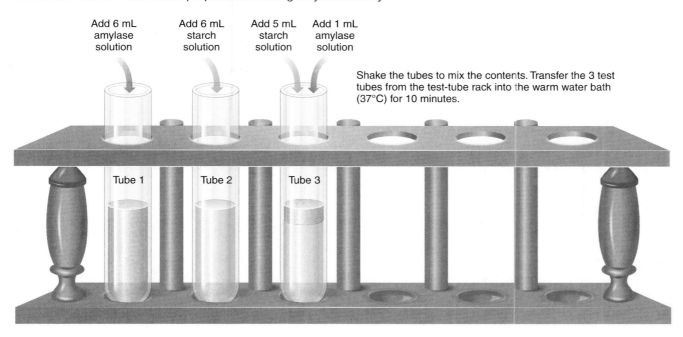

Add 6 mL amylase solution

Add 6 mL starch solution

Add 5 mL starch solution

Add 1 mL amylase solution

Shake the tubes to mix the contents. Transfer the 3 test tubes from the test-tube rack into the warm water bath (37°C) for 10 minutes.

Tube 1   Tube 2   Tube 3

b. Next add one drop of iodine-potassium-iodide solution, and note the color of the mixture. If the solution becomes blue-black, starch is present.

c. Record the results in Part A of Laboratory Report 55.

6. Test the contents of each tube for the presence of sugar (disaccharides in this instance). To do this, follow these steps:

a. Place 1 mL of the solution to be tested in a clean test tube.

b. Add 1 mL of Benedict's solution.

c. Place the test tube with a test-tube clamp in a beaker of boiling water for 2 minutes.

d. Note the color of the liquid. If the solution becomes green, yellow, orange, or red, sugar is present. Blue indicates a negative test, whereas green indicates a positive test with the least amount of sugar, and red indicates the greatest amount of sugar present.

e. Record the results in Part A of the laboratory report.

7. Complete Part A of the laboratory report.

Procedure B—Effect of Heat

1. Mark three clean test tubes as *tubes 4, 5,* and *6.*

2. Add 1 mL of amylase solution to each of the tubes, and expose each solution to a different test temperature for 3 minutes as follows:

Tube 4: **Place in beaker of ice water (about 0°C [32°F]).**

Tube 5: **Place in warm water bath (about 37°C [98.6°F]).**

Tube 6: **Place in beaker of boiling water (about 100°C [212°F]). Use a test-tube clamp.**

3. Add 5 mL of starch solution to each tube, shake to mix the contents, and return the tubes to their respective test temperatures for 10 minutes. It is important that the 5 mL of starch solution added to tube 4 be at ice-water temperature before it is added to the 1 mL of amylase solution.

4. At the end of the 10 minutes, test the contents of each tube for the presence of starch and the presence of sugar by following the directions in Procedure A.

5. Complete Part B of the laboratory report.

Learning Extension Activity

Devise an experiment to test the effect of some other environmental factor on amylase activity. For example, you might test the effect of a strong acid by adding a few drops of concentrated hydrochloric acid to a mixture of starch and amylase solutions. Be sure to include a control in your experimental plan. That is, include a tube containing everything except the factor you are testing. Then you will have something with which to compare your results. *Carry out your experiment only if it has been approved by the laboratory instructor.*

NOTES

Laboratory Report

55

Name _____

Date _____

Section _____

The ⚠ corresponds to the Learning Outcome(s) listed at the beginning of the laboratory exercise.

Action of a Digestive Enzyme

Part A—Amylase Activity Assessments

1. Test results: 2

Tube	Starch	Sugar
1 Amylase solution		
2 Starch solution		
3 Starch-amylase solution		

2. Complete the following:

   a. Explain the reason for including tube 1 in this experiment. 1 _____

   _____

   _____

   _____

   b. What is the importance of tube 2? 1 _____

   _____

   _____

   _____

   c. What do you conclude from the results of this experiment? 1 _____

   _____

   _____

   _____

Part B—Effect of Heat Assessments

1. Test results: 3

Tube	Starch	Sugar
4 0°C (32°F)		
5 37°C (98.6°F)		
6 100°C (212°F)		

2. Complete the following:

   a. What do you conclude from the results of this experiment? 3 _____

   _____

   _____

   _____

   _____

   _____

   b. If digestion failed to occur in one of the tubes in this experiment, how can you tell if the amylase was destroyed by the factor being tested or if the amylase activity was simply inhibited by the test treatment? 3

   _____

   _____

   _____

   _____

   _____

Critical Thinking Activity

What test result would occur if the amylase you used contained sugar? _____ Would your results be as valid? Explain your answer.

_____

_____

_____

_____

Kidney Structure

Pre-Lab

1. Carefully read the introductory material and examine the entire lab content.
2. Be familiar with the structures and functions (from lecture or the textbook) of the kidney and the nephron.
3. Visit www.mhhe.com/martinseriesl for pre-lab questions and Anatomy & Physiology Revealed animations.

Materials Needed

Human torso model
Kidney model
Preserved pig (or sheep) kidney
Dissecting tray
Dissecting instruments
Long knife
Compound light microscope
Prepared microscope slide of a kidney section

Safety

▶ Wear disposable gloves when working on the kidney dissection.
▶ Dispose of the kidney and gloves as directed by your laboratory instructor.
▶ Wash the dissecting tray and instruments as instructed.
▶ Wash your laboratory table.
▶ Wash your hands before leaving the laboratory.

The two kidneys are the primary organs of the urinary system. They are located in the abdominal cavity, against the posterior wall and behind the parietal peritoneum (retroperitoneal). Masses of adipose tissue associated with the kidneys hold them in place at a vertebral level between T12 and L3. The right kidney is slightly more inferior due to the large mass of the liver near its superior border. Ureters force urine by means of peristaltic waves into the urinary bladder, which temporarily stores urine. The urethra conveys urine to the outside of the body.

Each kidney contains over 1 million nephrons, which serve as the basic structural and functional units of the kidney. A glomerular capsule, proximal convoluted tubule, nephron loop, and distal convoluted tubule compose the microscopic, multicellular structure of a relatively long nephron tubule, which drains into a collecting duct. Approximately 85% of the nephrons are cortical nephrons with short nephron loops, while the remaining represent juxtamedullary nephrons, with long nephron loops extending deeper into the renal medulla. An elaborate network of blood vessels surrounds the entire nephron. Glomerular filtration, tubular reabsorption, and tubular secretion represent three processes resulting in urine as the final product.

A variety of overall functions occur in the kidneys. They remove metabolic wastes from the blood; help regulate blood volume, blood pressure, and pH of blood; control water and electrolyte concentrations; and secrete renin and erythropoietin.

Purpose of the Exercise

To review the structure of the kidney, to dissect a kidney, and to observe the major structures of a nephron.

Learning Outcomes

After completing this exercise, you should be able to

1. Locate and identify the organs of the urinary system.
2. Locate and identify the major structures of a kidney.
3. Identify and sketch the major structures of a nephron.
4. Trace the path of filtrate through a renal nephron.
5. Trace the path of blood through the renal blood vessels.

Procedure A—Kidney Structure

1. Label figure 56.1.
2. Study figure 56.2.

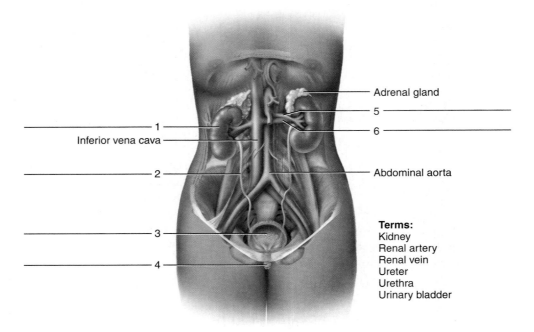

Adrenal gland

5

6

Inferior vena cava

Abdominal aorta

1

2

3

4

Terms:
Kidney
Renal artery
Renal vein
Ureter
Urethra
Urinary bladder

3. Observe the human torso model and the kidney model. Locate the following:

> **kidneys**
> **ureters**
> **urinary bladder**
> **urethra**
> **renal sinus**
> **renal pelvis**
>> major calyces
>> minor calyces
> **renal medulla**
>> renal pyramids
>> renal papillae
> **renal cortex**
> **renal columns**
> **nephrons**
>> cortical nephrons (85% of nephrons)
>> juxtamedullary nephrons (15% of nephrons)

4. Complete Part A of Laboratory Report 56.

5. To observe the structure of a kidney, follow these steps:

  a. Obtain a pig or sheep kidney and rinse it with water to remove as much of the preserving fluid as possible.

  b. Carefully remove any adipose tissue from the surface of the specimen.

  c. Locate the following features:

> **fibrous (renal) capsule**
> **hilum of kidney**
> **renal artery**
> **renal vein**
> **ureter**

  d. Use a long knife to cut the kidney in half longitudinally along the frontal plane, beginning on the convex border.

  e. Rinse the interior of the kidney with water, and using figure 56.2 as a reference, locate the following:

> **renal pelvis**
>> major calyces
>> minor calyces
> **renal cortex**
> **renal columns** (extensions of renal cortical tissue between renal pyramids)
> **renal medulla**
>> renal pyramids

6. Complete Part B of the laboratory report.

FIGURE 56.2 Frontal section of a pig kidney that has a triple injection of latex (*red* in the renal artery, *blue* in the renal vein, and *yellow* in the ureter and renal pelvis).

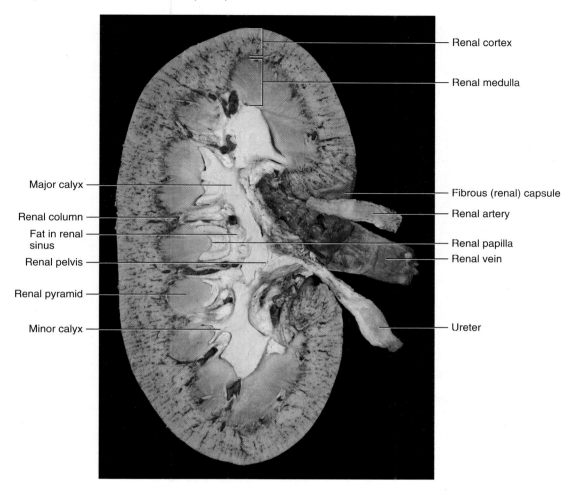

Labels (left side): Major calyx, Renal column, Fat in renal sinus, Renal pelvis, Renal pyramid, Minor calyx

Labels (right side): Renal cortex, Renal medulla, Fibrous (renal) capsule, Renal artery, Renal papilla, Renal vein, Ureter

Procedure B—The Renal Blood Vessels and Nephrons

1. Study figures 56.3 and 56.4.
2. Obtain a microscope slide of a kidney section, and examine it using low-power magnification. Locate the *renal capsule,* the *renal cortex* (which appears somewhat granular and may be more darkly stained than the other renal tissues), and the *renal medulla* (fig. 56.5).
3. Examine the renal cortex, using high-power magnification. Locate a *renal corpuscle.* These structures appear as isolated circular areas. Identify the *glomerulus,* the capillary cluster inside the corpuscle, and the *glomerular (Bowman's) capsule,* which appears as a clear area surrounding the glomerulus. Also note the numerous sections of renal tubules that occupy the spaces between renal corpuscles (fig. 56.5a).
4. Prepare a labeled sketch of a representative section of renal cortex in Part C of the laboratory report.
5. Examine the renal medulla, using high-power magnification. Identify longitudinal and cross sectional views of various collecting ducts. These ducts are lined with simple epithelial cells, which vary in shape from squamous to cuboidal (fig. 56.5b).
6. Prepare a labeled sketch of a representative section of renal medulla in Part C of the laboratory report.
7. Complete Part D of the laboratory report.

FIGURE 56.3 Renal blood vessels associated with cortical and juxtamedullary nephrons. The arrows indicate the flow of blood.

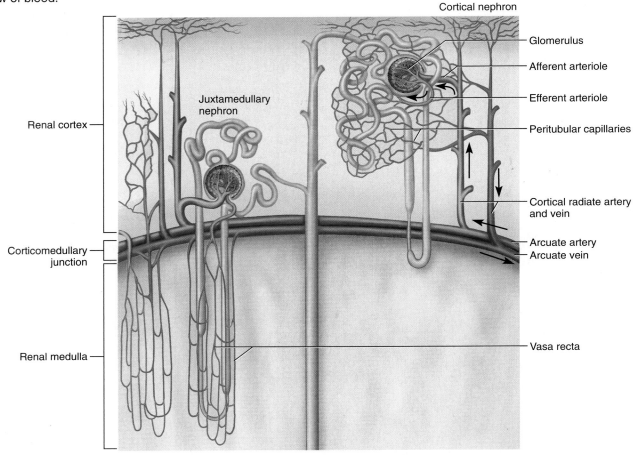

Cortical nephron

Glomerulus

Afferent arteriole

Efferent arteriole

Peritubular capillaries

Renal cortex

Juxtamedullary nephron

Cortical radiate artery and vein

Corticomedullary junction

Arcuate artery
Arcuate vein

Renal medulla

Vasa recta

FIGURE 56.4 Structure of a nephron with arrows indicating the flow of tubular fluid. The corticomedullary junction is typical for a cortical nephron.

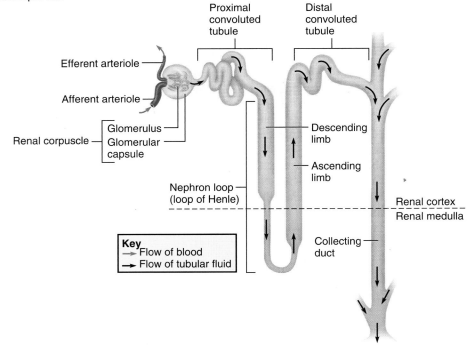

Proximal convoluted tubule

Distal convoluted tubule

Efferent arteriole

Afferent arteriole

Renal corpuscle
Glomerulus
Glomerular capsule

Descending limb

Ascending limb

Nephron loop (loop of Henle)

Renal cortex
Renal medulla

Collecting duct

Key
Flow of blood
Flow of tubular fluid

FIGURE 56.5 (*a*) Micrograph of a section of the renal cortex (220×). (*b*) Micrograph of a section of the renal medulla (80× micrograph enlarged to 200×).

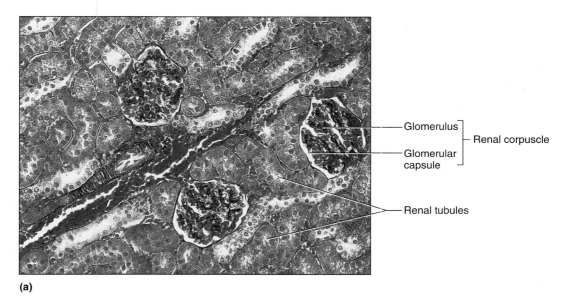

Glomerulus ⎤
⎥ Renal corpuscle
Glomerular capsule ⎦

Renal tubules

(a)

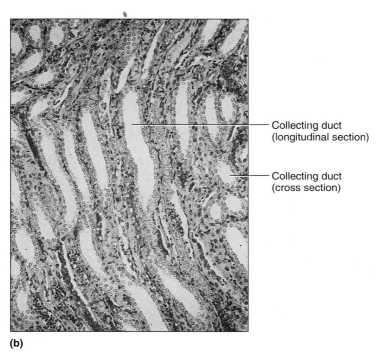

Collecting duct (longitudinal section)

Collecting duct (cross section)

(b)

Name _____

Date _____

Section _____

The ⬚ corresponds to the Learning Outcome(s) listed at the beginning of the laboratory exercise.

Kidney Structure

Part A Assessments

Label the features indicated in figure 56.6 of a kidney (frontal section).

FIGURE 56.6 Label the structures in the frontal section of a kidney. ⬚

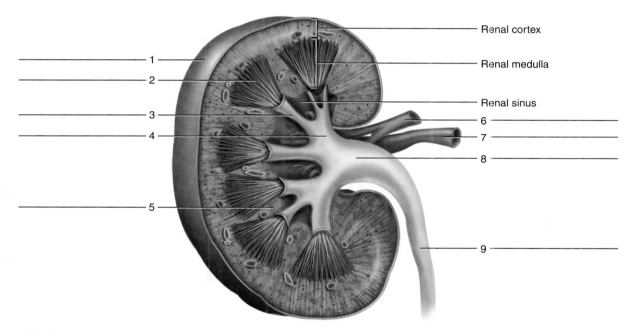

Renal cortex

Renal medulla

Renal sinus

Part B Assessments

Match the terms in column A with the descriptions in column B. Place the letter of your choice in the space provided. ⬚

Column A	Column B
a. Calyces	_____ **1.** Shell around the renal medulla
b. Hilum of kidney	_____ **2.** Branches of renal pelvis to renal papillae
c. Nephron	_____ **3.** Conical mass of tissue within renal medulla
d. Renal column	_____ **4.** Projection with tiny openings into a minor calyx
e. Renal cortex	_____ **5.** Hollow chamber within kidney
f. Renal papilla	_____ **6.** Microscopic functional unit of kidney
g. Renal pelvis	_____ **7.** Located between renal pyramids
h. Renal pyramid	_____ **8.** Superior funnel-shaped end of ureter inside the renal sinus
i. Renal sinus	_____ **9.** Medial depression for blood vessels and ureter to enter kidney chamber

Part C Assessments

Sketch a representative section of the renal cortex and the renal medulla. Label the glomerulus, glomerular capsule, and sections of renal tubules in the renal cortex. Label a longitudinal section and cross section of a collecting duct in the renal medulla. ⟋3⟍

Renal cortex (_____×)	Renal medulla (_____×)

Part D Assessments

Complete the following:

1. Distinguish between a renal corpuscle and a renal tubule. ⟋3⟍ _____

    _____

    _____

2. Number the following structures to indicate their respective positions in relation to the nephron. Assign the number 1 to the structure nearest the glomerulus. ⟋4⟍

    _____ Ascending limb of nephron loop   _____ Glomerular capsule

    _____ Collecting duct   _____ Proximal convoluted tubule

    _____ Descending limb of nephron loop   _____ Renal papilla

    _____ Distal convoluted tubule

3. Number the following structures (the list does not include all possible blood vessels) to indicate their respective positions in the blood pathway within the kidney. Assign the number 1 to the vessel nearest the abdominal aorta. ⟋5⟍

    _____ Afferent arteriole   _____ Efferent arteriole

    _____ Arcuate artery   _____ Glomerulus

    _____ Arcuate vein   _____ Peritubular capillary (or vasa recta)

    _____ Cortical radiate artery   _____ Renal artery

    _____ Cortical radiate vein   _____ Renal vein

Urinalysis

Pre-Lab

1. Carefully read the introductory material and examine the entire lab content.
2. Be familiar with normal urine components (from lecture or the textbook).
3. Visit www.mhhe.com/martinseries1 for pre-lab questions and Anatomy & Physiology Revealed animations.

Materials Needed

Normal and abnormal simulated urine specimens are suggested as a substitute for collected urine.
Disposable urine-collecting container
Paper towel
Urinometer cylinder
Urinometer hydrometer
Laboratory thermometer
pH test paper
Reagent strips (individual or combination strips such as Chemstrip or Multistix) to test for the presence of the following:
    Glucose
    Protein
    Ketones
    Bilirubin
    Hemoglobin/occult blood
Compound light microscope
Microscope slide
Coverslip
Centrifuge
Centrifuge tube
Graduated cylinder, 10 mL
Medicine dropper
Sedi-stain

Safety

▶ Consider using simulated urine samples available from various laboratory supply houses.
▶ Wear disposable gloves when working with body fluids.
▶ Work only with your own urine sample.
▶ Use an appropriate disinfectant to wash the laboratory table before and after the procedures.
▶ Place glassware in a disinfectant when finished.
▶ Dispose of contaminated items as directed by your laboratory instructor.
▶ Wash your hands before leaving the laboratory.

Urine is the product of three processes of the nephrons within the kidneys: glomerular filtration, tubular reabsorption, and tubular secretion. As a result of these processes, various waste substances are removed from the blood and body fluid and electrolyte balance are maintained. Consequently, the composition of urine varies considerably because of differences in dietary intake and physical activity from day to day. The normal urinary output ranges from 1.0–1.8 liters per day. Typically, urine consists of 95% water and 5% solutes. The volume of urine produced by the kidneys varies with such factors as fluid intake, environmental temperature, relative humidity, respiratory rate, and body temperature.

An analysis of urine composition and volume often is used to evaluate the functions of the kidneys and other organs. This procedure, called *urinalysis,* is a clinical assessment and a diagnostic tool for certain pathological conditions and general overall health. A urinalysis and a complete blood analysis complement each other for an evaluation of certain diseases and general health.

A urinalysis involves three aspects: physical characteristics, chemical analysis, and a microscopic examination. Physical characteristics of urine that are noted include volume, color, transparency, and odor. The chemical analysis of solutes in urine addresses urea and other nitrogenous wastes; electrolytes; pigments; as well as possible glucose, protein, ketones, bilirubin, and hemoglobin. The specific gravity and pH of urine are greatly influenced by the components and amounts of solutes. An examination of microscopic solids, including cells, casts, and crystals, assists in the diagnosis of injury, various diseases, and urinary infections.

Purpose of the Exercise

To perform the observations and tests commonly used to analyze the characteristics and composition of urine.

Learning Outcomes

After completing this exercise, you should be able to

1. Evaluate the color, transparency, and specific gravity of a urine sample.

2. Measure the pH of a urine sample.

3. Test a urine sample for the presence of glucose, protein, ketones, bilirubin, and hemoglobin.

4. Perform a microscopic study of urine sediment.

5. Summarize the results of these observations and tests.

 Warning

While performing the following tests, you should wear disposable latex gloves so that skin contact with urine is avoided. Observe all safety procedures listed for this lab. (Normal and abnormal simulated urine specimens could be used instead of real urine for this lab.)

Procedure—Urinalysis

1. Proceed to the restroom with a clean, disposable container. The first small volume of urine should not be collected because it contains abnormally high levels of microorganisms from the urethra. Collect a midstream sample of about 50 mL of urine. The best collections are the first specimen in the morning or one taken 3 hours after a meal. Refrigerate samples if they are not used immediately.

2. Place a sample of urine in a clean, transparent container. Describe the *color* of the urine. Normal urine varies from light yellow to amber, depending on the presence of urochromes, end-product pigments produced during the decomposition of hemoglobin. Dark urine indicates a high concentration of pigments.

   Abnormal urine colors include yellow-brown or green, due to elevated concentrations of bile pigments, and red to dark brown, due to the presence of blood. Certain foods, such as beets or carrots, and various drug substances also may cause color changes in urine, but in such cases the colors have no clinical significance. Enter the results of this and the following tests in Part A of Laboratory Report 57.

3. Evaluate the *transparency* of the urine sample (judge whether the urine is clear, slightly cloudy, or very cloudy). Normal urine is clear enough to see through.

You can read newsprint through slightly cloudy urine; you can no longer read newsprint through cloudy urine. Cloudy urine indicates the presence of various substances that may include mucus, bacteria, epithelial cells, fat droplets, or inorganic salts.

4. Determine the *specific gravity* of the urine sample, which indicates the solute concentration. Specific gravity is the ratio of the weight of something to the weight of an equal volume of pure water. For example, mercury (at 15°C) weighs 13.6 times as much as an equal volume of water; thus, it has a specific gravity of 13.6. Although urine is mostly water, it has substances dissolved in it and is slightly heavier than an equal volume of water. Thus, urine has a specific gravity of more than 1.000. Actually, the specific gravity of normal urine varies from 1.003 to 1.035. If the specific gravity is too low, the urine contains few solutes and represents dilute urine, a likely result of excessive fluid intake or the use of diuretics. A specific gravity above the normal range represents a higher concentration of solutes, likely from a limited fluid intake. Concentrated urine over an extended time increases the risk of the formation of kidney stones.

   To determine the specific gravity of a urine sample, follow these steps:

   a. Pour enough urine into a clean urinometer cylinder to fill it about three-fourths full. Any foam that appears should be removed with a paper towel.

   b. Use a laboratory thermometer to measure the temperature of the urine.

   c. Gently place the urinometer hydrometer into the urine, *and make sure that the float is not touching the sides or the bottom of the cylinder* (fig. 57.1).

   d. Position your eye at the level of the urine surface. Determine which line on the stem of the hydrometer intersects the lowest level of the concave surface (meniscus) of the urine.

   e. Liquids tend to contract and become denser as they are cooled, or to expand and become less dense as they are heated, so it may be necessary to make a temperature correction to obtain an accurate specific gravity measurement. To do this, add 0.001 to the hydrometer reading for each 3 degrees of urine temperature above 25°C or subtract 0.001 for each 3 degrees below 25°C. Enter this calculated value in the table of the laboratory report as the test result.

5. Reagent strips can be used to perform a variety of urine tests. In each case, directions for using the strips are found on the strip container. *Be sure to read them.*

   To perform each test, follow these steps:

   a. Obtain a urine sample and the proper reagent strip.

   b. Read the directions on the strip container.

   c. Dip the strip in the urine sample.

   d. Remove the strip at an angle and let it touch the inside rim of the urine container to remove any excess liquid.

   e. Wait for the length of time indicated by the directions on the container before you compare the color

FIGURE 57.1 Float the hydrometer in the urine, making sure that it does not touch the sides or the bottom of the cylinder.

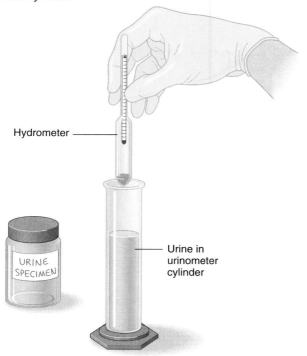

Hydrometer

Urine in urinometer cylinder

URINE SPECIMEN

of the test strip with the standard color scale on the side of the container. The value or amount represented by the matching color should be used as the test result and recorded in Part A of the laboratory report.

Alternative Procedure

If combination reagent strips (Chemstrip or Multistix) are being used, locate the appropriate color chart for each test being evaluated. Wait the designated time for each color reaction before the comparison is made to the standard color scale.

6. Perform the *pH test.* The pH of normal urine varies from 4.6 to 8.0, but most commonly, it is near 6.0 (slightly acidic). The pH of urine may decrease as a result of a diet high in protein, or it may increase with a vegetarian diet. Significant daily variations within the broad normal range are results of concentrations of excesses from variable diets.

7. Perform the *glucose test.* Normally, there is no glucose in urine. However, glucose may appear in the urine temporarily following a meal high in carbohydrates. Glucose also may appear in the urine as a result of uncontrolled diabetes mellitus.

8. Perform the *protein test.* Normally, proteins of large molecular size are not present in urine. However, those of small molecular sizes, such as albumins, may appear in trace amounts, particularly following strenuous exercise. Increased amounts of proteins also may appear as a result of kidney diseases in which the glomeruli are damaged or as a result of high blood pressure.

9. Perform the *ketone test.* Ketones are products of fat metabolism. Usually they are not present in urine. However, they may appear in the urine if the diet fails to provide adequate carbohydrate, as in the case of prolonged fasting or starvation or as a result of insulin deficiency (diabetes mellitus).

10. Perform the *bilirubin test.* Bilirubin, which results from hemoglobin decomposition in the liver, normally is absent in urine. It may appear, however, as a result of liver disorders that cause obstructions of the biliary tract. Urochrome, a normal yellow component of urine, is a result of additional breakdown of bilirubin.

11. Perform the *hemoglobin/occult blood test.* Hemoglobin occurs in the red blood cells, and because such cells normally do not pass into the renal tubules, hemoglobin is not found in normal urine. Its presence in urine usually indicates a disease process, a transfusion reaction, an injury to the urinary organs, or menstrual blood.

12. Complete Part A of the laboratory report.

13. A urinalysis usually includes a study of urine sediment—the microscopic solids present in a urine sample. This sediment normally includes mucus, certain crystals, and a variety of cells, such as the epithelial cells that line the urinary tubes and an occasional white blood cell. Other types of solids, such as casts or red blood cells, may indicate a disease or injury if they are present in excess. (Casts are cylindrical masses of cells or other substances that form in the renal tubules and are flushed out by the flow of urine.)

    To observe urine sediment, follow these steps:

    a. Thoroughly stir or shake a urine sample to suspend the sediment, which tends to settle to the bottom of the container.

    b. Pour 10 mL of urine into a clean centrifuge tube and centrifuge it for 5 minutes at slow speed (1,500 rpm). Be sure to balance the centrifuge with an even number of tubes filled to the same levels.

    c. Carefully decant 9 mL (leave 1 mL) of the liquid from the sediment in the bottom of the centrifuge tube, as directed by your laboratory instructor. Resuspend the 1 mL of sediment.

    d. Use a medicine dropper to remove some of the sediment and place it on a clean microscope slide.

    e. Add a drop of Sedi-stain to the sample, and add a coverslip.

    f. Examine the sediment with low-power (reduce the light when using low power) and high-power magnifications.

    g. With the aid of figure 57.2, identify the types of solids present.

    h. In Part B of the laboratory report, make a sketch of each type of sediment that you observed.

14. Complete Part B of the laboratory report.

FIGURE 57.2 Types of urine sediment. Healthy individuals lack many of these sediments and possess only occasional to trace amounts of others. (*Note:* Shades of white to purple sediments are most characteristic when using Sedi-stain.)

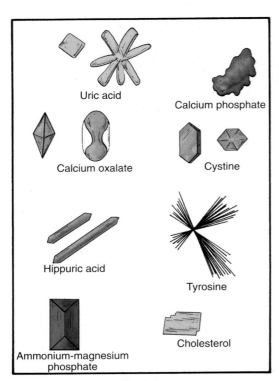

Crystals

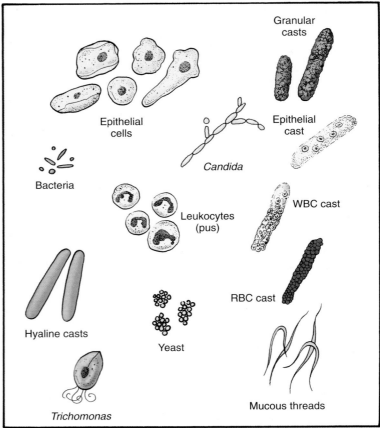

Cells and casts

Laboratory Report

57

Name _____

Date _____

Section _____

The ⚠ corresponds to the Learning Outcome(s) listed at the beginning of the laboratory exercise.

Urinalysis

Part A Assessments

1. Enter your observations, test results, and evaluations in the following table: ⚠1 ⚠2 ⚠3

Urine Characteristics	Observations and Test Results	Normal Values	Evaluations
Color		Light yellow to amber	
Transparency		Clear	
Specific gravity (corrected for temperature)		1.003–1.035	
pH		4.6–8.0	
Glucose		0 (negative)	
Protein		0 to trace	
Ketones		0	
Bilirubin		0	
Hemoglobin/occult blood		0	
(Other)			
(Other)			

2. Summarize the results of the urinalysis. **5** _____

_____

_____

_____

_____

_____

_____

Critical Thinking Activity

Why do you think it is important to refrigerate a urine sample if an analysis cannot be performed immediately after collecting it?

_____

_____

_____

_____

Part B Assessments

1. Make a sketch for each type of sediment you observed. Label any from those shown in figure 57.2. **4**

2. Summarize the results of the urine sediment study. **5** _____

_____

_____

_____

_____

Laboratory Exercise 58

Male Reproductive System

Materials Needed

Human torso model
Model of the male reproductive system
Anatomical chart of the male reproductive system
Compound light microscope
Prepared microscope slides of the following:
 Testis section
 Epididymis, cross section
 Penis, cross section

The organs of the male reproductive system are specialized to produce and maintain the male sex cells, to transport these cells together with supporting fluids to the female reproductive tract, and to produce and secrete male sex hormones. These organs includes the testes and sets of internal and external genitalia.

The testes originate internally near the kidneys but descend through an inguinal canal to a position in the scrotum, which provides a lower temperature necessary for sperm production and storage. Within the testes, numerous seminiferous tubules produce sperm (spermatozoa) by spermatogenesis, and interstitial cells between seminiferous tubules produce testosterone. Testosterone is secreted by the interstitial cells and then transported in the bloodstream. Testosterone influences the sex drive and development of the secondary sex characteristics.

The sperm produced in the seminiferous tubules enter a tubular network, the rete testis, which joins the tightly coiled epididymis. Within head, body, and tail of the epididymis, immature sperm are stored and nourished during their maturation. A muscular ductus deferens, part of the spermatic cord, passes through the inguinal canal into the pelvic cavity and posterior to the urinary bladder, where it unites with ducts from seminal vesicles to form the ejaculatory ducts. An alkaline fluid containing nutrients and prostaglandins is secreted by the seminal vesicles. The ejaculatory ducts unite with the urethra within the prostate gland. Additional secretions are added from the prostate gland and two bulbourethral glands. The combination of sperm and the secretions from the accessory glands is semen.

The urethra, a common tube to convey semen and urine, passes through the penis within the corpus spongiosum to the external urethral orifice. The corpus spongiosum plus two corpora cavernosa provide vascular spaces that become engorged with blood during an erection. A loose fold of skin, forming a cuff over the glans penis, is called the prepuce, and is frequently removed in a procedure called a circumcision shortly after birth.

Purpose of the Exercise

To review the structure and functions of the male reproductive organs and to examine some of these organs.

Learning Outcomes

After completing this exercise, you should be able to

1 Locate and identify the organs of the male reproductive system.
2 Match the structures and functions of these organs.
3 Sketch and label the major features of microscopic sections of the testis, epididymis, and penis.

Procedure A—Male Reproductive Organs

1. Study and label figures 58.1 and 58.2.
2. Observe the human torso model, the model of the male reproductive system, and the anatomical chart

FIGURE 58.1 Label the major structures of the male reproductive system in this sagittal view, using the terms provided. ⚠

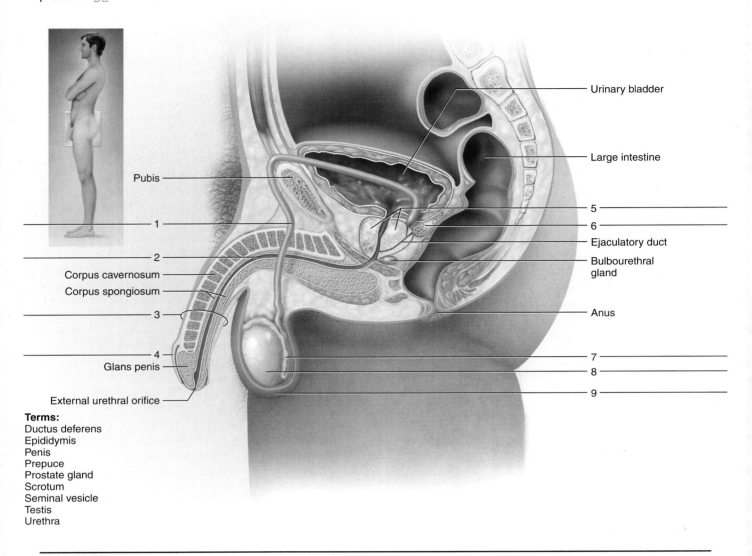

Urinary bladder

Large intestine

Pubis

5

6

Ejaculatory duct

Bulbourethral gland

Corpus cavernosum

Corpus spongiosum

Anus

Glans penis

7

8

9

External urethral orifice

Terms:
Ductus deferens
Epididymis
Penis
Prepuce
Prostate gland
Scrotum
Seminal vesicle
Testis
Urethra

1
2
3
4

FIGURE 58.2 Label the diagram of a sagittal section of a testis and associated structures by placing the correct numbers in the spaces provided. ⚠

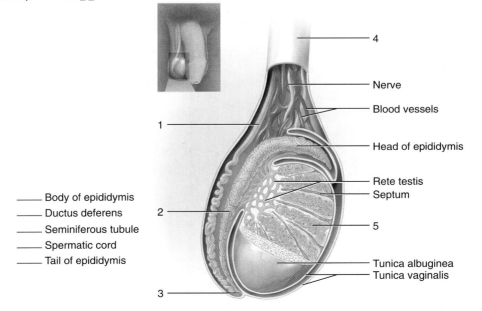

4

Nerve

Blood vessels

Head of epididymis

Rete testis
Septum

5

Tunica albuginea
Tunica vaginalis

1

2

3

_____ Body of epididymis
_____ Ductus deferens
_____ Seminiferous tubule
_____ Spermatic cord
_____ Tail of epididymis

486

of the male reproductive system. Locate the following features:

testes
    seminiferous tubules
    interstitial cells (Leydig cells)

inguinal canal

spermatic cord

epididymis

ductus deferens (vas deferens)

ejaculatory duct

seminal vesicles (seminal glands)

prostate gland

bulbourethral glands

scrotum

penis
    corpora cavernosa
    corpus spongiosum
    tunica albuginea
    glans penis
    external urethral orifice (external urethral meatus)
    prepuce (foreskin)

3. Complete Part A of Laboratory Report 58.

Procedure B—Microscopic Anatomy

1. Obtain a microscope slide of a human testis section and examine it, using low-power magnification (fig. 58.3). Locate the thick *fibrous capsule* (tunica albuginea) on the surface and the numerous sections of *seminiferous tubules* inside.

2. Focus on some of the seminiferous tubules, using high-power magnification (fig. 58.4). Locate the *basement membrane* and the layer of *spermatogonia* just beneath the basement membrane. Identify some *sustentacular cells* (supporting cells; Sertoli cells), which have pale, oval-shaped nuclei, and some *spermatogenic cells,* which have smaller, round nuclei. Spermatogonia give rise to spermatogenic cells that are in various stages of spermatogenesis as they are forced toward the lumen. Near the lumen of the tube, find some darkly stained, elongated heads of developing sperm (spermatozoa). In the spaces between adjacent seminiferous tubules, locate some isolated *interstitial cells* (Leydig cells) of the endocrine system. Interstitial cells produce the hormone testosterone, transported by the blood.

3. Prepare a labeled sketch of a representative section of the testis in Part B of the laboratory report.

FIGURE 58.3   Micrograph of a human testis (1.7×).

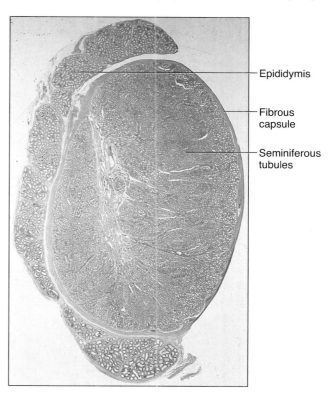

FIGURE 58.4   Micrograph of seminiferous tubules (50× micrograph enlarged to 135×).

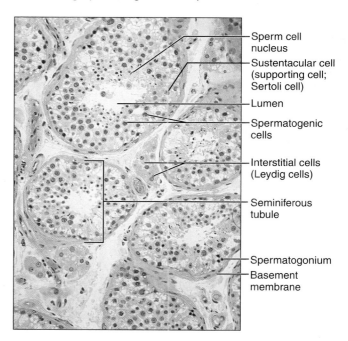

4. Obtain a microscope slide of a cross section of *epididymis* (fig. 58.5). Examine its wall, using high-power magnification. Note the elongated, *pseudostratified columnar epithelial cells* that comprise most of the inner lining. These cells have nonmotile stereocilia (microvilli) on their free surfaces. Also note the thin layer of smooth muscle and connective tissue surrounding the tube.

5. Prepare a labeled sketch of the epididymis wall in Part B of the laboratory report.

6. Obtain a microscope slide of a *penis* cross section, and examine it with low-power magnification (fig. 58.6). Identify the following features:

> **corpora cavernosa**
>
> **corpus spongiosum**
>
> **tunica albuginea**
>
> **urethra**
>
> **skin**

7. Prepare a labeled sketch of a penis cross section in Part B of the laboratory report.

8. Complete Part B of the laboratory report.

FIGURE 58.5 Micrograph of a cross section of a human epididymis (50× micrograph enlarged to 145×).

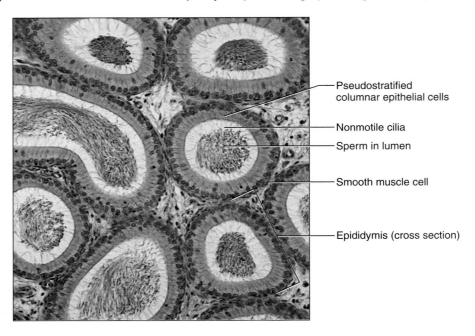

Pseudostratified columnar epithelial cells

Nonmotile cilia

Sperm in lumen

Smooth muscle cell

Epididymis (cross section)

FIGURE 58.6 Micrograph of a cross section of the body of the penis (5×).

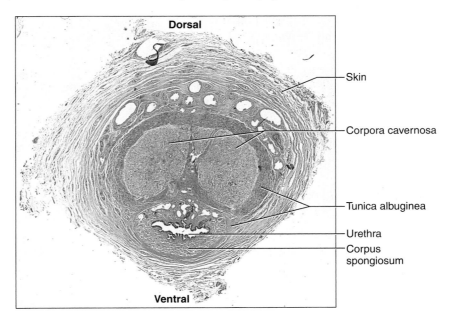

Dorsal

Skin

Corpora cavernosa

Tunica albuginea

Urethra

Corpus spongiosum

Ventral

Name _____

Date _____

Section _____

The Ⓐ corresponds to the Learning Outcome(s) listed at the beginning of the laboratory exercise.

Male Reproductive System

Part A Assessments

Match the terms in column A with the descriptions in column B. Place the letter of your choice in the space provided. ②

Column A		Column B
a. Bulbourethral glands	_____	**1.** Process by which sperm are formed
b. Ductus deferens	_____	**2.** Common tube for urine and semen
c. Ejaculation	_____	**3.** Located within spermatic cord
d. Epididymis	_____	**4.** Produce male hormones
e. Interstitial cells	_____	**5.** Location of spermatogenesis
f. Penis		
g. Prepuce	_____	**6.** Process by which semen is forced out through urethra
h. Prostate	_____	**7.** Contains sperm and secretions from accessory glands
i. Semen	_____	**8.** Paired glands that secrete into urethra
j. Seminiferous tubules	_____	**9.** Stores immature sperm
k. Spermatogenesis	_____	**10.** Surrounds the urethra and secretes a milky fluid
l. Urethra	_____	**11.** Contains three columns of erectile tissue
	_____	**12.** Often removed in circumcision

Part B Assessments

1. Sketch and label a representative section of the testis. ③

2. Sketch and label a region of the epididymis. △3

3. Sketch and label a penis cross section. △3

4. Describe the function of each of the following: △2

    a. Scrotum

        _____

        _____

    b. Spermatogenic cell

        _____

        _____

    c. Interstitial cell (Leydig cell)

        _____

        _____

    d. Epididymis

        _____

        _____

    e. Corpora cavernosa and corpus spongiosum

        _____

        _____

Female Reproductive System

Pre-Lab

1. Carefully read the introductory material and examine the entire lab content.
2. Be familiar with the structures and functions (from lecture or the textbook) of the female reproductive organs.
3. Visit www.mhhe.com/martinseriesl for pre-lab questions and Anatomy & Physiology Revealed animations.

Materials Needed

Human torso model
Model of the female reproductive system
Anatomical chart of the female reproductive system
Compound light microscope
Prepared microscope slides of the following:
    Ovary section with maturing follicles
    Uterine tube, cross section
    Uterine wall section

For Demonstration Activity:
Prepared microscope slides of the following:
    Uterine wall, early proliferative phase
    Uterine wall, secretory phase
    Uterine wall, early menstrual phase

The organs of the female reproductive system are specialized to produce and maintain the female sex cells, to transport these cells to the site of fertilization, to provide a favorable environment for a developing offspring, to move the offspring to the outside, and to produce female sex hormones.

These organs include the ovaries, which produce the egg cells and female sex hormones, and sets of internal and external genitalia (accessory organs). The internal genitalia include the uterine tubes, uterus, and vagina. The external genitalia are the labia majora, labia minora, clitoris, and vestibular glands.

Within the cortex region of an ovary, numerous follicles are in some stage of development to produce ova (eggs) by the process called oogenesis (meiosis). Numerous primordial follicles formed during prenatal development. During the period from puberty to menopause, reproductive cycles occur as individual follicles complete maturation. A primary follicle contains a primary oocyte that undergoes oogenesis and develops into a secondary follicle, and finally a mature follicle shortly before ovulation of a secondary oocyte. The ruptured follicle becomes a corpus luteum, which secretes estrogens and progesterone as it continues to degenerate into the corpus albicans composed of connective tissues.

The uterine tube has fingerlike fimbriae partially around the ovary. After ovulation, the secondary oocyte is conveyed through the uterine tube by the action of cilia and peristaltic waves. If fertilization occurs, early cleavage divisions take place during the passage to the uterus during the next several days. The uterus has three regions: fundus, body, and cervix. There are three layers to its wall: perimetrium, myometrium, and endometrium. Development of the embryo and the fetus take place within the uterus until birth. The vagina has a mucosal lining, and near the vaginal orifice a membrane (hymen) partially closes the orifice unless broken, often during the first intercourse.

The external genitalia (vulva) are within a diamond-shaped perineum with boundaries established by the pubis, ischial tuberosities, and coccyx. Larger rounded folds of skin, labia majora, protect the labia minora and other external genitalia. Between the labia minora is a space (vestibule) that encloses the urethral and vaginal openings. A clitoris at the anterior vulva has similar structures to a penis, including a prepuce, glans, and corpora cavernosa.

The breasts contain the mammary glands, which are comprised of many lobes, ducts, ligaments, and adipose tissue. At puberty, estrogens stimulate breast development, and at the birth of a baby, various hormones promote alveolar gland development and milk production. Lactiferous ducts terminate in 15–20 openings on the nipple in the center of the pigmented areola.

Purpose of the Exercise

To review the structure and functions of the female reproductive organs and to examine some of their features.

Learning Outcomes

After completing this exercise, you should be able to

① Locate and identify the organs of the female reproductive system.

② Match the structures and functions of these organs.

③ Sketch and label the major features of microscopic sections of the ovary, uterine tube, and uterine wall.

Procedure A—Female Reproductive Organs

1. Study and label figures 59.1, 59.2, 59.3, and 59.4.
2. Observe the human torso model, the model of the female reproductive system, and the anatomical chart of the female reproductive system. Locate the following features:

> **ovaries**
>> medulla
>> cortex

ligaments
> broad ligament
> suspensory ligament of ovary
> ovarian ligament
> round ligament of uterus

uterine tubes (oviducts; fallopian tubes)
> infundibulum
> fimbriae

uterus
> fundus of uterus
> body of uterus
> cervix of uterus
>> cervical canal
>> external os
> uterine wall
>> endometrium
>> myometrium
>> perimetrium (serosa; serous coat)

FIGURE 59.1   Label the structures of the female reproductive system, using the terms provided. ⚟

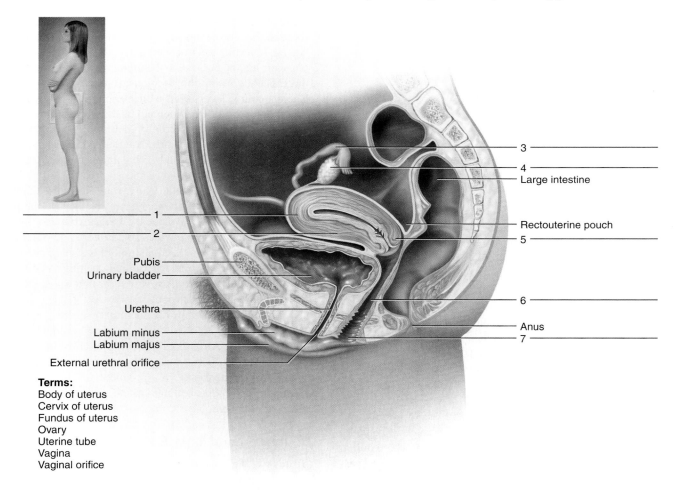

3 ——————
4 ——————
— Large intestine

1 ——————
2 ——————

Pubis —
Urinary bladder —

Urethra —

Labium minus —
Labium majus —

External urethral orifice —

— Rectouterine pouch
5 ——————

6 ——————

— Anus
7 ——————

Terms:
Body of uterus
Cervix of uterus
Fundus of uterus
Ovary
Uterine tube
Vagina
Vaginal orifice

FIGURE 59.2 Label the female reproductive organs in this posterior view, using the terms provided. The left-side organs are shown in section and the ligaments are shown on the right side. ⚠

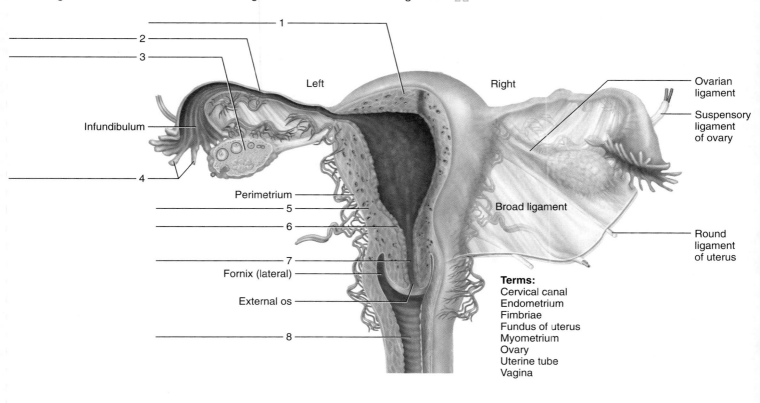

Infundibulum

Left

Right

Ovarian ligament

Suspensory ligament of ovary

Broad ligament

Round ligament of uterus

Perimetrium

Fornix (lateral)

External os

Terms:
Cervical canal
Endometrium
Fimbriae
Fundus of uterus
Myometrium
Ovary
Uterine tube
Vagina

FIGURE 59.3 Label the structures associated with the female perineum by placing the correct numbers in the spaces provided. The external genitalia (vulva) are within the perineum. ⚠

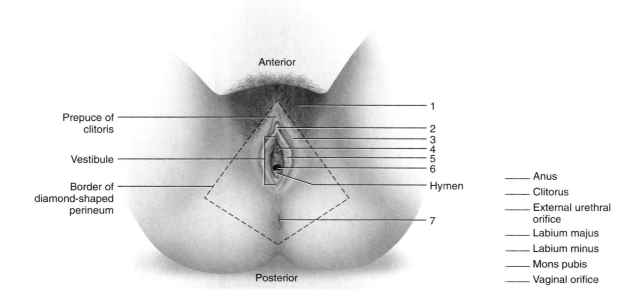

Anterior

Prepuce of clitoris

Vestibule

Border of diamond-shaped perineum

Hymen

Posterior

_____ Anus

_____ Clitorus

_____ External urethral orifice

_____ Labium majus

_____ Labium minus

_____ Mons pubis

_____ Vaginal orifice

FIGURE 59.4 Label the structures of a lactating breast (anterior view), using the terms provided. ⚠

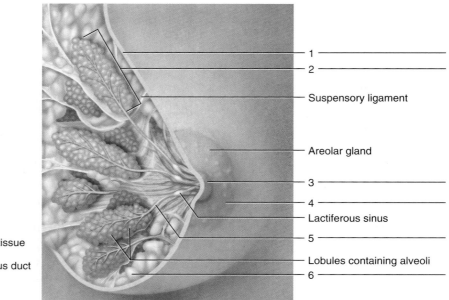

1 _____
2 _____

Suspensory ligament

Areolar gland

3 _____
4 _____

Lactiferous sinus

5 _____

Lobules containing alveoli

6 _____

Terms:
Adipose tissue
Areola
Lactiferous duct
Lobe
Nipple
Skin

rectouterine pouch

vagina

    fornices

    vaginal orifice

    hymen

    mucosal layer

external genitalia (vulva; pudendum)

    mons pubis

    labia majora

    labia minora

    vestibule

    vestibular glands

    clitoris

breasts

    nipple

    areola

    alveolar glands (compose milk-producing parts of mammary glands)

    lactiferous ducts

    adipose tissue

3. Complete Part A of Laboratory Report 59.

Procedure B—Microscopic Anatomy

1. Obtain a microscope slide of an ovary section with maturing follicles, and examine it with low-power magnification (fig. 59.5). Locate the outer layer, or *cortex,* composed of densely packed cells, and the inner layer, or *medulla,* which largely consists of loose connective tissue.

2. Focus on the cortex of the ovary, using high-power magnification (fig. 59.6). Note the thin layer of small cuboidal cells on the free surface. These cells comprise the *germinal epithelium.* Also locate some *primordial follicles* just beneath the germinal epithelium. Each follicle consists of a single, relatively large *primary oocyte* with a prominent nucleus and a covering of *follicular cells.*

3. Use low-power magnification to search the ovarian cortex for maturing follicles in various stages of development. Prepare three labeled sketches in Part B of the laboratory report to illustrate the changes that occur in a follicle as it matures.

4. Obtain a microscope slide of a cross section of a uterine tube. Examine it, using low-power magnification (fig. 59.7). The shape of the lumen is very irregular.

5. Focus on the inner lining of the uterine tube, using high-power magnification. The lining is composed of *simple columnar epithelium,* and some of the epithelial cells are ciliated on their free surfaces.

6. Prepare a labeled sketch of a representative region of the wall of the uterine tube in Part B of the laboratory report.

7. Obtain a microscope slide of the uterine wall section (fig. 59.8). Examine it, using low-power magnification, and locate the following:

    endometrium (inner mucosal layer)

    myometrium (middle, thick muscular layer)

    perimetrium (outer serosal layer)

8. Prepare a labeled sketch of a representative section of the uterine wall in Part B of the laboratory report.

9. Complete Part C of the laboratory report.

FIGURE 59.5 Micrograph of the ovary (30× micrograph enlarged to 80×).

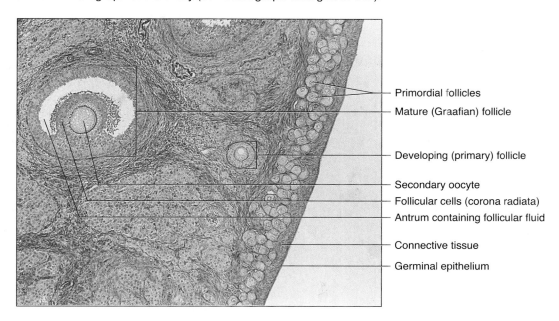

- Primordial follicles
- Mature (Graafian) follicle
- Developing (primary) follicle
- Secondary oocyte
- Follicular cells (corona radiata)
- Antrum containing follicular fluid
- Connective tissue
- Germinal epithelium

FIGURE 59.6 Micrograph of the ovarian cortex (100× micrograph enlarged to 200×).

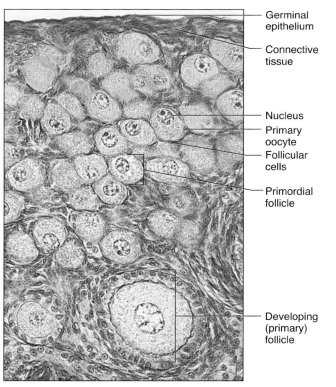

- Germinal epithelium
- Connective tissue
- Nucleus
- Primary oocyte
- Follicular cells
- Primordial follicle
- Developing (primary) follicle

FIGURE 59.7 Micrograph of a cross section of the uterine tube (8×).

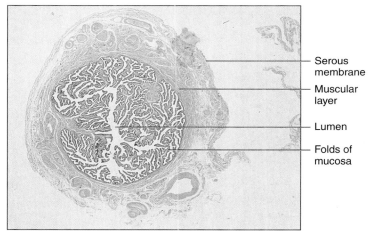

- Serous membrane
- Muscular layer
- Lumen
- Folds of mucosa

FIGURE 59.8 Micrograph of the uterine wall (10× micrograph enlarged to 35×).

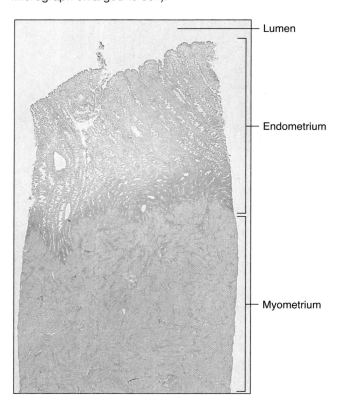

- Lumen
- Endometrium
- Myometrium

Demonstration Activity

Observe the slides in the demonstration microscopes. Each slide contains a section of uterine mucosa taken during a different phase in the reproductive cycle. In the *early proliferative phase,* note the simple columnar epithelium on the free surface of the mucosa and the many sections of tubular uterine glands in the tissues beneath the epithelium. In the *secretory phase,* the endometrium is thicker and the uterine glands appear more extensive and they are coiled. In the *early menstrual phase,* the endometrium is thinner because its surface layer has been lost. Also the uterine glands are less apparent, and the spaces between the glands contain many leukocytes. What is the significance of these changes?

_____

_____

_____

Name _____

Date _____

Section _____

The ⒶÅ corresponds to the Learning Outcome(s) listed at the beginning of the laboratory exercise.

Female Reproductive System

Part A Assessments

Match the terms in column A with the descriptions in column B. Place the letter of your choice in the space provided. Ⓐ2

Column A	Column B
a. Cervix	_____ **1.** Rounded end of uterus near uterine tubes
b. Cilia	_____ **2.** Smooth muscle that contracts with force during childbirth
c. Endometrium	_____ **3.** Thin membrane that partially closes vaginal orifice
d. Fimbriae	_____ **4.** Contains the vulva
e. Fundus	_____ **5.** Process by which a secondary oocyte is released from the ovary
f. Hymen	_____ **6.** Portion of uterus extending into superior portion of vagina
g. Lactiferous duct	_____ **7.** Inner mucosal lining of uterus
h. Myometrium	_____ **8.** Transports milk from milk-producing lobes
i. Ovulation	_____ **9.** Help move secondary oocyte through the uterine tube toward uterus
j. Perineum	_____ **10.** Fingerlike projections of uterine tube near ovary

Part B Assessments

1. Sketch and label a series of three changes to illustrate follicular maturation. Ⓐ3

2. Sketch and label a representative section of the wall of a uterine tube. **3**

3. Sketch and label a representative section of uterine wall. **3**

Part C Assessments

Identify the numbered features in figure 59.9 of a sectioned ovary that represent structures present at some stage during a reproductive cycle.

FIGURE 59.9   Label this ovary by placing the correct numbers in the spaces provided. The arrows indicate development over time; follicles do not migrate through an ovary. **1**

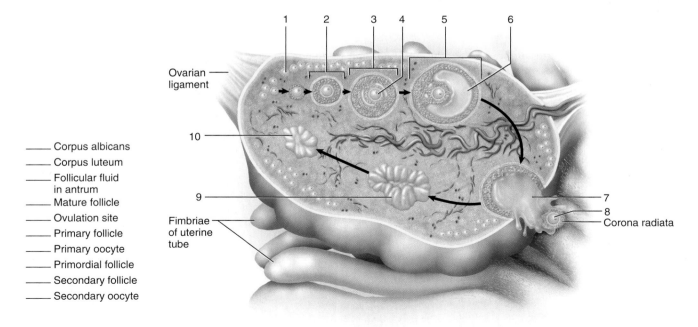

_____ Corpus albicans
_____ Corpus luteum
_____ Follicular fluid
      in antrum
_____ Mature follicle
_____ Ovulation site
_____ Primary follicle
_____ Primary oocyte
_____ Primordial follicle
_____ Secondary follicle
_____ Secondary oocyte

Laboratory Exercise 60

Fertilization and Early Development

Pre-Lab

1. Carefully read the introductory material and examine the entire lab content.
2. Be familiar with fertilization and early embryonic development (from lecture or the textbook).
3. Visit www.mhhe.com/martinseries1 for pre-lab questions and Anatomy & Physiology Revealed animations.

Materials Needed

Sea urchin egg suspension[*]
Sea urchin sperm suspension[*]
Compound light microscope
Depression microscope slide
Coverslip
Medicine droppers
Models of human embryos
Prepared microscope slide of the following:
  Sea urchin embryos (early and late cleavage)

For Learning Extension Activity:
Vaseline
Toothpick

For Demonstration Activity:
Preserved mammalian embryos

[*]See the Instructor's Manual for a source of materials.

Ovulation occurs when a secondary oocyte, surrounded by a zona pellucida, is expelled from the ovary into the uterine tube. Before fertilization can occur, semen must be deposited in the vagina; some of the sperm must pass through the uterus into the uterine tubes and contact a secondary oocyte. Upon penetration of a sperm, the second meiotic division takes place, forming a second polar body. Fertilization is the process by which the pronuclei of the ovum (egg) and sperm come together and combine their chromosomes, forming a single diploid cell called a *zygote.*

Shortly after fertilization, the zygote undergoes cell division (mitosis and cytokinesis) to form two cells. These two cells become four, then in turn divide into eight, and so forth. A solid cluster of cells forms, the morula, which is still surrounded by the zona pellucida. The cells of the morula continue cell cycles and the hollow *blastocyst* develops. The blastocyst contains an inner cell mass, which develops into the embryo, and an outer trophoblast layer, which develops into the extraembryonic membranes and a portion of the *placenta.* These early cleavage divisions, resulting in an increasing number of cells that are smaller than the zygote, continue for most of the first week of development. The blastocyst implants into the endometrium about 6–7 days after fertilization. Most of the early stages of development are similar in other animals, including the sea urchin.

In human development, the embryonic stage lasts through the first eight weeks, followed by the fetal stage until birth. The blastocyst develops into a gastrula containing three primary germ layers: the ectoderm, endoderm, and mesoderm, which give rise to particular tissues and body systems. The extraembryonic membranes include the amnion, yolk sac, allantois, and chorion. The amnion extends around the embryo and is filled with amniotic fluid, providing a protective cushion and constant environment. The yolk sac forms a portion of the embryonic digestive tube and stem cells of the bone marrow; however, the placental region, not the yolk, provides the nutritive functions. The allantois produces early blood cells, gives rise to the umbilical blood vessels, and forms part of the urinary bladder. The chorion extends around the embryo and the other membranes, and it eventually fuses with the amnion to form the amniochorionic membrane.

The placenta is comprised of maternal and fetal components. The maternal component is a region of the endometium (decidua basalis); the fetal portion contains an elaborate series of chorionic villi formed from the chorion. An examination of the placenta upon birth (the afterbirth) reveals a rough surface of the maternal portion, and a smooth surface with the attached umbilical cord of the fetal part.

Purpose of the Exercise

To review the process of fertilization, to observe sea urchin eggs being fertilized, and to examine embryos in early stages of development.

Learning Outcomes

After completing this exercise, you should be able to

(1) Sketch the early developmental stages of a sea urchin.

(2) Distinguish the major features of human embryos.

(3) Describe fertilization and the early developmental stages of a human.

Procedure A—Fertilization

1. Study figure 60.1.

2. Although it is difficult to observe fertilization in animals since the process occurs internally, it is possible to view forms of external fertilization. For example, egg and sperm cells can be collected from sea urchins, and the process of fertilization can be observed microscopically. To make this observation, follow these steps:

   a. Place a drop of sea urchin egg-cell suspension in the chamber of a depression slide, and add a coverslip.

   b. Examine the egg cells, using low-power magnification.

   c. Focus on a single egg cell with high-power magnification, and sketch the cell in Part A of Laboratory Report 60.

   d. Remove the coverslip and add a drop of sea urchin sperm-cell suspension to the depression slide.

Replace the coverslip, and observe the sperm cells with high-power magnification as they cluster around the egg cells. This attraction is stimulated by gamete secretions.

   e. Observe the egg cells with low-power magnification once again, and watch for the appearance of *fertilization membranes*. Such a membrane forms as soon as an egg cell is penetrated by a sperm cell; it looks like a clear halo surrounding the egg cell.

   f. Focus on a single fertilized egg cell, and sketch it in Part A of the laboratory report.

Learning Extension Activity

Use a toothpick to draw a thin line of Vaseline around the chamber of the depression slide containing the fertilized sea urchin egg cells. Place a coverslip over the chamber, and gently press it into the Vaseline to seal the chamber and prevent the liquid inside from evaporating. Keep the slide in a cool place so that the temperature never exceeds 22°C (72°F). Using low-power magnification, examine the slide every 30 minutes, and look for the appearance of two-, four-, and eight-cell stages of developing sea urchin embryos.

FIGURE 60.1 Fertilization and early embryonic development in the uterine tube and implantation of blastocyst stage in endometrium of uterus.

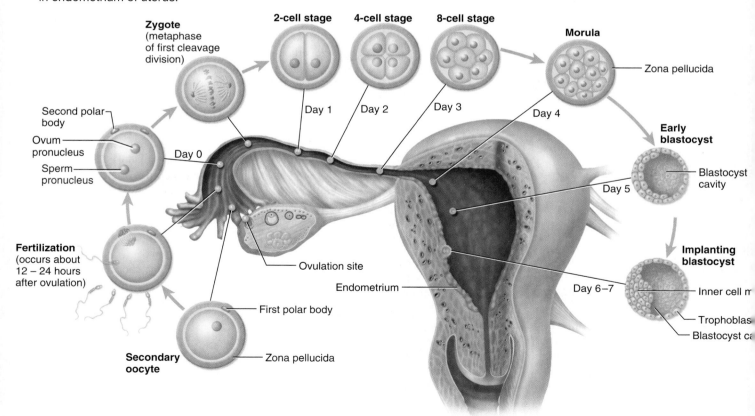

Procedure B—Sea Urchin Early Development

1. Obtain a prepared microscope slide of developing sea urchin embryos. This slide contains embryos in various stages of cleavage. Search the slide, using low-power magnification, and locate embryos in two-, four-, eight-cell, and morula stages. Observe that cleavage results in an increase of cell numbers; however, the cells get progressively smaller.
2. Prepare a sketch of each stage in Part B of the laboratory report.

Procedure C—Human Early Development

1. Study figures 60.2 and 60.3.
2. Observe the models of human embryos, and identify the following features:

   blastocyst
   blastocyst cavity (blastocoel)
   inner cell mass (embryoblast)
   trophoblast

primary germ layers
   ectoderm
   endoderm
   mesoderm
chorion
chorionic villi
amnion
amniotic fluid
umbilical cord
   umbilical arteries (2)
   umbilical vein (1)
yolk sac
allantois
placenta

3. Complete Parts C and D of the laboratory report.

FIGURE 60.2 Structures associated with this 12-week-old fetus: (a) fetus, membranes, and uterus; (b) fetal and maternal vasculature of placenta.

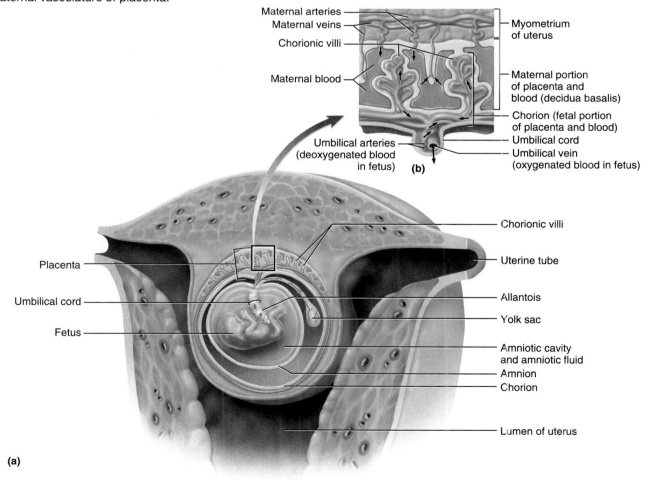

FIGURE 60.3 Human placenta (afterbirth): (*a*) fetal surface; (*b*) maternal (uterine) surface. At full term, the placenta is about 18 cm in diameter, 4 cm thickness, and weighs about 450 g (one pound).

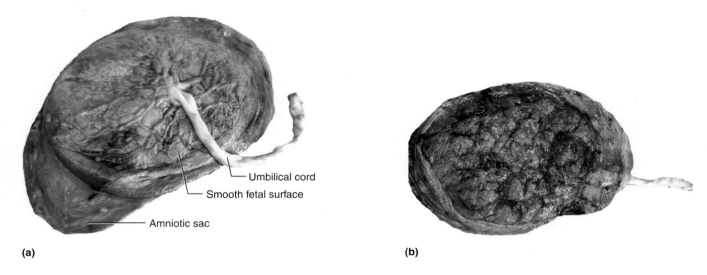

(a)
    — Umbilical cord
    — Smooth fetal surface
    — Amniotic sac

(b)

 Demonstration Activity

Observe the preserved mammalian embryos that are on display. In addition to observing the developing external body structures, identify such features as the chorion, chorionic villi, amnion, yolk sac, umbilical cord, and placenta. What special features provide clues to the types of mammals these embryos represent?

_____

_____

_____

Name _____

Date _____

Section _____

The ▲ corresponds to the Learning Outcome(s) listed at the beginning of the laboratory exercise.

Fertilization and Early Development

Part A Assessments

Prepare sketches of the following: ▲1

Single sea urchin egg (_____×)	Fertilized sea urchin egg (_____×)

Part B Assessments

Prepare sketches of the following: ▲1

Two-cell sea urchin embryo (_____×)	Four-cell sea urchin embryo (_____×)
Eight-cell sea urchin embryo (_____×)	Morula sea urchin embryo (_____×)

Part C Assessments

Match the terms in column A with the descriptions in column B. Place the letter of your choice in the space provided. /2\

Column A	Column B
a. Blastocyst	_____ **1.** Gel layer around oocyte and early cleavage stages
b. Chorionic villi	_____ **2.** Hollow ball of cells
c. Endometrium	_____ **3.** Develops into extraembryonic membranes
d. Gastrula	_____ **4.** Solid ball of about sixteen cells
e. Morula	_____ **5.** Forms fetal portion of placenta
f. Polar bodies	_____ **6.** Forms maternal portion of placenta
g. Trophoblast	_____ **7.** Contains the ectoderm, endoderm, and mesoderm
h. Zona pellucida	_____ **8.** Small cells formed during meiosis that degenerate

Part D Assessments

Complete the following statements:

1. The umbilical cord contains three blood vessels, two of which are _____. /3\

2. _____ fluid protects the embryo from jarred movements and provides a watery environment for development. /3\

3. The cell resulting from fertilization is called a(an) _____. /3\

4. The cell resulting from fertilization divides by the process of _____. /3\

5. _____ is the phase of development during which cellular divisions result in smaller and smaller cells. /3\

6. _____ on the inner lining of the uterine tube aid in moving a developing embryo. /3\

7. A human offspring is called a(an) _____ until the end of the eighth week of development. /3\

8. After the eighth week, a developing human is called a(an) _____ until the time of birth. /3\

9. Summarize your observations of any similarities noted between sea urchin and human early development stages. /3\

_____

_____

_____

_____

_____

Laboratory Exercise 61

Genetics

Pre-Lab

1. Carefully read the introductory material and examine the entire lab content.
2. Be familiar with genetic terminology and genetic problems (from lecture or the textbook).
3. Visit www.mhhe.com/martinseries1 for pre-lab questions.

Materials Needed

Pennies (or other coins)
Dice
PTC paper
Astigmatism chart
Ichikawa's or Ishihara's color plates for color blindness test

Many examples of human genetics in this laboratory exercise are basic, external features to observe. Some traits are based upon simple Mendelian genetics. As our knowledge of genetics continues to develop, traits such as tongue roller and free earlobe may no longer be considered examples of simple Mendelian genetics. We may need to continue to abandon some simple Mendelian models. New evidence may involve polygenic inheritance, effects of other genes, or environmental factors as more appropriate explanations. Therefore, the analysis of family genetics is not appropriate nor is it the purpose of this laboratory exercise.

Genetics is the study of the inheritance of characteristics. The genes that transmit this information are coded in segments of DNA in chromosomes. Homologous chromosomes possess the same gene at the same *locus*. These genes may exist in variant forms, called *alleles*. If a person pos-

sesses two identical alleles, the condition is *homozygous*. If a person possesses two different alleles, the condition is *heterozygous*. The particular combination of these gene variants (alleles) represents the person's *genotype;* the appearance of the individual that develops as a result of the way the genes are expressed represents the person's *phenotype.*

If one allele determines the phenotype by masking the expression of the other allele in a heterozygous individual, the allele is termed *dominant.* The allele whose expression is masked is termed *recessive.* If the heterozygous condition determines an intermediate phenotype, the inheritance represents *incomplete dominance.* However, different alleles are codominant if both are expressed in the heterozygous condition. Some characteristics inherited on the sex chromosomes result in phenotype frequencies that might be more prevalent in males or females. Such characteristics are called sex-linked (X-linked or Y-linked) characteristics.

As a result of meiosis during the formation of eggs and sperm, a mother and father each transmit an equal number of chromosomes (the haploid number 23) to form the zygote (diploid number 46). An offspring will receive one allele from each parent. These gametes combine randomly in the formation of each offspring. Hence, the *laws of probability* can be used to predict possible genotypes and phenotypes of offspring. A genetic tool called a *Punnett square* simulates all possible combinations (probabilities) that can occur in offspring genotypes and resulting phenotypes.

Purpose of the Exercise

To observe some selected human traits, to use pennies and dice to demonstrate laws of probability, and to solve some genetic problems using a Punnett square.

Learning Outcomes

After completing this exercise, you should be able to

1. Examine and record twelve genotypes and phenotypes of selected human traits.
2. Demonstrate the laws of probability using tossed pennies and dice and interpret the results.
3. Predict genotypes and phenotypes of complete dominance, codominance, and sex-linked problems using Punnett squares.

Procedure A—Human Genotypes and Phenotypes

A complete set of genetic instructions in one human cell constitutes one's *genome*. The human genome contains about 2.9 billion base pairs representing approximately 20,500 protein-encoding genes. These instructions represent our genotypes and are expressed as phenotypes sometimes clearly observable on our bodies. Some of these traits are listed in table 61.1 and are discernible in figure 61.1.

A dominant trait might be homozygous or heterozygous, only one capital letter is used along with a blank for the possible second dominant or recessive allele. For a recessive trait, two lowercase letters represent the homozygous recessive genotype for that characteristic. Dominant does not always correlate with the predominance of the allele in the gene pool; dominant means one allele will determine the appearance of the phenotype.

1. **Tongue roller/nonroller:** The dominant allele (R) determines the person's ability to roll the tongue into a U-shaped trough. The homozygous recessive condition (rr) prevents this tongue rolling (fig. 61.1). Record your results in the table in Part A of Laboratory Report 61.
2. **Freckles/no freckles:** The dominant allele (F) determines the appearance of freckles. The homozygous recessive condition (ff) does not produce freckles (fig. 61.1). Record your results in the table in Part A of the laboratory report.
3. **Widow's peak/straight hairline:** The dominant allele (W) determines the appearance of a hairline above the forehead that has a distinct downward point in the center, called a widow's peak. The homozygous recessive condition (ww) produces a straight hairline (fig. 61.1). A receding hairline would prevent this phenotype determination. Record your results in the table in Part A of the laboratory report.
4. **Dimples/no dimples:** The dominant allele (D) determines the appearance of a distinct dimple in one or both cheeks upon smiling. The homozygous recessive condition (dd) results in the absence of dimples (fig. 61.1). Record your results in the table in Part A of the laboratory report.
5. **Free earlobe/attached earlobe:** The dominant allele (E) codes for the appearance of an inferior earlobe that hangs freely below the attachment to the head. The homozygous recessive condition (ee) determines the earlobe attaching directly to the head at its inferior border (fig. 61.1). Record your results in the table in Part A of the laboratory report.
6. **Normal skin coloration/albinism:** The dominant allele (M) determines the production of some melanin, producing normal skin coloration. The homozygous recessive condition (mm) determines albinism due to the inability to produce or use the enzyme tyrosinase in pigment cells. An albino does not produce melanin in the skin, hair, or the middle tunic (choroid, ciliary body, and iris) of the eye. The absence of melanin in the middle tunic allows the pupil to appear slightly red to nearly black. Remember that the pupil is an opening in the iris filled with transparent aqueous humor. An albino human has pale white skin, flax-white hair, and a pale blue iris. Record your results in the table in Part A of the laboratory report.
7. **Astigmatism/normal vision:** The dominant allele (A) results in an abnormal curvature to the cornea or the lens. As a consequence, some portions of the image projected on the retina are sharply focused, and other portions are blurred. The homozygous recessive condition (aa) generates normal cornea and lens shapes and normal vision. Use the astigmatism chart and directions to assess this possible defect described in Laboratory Exercise 36. Other eye defects, such as nearsightedness (myopia) and

TABLE 61.1 Examples of Some Common Human Phenotypes

Dominant Traits and Genotypes	Recessive Traits and Genotypes
Tongue roller (R__)	Nonroller (rr)
Freckles (F__)	No freckles (ff)
Widow's peak (W__)	Straight hairline (ww)
Dimples (D__)	No dimples (dd)
Free earlobe (E__)	Attached earlobe (ee)
Normal skin coloration (M__)	Albinism (mm)
Astigmatism (A__)	Normal vision (aa)
Natural curly hair (C__)	Natural straight hair (cc)
PTC taster (T__)	Nontaster (tt)
Blood type A (I^A __), B (I^B __), or AB ($I^A I^B$)	Blood type O (ii)
Normal color vision ($X^C X^C$), ($X^C X^c$), or ($X^C Y$)	Red-green color blindness ($X^c X^c$) or ($X^c Y$)

FIGURE 61.1 Representative genetic traits comparing dominant and recessive phenotypes: (*a*) tongue roller; (*b*) nonroller; (*c*) freckles; (*d*) no freckles; (*e*) widow's peak; (*f*) straight hairline; (*g*) dimples; (*h*) no dimples; (*i*) free earlobe; (*j*) attached earlobe.

Dominant Traits   **Recessive Traits**

(a) Tongue roller

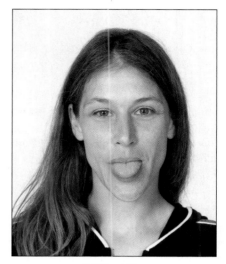

(b) Nonroller

(c) Freckles

(d) No freckles

(e) Widow's peak

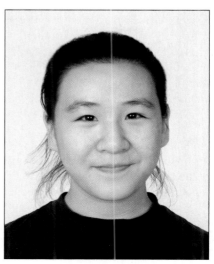

(f) Straight hairline

FIGURE 61.1 *(Continued).*

Dominant Traits

Recessive Traits

(g) Dimples

(h) No dimples

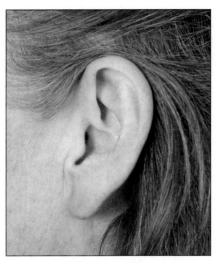

(i) Free earlobe

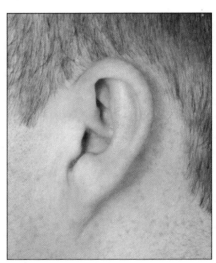

(j) Attached earlobe

farsightedness (hyperopia), are different genetic traits due to genes at other locations. Record your results in the table in Part A of the laboratory report.

8. **Curly hair/straight hair:** The dominant allele (*C*) determines the appearance of curly hair. Curly hair is somewhat flattened in cross section, as the hair follicle of a similar shape served as a mold for the root of the hair during its formation. The homozygous recessive condition (*cc*) produces straight hair. Straight hair is nearly round in cross section from being molded into this shape in the hair follicle. In some populations (Caucasians) the heterozygous condition (*Cc*) expresses the intermediate wavy hair phenotype (incomplete dominance). This trait determination assumes no permanents or hair straightening procedures have been performed.

Such hair alterations do not change the hair follicle shape, and future hair growth results in original genetic hair conditions. Record your unaltered hair appearance in the table in Part A of the laboratory report.

9. **PTC taster/nontaster:** The dominant allele (*T*) determines the ability to experience a bitter sensation when PTC paper is placed on the tongue. About 70% of people possess this dominant gene. The homozygous recessive condition (*tt*) makes a person unable to notice the substance. Place a piece of PTC (phenylthiocarbamide) paper on the upper tongue surface and chew it slightly to see if you notice a bitter sensation from this harmless chemical. The nontaster of the PTC paper does not detect any taste at all from this substance. Record your results in the table in Part A of the laboratory report.

TABLE 61.2   Genotypes and Phenotypes (Blood Types)

Genotypes	Phenotypes (Blood Types)
$I^A I^A$ or $I^A i$	A
$I^B I^B$ or $I^B i$	B
$I^A I^B$ (codominant)	AB
ii	O

10. **Blood type A, B, or AB/blood type O:** There are three alleles (I^A, I^B, and i) in the human population affecting RBC membrane structure. These alleles are located on a single pair of homologous chromosomes, so a person could possess either two of the three alleles or two of the same allele. All of the possible combinations of these alleles of genotypes and the resulting phenotypes are depicted in table 61.2. The expression of the blood type AB is a result of both codominant alleles located in the same individual. Possibly you have already determined your blood type in Laboratory Exercise 43 or have it recorded on a blood donor card. (If simulated blood-typing kits were used for Laboratory Exercise 43, those results would not be valid for your genetic factors.) Record your results in the table in Part A of the laboratory report.

11. **Sex determination:** A person with sex chromosomes XX displays a female phenotype. A person with sex chromosomes XY displays a male phenotype. Record your results in the table in Part A of the laboratory report.

12. **Normal color vision/red-green color blindness:** This condition is a sex-linked (X-linked) characteristic. The alleles for color vision are also on the X chromosome, but absent on the Y chromosome. As a result, a female might possess both alleles (C and c), one on each of the X chromosomes. The dominant allele (C) determines normal color vision; the homozygous recessive condition (cc) results in red-green color blindness. However, a male would possess only one of the two alleles for color vision because there is only a single X chromosome in a male. Hence a male with even a single recessive gene for color blindness possesses the defect. Note all the possible genotypes and phenotypes for this condition (table 61.1). Review the color vision test in Laboratory Exercise 36 using the color plates in figure 36.4 and Ichikawa's or Ishihara's book. Record your results in the table in Part A of the laboratory report.

13. Complete Part A of Laboratory Report 61.

Procedure B—Laws of Probability

The laws of probability provide a mathematical way to determine the likelihood of events occurring by chance. This prediction is often expressed as a ratio of the number of results from experimental events to the number of results considered possible. For example, when tossing a coin there is an equal chance of the results displaying heads or tails. Hence the probability is one-half of obtaining either a heads or a tails (there are two possibilities for each toss). When all of the probabilities of all possible outcomes are considered for the result, they will always add up to a 1. To predict the probability of two or more events occurring in succession, multiply the probabilities of each individual event. For example, the probability of tossing a die and displaying a 4 two times in a row is $1/6 \times 1/6 = 1/36$ (there are six possibilities for each toss). Each toss in a sequence is an *independent event* (chance has no memory). The same laws apply when parents have multiple children (each fertilization is an independent event). Perform the following experiments to demonstrate the laws of probability:

1. Use a single penny (or other coin) and toss it 20 times. Predict the number of heads and tails that would occur from the 20 tosses. Record your prediction and the actual results observed in Part B of the laboratory report.

2. Use a single die (*pl.* dice) and toss it 24 times. Predict the number of times a number below 3 (numbers 1 and 2) would occur from the 24 tosses. Record your prediction and the actual results in Part B of the laboratory report.

3. Use two pennies and toss them simultaneously 32 times. Predict the number of times two heads, a heads and a tails, and two tails occur. Record your prediction and the actual results in Part B of the laboratory report.

4. Use a pair of dice and toss them simultaneously 32 times. Predict the number of times for both dice coming up with odd numbers, one die an odd and the other an even number, and both dice coming up with even numbers. Record your prediction and the actual results in Part B of the laboratory report.

5. Obtain class totals for all of the coins and dice tossed by adding your individual results to a class tally location as on the blackboard.

6. A Punnett square can be used for a visual representation to demonstrate the probable results for two pennies tossed simultaneously. For the purpose of a genetic comparison, an *h* (heads) will represent one "allele" on the coin; a *t* (tails) will represent a different "allele" on the coin.

Coin #1 Possibilities

	h	*t*	
h	*hh*	*ht*	Possible Combinations
t	*ht*	*tt*	of Two Tossed Pennies (in Boxes)

Coin #2 Possibilities

7. Complete Part B of the laboratory report.

Procedure C—Genetic Problems

1. A Punnett square can be constructed to demonstrate a visual display of the predicted offspring from parents with known genotypes. Recall that in complete dominance, a

dominant allele is expressed in the phenotype, as it can mask the other recessive allele on the homologous chromosome pair. Recall also that during meiosis, the homologous chromosomes with their alleles separate (Mendel's Law of Segregation) into different gametes. An example of such a cross might be a homozygous dominant mother for dimples (*DD*) has offspring with a father homozygous recessive (*dd*) for the same trait. The results of such a cross, according to the laws of probability, would be represented by the following Punnett square:

Female Gametes

		D	*D*	
Male Gametes	*d*	*Dd*	*Dd*	Possible Genotypes of Offspring (in Boxes)
	d	*Dd*	*Dd*	

Results: Genotypes: 100% *Dd* (all heterozygous)

Phenotypes: 100% dimples

In another example, assume that both parents are heterozygous (*Dd*) for dimples. The results of such a cross, according to the laws of probability, would be represented by the following Punnett square:

Female Gametes

		D	*D*	
Male Gametes	*D*	*DD*	*Dd*	Possible Genotypes of Offspring
	d	*Dd*	*dd*	

Results: Genotypes: 25% *DD* (homozygous dominant); 50% *Dd* (heterozygous); 25% *dd* (homozygous recessive) (1:2:1 genotypic ratio)

Phenotypes: 75% dimples; 25% no dimples (3:1 phenotypic ratio)

2. Work the genetic problems 1 and 2 in Part C of the laboratory report.

3. The ABO blood type inheritance represents an example of codominance. Review table 61.2 for the genotypes and phenotypes for the expression of this trait. A similar technique completed previously can be used to predict the offspring of parents of known genotypes. In this example, assume the genotype of the mother is $I^A I^B$, and the father is *ii*. The results of such a cross would be represented by the following Punnett square:

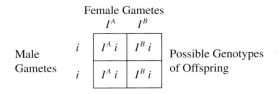

Results: Genotypes: 50% $I^A i$ (heterozygous for A); 50% $I^B i$ (heterozygous for B) (1:1 genotypic ratio)

Phenotypes: 50% blood type A; 50% blood type B (1:1 phenotypic ratio)

Note: In this particular cross, all of the children would have blood types unlike either parent.

4. Work the genetic problems 3 and 4 in Part C of the laboratory report.

5. Review the inheritance of red-green color blindness, an X-linked characteristic, in table 61.1. A similar technique used to identify complete dominance can be performed to predict the offspring of parents of known genotypes. In this example, assume the genotype of the mother is heterozygous $X^C X^c$ (normal color vision, but a carrier for the color blindness defect), and the father is X^C Y (normal color vision; no allele on the Y chromosome). The results of such a cross would be represented by the following Punnett square:

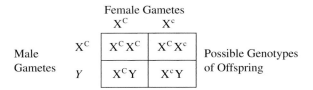

Results: Genotypes: 25% $X^C X^C$; 25% $X^C X^c$; 25% $X^C Y$; 25% $X^c Y$ (1:1:1:1 genotypic ratio)

Phenotypes for sex determination: 50% females; 50% males (1:1 phenotypic ratio)

Phenotypes for color vision: Females 100% normal color vision (however, 50% are heterozygous carriers for color blindness). Males 50% normal; 50% with red-green color blindness (1:1 phenotypic ratio)

Note: In X-linked inheritance, males with color blindness received the recessive gene from their mothers.

6. Complete Part C of the laboratory report.

Name _____

Date _____

Section _____

The ⒜ corresponds to the Learning Outcome(s) listed at the beginning of the laboratory exercise.

Genetics

Part A Assessments

1. Enter your test results for genotypes and phenotypes in the table. Circle your particular phenotype and genotype for each of the twelve traits. ⒤

Trait	Dominant Phenotype	Genotype	Recessive Phenotype	Genotype
Tongue movement	Roller	R__	Nonroller	rr
Freckles	Freckles	F__	No freckles	ff
Hairline	Widow's peak	W__	Straight	ww
Dimples	Dimples	D__	No dimples	dd
Earlobe	Free	E__	Attached	ee
Skin coloration	Normal (some melanin)	M__	Albinism	mm
Vision	Astigmatism	A__	Normal	aa
Hair shape*	Curly	C__	Straight	cc
Taste	PTC taster	T__	Nontaster for PTC	tt
Blood type	A, B, or AB	I^A__; I^B__; or $I^A I^B$	O	ii
Sex		XX or XY		
Color vision	Normal	$X^C X$__ or $X^C Y$	Red-green color blindness	$X^c X^c$ or $X^c Y$

*In some populations (Caucasians) the heterozygous condition (Cc) results in the appearance of wavy hair, which actually represents an example of incomplete dominance for this trait.

2. Analyze all the genotypes that you circled for the dominant phenotypes. If it is feasible to observe your biological parents and siblings for any of these traits, are you able to determine if any of your dominant genotypes are homozygous dominant or heterozygous? _____ If so, which ones? Explain the rationale for your response.

_____

_____

_____

_____

_____

Part B Assessments

1. Single penny tossed 20 times and counting heads and tails: /2\

   Probability (prediction): ___/20 heads ___/20 tails

   (Note: Traditionally, probabilities are converted to the lowest fractional representation.)

   Actual results: _____ heads _____ tails

   Class totals: _____ heads _____ tails

2. Single die tossed 24 times and counting the number of times a number below 3 occurs: /2\

   Probability: ____/24 number below 3 (numbers 1 and 2)

   Actual results: ____ number below 3

   Class totals: ____ number below 3 ____ total tosses by class members

3. Two pennies tossed simultaneously 32 times and counting the number of two heads, a heads and a tails, and two tails: /2\

   Probability: ____/32 of two heads ____/32 of a heads and a tails ____/32 of two tails

   Actual results: ____two heads ____heads and tails ____two tails

   Class totals: ____two heads ____heads and tails ____two tails

4. Two dice tossed simultaneously 32 times and counting the number of two odd numbers, an odd and an even number, and two even numbers: /2\

   Probability: ____/32 of two odd numbers ____/32 of an odd and an even number ____/32 of two even numbers

   Actual results: ____ two odd numbers ____ an odd and an even number ____ two even numbers

   Class totals: ____ two odd numbers ____ an odd and an even number ____ two even numbers

5. Use the example of the two dice tossed 32 times and construct a Punnett square to represent the possible combinations that could be used to determine the probability (prediction) of odd and even numbers for the resulting tosses. Your construction should be similar to the Punnett square for the two coins tossed that is depicted in Procedure B of the laboratory exercise. /2\

6. Complete the following:

   a. Are the class totals closer to the predicted probabilities than your results? _____ Explain your response. /2\

      _____

      _____

   b. Does the first toss of the penny or the first toss of the die have any influence on the next toss? _____
      Explain your response. /2\

      _____

   c. Assume a family has two boys or two girls. They wish to have one more child, but hope for the child to be of the
      opposite sex from the two they already have. What is the probability that the third child will be of the opposite sex?
      _____ Explain your response. /2\

      _____

d. What is the probability (prediction) that a couple without children will eventually have four children, all girls?

_____ Explain your response. **2**

_____

_____

Part C Assessments

For each of the genetic problems, (a) determine the parents' genotypes, (b) determine the possible gametes for each parent, (c) construct a Punnett square, and (d) record the resulting genotypes and phenotypes as ratios from the cross. Problems 1 and 2 involve examples of complete dominance; problems 3 and 4 are examples of codominance; problem 5 is an example of sex-linked (X-linked) inheritance.

1. Determine the results from a cross of a mother who is heterozygous (*Rr*) for tongue rolling with a father who is homozygous recessive (*rr*). **3**

2. Determine the results from a cross of a mother and a father who are both heterozygous for freckles. **3**

3. Determine the results from a mother who is heterozygous for blood type B and a father who is homozygous dominant for blood type A. **3**

4. Determine the results from a mother who is heterozygous for blood type A and a father who is heterozygous for blood type B.

5. Color blindness is an example of X-linked inheritance. Hemophilia is another example of X-linked inheritance, also from a recessive allele (*h*). The dominant allele (*H*) determines whether the person possesses normal blood clotting. A person with hemophilia has a permanent tendency for hemorrhaging due to a deficiency of one of the clotting factors (VIII—antihemophilic factor). Determine the offspring from a cross of a mother who is a carrier (heterozygous) for the disease and a father with normal blood coagulation.

Critical Thinking Activity

Assume that the genes for hairline and earlobes are on different pairs of homologous chromosomes. Determine the genotypes and phenotypes of the offspring from a cross if both parents are heterozygous for both traits. (1) First determine the genotypes for each parent. (2) Determine the gametes, but remember each gamete has one allele for each trait (gametes are haploid). (3) Construct a Punnett square with 16 boxes that has four different gametes from each parent along the top and the left edges. (This is an application to demonstrate Mendel's Law of Independent Assortment.) (4) List the results of genotypes and phenotypes as ratios.

Cat Dissection: Musculature

Pre-Lab

1. Carefully read the introductory material and examine the entire lab content.
2. Be familiar with the location, origin, insertion, and action of muscles of the human body (from lecture, the textbook, or laboratory exercises).

Materials Needed

Preserved cat (double injection)
Dissecting tray
Dissecting instruments
Large plastic bag
Identification tag
Disposable gloves
Bone shears
Human torso model
Human upper and lower limb models

For Demonstration Activity:
Cat skeleton

Safety

▶ Wear disposable gloves when working on the cat dissection.
▶ Dispose of tissue remnants and gloves as instructed.
▶ Wash the dissecting tray and instruments as instructed.
▶ Wash your laboratory table.
▶ Wash your hands before leaving the laboratory.

Although the aim of this exercise is to become more familiar with the human musculature, human cadavers are not always available for dissection. Instead, preserved cats often are used for dissection because they are relatively small

and can be purchased from biological suppliers. Also, as mammals, cats have many features in common with humans, including similar skeletal muscles (with similar names).

On the other hand, cats make use of four limbs for support, whereas humans use only two limbs. Because the musculature of each type of organism is adapted to provide for its special needs, comparisons of the muscles of cats and humans may not be precise.

As you continue your dissection of various systems of the cat, many anatomical similarities between cats and humans will be observed. These fundamental similarities are called homologous structures. Although homologous structures have a similar structure and embryological origin, the functions are sometimes different.

Purpose of the Exercise

To observe the musculature of the cat and to compare it with that of the human.

Learning Outcomes

After completing this exercise, you should be able to

1. Name and locate the major skeletal muscles of the cat.
2. Name the origins, insertions, and actions of the muscles designated by the laboratory instructor.
3. Name and locate the corresponding muscles of the human.

Muscle Dissection Techniques

Dissect:	to expose the entire length of the muscle from origin to insertion. Most of the procedures are accomplished with blunt probes used to separate various connective tissues that hold adjacent structures together. It does not mean to remove or to cut into the muscles or other organs.
Transect:	to cut through the muscle near its midpoint. The cut is perpendicular to the muscle fibers.
Reflect:	to lift a transected muscle aside to expose deeper muscles or other organs.

Procedure A—Skin Removal

1. Cats usually are preserved with a solution that prevents microorganisms from causing the tissues to decompose. However, because fumes from this preserving fluid may be annoying and may irritate skin, be sure to work in a well-ventilated room and wear disposable gloves to protect your hands.

2. Obtain a preserved cat, a dissecting tray, a set of dissecting instruments, a large plastic bag, and an identification tag.

3. If the cat is in a storage bag, dispose of any excess preserving fluid in the bag, as directed by the laboratory instructor. However, you may want to save some of the preserving fluid to keep the cat moist until you have completed your work later in the year.

4. If the cat is too wet with preserving fluid, blot the cat dry with paper towels.

5. Remove the skin from the cat. To do this, follow these steps:
   a. Place the cat in the dissecting tray with its ventral surface down.
   b. Use a sharp scalpel to make a short, shallow incision through the skin in the dorsal midline at the base of the cat's neck.
   c. Insert the pointed blade of a dissecting scissors through the opening in the skin and cut along the dorsal midline, following the tips of the vertebral spinous processes to the base of the tail (fig. 62.1).
   d. Use the scissors to make an incision encircling the region of the tail, anus, and genital organs. Do not remove the skin from this area.
   e. Use bone shears or a bone saw to sever the tail at its base and discard it (optional procedure).

FIGURE 62.1   Incisions to be made for removing the skin of the cat.

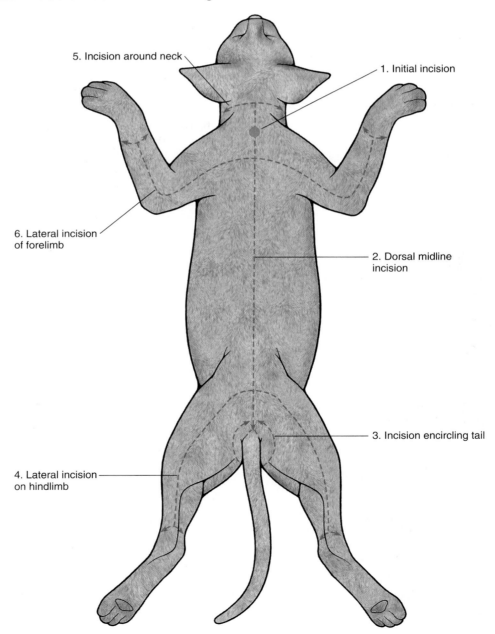

5. Incision around neck

1. Initial incision

6. Lateral incision of forelimb

2. Dorsal midline incision

3. Incision encircling tail

4. Lateral incision on hindlimb

516

f. From the incision at the base of the tail, cut along the lateral surface of each hindlimb to the paw, and encircle the ankle. Also clip off the claws from each paw to prevent them from tearing the plastic storage bag and scratching your skin.

g. Make an incision from the initial cut at the base of the neck to encircle the neck.

h. From the incision around the neck, cut along the lateral surface of each forelimb to the paw and encircle the wrist (fig. 62.1).

i. Grasp the skin on either side of the dorsal midline incision, and carefully pull the skin laterally away from the body. At the same time, use your fingers or a blunt probe to help remove the skin from the underlying muscles by separating the loose connective tissue (superficial fascia). As you pull the skin away, you may note a thin sheet of muscle tissue attached to it. This muscle is called the *cutaneous maximus,* and it functions to move the cat's skin. Humans lack the cutaneous maximus, but a similar sheet of muscle (platysma) is present in the neck of a human.

j. As you remove the skin, work toward the ventral surface, then work toward the head, and finally work toward the tail. Pull the skin over each limb as if you were removing a glove.

FIGURE 62.2 Mammary glands on the ventral surface of the cat's thorax and abdomen.

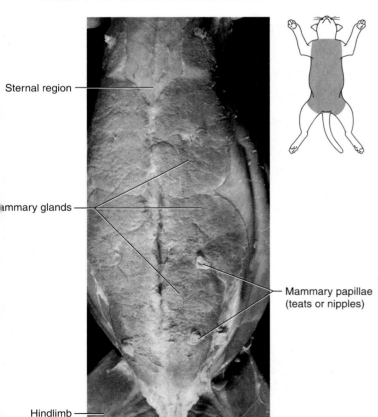

Sternal region

mammary glands

Mammary papillae (teats or nipples)

Hindlimb

k. If the cat is a female, note the mammary glands on the ventral surface of the thorax and abdomen (fig. 62.2). These glands will adhere to the skin and can be removed from it and discarded.

l. After the skin has been pulled away, carefully remove as much of the remaining connective tissue as possible to expose the underlying skeletal muscles. The muscles should appear light brown and fibrous.

6. After skinning the cat, follow these steps:

a. Discard the tissues you have removed as directed by the laboratory instructor.

b. Wrap the skin around the cat to help keep its body moist, and place it in a plastic storage bag.

c. Write your name in pencil on an identification tag, and tie the tag to the storage bag so that you can identify your specimen.

7. Observe the recommended safety procedures for the conclusion of a laboratory session.

Procedure B—Skeletal Muscle Dissection

1. The purpose of a skeletal muscle dissection is to separate the individual muscles from any surrounding tissues and thus expose the muscles for observation. To do a muscle dissection, follow these steps:

a. Use the appropriate figure as a guide and locate the muscle to be dissected in the specimen.

b. Use a blunt probe or a finger to separate the muscle from the surrounding connective tissue along its natural borders. The muscle should separate smoothly. If the border appears ragged, you probably have torn the muscle fibers.

2. If it is necessary to cut through (transect) a superficial muscle to observe a deeper one, use scissors to transect the muscle about halfway between its origin and insertion. Then, lift aside (reflect) the cut ends, leaving their attachments intact.

3. The following procedures will instruct you to dissect some of the larger and more easily identified muscles of the cat. In each case, the procedure will include the names of the muscles to be dissected, figures illustrating their locations, and tables listing the origins, insertions, and actions of the muscles. (More detailed dissection instructions can be obtained by consulting an additional guide or atlas for cat dissection.) As you dissect each muscle, review the muscle figures and tables in the lab manual and identify any corresponding (homologous) muscles of the human body. Also locate these homologous muscles in the human torso model or models of the human upper and lower limbs.

Demonstration Activity

Refer to the cat skeleton to help you to locate origins and insertions of the cat's muscles. Reference to the cat skeleton will also help you visualize the actions of these muscles, if you remember that a muscle's insertion is pulled toward its origin.

Procedure C—Muscles of the Head and Neck

1. Place the cat in the dissecting tray with its ventral surface up.
2. Remove the skin and underlying connective tissues from one side of the neck, forward to the chin, and up to the ear.

3. Study figures 62.3 and 62.4, and then locate and dissect the following muscles:

   sternomastoid
   sternohyoid
   digastric
   masseter
   mylohyoid

FIGURE 62.3   Muscles on the ventral surface of the cat's head and neck.

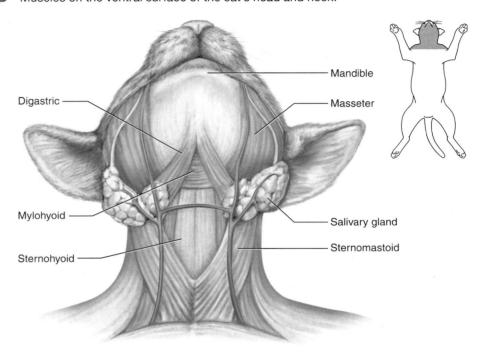

FIGURE 62.4   Deep muscles on the right ventral surface of the cat's neck.

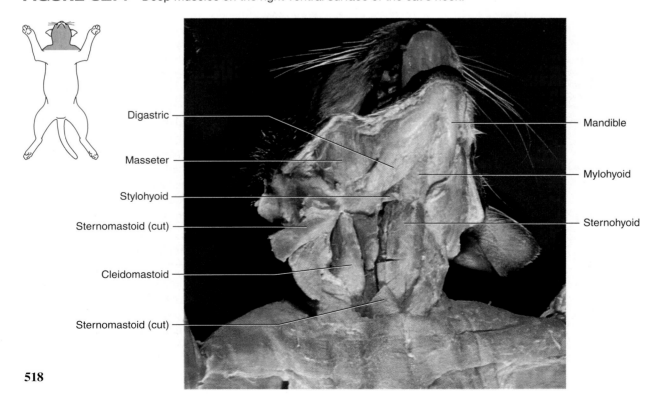

518

4. To find the deep muscles of the cat's neck and chin, carefully remove the sternomastoid, sternohyoid, and mylohyoid muscles from the right side.

5. Study figure 62.5, and locate the following muscles:

hyoglossus

cleidomastoid

styloglossus

thyrohyoid

geniohyoid

sternothyroid

6. See table 62.1 for the origins, insertions, and actions of these head and neck muscles.

7. Complete Part A of Laboratory Report 62.

FIGURE 62.5   Deep muscles of the cat's neck and chin regions.

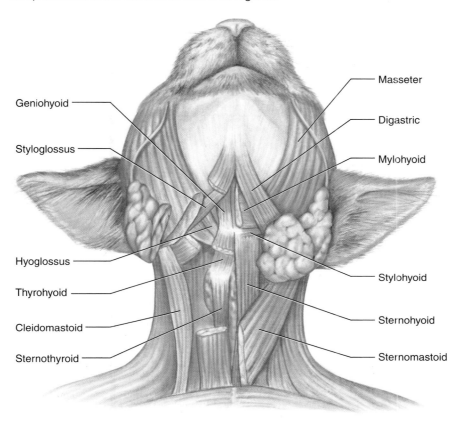

TABLE 62.1   Muscles of the Head and Neck

Muscle	Origin	Insertion	Action
Sternomastoid	Manubrium of sternum	Mastoid process of temporal bone	Turns and depresses head
Sternohyoid	Costal cartilage	Hyoid bone	Depresses hyoid bone
Digastric	Mastoid process and occipital bone	Ventral border of mandible	Depresses mandible
Mylohyoid	Inner surface of mandible	Hyoid bone	Raises floor of mouth
Masseter	Zygomatic arch	Coronoid fossa of mandible	Elevates mandible
Hyoglossus	Hyoid bone	Tongue	Pulls tongue back
Styloglossus	Styloid process	Tip of tongue	Pulls tongue back
Geniohyoid	Inner surface of mandible	Hyoid bone	Pulls hyoid bone forward
Cleidomastoid	Clavicle	Mastoid process	Turns head to side
Thyrohyoid	Thyroid cartilage	Hyoid bone	Raises larynx
Sternothyroid	Sternum	Thyroid cartilage	Pulls larynx back

Procedure D—Muscles of the Thorax

1. Place the cat in the dissecting tray with its ventral surface up.
2. Remove any remaining fat and connective tissue to expose the muscles in the walls of the thorax and abdomen.
3. Study figures 62.6 and 62.7. Transect the superficial pectoantebrachialis to expose the entire pectoralis major. Locate and dissect the following pectoral muscles:

   pectoantebrachialis
   pectoralis major
   pectoralis minor
   xiphihumeralis

4. To find the deep thoracic muscles, transect the pectoralis minor, pectoralis major, and pectoantebrachialis muscles, and reflect their cut edges to the sides. Remove the underlying fat and connective tissues to expose the following muscles, as shown in figure 62.8:

   scalenus
   serratus ventralis
   levator scapulae
   transversus costarum

5. See table 62.2 for the origins, insertions, and actions of these muscles.

FIGURE 62.6 Muscles of the cat's thorax and abdomen.

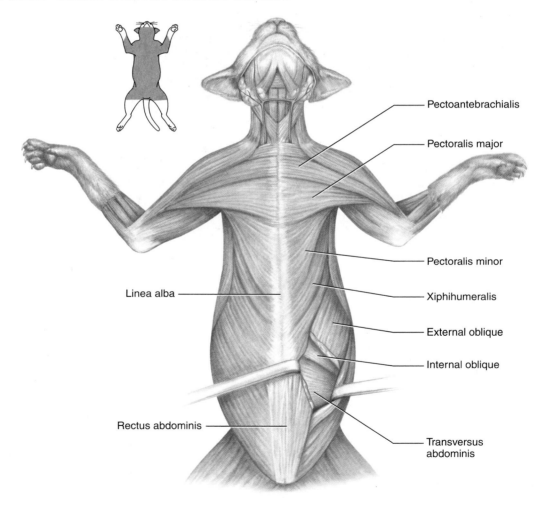

Pectoantebrachialis

Pectoralis major

Pectoralis minor

Xiphihumeralis

External oblique

Internal oblique

Linea alba

Rectus abdominis

Transversus abdominis

FIGURE 62.7 Superficial muscles of the cat's thorax.

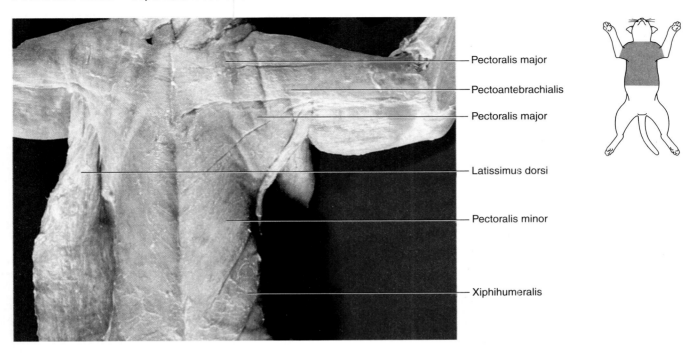

— Pectoralis major

— Pectoantebrachialis

— Pectoralis major

— Latissimus dorsi

— Pectoralis minor

— Xiphihumeralis

FIGURE 62.8 Superficial and deep muscles of the cat's thorax.

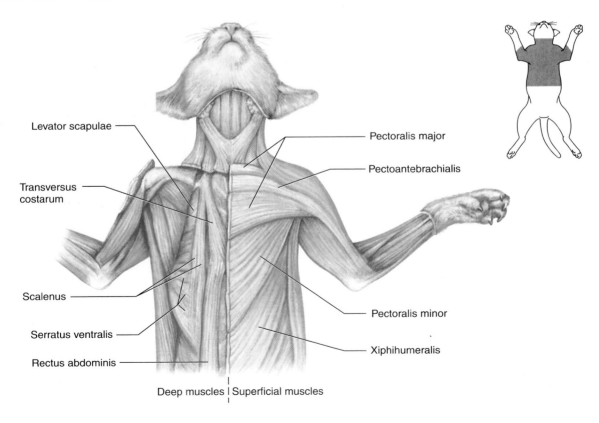

Levator scapulae

Transversus costarum

Scalenus

Serratus ventralis

Rectus abdominis

Pectoralis major

Pectoantebrachialis

Pectoralis minor

Xiphihumeralis

Deep muscles | Superficial muscles

TABLE 62.2 Muscles of the Thorax

Muscle	Origin	Insertion	Action
Pectoantebrachialis	Manubrium of sternum	Fascia of forelimb	Adducts forelimb
Pectoralis major	Manubrium of sternum and costal cartilages	Greater tubercle of humerus	Adducts and rotates forelimb
Pectoralis minor	Sternum	Proximal end of humerus	Adducts and rotates forelimb
Xiphihumeralis	Xiphoid process of sternum	Proximal end of humerus	Adducts forelimb
Scalenus	Ribs	Cervical vertebrae	Flexes neck
Serratus ventralis	Ribs	Scapula	Pulls scapula ventrally and posteriorly
Levator scapulae	Cervical vertebrae	Scapula	Pulls scapula ventrally and anteriorly
Transversus costarum	Sternum	First rib	Pulls sternum toward head

Procedure E—Muscles of the Abdominal Wall

1. Study figures 62.6 and 62.9
2. Locate the *external oblique* muscle in the abdominal wall.
3. Make a shallow, longitudinal incision through the external oblique. Reflect the cut edge and expose the *internal oblique* muscle beneath (fig. 62.6). The fibers of the internal oblique run at a right angle to those of the external oblique.
4. Make a longitudinal incision through the internal oblique. Reflect the cut edge and expose the *transversus abdominis* muscle.
5. Expose the *rectus abdominis* muscle on one side of the midventral line. This muscle lies beneath an aponeurosis.
6. See table 62.3 for the origins, insertions, and actions of these muscles.
7. Complete Part B of the laboratory report.

Procedure F—Muscles of the Shoulder and Back

1. Place the cat in the dissecting tray with its dorsal surface up.
2. Remove any remaining fat and connective tissue to expose the muscles of the shoulder and back.

FIGURE 62.9 Muscles of the cat's thorax and abdomen.

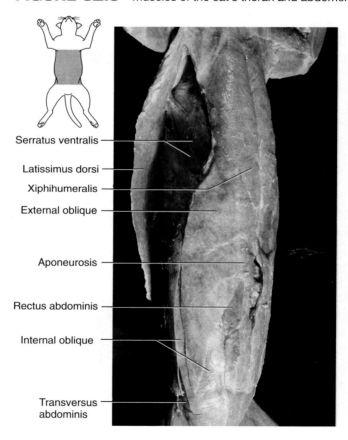

Serratus ventralis
Latissimus dorsi
Xiphihumeralis
External oblique
Aponeurosis
Rectus abdominis
Internal oblique
Transversus abdominis

TABLE 62.3 Muscles of the Abdominal Wall

Muscle	Origin	Insertion	Action
External oblique	Ribs and fascia of back	Linea alba	Compresses abdominal wall
Internal oblique	Fascia of back	Linea alba	Compresses abdominal wall
Transversus abdominis	Lower ribs	Linea alba	Compresses abdominal wall
Rectus abdominis	Pubic symphysis	Sternum and costal cartilages	Compresses abdominal wall and flexes trunk

3. Study figures 62.10 and 62.11, and then locate and dissect the following superficial muscles:

clavotrapezius
acromiotrapezius
spinotrapezius
clavodeltoid
acromiodeltoid
spinodeltoid
latissimus dorsi

Note: The clavodeltoid (clavobrachialis) appears as a continuation of the clavotrapezius.

4. Using scissors, transect the latissimus dorsi and the group of trapezius muscles. Reflect their cut edges and remove any underlying fat and connective tissue. Study figures 62.10 and 62.12, and then locate and dissect the following deep muscles of the shoulder and back:

supraspinatus
teres major
rhomboid capitis

infraspinatus
rhomboid
splenius

5. See table 62.4 for the origins, insertions, and actions of these muscles in the shoulder and back.
6. Review the locations, origins, insertions, and actions designated by the laboratory instructor without the aid of the figures and tables. ⚠️1 ⚠️2
7. Complete Part C of the laboratory report.

Procedure G—Muscles of the Forelimb

1. Place the cat in the dissecting tray with its ventral surface up.
2. Remove any remaining fat and connective tissue from a forelimb to expose the muscles.
3. Study figure 62.13a, and then locate and dissect the following muscles from the medial surface of the upper forelimb:

epitrochlearis
triceps brachii

FIGURE 62.10 Superficial muscles of the cat's shoulder and back (*right side*); deep muscles of the shoulder and back (*left side*).

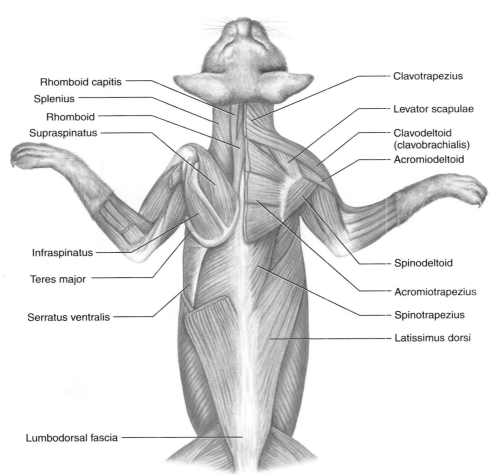

Rhomboid capitis
Splenius
Rhomboid
Supraspinatus
Infraspinatus
Teres major
Serratus ventralis
Lumbodorsal fascia

Clavotrapezius
Levator scapulae
Clavodeltoid (clavobrachialis)
Acromiodeltoid
Spinodeltoid
Acromiotrapezius
Spinotrapezius
Latissimus dorsi

FIGURE 62.11 Muscles of the cat's shoulder and back.

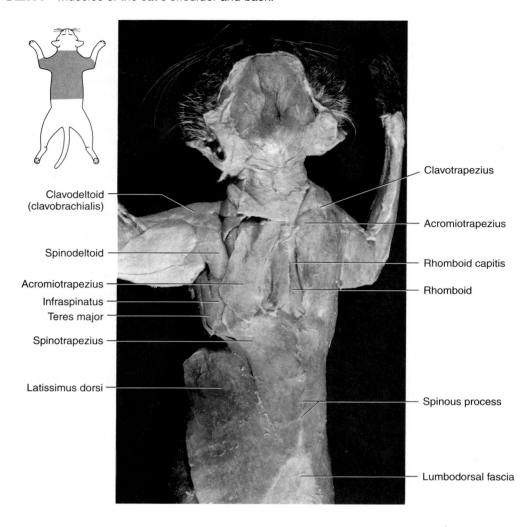

Clavotrapezius

Clavodeltoid (clavobrachialis)

Acromiotrapezius

Spinodeltoid

Rhomboid capitis

Acromiotrapezius

Rhomboid

Infraspinatus

Teres major

Spinotrapezius

Latissimus dorsi

Spinous process

Lumbodorsal fascia

FIGURE 62.12 Deep muscles of the cat's back.

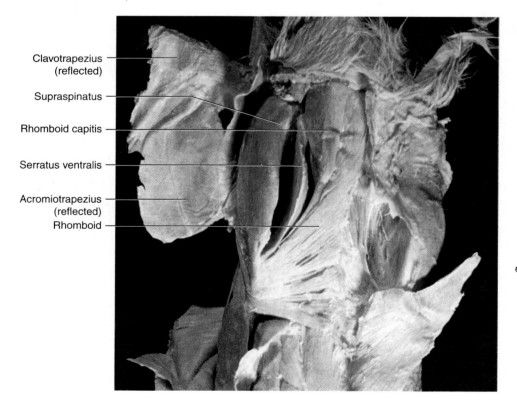

Clavotrapezius (reflected)

Supraspinatus

Rhomboid capitis

Serratus ventralis

Acromiotrapezius (reflected)

Rhomboid

TABLE 62.4 Muscles of the Shoulder and Back

Muscle	Origin	Insertion	Action
Clavotrapezius	Occipital bone of skull and spines of cervical vertebrae	Clavicle	Pulls clavicle upward
Acromiotrapezius	Spines of cervical and thoracic vertebrae	Spine and acromion of scapula	Pulls scapula upward
Spinotrapezius	Spines of thoracic vertebrae	Spine of scapula	Pulls scapula upward and back
Clavodeltoid	Clavicle	Ulna near semilunar notch	Flexes forelimb
Acromiodeltoid	Acromion of scapula	Proximal end of humerus	Flexes and rotates forelimb
Spinodeltoid	Spine of scapula	Proximal end of humerus	Flexes and rotates forelimb
Latissimus dorsi	Thoracic and lumbar vertebrae	Proximal end of humerus	Pulls forelimb upward and back
Supraspinatus	Fossa above spine of scapula	Greater tubercle of humerus	Extends forelimb
Infraspinatus	Fossa below spine of scapula	Greater tubercle of humerus	Rotates forelimb
Teres major	Posterior border of scapula	Proximal end of humerus	Rotates forelimb
Rhomboid	Spines of cervical and thoracic vertebrae	Medial border of scapula	Pulls scapula back
Rhomboid capitis	Occipital bone of skull	Medial border of scapula	Pulls scapula forward
Splenius	Fascia of neck	Occipital bone	Turns and raises head

FIGURE 62.13 (a) Superficial muscles of the cat's left medial shoulder and forelimb. (b) Deep muscles of the left medial shoulder and forelimb.

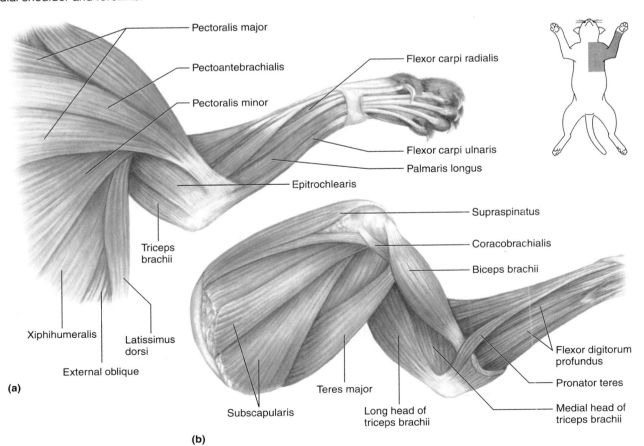

FIGURE 62.14 Deep muscles of the cat's right medial forelimb.

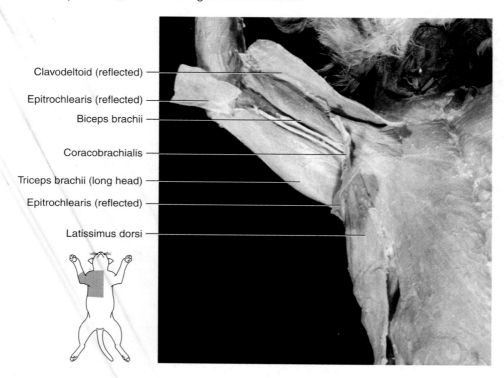

Clavodeltoid (reflected)

Epitrochlearis (reflected)

Biceps brachii

Coracobrachialis

Triceps brachii (long head)

Epitrochlearis (reflected)

Latissimus dorsi

FIGURE 62.15 (a) Superficial muscles of the cat's left lateral shoulder and forelimb. (b) Deep muscles of the left lateral shoulder and forelimb.

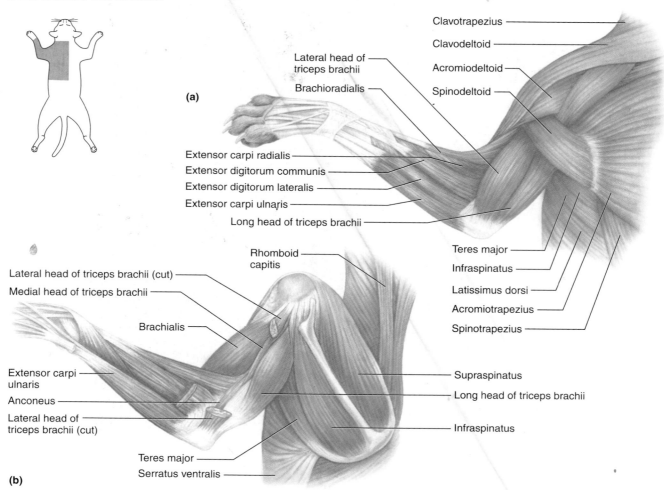

(a)

Lateral head of triceps brachii

Brachioradialis

Extensor carpi radialis

Extensor digitorum communis

Extensor digitorum lateralis

Extensor carpi ulnaris

Long head of triceps brachii

Clavotrapezius

Clavodeltoid

Acromiodeltoid

Spinodeltoid

Teres major

Infraspinatus

Latissimus dorsi

Acromiotrapezius

Spinotrapezius

Lateral head of triceps brachii (cut)

Medial head of triceps brachii

Brachialis

Extensor carpi ulnaris

Anconeus

Lateral head of triceps brachii (cut)

Teres major

Serratus ventralis

Rhomboid capitis

Supraspinatus

Long head of triceps brachii

Infraspinatus

(b)

4. Transect the pectoral muscles (pectoantebrachialis, pectoralis major, and pectoralis minor) and reflect their cut edges.

5. Study figures 62.13b and 62.14, and then locate and dissect the following muscles:

> **coracobrachialis**
> **biceps brachii**
> **long head of the triceps brachii**
> **medial head of the triceps brachii**

6. Study figures 62.15a and 62.16, which show the lateral surface of the upper forelimb.

7. Locate the lateral head of the triceps muscle, transect it, and reflect its cut edges.

8. Study figure 62.15b. Locate and dissect the following deeper muscles:

> **anconeus**
> **brachialis**

9. Separate the muscles in the lower forelimb into an *extensor group* (fig. 62.15a) and a *flexor group* (fig. 62.13a). Generally, the extensors are located on the posterior surface of the lower forelimb, and the flexors are located on the anterior surface. A thick, tough layer of fascia (antebrachial fascia) surrounds the forelimb. Remove this fascia to expose and dissect the following muscles:

> **brachioradialis**
> **extensor carpi radialis**
> **extensor digitorum communis**
> **extensor digitorum lateralis**
> **extensor carpi ulnaris**
> **flexor carpi ulnaris**
> **palmaris longus**
> **flexor carpi radialis**
> **pronator teres**

FIGURE 62.16 Superficial muscles of the cat's left lateral shoulder and forelimb, dorsal view.

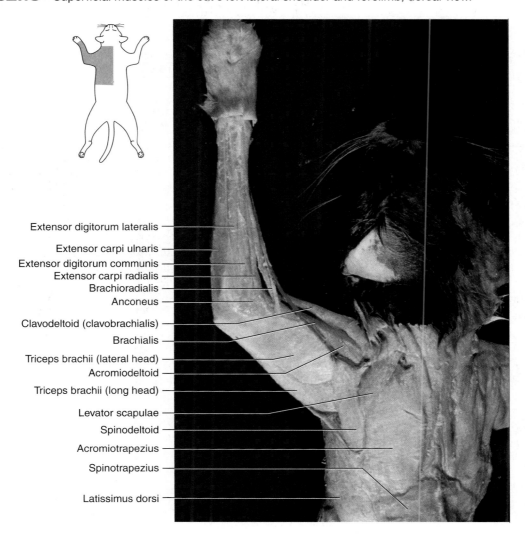

TABLE 62.5 Muscles of the Forelimb

Muscle	Origin	Insertion	Action
Coracobrachialis	Coracoid process of scapula	Proximal end of humerus	Adducts forelimb
Epitrochlearis	Fascia of latissimus dorsi	Olecranon process of ulna	Extends forelimb
Biceps brachii	Border of glenoid cavity of scapula	Radial tuberosity of radius	Flexes forelimb
Triceps brachii			
Lateral head	Deltoid tuberosity of humerus	Olecranon process of ulna	Extends forelimb
Long head	Border of glenoid cavity of scapula	Olecranon process of ulna	Extends forelimb
Medial head	Shaft of humerus	Olecranon process of ulna	Extends forelimb
Anconeus	Lateral epicondyle of humerus	Olecranon process of ulna	Extends forelimb
Brachialis	Lateral surface of humerus	Proximal end of ulna	Flexes forelimb
Brachioradialis	Shaft of humerus	Distal end of radius	Supinates lower forelimb
Extensor carpi radialis	Lateral surface of humerus	Second and third metacarpals	Extends wrist
Extensor digitorum communis	Lateral surface of humerus	Digits two to five	Extends digits
Extensor digitorum lateralis	Lateral surface of humerus	Digits two to five	Extends digits
Extensor carpi ulnaris	Lateral epicondyle of humerus and ulna	Fifth metacarpal	Extends wrist
Flexor carpi ulnaris	Medial epicondyle of humerus	Carpals	Flexes wrist
Palmaris longus	Medial epicondyle of humerus	Digits	Flexes digits
Flexor carpi radialis	Medial epicondyle of humerus	Second and third metacarpals	Flexes metacarpals
Pronator teres	Medial epicondyle of humerus	Radius	Pronates lower forelimb
Flexor digitorum profundus	Radius, ulna, and medial epicondyle of humerus	Distal phalanges	Flexes digits

10. Transect the flexor carpi ulnaris, palmaris longus, and flexor carpi radialis to expose the flexor digitorum profundus.
11. See table 62.5 for the origins, insertions, and actions of these muscles of the forelimb.
12. Complete Part D of the laboratory report.

Procedure H—Muscles of the Hip and Hindlimb

1. Place the cat in the dissecting tray with its ventral surface up.
2. Remove any remaining fat and connective tissue from the hip and hindlimb to expose the muscles.
3. Study figures 62.17a and 62.18, and then locate and dissect the following muscles from the medial surface of the thigh:

    sartorius

    gracilis

4. Using scissors, transect the sartorius and gracilis, and reflect their cut edges to observe the deeper muscles of the thigh.
5. Study figures 62.17b and 62.19, and then locate and dissect the following muscles:

    rectus femoris

    vastus medialis

    adductor longus

    adductor femoris

    semimembranosus

    semitendinosus

    tensor fasciae latae

6. Transect the tensor fasciae latae and rectus femoris muscles and reflect their cut edges. Locate the *vastus*

FIGURE 62.17 (a) Superficial muscles of the cat's left medial hindlimb. (b) Deep muscles of the left medial hindlimb.

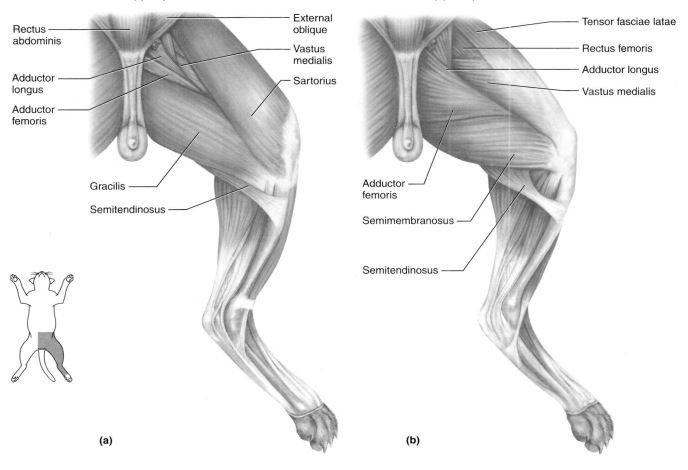

Rectus abdominis
Adductor longus
Adductor femoris
Gracilis
Semitendinosus
External oblique
Vastus medialis
Sartorius

Adductor femoris
Semimembranosus
Semitendinosus
Tensor fasciae latae
Rectus femoris
Adductor longus
Vastus medialis

(a)

(b)

FIGURE 62.18 Superficial muscles of the cat's hindlimb, medial view.

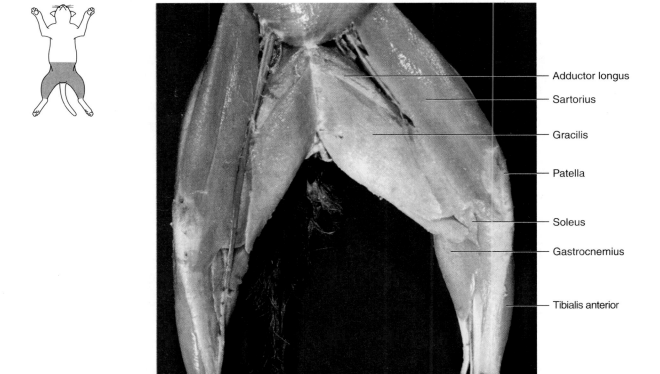

Adductor longus
Sartorius
Gracilis
Patella
Soleus
Gastrocnemius
Tibialis anterior

FIGURE 62.19 Deep muscles of the cat's left hindlimb, medial view.

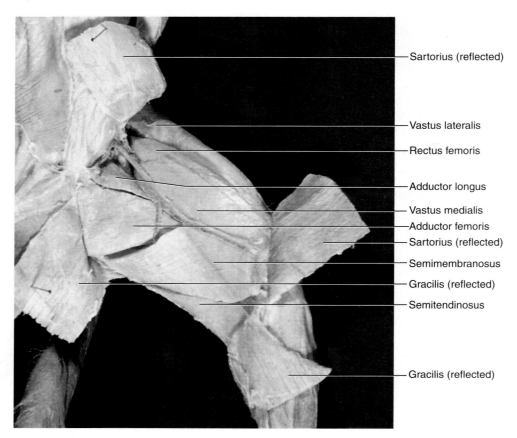

Sartorius (reflected)

Vastus lateralis

Rectus femoris

Adductor longus

Vastus medialis
Adductor femoris
Sartorius (reflected)
Semimembranosus
Gracilis (reflected)
Semitendinosus

Gracilis (reflected)

FIGURE 62.20 Other deep muscles of the cat's left thigh, medial view.

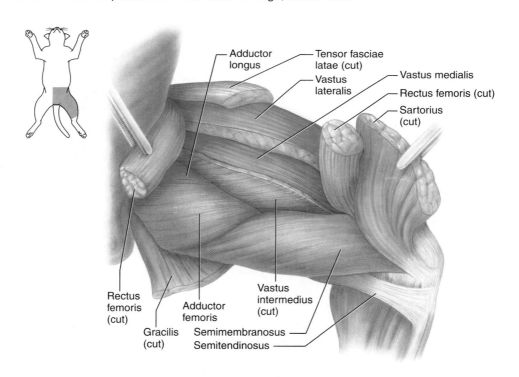

Adductor
longus

Tensor fasciae
latae (cut)

Vastus
lateralis

Vastus medialis

Rectus femoris (cut)

Sartorius
(cut)

Rectus
femoris
(cut)

Adductor
femoris

Gracilis
(cut)

Vastus
intermedius
(cut)

Semimembranosus

Semitendinosus

intermedius and *vastus lateralis* muscles beneath (fig. 62.20). (*Note:* In some specimens, the vastus intermedius, vastus lateralis, and vastus medialis muscles are closely united by connective tissue and are difficult to separate.)

7. Study figures 62.21*a* and 62.22, and then locate and dissect the following muscles from the lateral surface of the hip and thigh:

> **biceps femoris**
> **caudofemoralis**
> **gluteus maximus**

8. Using scissors, transect the tensor fasciae latae and biceps femoris, and reflect their cut edges to observe the deeper muscles of the thigh.

9. Locate and dissect the following muscles:

> **tenuissimus**
> **gluteus medius**

(*Note:* The tenuissimus muscle and the sciatic nerve may adhere tightly to the dorsal surface of the biceps femoris muscle. Take care to avoid cutting these structures.)

FIGURE 62.21   (*a*) Superficial muscles of the cat's left lateral hip and hindlimb. (*b*) Deep muscles of the left lateral hip and hindlimb.

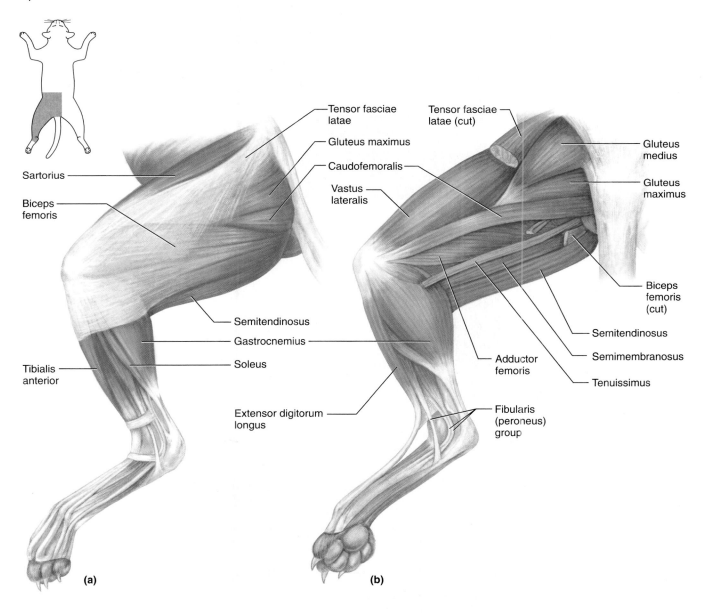

FIGURE 62.22   Muscles of the cat's left hip and hindlimb, lateral view.

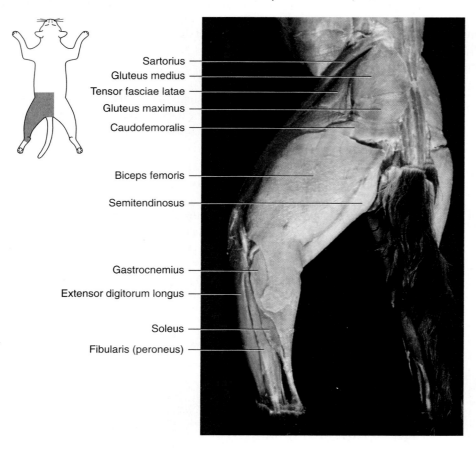

Sartorius
Gluteus medius
Tensor fasciae latae
Gluteus maximus
Caudofemoralis

Biceps femoris

Semitendinosus

Gastrocnemius

Extensor digitorum longus

Soleus

Fibularis (peroneus)

FIGURE 62.23   Deep muscles of the cat's left hindlimb, lateral view.

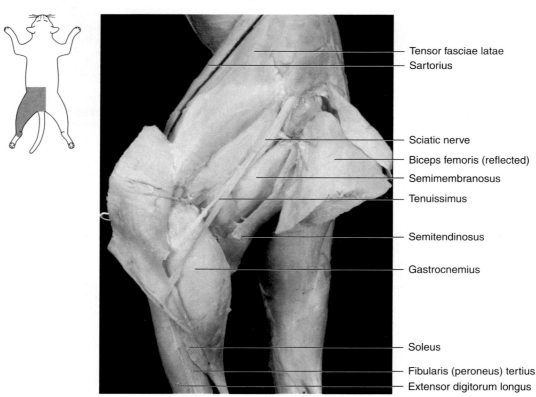

Tensor fasciae latae
Sartorius

Sciatic nerve

Biceps femoris (reflected)

Semimembranosus

Tenuissimus

Semitendinosus

Gastrocnemius

Soleus

Fibularis (peroneus) tertius
Extensor digitorum longus

10. On the lateral surface of the lower hindlimb (figs. 62.21 and 62.23), locate and dissect the following muscles:

> **gastrocnemius**
>
> **soleus**
>
> **tibialis anterior**
>
> **fibularis (peroneus) group**

11. See table 62.6 for the origins, insertions, and actions of these muscles of the hip and hindlimb.

12. Review the locations, origins, insertions, and actions designated by the laboratory instructor without the aid of the figures and tables. 🔺1 🔺2

13. Complete Part E of the laboratory report.

TABLE 62.6   Muscles of the Hip and Hindlimb

Muscle	Origin	Insertion	Action
Sartorius	Crest of ilium	Fascia of knee and proximal end of tibia	Rotates and extends hindlimb
Gracilis	Pubic symphysis and ischium	Proximal end of tibia	Adducts hindlimb
Quadriceps femoris			
Vastus lateralis	Shaft of femur and greater trochanter	Patella	Extends hindlimb
Vastus intermedius	Shaft of femur	Patella	Extends hindlimb
Rectus femoris	Ilium	Patella	Extends hindlimb
Vastus medialis	Shaft of femur	Patella	Extends hindlimb
Adductor longus	Pubis	Proximal end of femur	Adducts hindlimb
Adductor femoris	Pubis and ischium	Shaft of femur	Adducts hindlimb
Semimembranosus	Ischial tuberosity	Medial epicondyle of femur	Extends thigh
Tensor fasciae latae	Ilium	Fascia of thigh	Extends thigh
Biceps femoris	Ischial tuberosity	Patella and tibia	Abducts thigh and flexes lower hindlimb
Tenuissimus	Second caudal vertebra	Tibia and fascia of biceps femoris	Abducts thigh and flexes lower hindlimb
Semitendinosus	Ischial tuberosity	Crest of tibia	Flexes lower hindlimb
Caudofemoralis	Caudal vertebrae	Patella	Abducts thigh and extends lower hindlimb
Gluteus maximus	Sacral and caudal vertebrae	Greater trochanter of femur	Abducts thigh
Gluteus medius	Ilium, and sacral and caudal vertebrae	Greater trochanter of femur	Abducts thigh
Gastrocnemius	Lateral and medial epicondyles of femur	Calcaneus	Extends foot
Soleus	Proximal end of fibula	Calcaneus	Extends foot
Tibialis anterior	Proximal end of tibia and fibula	First metatarsal	Flexes foot
Fibularis (peroneus) group	Shaft of fibula	Metatarsals	Flexes foot

Laboratory Report

62

Name _____

Date _____

Section _____

The ⒶＡ corresponds to the Learning Outcome(s) listed at the beginning of the laboratory exercise.

Cat Dissection: Musculature

Part A Assessments

Complete the following statements:

1. The _____ muscle of the human is homologous to the sternomastoid muscle of the cat. **3**

2. The _____ muscle elevates the mandible in the human and in the cat. **2** **3**

3. Two muscles of the cat that are inserted on the hyoid bone are the _____ and the
_____ . **1** **2**

Part B Assessments

Complete the following:

1. Name two muscles that are found in the thoracic wall of the cat but are absent in the human. **1** **3**

   _____   _____

2. Name two muscles that are found in the thoracic wall of the cat and the human. **3**

   _____   _____

3. Name four muscles that are found in the abdominal wall of the cat and the human. **3**

   _____   _____

   _____   _____

Part C Assessments

Complete the following:

1. Name three muscles of the cat that together correspond to the trapezius muscle in the human. **1** **3**

   _____   _____   _____

2. Name three muscles of the cat that together correspond to the deltoid muscle in the human. **1** **3**

   _____   _____   _____

3. Name the muscle in the cat and in the human that occupies the fossa above the spine of the scapula. **1** **3**

   _____

4. Name the muscle in the cat and in the human that occupies the fossa below the spine of the scapula. ◢1◣ ◢3◣

_____

5. Name two muscles found in the cat and in the human that can rotate the forelimb. ◢2◣ ◢3◣

_____   _____

Part D Assessments

Complete the following:

1. Name two muscles found in the cat and in the human that can flex the forelimb. ◢2◣ ◢3◣

_____   _____

2. Name a muscle found in the cat but absent in the human that can extend the forelimb. ◢2◣ ◢3◣

_____

3. Name a muscle that has three heads, can extend the forelimb, and is found in the cat and in the human. ◢2◣ ◢3◣

_____

Part E Assessments

Observe the muscles on the human torso model and the upper and lower limb models. Each of the muscles in figure 62.24 is found both in the cat and in the human. Identify each of the numbered muscles in the figure by placing its name in the space next to its number.

FIGURE 62.24   Identify the numbered muscles that occur in both the cat and the human. ◢3◣

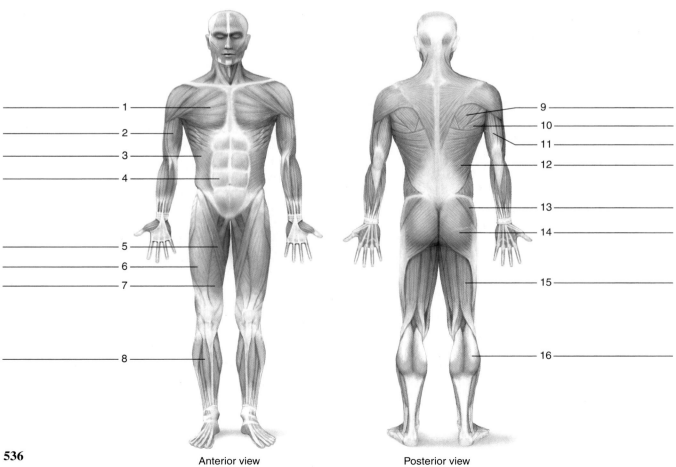

Anterior view   Posterior view

Cat Dissection: Cardiovascular System

Pre-Lab

1. Carefully read the introductory material and examine the entire lab content.
2. Be familiar with the structures and functions of the human cardiovascular system (from lecture, the textbook, or laboratory exercises).

Materials Needed

Preserved cat
Dissecting tray
Dissecting instruments
Disposable gloves
Human torso model

Safety

▶ Wear disposable gloves when working on the cat dissection.
▶ Dispose of tissue remnants and gloves as instructed.
▶ Wash the dissecting tray and instruments as instructed.
▶ Wash your laboratory table.
▶ Wash your hands before leaving the laboratory.

In this laboratory exercise, you will dissect the major organs of the cardiovascular system of the cat. As before, while you are examining the organs of the cat, compare them with the corresponding organs of the human torso model.

If the cardiovascular system of the cat has been injected, the arteries will be filled with red latex, and the veins will be filled with blue latex. This will make it easier for you to trace the vessels as you dissect them.

Purpose of the Exercise

To examine the major organs of the cardiovascular system of the cat and to compare them with the corresponding organs of the human.

Learning Outcomes

After completing this exercise, you should be able to

1. Locate and identify the major organs of the cardiovascular system of the cat.
2. Compare the features of the cardiovascular system of the cat with those of the human.
3. Identify the corresponding organs in the human torso model and a cadaver.

Procedure A—the Arterial System

1. Place the preserved cat in the dissecting tray with its ventral surface up.
2. Open the thoracic cavity, and expose its contents. To do this, follow these steps:
   a. Make a longitudinal incision passing anteriorly from the diaphragm along one side of the sternum. Continue the incision through the neck muscles to the mandible. Try to avoid damaging the internal organs as you cut.
   b. Make a lateral cut on each side along the anterior surface of the diaphragm, and cut the diaphragm loose from the thoracic wall.
   c. Spread the sides of the thoracic wall outward, and use a scalpel to make a longitudinal cut along each side of the inner wall of the rib cage to weaken the ribs. Continue to spread the thoracic wall laterally to break the ribs so that the flaps of the wall will remain open (figs. 63.1 and 63.2).
3. Note the location of the heart and the large blood vessels associated with it. Slit the outer *parietal pericardium* that surrounds the heart by cutting with scissors along the midventral line. Note how this membrane is connected to the *visceral pericardium* that is attached to the surface of the heart. Locate the *pericardial cavity,* the space between the two layers of the pericardium.

FIGURE 63.1 Arteries of the cat's thorax, neck, and forelimb.

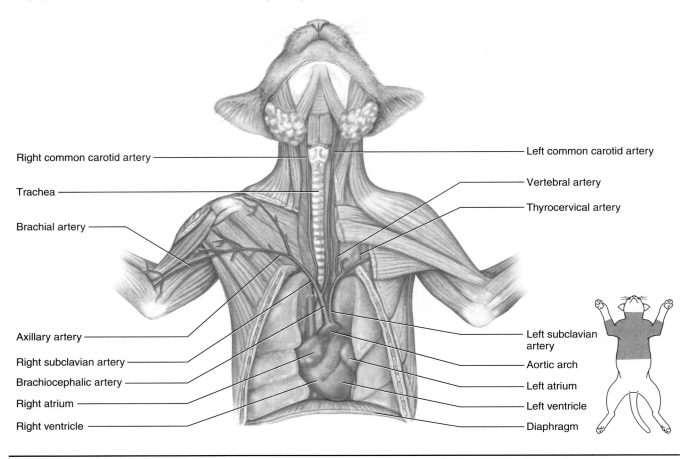

Right common carotid artery

Trachea

Brachial artery

Axillary artery

Right subclavian artery

Brachiocephalic artery

Right atrium

Right ventricle

Left common carotid artery

Vertebral artery

Thyrocervical artery

Left subclavian artery

Aortic arch

Left atrium

Left ventricle

Diaphragm

FIGURE 63.2 Features of the cat's neck and thoracic cavity.

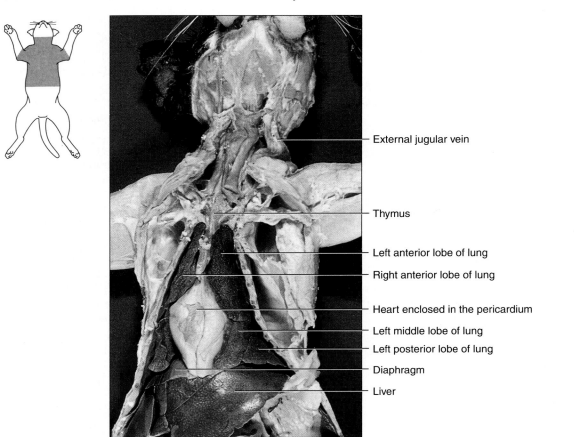

External jugular vein

Thymus

Left anterior lobe of lung

Right anterior lobe of lung

Heart enclosed in the pericardium

Left middle lobe of lung

Left posterior lobe of lung

Diaphragm

Liver

4. Examine the heart (fig. 63.3*a*). The arrangement of blood vessels coming off the aortic arch is different in the cat than in the human body (fig. 63.3*b*). Locate the following:

>**right atrium**
>**left atrium**
>**right ventricle**

left ventricle
pulmonary trunk
aorta
coronary arteries

5. Use a scalpel to open the heart chambers by making a cut along the frontal plane from its apex to its base. Remove any remaining latex from the chambers. Examine the

---

FIGURE 63.3 (*a*) Heart and associated arteries and veins of the cat, ventral view. (*b*) Heart and associated arteries and veins of a cadaver, anterior view.

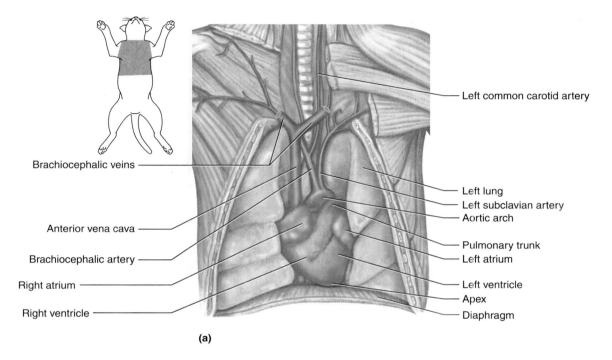

(a)

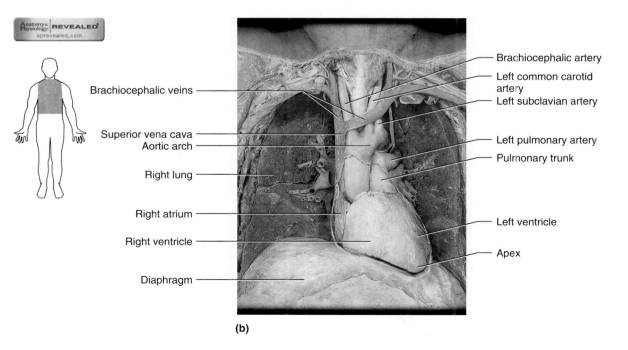

(b)

valves between the chambers, and note the relative thicknesses of the chamber walls (fig. 63.4). Do not remove the heart as it is needed for future examination of the relationship between major blood vessels.

6. Using figures 63.1, 63.4, and 63.5 as guides, locate and dissect the following arteries of the thorax and neck:

   aortic arch

   brachiocephalic artery (on right only; first branch of the aortic arch)

   right subclavian artery

   left subclavian artery

   right common carotid artery

   left common carotid artery

7. Trace the right subclavian artery into the forelimb, and locate the following arteries:

   vertebral artery

   thyrocervical artery

   subscapular artery

   axillary artery

   brachial artery

   radial artery

   ulnar artery

8. Open the abdominal cavity. To do this, follow these steps:

   a. Use scissors to make an incision through the body wall along the midline from the pubic symphysis to the diaphragm.

   b. Make a lateral incision through the body wall along either side of the inferior border of the diaphragm and along the bases of the thighs.

   c. Reflect the flaps created in the body wall as you would open a book, and expose the contents of the abdominal cavity.

   d. Note the *parietal peritoneum* that forms the inner lining of the abdominal wall. Also note the *greater omentum,* a structure composed of a double layer of peritoneum that hangs from the border of the stomach

---

FIGURE 63.4 Blood vessels of the cat's neck and thorax.

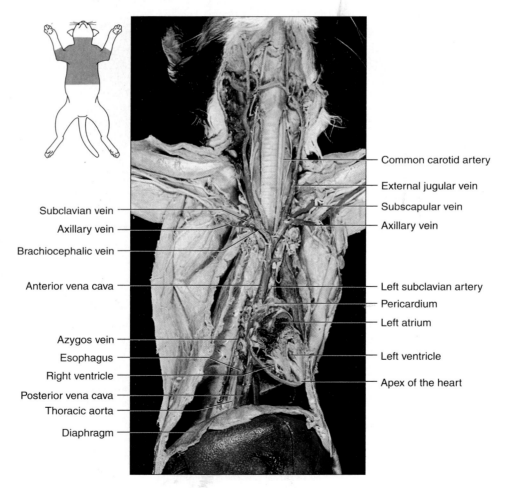

Subclavian vein

Axillary vein

Brachiocephalic vein

Anterior vena cava

Azygos vein

Esophagus

Right ventricle

Posterior vena cava

Thoracic aorta

Diaphragm

Common carotid artery

External jugular vein

Subscapular vein

Axillary vein

Left subclavian artery

Pericardium

Left atrium

Left ventricle

Apex of the heart

FIGURE 63.5   Blood vessels of the cat's upper forelimb and neck.

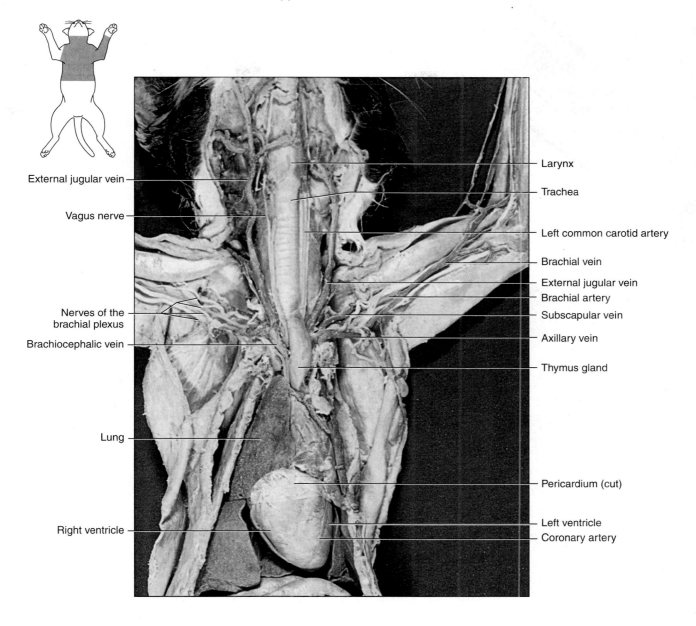

External jugular vein

Vagus nerve

Nerves of the brachial plexus

Brachiocephalic vein

Lung

Right ventricle

Larynx

Trachea

Left common carotid artery

Brachial vein

External jugular vein
Brachial artery

Subscapular vein

Axillary vein

Thymus gland

Pericardium (cut)

Left ventricle
Coronary artery

and covers the lower abdominal organs like a fatty apron (fig. 63.6).

9. As you expose and dissect blood vessels, try not to destroy other visceral organs needed for future studies. Using figures 63.7, 63.8, and 63.9 as guides, locate and dissect the following arteries of the abdomen:

abdominal aorta (unpaired)
celiac artery (unpaired)
hepatic artery (unpaired)
gastric artery (unpaired)
splenic artery (unpaired)

anterior mesenteric artery (corresponds to superior mesenteric artery)
renal arteries (paired)
adrenolumbar arteries (paired)
gonadal arteries (paired)
posterior mesenteric artery (corresponds to inferior mesenteric artery)
iliolumbar arteries (paired)
external iliac arteries (paired)
internal iliac arteries (paired)
caudal artery (unpaired)

FIGURE 63.6   Organs of the cat's thoracic and abdominal cavities.

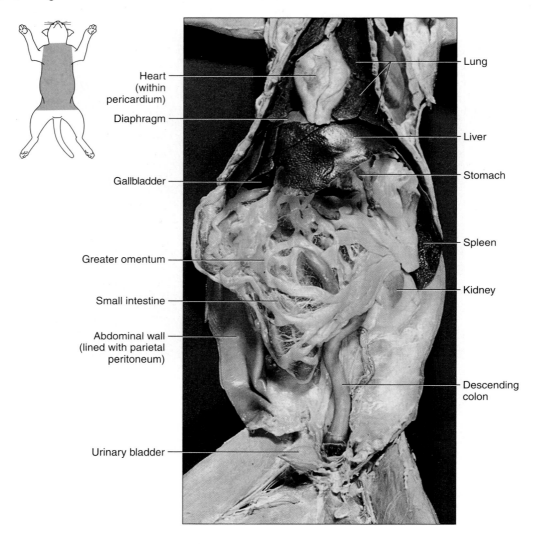

FIGURE 63.7 Arteries of the cat's abdomen and hindlimb.

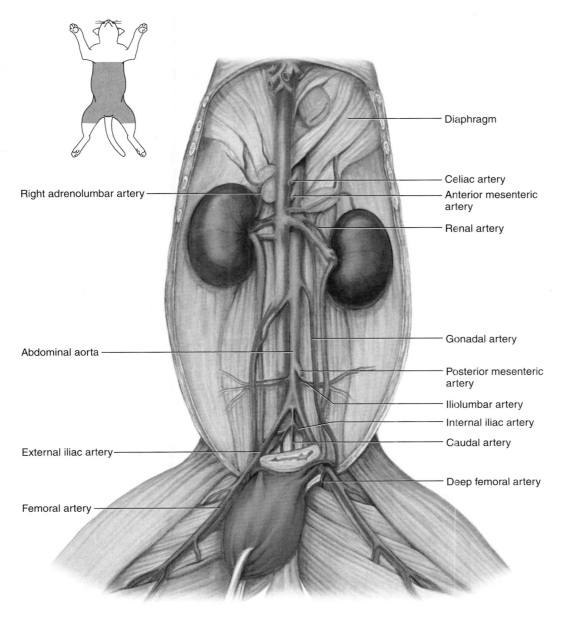

Diaphragm

Celiac artery

Anterior mesenteric artery

Right adrenolumbar artery

Renal artery

Gonadal artery

Abdominal aorta

Posterior mesenteric artery

Iliolumbar artery

Internal iliac artery

External iliac artery

Caudal artery

Deep femoral artery

Femoral artery

FIGURE 63.8 Arteries of the cat's abdomen, left lateral view.

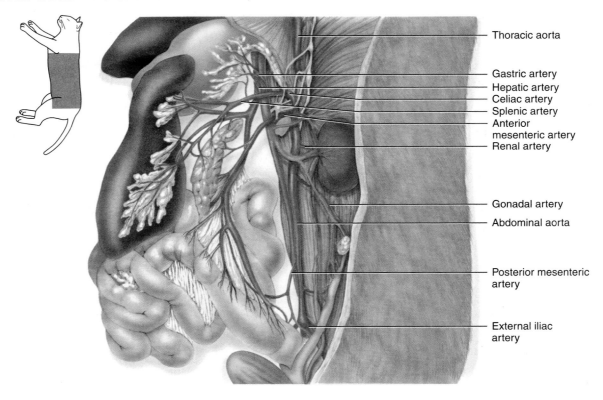

Thoracic aorta

Gastric artery
Hepatic artery
Celiac artery
Splenic artery
Anterior mesenteric artery
Renal artery

Gonadal artery

Abdominal aorta

Posterior mesenteric artery

External iliac artery

FIGURE 63.9 Blood vessels of the cat's abdominal cavity, with the digestive organs removed.

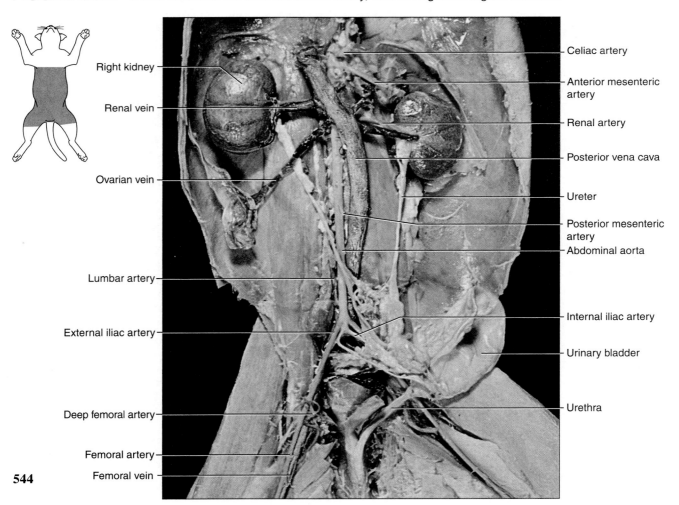

Right kidney

Renal vein

Ovarian vein

Lumbar artery

External iliac artery

Deep femoral artery

Femoral artery
Femoral vein

Celiac artery

Anterior mesenteric artery

Renal artery

Posterior vena cava

Ureter

Posterior mesenteric artery
Abdominal aorta

Internal iliac artery

Urinary bladder

Urethra

10. Trace the external iliac artery into the right hindlimb (fig. 63.10), and locate the following:

> **femoral artery**
> **deep femoral artery**

11. Review the locations of the heart structures and arteries without the aid of figures. ⚠

12. Complete Part A of Laboratory Report 63.

Procedure B—The Venous System

1. Examine the heart again, and locate the following veins:

> **anterior vena cava** (corresponds to superior vena cava)
> **posterior vena cava** (corresponds to inferior vena cava)
> **pulmonary veins**

2. Using figures 63.4, 63.5, and 63.11 as guides, locate and dissect the following veins in the thorax and neck:

> **right brachiocephalic vein**
> **left brachiocephalic vein**
> **right subclavian vein**
> **left subclavian vein**
> **internal jugular vein**
> **external jugular vein**

3. Trace the right subclavian vein into the forelimb, and locate the following veins:

> **axillary vein**
> **subscapular vein**
> **brachial vein**

FIGURE 63.10 Blood vessels of the cat's upper hindlimb.

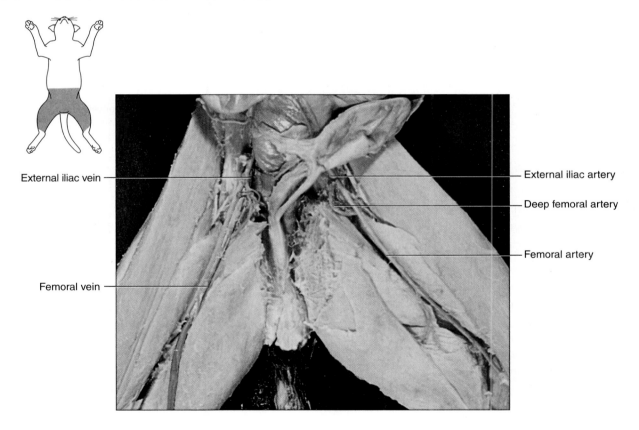

External iliac vein

Femoral vein

External iliac artery

Deep femoral artery

Femoral artery

FIGURE 63.11 Veins of the cat's thorax, neck, and forelimb.

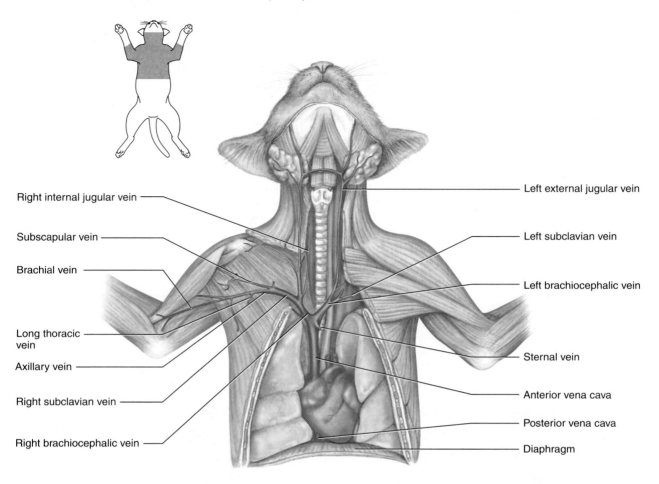

Right internal jugular vein

Subscapular vein

Brachial vein

Long thoracic vein

Axillary vein

Right subclavian vein

Right brachiocephalic vein

Left external jugular vein

Left subclavian vein

Left brachiocephalic vein

Sternal vein

Anterior vena cava

Posterior vena cava

Diaphragm

4. Using figures 63.9, 63.12, and 63.13 as guides, locate and dissect the following veins in the abdomen:

> **posterior vena cava**
>
> **adrenolumbar vein**
>
> **anterior mesenteric vein** (corresponds to superior mesenteric vein)
>
> **posterior mesenteric vein** (corresponds to inferior mesenteric vein)
>
> **hepatic portal vein**
>
> **splenic vein**
>
> **renal vein**
>
> **gonadal vein**

> **iliolumbar vein**
>
> **common iliac vein**
>
> **internal iliac vein**
>
> **external iliac vein**

5. Trace the external iliac vein into the hindlimb (fig. 63.10), and locate the following veins:

> **femoral vein**
>
> **deep femoral vein**

6. Review the locations of the veins without the aid of figures. ⚠**1**

7. Complete Parts B and C of the laboratory report.

FIGURE 63.12 Veins of the cat's abdomen and hindlimb.

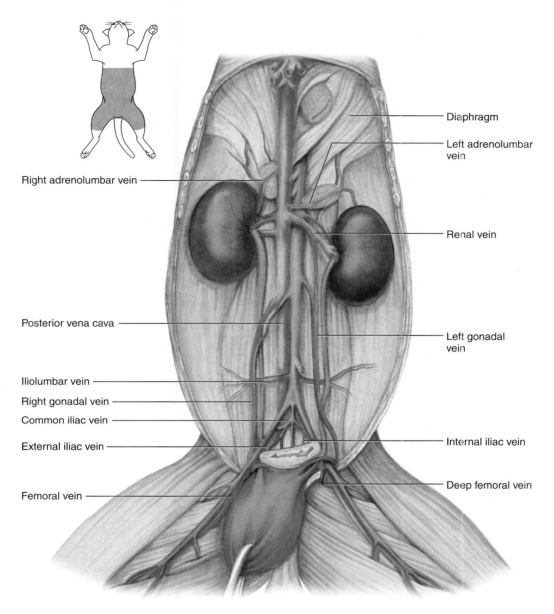

Diaphragm

Left adrenolumbar vein

Right adrenolumbar vein

Renal vein

Posterior vena cava

Left gonadal vein

Iliolumbar vein

Right gonadal vein

Common iliac vein

External iliac vein

Internal iliac vein

Deep femoral vein

Femoral vein

FIGURE 63.13 Veins of the cat's abdomen, left lateral view.

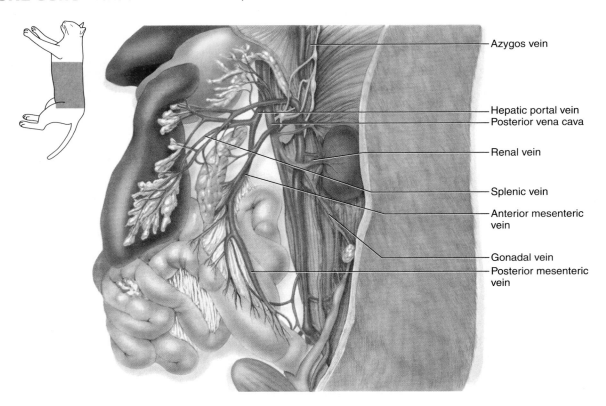

Azygos vein

Hepatic portal vein
Posterior vena cava

Renal vein

Splenic vein

Anterior mesenteric
vein

Gonadal vein
Posterior mesenteric
vein

Name _____

Date _____

Section _____

The ⚠ corresponds to the Learning Outcome(s) listed at the beginning of the laboratory exercise.

Cat Dissection: Cardiovascular System

Part A Assessments

Complete the following:

1. Describe the position and attachments of the parietal pericardium of the heart of the cat. ⚠ _____

   _____

   _____

2. Describe the relative thicknesses of the walls of the heart chambers of the cat. ⚠ _____

   _____

   _____

3. Explain how the wall thicknesses are related to the functions of the chambers. ⚠ _____

   _____

   _____

4. Compare the origins of the common carotid arteries of the cat with those of the human. ⚠ _____

   _____

   _____

5. Compare the origins of the iliac arteries of the cat with those of the human. ⚠ _____

   _____

   _____

   _____

Part B Assessments

Complete the following:

1. Compare the origins of the brachiocephalic veins of the cat with those of the human. ⚠ _____

   _____

   _____

2. Compare the relative sizes of the external and internal jugular veins of the cat with those of the human. **2** _____

_____

_____

_____

3. List twelve veins that cats and humans have in common. **2** **3**

_____

_____

_____

Part C Assessments

Observe the human torso model and figures 63.7, 63.9, and 63.12 of a cat. Locate the labeled features and identify the numbered features in figure 63.14 of the cadaver that were also identified in the cat dissection.

FIGURE 63.14 Identify the arteries and veins indicated on this anterior view of the abdomen of a cadaver, using the terms provided. **2** **3**

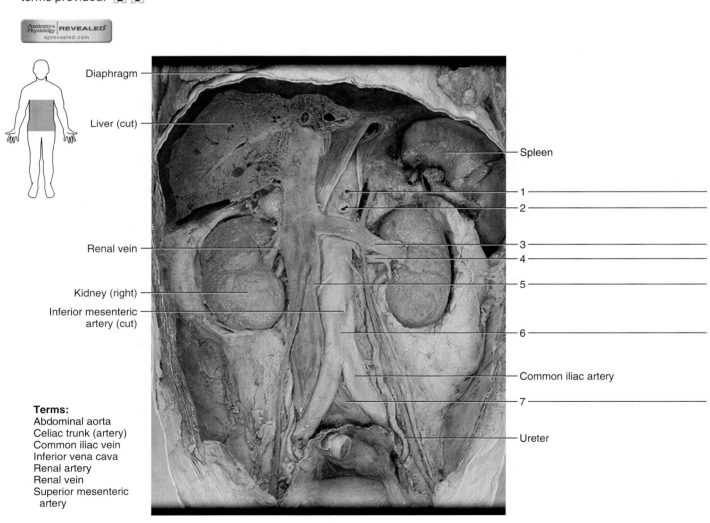

Terms:
Abdominal aorta
Celiac trunk (artery)
Common iliac vein
Inferior vena cava
Renal artery
Renal vein
Superior mesenteric
 artery

64

Cat Dissection: Respiratory System

Pre-Lab

1. Carefully read the introductory material and examine the entire lab content.
2. Be familiar with the structures and functions of the human respiratory system (from lecture, the textbook, or laboratory exercises).

Materials Needed

Preserved cat
Dissecting tray
Dissecting instruments
Bone cutter
Disposable gloves
Human torso model

Safety

▶ Wear disposable gloves when working on the cat dissection.
▶ Dispose of tissue remnants and gloves as instructed.
▶ Wash the dissecting tray and instruments as instructed.
▶ Wash your laboratory table.
▶ Wash your hands before leaving the laboratory.

In this laboratory exercise, you will dissect the major respiratory organs of a cat. As you observe these structures in the cat, compare them with those of the human torso model.

Purpose of the Exercise

To examine the major respiratory organs of the cat and to compare these organs with those of the human.

Learning Outcomes

After completing this exercise, you should be able to

1. Locate and identify the major respiratory organs of the cat.
2. Compare the respiratory system of the human with that of the cat.
3. Identify the corresponding organs in the human torso model and a cadaver.

Procedure—Respiratory System Dissection

1. Place the preserved cat in a dissecting tray with its ventral surface up.
2. Examine the *external nares* (*nostrils*) and *nasal septum.*
3. Open the oral cavity. To do this, follow these steps:
   a. Use scissors to cut through the soft tissues at the angle of the mouth.
   b. When you reach the bone of the jaw, use a bone cutter to cut through the bone, thus freeing the mandible.
4. Open the oral cavity wide to expose the *hard palate* and the *soft palate* (fig. 64.1). Cut through the tissues of the soft palate, and observe the *nasopharynx* above it. Locate the small openings of the *auditory tubes* in the lateral walls of the nasopharynx. Insert a probe into an opening of an auditory tube.
5. Pull the tongue forward, and locate the *epiglottis* near its base. Also identify the *glottis,* which is the opening between the *vocal folds* (*vocal cords*) into the larynx. Locate the *esophagus,* which is dorsal to the larynx (figs. 64.1 and 64.2).
6. Open the thoracic cavity, and expose its contents.
7. Dissect the *trachea* in the neck, and expose the *larynx* at the anterior end near the base of the tongue. Note the *tracheal rings,* and locate the lobes of the *thyroid gland* on each side of the trachea, just posterior to the larynx. Also note the *thymus gland,* a rather diffuse mass of glandular tissue extending along the ventral surface of the trachea into the thorax (figs. 64.1 and 64.2).
8. Examine the larynx, and identify the *thyroid cartilage* and *cricoid cartilage* in its wall (fig. 64.3*a*). Make a

FIGURE 64.1 Features of the cat's neck and oral cavity, ventral view.

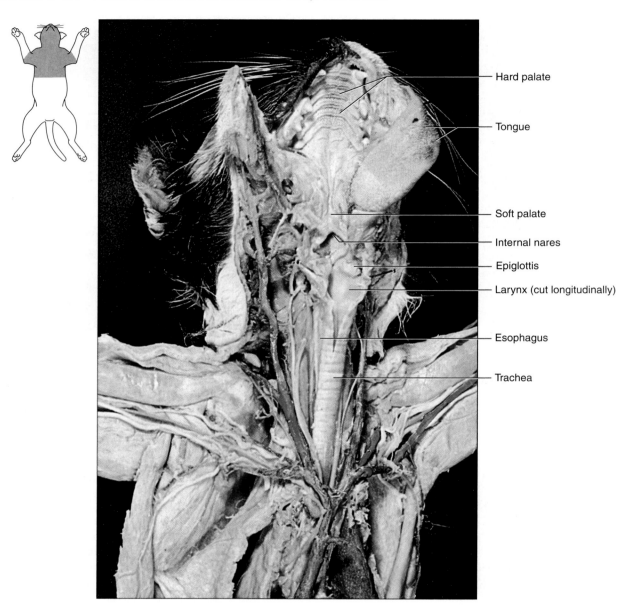

Hard palate

Tongue

Soft palate

Internal nares

Epiglottis

Larynx (cut longitudinally)

Esophagus

Trachea

longitudinal incision through the ventral wall of the larynx, and locate the vocal folds inside. Compare the structures of the larynx of the cat with those of a human (fig. 64.3*b*).

9. Trace the trachea posteriorly to where it divides into *main (primary) bronchi,* which pass into the lungs.

10. Examine the *lungs,* each of which is subdivided into three lobes—an *anterior (cranial),* a *middle,* and a *posterior (caudal) lobe.* The right lung of the cat has a small fourth *accessory (mediastinal; cardiac) lobe* associated with the right posterior lobe. Notice the thin membrane, the *visceral pleura,* on the surface of each lung. Also notice the parietal pleura, which forms the inner lining of the thoracic wall, and locate the spaces of the *pleural cavities* (fig. 64.2).

11. Make an incision through a lobe of a lung, and examine its interior. Note the branches of the smaller air passages.

12. Examine the *diaphragm,* and note that its central portion is tendinous.

13. Locate the *phrenic nerve.* This nerve appears as a white thread passing along the side of the heart to the diaphragm.

14. Examine the organs located in the *mediastinum.* Note that the mediastinum separates the right and left lungs and pleural cavities within the thorax.

15. Review the locations of the respiratory organs without the aid of figures. ⚠

16. Complete Parts A and B of Laboratory Report 64.

FIGURE 64.2 Ventral view of the cat's thoracic cavity.

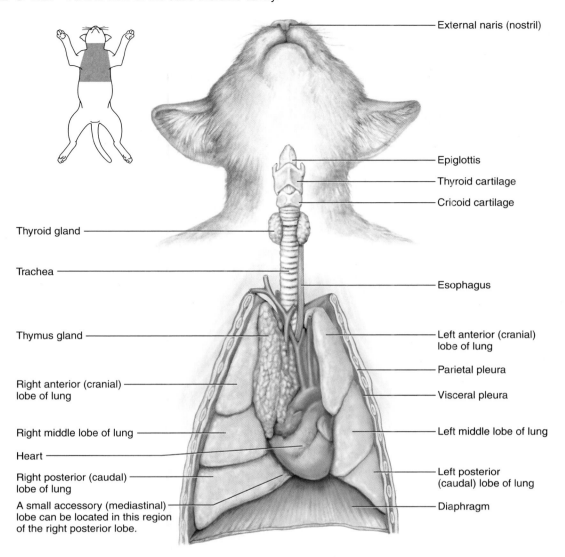

External naris (nostril)

Epiglottis

Thyroid cartilage

Cricoid cartilage

Thyroid gland

Trachea

Esophagus

Thymus gland

Left anterior (cranial) lobe of lung

Parietal pleura

Right anterior (cranial) lobe of lung

Visceral pleura

Right middle lobe of lung

Left middle lobe of lung

Heart

Right posterior (caudal) lobe of lung

Left posterior (caudal) lobe of lung

A small accessory (mediastinal) lobe can be located in this region of the right posterior lobe.

Diaphragm

FIGURE 64.3 (a) Larynx and associated structures of the cat, ventral view. (b) Larynx structures of a cadaver, anterior view.

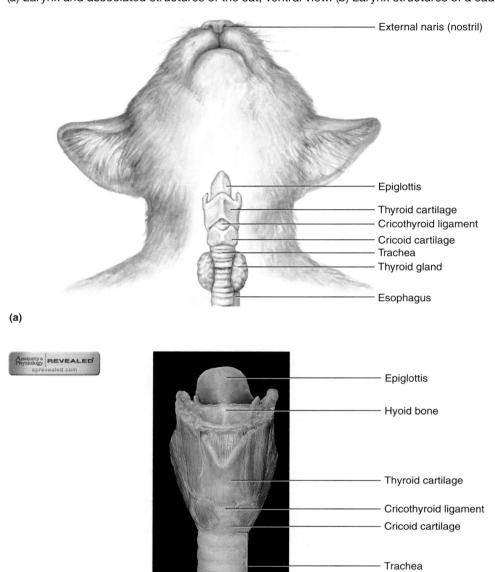

(a)

(b)

Name _____

Date _____

Section _____

The ⚠ corresponds to the Learning Outcome(s) listed at the beginning of the laboratory exercise.

Cat Dissection: Respiratory System

Part A Assessments

Complete the following:

1. What is the purpose of the auditory tubes opening into the nasopharynx? ⚠ _____

_____

_____

_____

2. Distinguish between the glottis and the epiglottis. ⚠ _____

_____

_____

_____

3. Are the tracheal rings of the cat complete or incomplete? ⚠ _____ How does this feature

   compare with that of the human? ⚠2 _____

_____

_____

4. How does the structure of the main (primary) bronchi compare with that of the trachea? ⚠ _____

_____

_____

5. Compare the number of lobes in the human lungs with the number of lobes in the cat. ⚠2 _____

_____

_____

_____

6. Describe the attachments of the diaphragm. **1** _____

_____

_____

7. What major structures are located within the mediastinum of the cat? **1** _____

_____

How does this compare to the human mediastinum? **2** _____

_____

Part B Assessments

Observe the human torso model and figures 64.3*b* and 64.4 of a cadaver. Locate the labeled features and identify the numbered features in figure 64.4 that were also identified in the cat dissection.

FIGURE 64.4   Identify the respiratory features indicated on this anterior view of the thorax, using the terms provided. **2** **3**

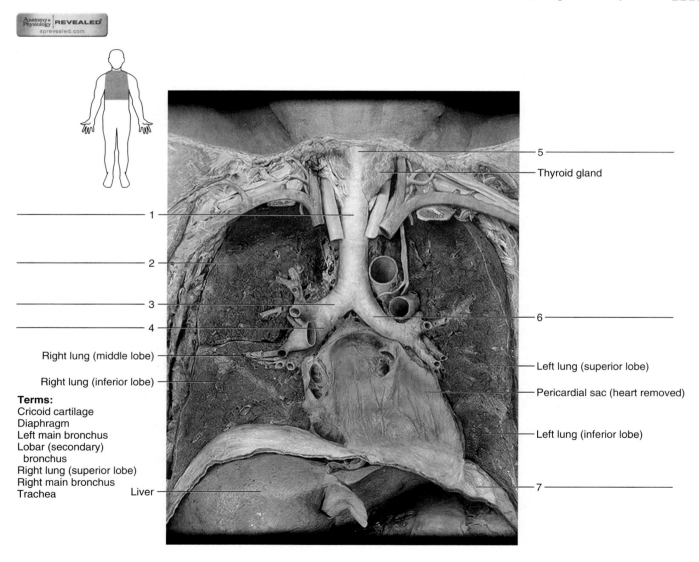

Right lung (middle lobe)
Right lung (inferior lobe)

Terms:
Cricoid cartilage
Diaphragm
Left main bronchus
Lobar (secondary)
  bronchus
Right lung (superior lobe)
Right main bronchus
Trachea

Liver

Thyroid gland

Left lung (superior lobe)
Pericardial sac (heart removed)
Left lung (inferior lobe)

Cat Dissection: Digestive System

Pre-Lab

1. Carefully read the introductory material and examine the entire lab content.
2. Be familiar with the structures and functions of the human digestive system (from lecture, the textbook, or laboratory exercises).

Materials Needed

Preserved cat
Dissecting tray
Dissecting instruments
Bone cutter
Disposable gloves
Human torso model
Hand lens

Safety

▶ Wear disposable gloves when working on the cat dissection.
▶ Dispose of tissue remnants and gloves as instructed.
▶ Wash the dissecting tray and instruments as instructed.
▶ Wash your laboratory table.
▶ Wash your hands before leaving the laboratory.

In this laboratory exercise, you will dissect the major digestive organs of the cat. As you observe these organs, compare them with those of the human by observing the parts of the human torso model; however, keep in mind that the cat is a carnivore (flesh-eating mammal) and that the organs of its digestive system are adapted to capturing, holding, eating, and digesting the bodies of other animals. In contrast, humans are omnivores (eat both plant and animal

substances). Because humans are adapted to eating and digesting a greater variety of foods, comparisons of the cat and the human digestive organs may not be precise.

Purpose of the Exercise

To examine the major digestive organs of the cat and to compare these organs with those of the human.

Learning Outcomes

After completing this exercise, you should be able to

1. Locate and identify the major digestive organs of the cat.
2. Compare the digestive system of the cat with that of the human.
3. Identify the corresponding organs in the human torso model and a cadaver.

Procedure—Digestive System Dissection

1. Place the preserved cat in the dissecting tray on its left side.
2. Locate the major salivary glands on one side of the head (fig. 65.1a). To do this, follow these steps:
   a. Clear away any remaining fascia and other connective tissue from the region below the ear and near the joint of the mandible.
   b. Identify the *parotid gland,* a relatively large mass of glandular tissue just below the ear. Note the parotid duct (Stensen's duct) that passes over the surface of the masseter muscle and opens into the oral cavity.
   c. Look for the *submandibular gland* just below the parotid gland, near the angle of the jaw.
   d. Locate the *sublingual gland* that is adjacent and medial to the submandibular gland.
   e. Compare the location of the salivary glands of the cat with those of a human (fig. 65.1b).
3. Open the oral cavity. To do this, follow these steps:
   a. Use scissors to cut through the soft tissues at the angle of the mouth if not previously accomplished.
   b. When you reach the bone of the jaw, use a bone cutter to cut through the bone, thus freeing the mandible.

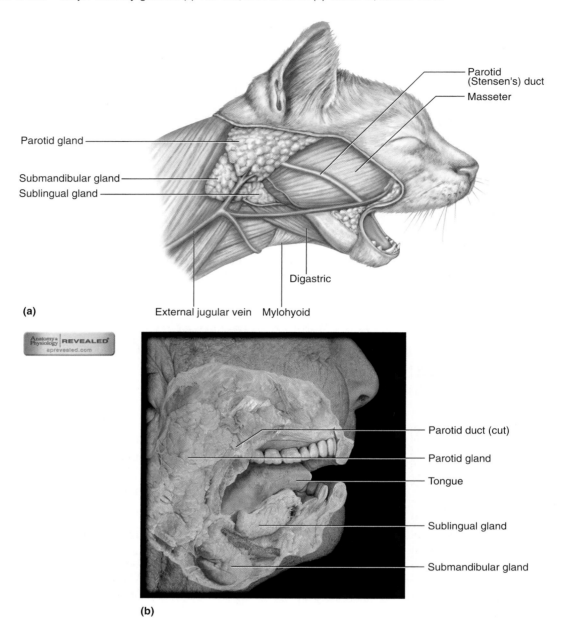

(a)

Parotid (Stensen's) duct
Masseter
Parotid gland
Submandibular gland
Sublingual gland
Digastric
External jugular vein   Mylohyoid

Anatomy & Physiology REVEALED
aprevealed.com

Parotid duct (cut)
Parotid gland
Tongue
Sublingual gland
Submandibular gland

(b)

c. Open the mouth wide, and locate the following features:

   cheek

   lip

   vestibule

   palate

     hard palate

     soft palate

   palatine tonsils (small, rounded masses of glandular tissue in the lateral wall of the soft palate)

   tongue

     frenulum

     papillae (examine with a hand lens)

4. Examine the teeth of the upper jaw. The adult cat has 6 incisors, 2 canines, 8 premolars, and 2 molars. The lower jaw teeth are similar, except 4 premolars are present.

5. Complete Part A of Laboratory Report 65.

6. Examine organs in the abdominal cavity with the cat positioned with its ventral surface up. Review the normal position and function of the greater omentum.

7. Examine the *liver,* which is located just beneath the diaphragm and is attached to the central portion of the diaphragm by the *falciform ligament.* Also, locate the *spleen,* which is posterior to the liver on the left side (fig. 65.2). The liver has five lobes—a right medial, right lateral (which is subdivided into two parts by a deep cleft), left medial, left lateral, and caudate lobe. The caudate lobe is the smallest lobe and is located in the median

FIGURE 65.2 Ventral view of abdominal organs of the cat.

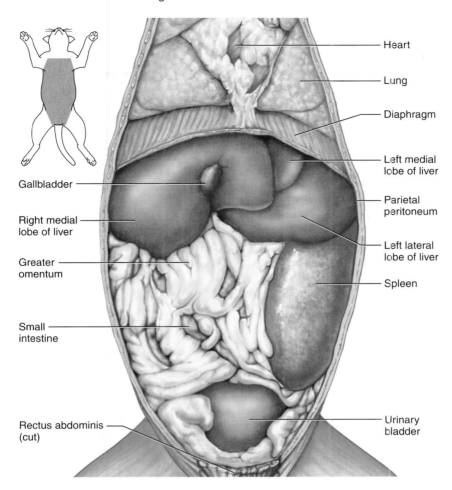

line, where it projects into the curvature of the stomach. The caudate lobe is covered by a sheet of mesentery, the *lesser omentum,* that connects the liver to the stomach (figs. 65.3 and 65.4). Find the greenish *gallbladder* on the inferior surface of the liver on the right side. Also note the *cystic duct,* by which the gallbladder is attached to the *bile duct,* and the *hepatic duct,* which originates in the liver and attaches to the cystic duct. Trace the bile duct to its connection with the duodenum.

8. Locate the *stomach* in the left side of the abdominal cavity. At its anterior end, note the union with the *esophagus,* which passes through the diaphragm. Identify the *cardia, fundus, body,* and *pyloric part* (listed from entrance to exit) of the stomach. Use scissors to make an incision along the convex border of the stomach from the cardia region to the pyloric part. The lining of the stomach has numerous *gastric folds (rugae).* Examine the *pyloric sphincter,* which creates a constriction between the stomach and small intestine.

9. Locate the *pancreas.* It appears as a grayish, two-lobed elongated mass of glandular tissue. One lobe lies dorsal to the stomach and extends across to the duodenum. The

other lobe is enclosed by the *mesentery,* which supports the duodenum (figs. 65.3 and 65.4).

10. Trace the *small intestine,* beginning at the pyloric sphincter. The first portion, the *duodenum,* travels posteriorly for several centimeters. Then it loops back around a lobe of the pancreas. The proximal half of the remaining portion of the small intestine is the *jejunum,* and the distal half is the *ileum.* Open the duodenum and note the velvety appearance of the villi. Note how the mesentery supports the small intestine from the dorsal body wall. The small intestine terminates on the right side, where it joins the large intestine.

11. Locate the *large intestine,* and identify the *cecum, ascending colon, transverse colon,* and *descending colon.* Also locate the *rectum,* which extends through the pelvic cavity to the *anus.* Make an incision at the junction between the ileum and cecum, and look for the *ileocecal sphincter.*

12. Review the locations of the digestive organs without the aid of figures. 🔟

13. Complete Parts B and C of the laboratory report.

FIGURE 65.3 Ventral view of the cat's abdominal organs with intestines reflected to the right side.

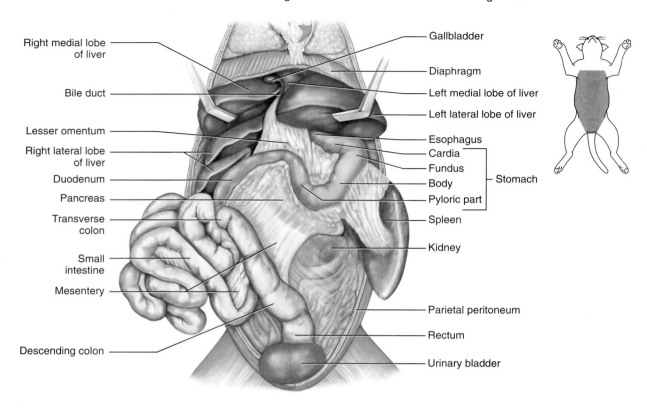

Right medial lobe of liver

Bile duct

Lesser omentum

Right lateral lobe of liver

Duodenum

Pancreas

Transverse colon

Small intestine

Mesentery

Descending colon

Gallbladder

Diaphragm

Left medial lobe of liver

Left lateral lobe of liver

Esophagus

Cardia

Fundus

Body

Pyloric part

Stomach

Spleen

Kidney

Parietal peritoneum

Rectum

Urinary bladder

FIGURE 65.4 Abdominal organs of the cat with the greater omentum removed. The left liver lobes have been reflected to the right side.

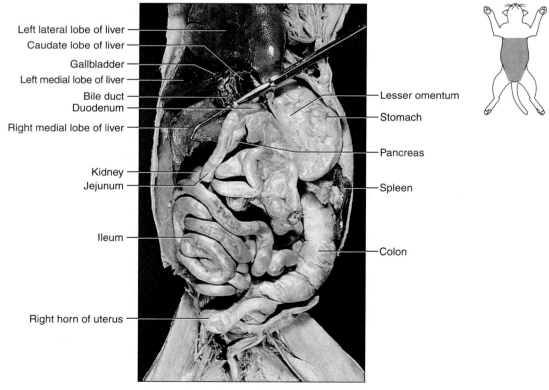

Left lateral lobe of liver

Caudate lobe of liver

Gallbladder

Left medial lobe of liver

Bile duct

Duodenum

Right medial lobe of liver

Kidney

Jejunum

Ileum

Right horn of uterus

Lesser omentum

Stomach

Pancreas

Spleen

Colon

Laboratory Report

65

Name _____

Date _____

Section _____

The ⬛A⬛ corresponds to the Learning Outcome(s) listed at the beginning of the laboratory exercise.

Cat Dissection: Digestive System

Part A Assessments

Complete the following:

1. Compare the locations of the major salivary glands of the human with those of the cat. **2** _____

_____

_____

2. Compare the types and numbers of teeth present in the cat's upper and lower jaws with those of the human. **2** _____

_____

_____

_____

_____

3. In what ways do the cat's teeth seem to be adapted to a special diet? **1** _____

_____

_____

4. What part of the human soft palate is lacking in the cat? **2** _____

5. What do you think is the function of the transverse ridges (rugae) in the hard palate of the cat? **1** _____

_____

_____

6. How do the papillae on the surface of the human tongue compare with those of the cat? **2** _____

_____

_____

Part B Assessments

Complete the following:

1. Describe how the peritoneum and mesenteries are associated with the organs in the abdominal cavity. **1** _____

   _____

   _____

   _____

   _____

2. Describe the inner lining of the stomach. **1** _____

   _____

3. Compare the structure of the human liver with that of the cat. **2** _____

   _____

   _____

   _____

4. Compare the structure and location of the human pancreas with those of the cat. **2** _____

   _____

   _____

5. What feature of the human cecum is lacking in the cat? **2** _____

   _____

Part C Assessments

Observe the human torso model and figures 65.1*b* and 65.5 of a cadaver. Locate the labeled features, and identify the numbered features in figure 65.5 that were also identified in the cat dissection.

FIGURE 65.5   Identify the digestive features indicated on this anterior view of the abdomen of a cadaver, using the terms provided. /2\ /3\

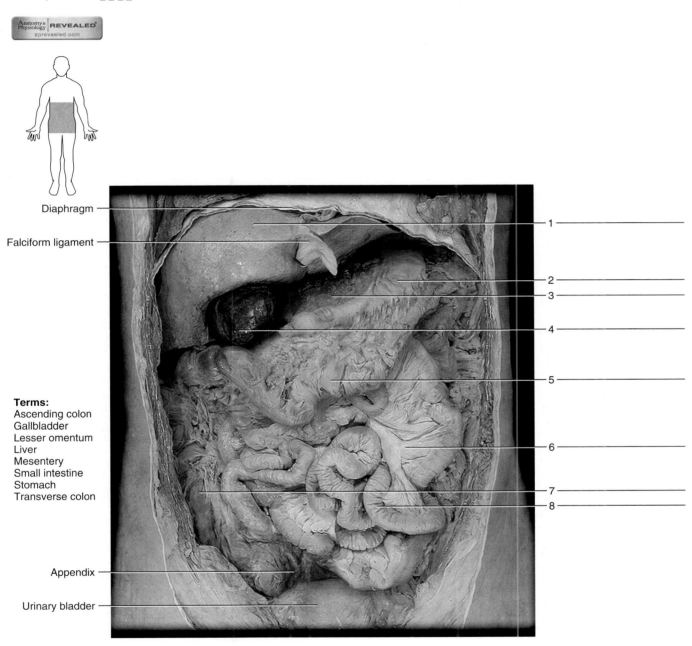

Diaphragm

Falciform ligament

1

2

3

4

5

Terms:
Ascending colon
Gallbladder
Lesser omentum
Liver
Mesentery
Small intestine
Stomach
Transverse colon

6

7

8

Appendix

Urinary bladder

Cat Dissection: Urinary System

Pre-Lab

1. Carefully read the introductory material and examine the entire lab content.
2. Be familiar with the structures and functions of the human urinary system (from lecture, the textbook, or laboratory exercises).

Materials Needed

Human torso model
Preserved cat
Dissecting tray
Dissecting instruments
Disposable gloves

Safety

▶ Wear disposable gloves when working on the cat dissection.
▶ Dispose of tissue remnants and gloves as instructed.
▶ Wash the dissecting tray and instruments as instructed.
▶ Wash your laboratory table.
▶ Wash your hands before leaving the laboratory.

In this laboratory exercise, you will dissect the urinary organs of the cat. As you observe these structures, compare them with the corresponding human organs by observing the parts of the human torso model.

Purpose of the Exercise

To examine the urinary organs of the cat and to compare them with those of the human.

Learning Outcomes

After completing this exercise, you should be able to

① Locate and identify the urinary organs of the cat.
② Compare the urinary organs of the cat with those of the human.
③ Identify the corresponding organs in the human torso model and a cadaver.

Procedure—Urinary System Dissection

1. Place the preserved cat in a dissecting tray with its ventral surface up.
2. Open the abdominal cavity, and remove the liver, stomach, and spleen (fig. 66.1).
3. Push the intestines to one side, and locate the *kidneys* in the dorsal abdominal wall on either side of the vertebral column. The kidneys are located dorsal to the *parietal peritoneum* (retroperitoneal).
4. Carefully remove the parietal peritoneum and the adipose tissue surrounding the kidneys. Locate the following, using figures 66.1 and 66.2 as guides:

   ureters
   renal arteries
   renal veins

5. Locate the *adrenal glands,* which lie medially and above the kidneys and are surrounded by connective tissues. The adrenal glands are attached to blood vessels and are separated by the abdominal aorta and posterior vena cava.
6. Expose the ureters by cleaning away the connective tissues along their lengths. They enter the *urinary bladder* on the dorsal surface.
7. Examine the urinary bladder, attached to the abdominal wall by folds of peritoneum that form ligaments. Locate the *medial ligament* on the ventral surface of the bladder that connects it with the linea alba and the *two lateral ligaments* on the dorsal surface.
8. Remove the connective tissue from around the urinary bladder. Use a sharp scalpel to open the bladder, and

FIGURE 66.1 Urinary system of the female cat's abdominal cavity, with the digestive organs removed.

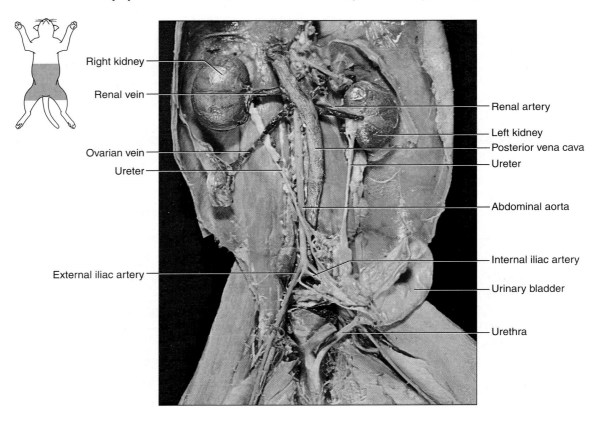

Right kidney
Renal vein
Ovarian vein
Ureter
External iliac artery

Renal artery
Left kidney
Posterior vena cava
Ureter
Abdominal aorta
Internal iliac artery
Urinary bladder
Urethra

FIGURE 66.2 The urinary organs of a female cat.

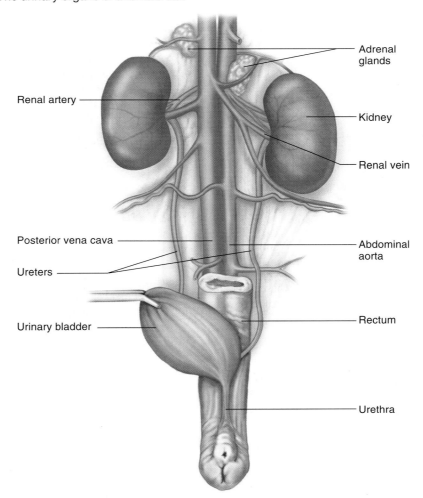

Renal artery
Posterior vena cava
Ureters
Urinary bladder

Adrenal glands
Kidney
Renal vein
Abdominal aorta
Rectum
Urethra

examine its interior. Locate the openings of the ureters and the urethra on the inside.

9. Expose the *urethra* at the posterior of the urinary bladder (figs. 66.1 and 66.3).

10. Remove one kidney, and section it longitudinally (figs. 66.3 and 66.4*a*). Identify the following features:

   fibrous capsule
   renal cortex (superficial lighter region)
   renal medulla (deeper darker region)
   renal pyramid (single in cat)
   renal papilla (single in cat)

renal sinus
renal pelvis
hilum of kidney

11. Compare the structures of the kidney of the cat with those of a human (fig. 66.4*b*).

12. Review the locations of the urinary organs without the aid of figures.

13. Discard the organs and tissues that were removed from the cat, as directed by the laboratory instructor.

14. Complete Parts A and B of Laboratory Report 66.

FIGURE 66.3 Female urinary and reproductive systems of a pregnant cat.

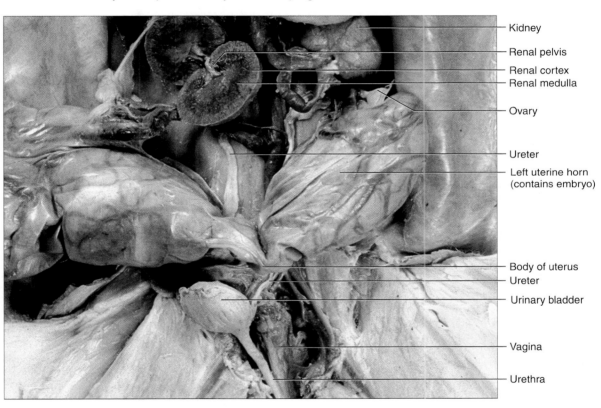

FIGURE 66.4   Longitudinal section of a kidney. (*a*) The cat, (*b*) cadaver.

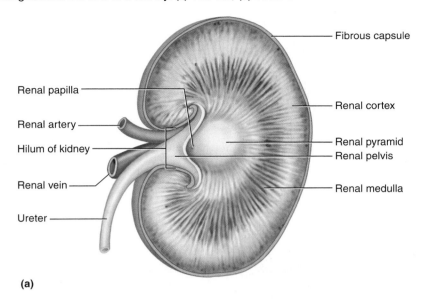

Fibrous capsule

Renal papilla

Renal artery

Hilum of kidney

Renal vein

Ureter

Renal cortex

Renal pyramid
Renal pelvis

Renal medulla

(a)

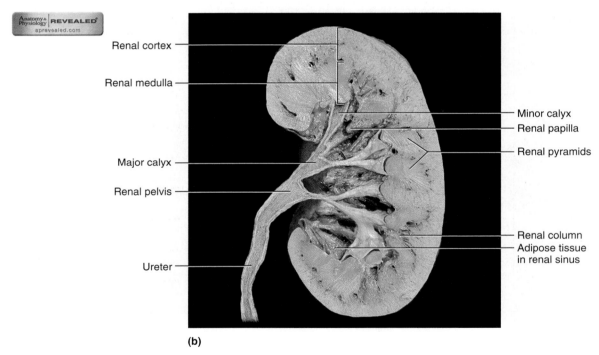

Renal cortex

Renal medulla

Major calyx

Renal pelvis

Ureter

Minor calyx
Renal papilla

Renal pyramids

Renal column
Adipose tissue
in renal sinus

(b)

Name _____

Date _____

Section _____

The ⬕ corresponds to the Learning Outcome(s) listed at the beginning of the laboratory exercise.

Cat Dissection: Urinary System

Part A Assessments

Complete the following:

1. Compare the positions of the kidneys in the cat with those in the human. ⬕2 _____

_____

_____

_____

2. Compare the locations of the adrenal glands in the cat with those in the human. ⬕2 _____

_____

_____

_____

_____

3. What structures of the cat urinary system are retroperitoneal? ⬕1 _____

_____

_____

_____

4. Describe the wall of the urinary bladder of the cat. ⬕1 _____

_____

_____

_____

5. Compare the renal pyramids and renal papillae of the human kidney with those of the cat. 2 _____

_____

_____

_____

Part B Assessments

Observe the human torso model and figures 66.4*b* and 66.5 of a cadaver. Locate the labeled features and identify the numbered features in figure 66.5 that were also identified in the cat dissection.

FIGURE 66.5   Identify the urinary features indicated on this anterior view of the abdomen of a cadaver, using the terms provided. 2 3

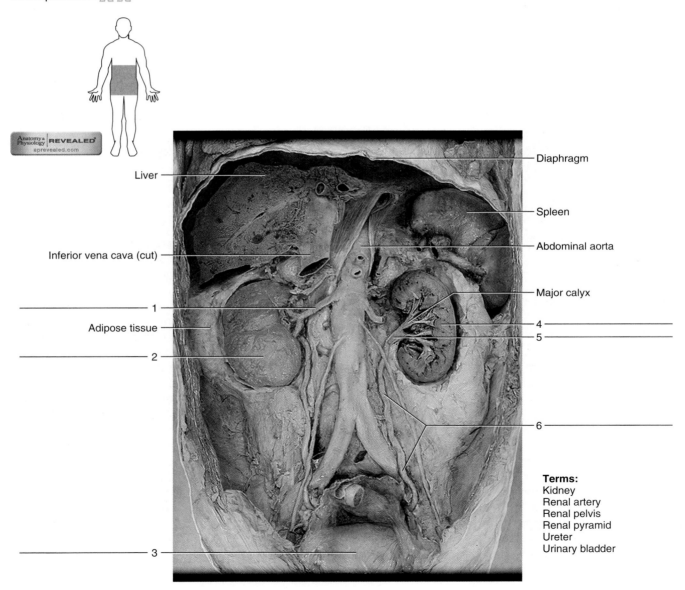

Liver

Inferior vena cava (cut)

1

Adipose tissue

2

3

Diaphragm

Spleen

Abdominal aorta

Major calyx

4

5

6

Terms:
Kidney
Renal artery
Renal pelvis
Renal pyramid
Ureter
Urinary bladder

Cat Dissection: Reproductive Systems

Pre-Lab

1. Carefully read the introductory material and examine the entire lab content.
2. Be familiar with the structures and functions of the human reproductive system (from lecture, the textbook, or laboratory exercises).

Materials Needed

Preserved cat
Dissecting tray
Dissecting instruments
Bone cutters
Disposable gloves
Models of human reproductive systems

Safety

► Wear disposable gloves when working on the cat dissection.
► Dispose of tissue remnants and gloves as instructed.
► Wash the dissecting tray and instruments as instructed.
► Wash your laboratory table.
► Wash your hands before leaving the laboratory.

In this laboratory exercise, you will dissect the reproductive system of the cat. If you have a female cat, begin with Procedure A. If you have a male cat, begin with Procedure B. After completing the dissection, exchange cats with someone who has dissected one of the opposite sex, and examine its reproductive organs.

As you observe the cat reproductive organs, compare them with the corresponding human organs by examining the models of the human reproductive systems.

Purpose of the Exercise

To examine the reproductive organs of the cat and to compare them with the corresponding organs of the human.

Learning Outcomes

After completing this exercise, you should be able to

1. Locate and identify the reproductive organs of a cat.
2. Compare the reproductive organs of the cat with those of the human.
3. Identify the corresponding organs in models of the human reproductive systems and a cadaver.

Procedure A—Female Reproductive System

1. Place the preserved female cat in a dissecting tray with its ventral surface up, and open its abdominal cavity.
2. Locate the small oval *ovaries* just posterior to the kidneys (figs. 67.1 and 67.2).
3. Examine an ovary, suspended from the dorsal body wall by a fold of peritoneum called the *mesovarium*. Another attachment, the *ovarian ligament,* connects the ovary to the tubular *uterine horn* (fig. 67.1).
4. Near the anterior end of the ovary, locate the funnel-shaped *infundibulum,* which is at the end of the *uterine tube* (*oviduct*). Note the tiny projections, or *fimbriae,* that create a fringe around the edge of the infundibulum. Trace the uterine tube around the ovary to its connection with the uterine horn near the attachment of the ovarian ligament.
5. Examine the uterine horn. It is suspended from the body wall by a fold of peritoneum, the *broad ligament,* that is continuous with the mesovarium of the ovary. Also note the fibrous *round ligament,* which extends from the uterine horn laterally and posteriorly to the body wall (figs. 67.1 and 67.2).

FIGURE 67.1 Ventral view of the female cat's reproductive system.

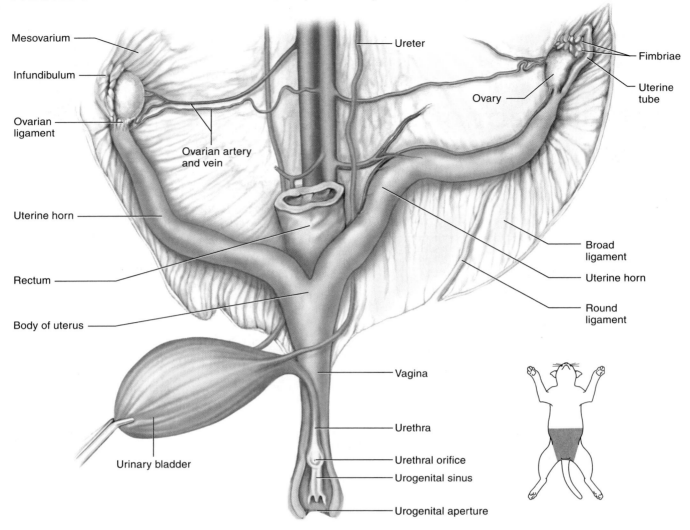

6. To observe the remaining organs of the reproductive system more easily, remove the left hindlimb by severing it near the hip with bone cutters. Then use the bone cutters to remove the left anterior portion of the pelvic girdle. Also remove the necessary muscles and connective tissue to expose the structures shown in figure 67.3.

7. Trace the uterine horns posteriorly. They unite to form the *uterine body,* which is located between the urethra and rectum. This Y-shaped uterus allows ample space for several offspring to develop at the same time in the uterine horns (fig. 67.2). The uterine body is continuous with the *vagina,* which leads to the outside.

8. Trace the *urethra* from the urinary bladder posteriorly. The urethra and the vagina open into a common chamber, called the *urogenital sinus.* The opening of this chamber, ventral to the anus, is called the *urogenital aperture.* Locate the *labia majora,* which are folds of skin on either side of the urogenital aperture.

9. Use scissors to open the vagina along its lateral wall, beginning at the urogenital aperture and continuing to the body of the uterus. Note the *urethral orifice* in the ventral wall of the urogenital sinus, and locate the small, rounded *cervix* of the uterus, which projects into the vagina at its deep end. The *clitoris* is located in the ventral wall of the urogenital sinus near its opening to the outside.

10. Review the locations of the reproductive organs without the aid of figures.

11. Complete Part A of Laboratory Report 67.

Procedure B—Male Reproductive System

1. Place the preserved male cat in a dissecting tray with its ventral surface up.

2. Locate the *scrotum* between the hindlimbs. Make an incision along the midline of the scrotum, and expose

FIGURE 67.2 Female urinary and reproductive systems of a pregnant cat.

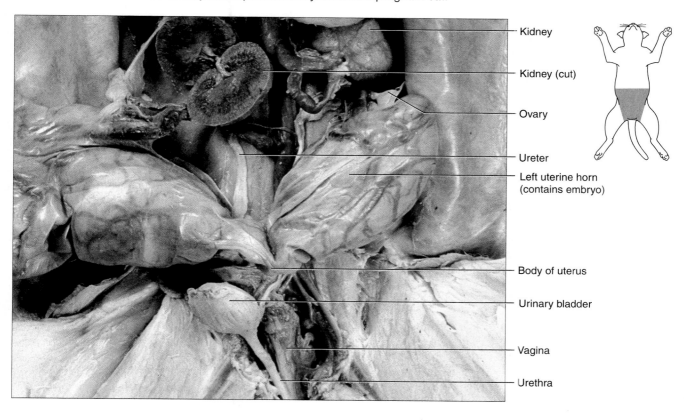

- Kidney
- Kidney (cut)
- Ovary
- Ureter
- Left uterine horn (contains embryo)
- Body of uterus
- Urinary bladder
- Vagina
- Urethra

FIGURE 67.3 Lateral view of the female cat's reproductive system.

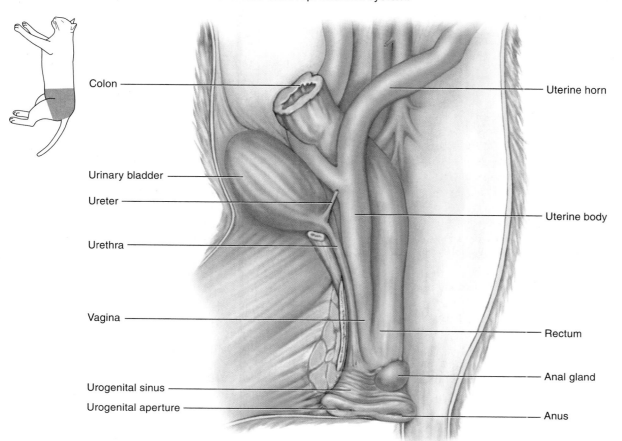

- Colon
- Urinary bladder
- Ureter
- Urethra
- Vagina
- Urogenital sinus
- Urogenital aperture
- Uterine horn
- Uterine body
- Rectum
- Anal gland
- Anus

the *testes* inside (fig. 67.4). The testes are separated by a septum and that each testis is enclosed in a sheath of connective tissue.

3. Remove the sheath surrounding the testes, and locate the convoluted *epididymis* on the dorsal surface of each testis (fig. 67.5).

4. Locate the *spermatic cord* on the right side, leading away from the testis. This spermatic cord contains the *ductus (vas) deferens,* which is continuous with the epididymis, as well as with the nerves and blood vessels that supply the testis on that side. Trace the spermatic cord to the body wall, where its contents pass through the *inguinal canal* and enter the pelvic cavity (fig. 67.5).

5. Locate the *penis,* and identify the *prepuce,* which forms a sheath around the penis. Make an incision through the skin of the prepuce, and expose the *glans penis.* The glans penis has minute spines on its surface.

6. To observe the remaining structures of the reproductive system more easily, remove the left hindlimb by severing it near the hip with bone cutters. Then use the bone cutters to remove the left anterior portion of the pelvic girdle. Also remove the necessary muscles and connective tissues to expose the organs shown in figure 67.6.

7. Trace the ductus deferens from the inguinal canal to the penis. Note that the ductus deferens loops over the ureter within the pelvic cavity and passes downward behind the urinary bladder to join the urethra. Locate the *prostate gland,* which appears as an enlargement at the junction of the ductus deferens and urethra.

8. Trace the urethra to the penis. Locate the *bulbourethral glands,* which form small swellings on either side at the proximal end of the penis.

9. Use a sharp scalpel to cut a transverse section of the penis. Identify the *corpora cavernosa,* each of which contains many blood spaces surrounded by a sheath of connective tissue.

10. Review the locations of the reproductive organs without the aid of figures.

11. Complete Parts B and C of the laboratory report.

FIGURE 67.4 Male reproductive system of the cat.

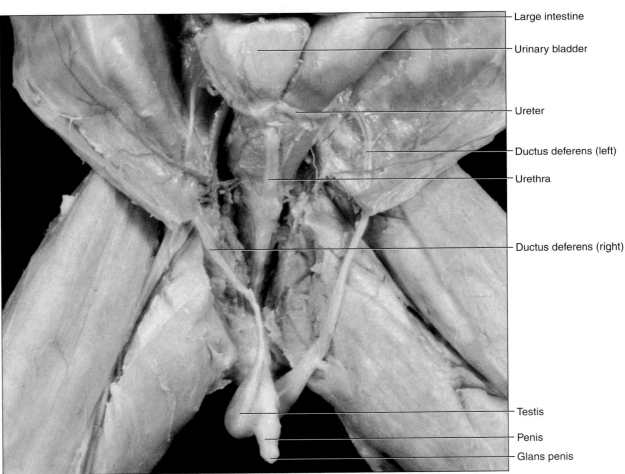

- Large intestine
- Urinary bladder
- Ureter
- Ductus deferens (left)
- Urethra
- Ductus deferens (right)
- Testis
- Penis
- Glans penis

FIGURE 67.5 Ventral view of the male cat's urinary and reproductive systems.

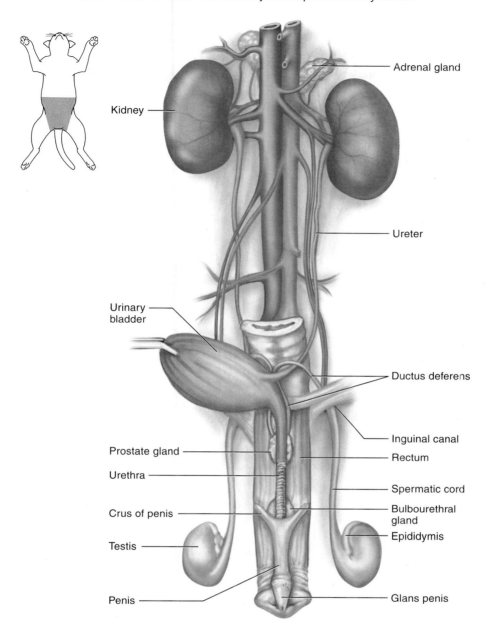

Adrenal gland

Kidney

Ureter

Urinary bladder

Ductus deferens

Inguinal canal

Prostate gland

Rectum

Urethra

Spermatic cord

Crus of penis

Bulbourethral gland

Epididymis

Testis

Penis

Glans penis

FIGURE 67.6 Lateral view of the male cat's reproductive system.

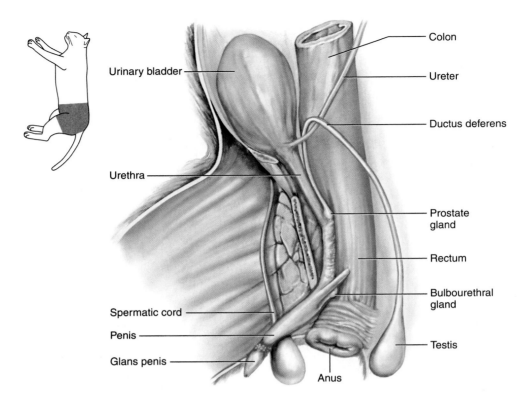

Urinary bladder

Urethra

Spermatic cord

Penis

Glans penis

Colon

Ureter

Ductus deferens

Prostate gland

Rectum

Bulbourethral gland

Testis

Anus

Name _____

Date _____

Section _____

The ⓐ corresponds to the Learning Outcome(s) listed at the beginning of the laboratory exercise.

Cat Dissection: Reproductive Systems

Part A Assessments

Complete the following:

1. Compare the relative lengths and paths of the uterine tubes (oviducts) of the cat and the human. **2** _____

   _____

   _____

   _____

2. How do the shape and structure of the uterus of the cat compare with that of the human? **2** _____

   _____

   _____

   _____

3. Assess the function of the uterine horns of the cat. **1** _____

   _____

   _____

4. Compare the relationship of the urethra and the vagina in the cat and in the human. **2** _____

   _____

   _____

   _____

Part B Assessments

Complete the following:

1. Compare the glans penis in the cat and in the human. /2\ _____

   _____

   _____

2. How do the location and the relative size of the prostate gland of the cat compare with that of the human? /2\ _____

   _____

   _____

3. What glands associated with the human ductus deferens are missing in the cat? /2\ _____

   _____

4. Compare the prepuce in the cat and in the human. /2\ _____

   _____

   _____

Part C Assessments

Observe the models of the human reproductive system and figure 67.7 of a cadaver. Locate the labeled features and identify the numbered features in figure 67.7 that were also identified in the cat dissection.

FIGURE 67.7   Sagittal (median; midsagittal) plane of a cadaver, (a) female, (b) male. Identify the reproductive features, using the terms provided. /2\ /3\

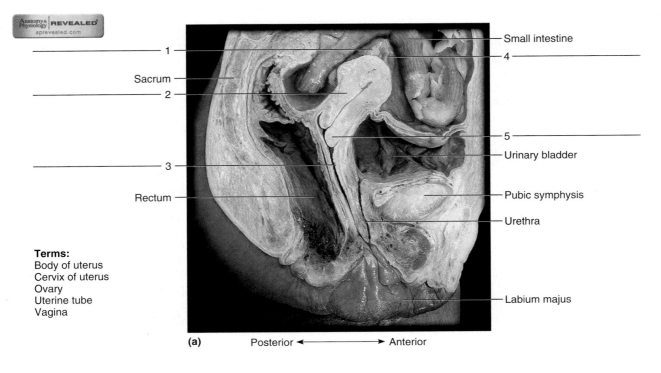

Terms:
Body of uterus
Cervix of uterus
Ovary
Uterine tube
Vagina

(a)   Posterior ◄————► Anterior

578

FIGURE 67.7 *Continued.*

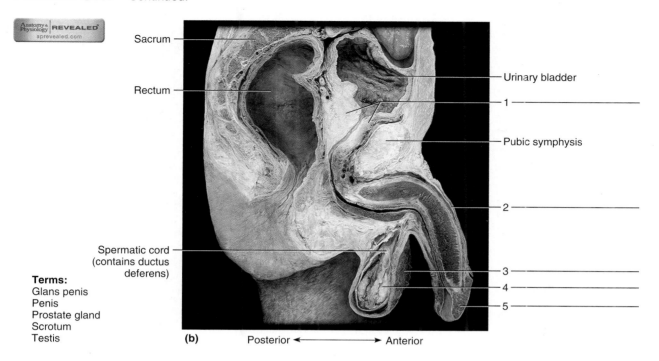

Sacrum

Rectum

Urinary bladder

1

Pubic symphysis

2

Spermatic cord
(contains ductus
deferens)

3
4
5

Terms:
Glans penis
Penis
Prostate gland
Scrotum
Testis

(b)   Posterior ◄─────────► Anterior

Preparation of Solutions

Amylase solution, 0.5%

Place 0.5 g of bacterial amylase in a graduated cylinder or volumetric flask. Add distilled water to the 100 mL level. Stir until dissolved. The amylase should be free of sugar for best results; a low-maltose solution of amylase yields good results. (Store amylase powder in a freezer until mixing this solution.)

Benedict's solution

Prepared solution is available from various suppliers.

Caffeine, 0.2%

Place 0.2 g of caffeine in a graduated cylinder or volumetric flask. Add distilled water to the 100 mL level. Stir until dissolved.

Calcium chloride, 2.0%

Place 2.0 g of calcium chloride in a graduated cylinder or volumetric flask. Add distilled water to the 100 mL level. Stir until dissolved.

Calcium hydroxide solution (limewater)

Add an excess of calcium hydroxide to 1 L of distilled water. Stopper the bottle and shake thoroughly. Allow the solution to stand for 24 hours. Pour the supernatant fluid through a filter. Store the clear filtrate in a stoppered container.

Epsom salt solution, 0.1%

Place 0.5 g of Epsom salt in a graduated cylinder or volumetric flask. Add distilled water to the 500 mL level. Stir until dissolved.

Glucose solutions

1. *1.0% solution.* Place 1 g of glucose in a graduated cylinder or volumetric flask. Add distilled water to the 100 mL level. Stir until dissolved.
2. *10% solution.* Place 10 g of glucose in a graduated cylinder or volumetric flask. Add distilled water to the 100 mL level. Stir until dissolved.

Iodine-potassium-iodide (IKI solution)

Add 20 g of potassium iodide to 1 L of distilled water, and stir until dissolved. Then add 4.0 g of iodine, and stir again until dissolved. Solution should be stored in a dark stoppered bottle.

Methylene blue

Dissolve 0.3 g of methylene blue powder in 30 mL of 95% ethyl alcohol. In a separate container, dissolve 0.01 g of potassium hydroxide in 100 mL of distilled water. Mix the two solutions. (Prepared solution is available from various suppliers.)

Monosodium glutamate (MSG) solution, 1%

Place 1 g of monosodium glutamate in a graduated cylinder or volumetric flask. Add distilled water to the 100 mL level. Stir until dissolved.

Physiological saline solution

Place 0.9 g of sodium chloride in a graduated cylinder or volumetric flask. Add distilled water to the 100 mL level. Stir until dissolved.

Potassium chloride, 5%

Place 5.0 g of potassium chloride in a graduated cylinder or volumetric flask. Add distilled water to the 100 mL level. Stir until dissolved.

Quinine sulfate, 0.5%

Place 0.5 g of quinine sulfate in a graduated cylinder or volumetric flask. Add distilled water to the 100 mL level. Stir until dissolved.

Ringer's solution (frog)

Dissolve the following salts in 1 L of distilled water:

6.50 g sodium chloride
0.20 g sodium bicarbonate
0.14 g potassium chloride
0.12 g calcium chloride

Sodium chloride solutions

1. *0.9% solution.* Place 0.9 g of sodium chloride in a graduated cylinder or volumetric flask. Add distilled water to the 100 mL level. Stir until dissolved.
2. *1.0% solution.* Place 1.0 g of sodium chloride in a graduated cylinder or volumetric flask. Add distilled water to the 100 mL level. Stir until dissolved.
3. *3.0% solution.* Place 3.0 g of sodium chloride in a graduated cylinder or volumetric flask. Add distilled water to the 100 mL level. Stir until dissolved.
4. *5.0% solution.* Place 5.0 g of sodium chloride in a graduated cylinder or volumetric flask. Add distilled water to the 100 mL level. Stir until dissolved.

Starch solutions

1. *0.5% solution.* Add 5 g of cornstarch to 1 L of distilled water. Heat until the mixture boils. Cool the liquid, and pour it through a filter. Store the filtrate in a refrigerator.
2. *1.0% solution.* Add 10 g of cornstarch to 1 L of distilled water. Heat until the mixture boils. Cool the liquid, and pour it through a filter. Store the filtrate in a refrigerator.
3. *10% solution.* Add 100 g of cornstarch to 1 L of distilled water. Heat until the mixture boils. Cool the liquid, and pour it through a filter. Store the filtrate in a refrigerator.

Sucrose, 5% solution

Place 5.0 g of sucrose in a graduated cylinder or volumetric flask. Add distilled water to the 100 mL level. Stir until dissolved.

Wright's stain

Prepared solution is available from various suppliers.

2

Assessments of Laboratory Reports

Many assessment models can be used for laboratory reports. A rubric, which can be used for performance assessments, contains a description of the elements (requirements or criteria) of success to various degrees. The term *rubric* originated from *rubrica terra,* which is Latin for the application of red earth to indicate anything of importance. A rubric used for assessment contains elements for judging student performance, with points awarded for varying degrees of success in meeting the learning outcomes. The content and the quality level necessary to attain certain points are indicated in the rubric. It is effective if the assessment tool is shared with the students before the laboratory exercise is performed.

Following are two sample rubrics that could easily be modified to meet the needs of a specific course. Some of the elements for these sample rubrics may not be necessary for every laboratory exercise. The generalized rubric needs to contain the possible assessment points that correspond to learning outcomes for a specific course. The point value for each element may vary. The specific rubric example contains performance levels for laboratory reports. The elements and the point values could easily be altered to meet the value placed on laboratory reports for a specific course.

Assessment: Generalized Laboratory Report Rubric

Element	Assessment Points Possible	Assessment Points Earned
1. Figures are completely and accurately labeled.		
2. Sketches are accurate, contain proper labels, and are of sufficient detail.		
3. Colored pencils were used extensively to differentiate structures on illustrations.		
4. Matching and fill-in-the-blank answers are completed and accurate.		
5. Short-answer/discussion questions contain complete, thorough, and accurate answers. Some elaboration is evident for some answers.		
6. Data collected are complete, accurately displayed, and contain a valid explanation.		

TOTAL POINTS: POSSIBLE _____ EARNED _____

Assessment: Specific Laboratory Report Rubric

Element	Excellent Performance (4 points)	Proficient Performance (3 points)	Marginal Performance (2 points)	Novice Performance (1 point)	Points Earned
Figure labels	Labels completed with ≥ 90% accuracy.	Labels completed with 80%–89% accuracy.	Labels completed with 70%–79% accuracy.	Labels < 70% accurate.	
Sketches	Accurate use of scale, details illustrated, and all structures labeled accurately.	Minor errors in sketches. Missing or inaccurate labels on one or more structures.	Sketch is not realistic. Missing or inaccurate labels on two or more structures.	Several missing or inaccurate labels.	
Matching and fill-in-the-blanks	All completed and accurate.	One to two errors or omissions.	Three to four errors or omissions.	Five or more errors or omissions.	
Short-answer and discussion questions	Answers are complete, valid, and contain some elaboration. No misinterpretations are noted.	Answers are generally complete and valid. Only minor inaccuracies were noted. Minimal elaboration exists.	Marginal answers to the questions and contains inaccurate information.	Many answers are incorrect or fail to address the topic. There may be misinterpretations.	
Data collection and analysis	Data are complete and displayed with a valid interpretation.	Only minor data missing or a slight misinterpretation exists.	Some omissions. Not displayed or interpreted accurately	Data are incomplete or show serious misinterpretations	

TOTAL POINTS EARNED _____

Appendix 3

Correlation of Laboratory Exercises and Ph.I.L.S. 3.0 Lab Simulations

LABORATORY EXERCISE	Ph.I.L.S. 3.0
Fundamentals of Human Anatomy and Physiology	
1 Scientific Method and Measurements	
2 Body Organization, Membranes, and Terminology	
3 Chemistry of Life	
4 Care and Use of the Microscope	
Cells	
5 Cell Structure and Function	3 Cyanide and Electron Transport System
Ph.I.L.S. 3.0 #2 Size and Basal Metabolic Rate	2 Size and Basal Metabolic Rate
6 Movements Through Cell Membranes	1 Varying Extracellular Concentration
Ph.I.L.S. 3.0 #1 Varying Extracellular Concentration	
7 Cell Cycle	
Tissues	
8 Epithelial Tissues	
9 Connective Tissues	
10 Muscle and Nervous Tissues	
Integumentary System	
11 Integumentary System	
Skeletal System	
12 Bone Structure and Classification	
13 Organization of the Skeleton	
14 Skull	
15 Vertebral Column and Thoracic Cage	
16 Pectoral Girdle and Upper Limb	
17 Pelvic Girdle and Lower Limb	
18 Fetal Skeleton	
19 Joint Structure and Movements	
Muscular System	
20 Skeletal Muscle Structure and Function	4 Stimulus-Dependent Force Generation
Ph.I.L.S. 3.0 #4 Stimulus-Dependent Force Generation	5 The Length-Tension Relationship
	6 Principles of Summation and Tetanus
	7 EMG and Twitch Amplitude
21 Electromyography: BIOPAC Exercise	
22 Muscles of the Head and Neck	
23 Muscles of the Chest, Shoulder, and Upper Limb	
24 Muscles of the Deep Back, Abdominal Wall, and Pelvic Outlet	
25 Muscles of the Hip and Lower Limb	
Surface Anatomy	
26 Surface Anatomy	

Continued

LABORATORY EXERCISE	Ph.I.L.S. 3.0
Fetal Pig Dissection Exercises (similar for cat version)	
62 Fetal Pig Dissection: Musculature	
63 Fetal Pig Dissection: Cardiovascular System	
64 Fetal Pig Dissection: Respiratory System	
65 Fetal Pig Dissection: Digestive System	
66 Fetal Pig Dissection: Urinary System	
67 Fetal Pig Dissection: Reproductive Systems	
Supplemental Laboratory Exercises*	
S-1 Skeletal Muscle Contractions	**6** Principles of Summation and Tetanus
Ph.I.L.S. 3.0 #6 Principles of Summation and Tetanus	
S-2 Nerve Impulse Stimulation	**10** The Compound Action Potential
Ph.I.L.S. 3.0 #10 The Compound Action Potential	
S-3 Factors Affecting the Cardiac Cycle	**18** Thermal and Chemical Effects
Ph.I.L.S. 3.0 #18 Thermal and Chemical Effects	

Appendix 4

Blood Cholesterol Level Determination

Materials Needed

Simulated blood kit for cholesterol determination*

* Kit contains all materials needed and is available from several biological supply companies.

Blood cholesterol is an important component needed for the structure of cellular membranes and the formation of steroid hormones and bile components. However, blood cholesterol higher than recommended levels increases the chances of cardiovascular diseases and risks of heart attacks and strokes. High cholesterol diets, lack of exercise, and heredity are among the contributing factors increasing the cholesterol levels. Low cholesterol diets, an increase in exercise, and various medications are some ways used to lower the levels of cholesterol.

The recommended total cholesterol level is less than 200 mg/dL (dL = deciliter), which decreases the probability of cardiovascular diseases. Total blood cholesterol of less than 100 (hypocholesterolemia) is associated with possible conditions as malnutrition, hyperthyroidism, and depression. Since cholesterol is not water soluble, it is transported in the plasma incorporated with a protein as a low-density lipoprotein (LDL) and a high-density lipoprotein (HDL). The **HDL**s are transported to the liver where they are degraded and removed from the body, and are therefore sometimes referred to as the "**H**ealthy/good" form. The LDLs are being transported to our tissue cells for cellular needs, but if the levels are excessive there is an increased occurrence of atherosclerosis. **LDL**s are sometimes referred to as the "**L**ousy/bad" cholesterol. The recommended levels of HDLs are > 40 mg/dL; recommended levels of LDLs are < 130 mg/dL.

Purpose of the Exercise

To determine the total cholesterol in simulated blood samples and the influence that medications or diets have on cholesterol levels.

Learning Outcomes

After completing this exercise, you should be able to

1. Test the total cholesterol level of simulated blood sample before and after a cholesterol reduction program.

2. Analyze the results of medications or diets on total cholesterol levels with a cholesterol reduction program.

Procedure—Blood Cholesterol Level Determination

Each kit contains cholesterol samples of simulated blood, test strips, color charts, and directions for the particular kit being used. Some available kits contain samples after using cholesterol-lowering medications while others might contain samples after diets of various cholesterol levels. Dip a test strip into a sample, then compare the strip to a color chart to determine the cholesterol level of the sample.

What was the cholesterol level of the sample before any cholesterol reduction program? 1 _____

What was the cholesterol level of each sample after a cholesterol reduction program was done? 1 _____

_____

_____

Assess the most beneficial method of lowering blood cholesterol from the samples provided. 2 _____

_____

_____

Credits

Lab Exercise 1
1.2: © McGraw-Hill Companies/ J. Womack Photography; **1.3:** © McGraw-Hill Companies/ J. Womack Photography

Lab Exercise 2
2.4: © McGraw-Hill Companies/ J. Womack Photography; **2.5:** © McGraw-Hill Companies/ J. Womack Photography; **2.6(1)–(4):** © McGraw-Hill Education, Inc./ Joe De Grandis, photographer

Lab Exercise 4
4.1: © McGraw-Hill Companies/ J. Womack Photography; **4.2:** © Stockbyte/Getty Images; **4.3(both), 4.4, 4.7:** © McGraw-Hill Companies/J. Womack Photography

Lab Exercise 5
5.3: © Ed Reschke; **5.5(both):** © Richard Rodewald/Biological Photo Service

Lab Exercise 6
6.1: © McGraw-Hill Companies/ Womack Photography, Ltd.; **6.6a–c:** © Dr. David M. Phillips/Visuals Unlimited

Lab Exercise 7
7.2(1)–(4), 7.4: © Ed Reschke; **7.5a:** Courtesy of the March of Dimes, Birth Defects Foundation; **7.5b:** CNRI/Photo Researchers, Inc.; **7.6a–d:** © Ed Reschke

Lab Exercise 8
8.1a,b: © McGraw-Hill Companies/ Al Telser, photographer; **8.1c:** © Manfred Kage/Peter Arnold; **8.1d:** © McGraw-Hill Companies/ Dennis Strete, photographer; **8.1e,f:** © McGraw-Hill Companies/ Al Telser, photographer; **8.2a–c:** © McGraw-Hill Companies/ Womack Photography, Ltd.

Lab Exercise 9
9.1a: © McGraw-Hill Companies/ Dennis Strete, photographer; **9.1b:** © Ed Reschke; **9.1c:** © McGraw-Hill Companies/Al Telser, photographer; **9.1d,e:** © McGraw-Hill Companies/ Dennis Strete, photographer; **9.1f:** © John D. Cunningham/Visuals Unlimited; **9.1g–i:** © McGraw-Hill Companies/Al Telser, photographer; **9.1j:** © McGraw-Hill Companies/ Dennis Strete, photographer; **9.1k:** © McGraw-Hill Companies/ Al Telser, photographer

Lab Exercise 10
10.1a–c: © McGraw-Hill Companies/Al Telser, photographer; **10.1d:** © Ed Reschke

Lab Exercise 11
11.2: © McGraw-Hill Companies/ Dennis Strete, photographer; **11.3a:** © Per H. Kjeldsen, University of Michigan, Ann Arbor; **11.3b:** © McGraw-Hill Companies/ Dennis Strete, photographer; **11.3c:** © Carolina Biological Supply Company/Phototake

Lab Exercise 12
12.3: © Victor B. Eichler; **12.4:** Courtesy of Utah Valley Regional Medical Center, Department of Radiology; **12.5a,b, 12.6a,b:** © McGraw-Hill Companies/ J. Womack Photography

Lab Exercise 13
13.2: © McGraw-Hill Companies/ J. Womack Photography

Lab Exercise 14
14.6–14.10: © McGraw-Hill Companies/J. Womack Photography

Lab Exercise 15
15.2a,b, 15.3a–c: © McGraw-Hill Companies/J. Womack Photography; **15.6:** © Dr. Kent M. Van de Graaff

Lab Exercise 16
16.6: © Martin M. Rotker; **16.7:** Courtesy of Eastman Kodak; **16.8:** © Martin M. Rotker; **16.9:** © McGraw-Hill Companies/ J. Womack Photography

Lab Exercise 17
17.6–17.9: © Martin M. Rotker

Lab Exercise 18
18.1–18.3, 18.4a,b, 18.5, 18.6: © McGraw-Hill Companies/ Womack Photography, Ltd.

Lab Exercise 19
19.4a–h, 19.5a–g: © McGraw-Hill Companies/Womack Photography, Ltd.

Lab Exercise 20
20.1: © Ed Reschke; **20.7:** © H. E. Huxley

Lab Exercise 22
22.6: © The McGraw-Hill Companies, Inc./APR; **22.7a–c:** © McGraw-Hill Companies/ J. Womack Photography

Lab Exercise 23
23.6, 23.7: © The McGraw-Hill Companies, Inc./APR; **23.8a–c:** © McGraw-Hill Companies/ J. Womack Photography

Lab Exercise 24
24.5: © The McGraw-Hill Companies, Inc./APR

Lab Exercise 25
25.7: © The McGraw-Hill Companies, Inc./APR; **25.8a,b:** © McGraw-Hill Companies/ J. Womack Photography

Lab Exercise 26
26.1a: © McGraw-Hill Companies/ Joe De Grandis, photographer; **26.1b,c:** © McGraw-Hill Companies/ Eric Wise, photographer; **26.2:** © Suza Scalora/PhotoDisc/ Getty Images; **26.3a:** © McGraw-Hill Companies/Joe De Grandis, photographer; **26.3b:** © Suza Scalora/PhotoDisc/Getty Images; **26.4, 26.5a,b:** © McGraw-Hill Companies/Joe De Grandis, photographer; **26.6:** © McGraw-Hill Companies/Eric Wise, photographer; **26.7, 26.8a–c:** © McGraw-Hill Companies/ Joe De Grandis, photographer; **26.9a,b:** © McGraw-Hill Companies/Eric Wise, photographer

Lab Exercise 27
27.3–27.5: © Ed Reschke; **27.6:** © Ed Reschke/Peter Arnold; **27.7:** © Ed Reschke

Lab Exercise 28
28.4: © The McGraw-Hill Companies, Inc./APR; **28.6:** © Per H. Kjeldsen, University of Michigan, Ann Arbor

Lab Exercise 29
29.2a–e: © McGraw-Hill Companies/J. Womack Photography

Lab Exercise 30
30.1, 30.5: © The McGraw-Hill Companies, Inc./APR; **30.6:** © Martin Rotker/Visuals Unlimited

Lab Exercise 32
32.2, 32.5: © McGraw-Hill Companies/J. Womack Photography

Lab Exercise 33
33.1, 33.2: © Ed Reschke

Lab Exercise 34
34.2: © Dwight Kuhn; **34.4:** © Ed Reschke

Lab Exercise 35
35.4: © Per H. Kjeldsen, University of Michigan, Ann Arbor; **35.5:** © McGraw-Hill Companies/ J. Womack Photography; **35.7:** © Lisa Klancher; **35.9, 35.10:** © McGraw-Hill Companies/ J. Womack Photography; **35.11a,b:** © Carroll Weiss/Camera M. D. Studios

Lab Exercise 36
36.3: © McGraw-Hill Companies/ J. Womack Photography

Lab Exercise 37
37.4: © McGraw-Hill Companies/ J. Womack Photography; **37.5:** © Biophoto Associates/Photo Researchers, Inc.; **37.6a,b, 37.7:** © McGraw-Hill Companies/ J. Womack Photography; **37.8:** © McGraw-Hill Companies/ Womack Photography, Ltd.; **37.9:** © John D. Cunningham/Visuals Unlimited

Lab Exercise 38
38.2, 38.4: © McGraw-Hill Companies/Womack Photography, Ltd.; **38.5:** © John D. Cunningham/ Visuals Unlimited

Lab Exercise 39
39.2: © Biophoto Associates/ Photo Researchers, Inc.; **39.3:** © John D. Cunningham/Visuals Unlimited; **39.4:** © Science VU/Visuals Unlimited; **39.6:** © Ed Reschke; **39.8:** © Biophoto Associates/Photo Researchers, Inc.; **39.10, 39.12:** © Ed Reschke

Lab Exercise 40
40.1, 40.2: © McGraw-Hill Companies/Al Telser, photographer; **40.3:** © Dr. F. C. Skvara/Visuals Unlimited

Lab Exercise 41
41.2(all): © McGraw-Hill Companies/Al Telser, photographer

Lab Exercise 42
42.2a–d: © McGraw-Hill Companies/ Photo by James Shaffer; **42.3a,b:** © McGraw-Hill Companies/ Womack Photography, Ltd.; **42.4:** © McGraw-Hill Companies/ J. Womack Photography

Lab Exercise 43
43.2: © Jean-Claude Revy/ISM/ Phototake

Lab Exercise 44

44.6, 44.7: © McGraw-Hill Companies/J. Womack Photography; **44.8:** © The McGraw-Hill Companies, Inc./APR; **44.9:** © McGraw-Hill Companies/J. Womack Photography

Lab Exercise 47

47.1: © Biophoto Associates/Photo Researchers, Inc.

Lab Exercise 49

49.4a: © McGraw-Hill Companies/Womack Photography, Ltd.; **49.4b:** © John Watney/Photo Researchers, Inc.; **49.5:** © John Cunningham/Visuals Unlimited; **49.6, 49.7:** © Biophoto Associates/Photo Researchers, Inc.

Lab Exercise 50

50.4: © Collection CNRI/Phototake; **50.5:** © Ed Reschke; **50.6:** © Biophoto Associates/Photo Researchers, Inc.; **50.8:** © imagingbody.com; **50.9:** © Dwight Kuhn; **50.10:** © The McGraw-Hill Companies, Inc./APR

Lab Exercise 51

51.1: © McGraw-Hill Companies/J. Womack Photography; **51.2:** © McGraw-Hill Companies/Womack Photography, Ltd.; **51.3:** © McGraw-Hill Companies/J. Womack Photography

Lab Exercise 54

54.4: © McGraw-Hill Companies/Dennis Strete; **54.5, 54.7:** © Ed Reschke; **54.11:** © McGraw-Hill Companies/J. Womack Photography; **54.12:** © Dennis Strete; **54.13:** © Ed Reschke/Peter Arnold; **54.14, 54.15:** © The McGraw-Hill Companies, Inc./APR

Lab Exercise 56

56.2: © McGraw-Hill Companies/J. Womack Photography; **56.5a:** © Biophoto Associates/Photo Researchers, Inc.; **56.5b:** © Manfred Kage/Peter Arnold

Lab Exercise 58

58.3: © Biophoto Associates/Photo Researchers, Inc.; **58.4, 58.5:** © Ed Reschke; **58.6:** © Michael Peres

Lab Exercise 59

59.5: © Ed Reschke/Peter Arnold; **59.6:** © Ed Reschke; **59.7, 59.8:** © Michael Peres

Lab Exercise 60

60.3a,b: © K. Benirschke

Lab Exercise 61

61.1(all): © McGraw-Hill Companies/J. Womack Photography

Lab Exercise 62

62.2–62.22: Courtesy of Dr. Sheril Burton; **62.23:** © McGraw-Hill Companies/Photo Ralph W. Stevens III, Ph.D

Lab Exercise 63

63.2: Courtesy of Dr. Sheril Burton; **63.3b:** © The McGraw-Hill Companies, Inc./APR; **63.4–63.10:** Courtesy of Dr. Sheril Burton; **63.14:** © The McGraw-Hill Companies, Inc./APR

Lab Exercise 64

64.1: Courtesy of Dr. Sheril Burton; **64.3b, 64.4:** © The McGraw-Hill Companies, Inc./APR

Lab Exercise 65

65.1b: © The McGraw-Hill Companies, Inc./APR; **65.4:** Courtesy of Dr. Sheril Burton; **65.5:** © The McGraw-Hill Companies, Inc./APR

Lab Exercise 66

66.1: Courtesy of Dr. Sheril Burton; **66.3:** © McGraw-Hill Companies/Photo Ralph W. Stevens III, Ph.D; **66.4b, 66.5:** © The McGraw-Hill Companies, Inc./APR

Lab Exercise 67

67.2, 67.4: © McGraw-Hill Companies/Photo Ralph W. Stevens III, Ph.D; **67.7a,b:** © The McGraw-Hill Companies, Inc./APR

Index

chambers
  of eye, 289, 291
  of heart, 367, 368f, 370, 370f, 373f, 374f
    in cat, 538f, 539–540, 539f, 540f
    in circulation, 398f
    in sheep, 371–372, 371f, 372f
cheek, of cat, 558
cheek cells, 41, 41f
chemistry, of life, 21–27
chemoreceptors, 281
Chemstrip, 481
chest, muscles of, 195–204, 196f, 203f–204f
chicken bone, 96, 96f
chief cells, 328, 329f, 456, 457f
chordae tendineae, 370, 370f, 374f
  in sheep, 372
chorion, 499, 501, 501f
chorionic villi, 501, 501f
choroid, 289, 290f, 291, 291f, 293, 294f, 295f
choroid plexuses, 254
chromaffin cells, 330f
chromatids, 62f, 63, 63f
chromatin, 39, 47f
chromosomes, 499, 505
  diploid number of, 505
  haploid number of, 505
  homologous, 505
  in meiosis, 505
  microscopy of, 63, 63f, 64f
  in mitosis, 62f
  sex (X and Y), 509
chyme, 453
cilia
  olfactory, 281–283, 282f, 283f
  respiratory system, 421, 424f
  uterine tube, 491
ciliary body, 289, 290f, 291, 293, 294f, 300–301
ciliary muscles, 291
ciliary processes, 291
circulation
  lymphatic, 415
  pulmonary, 395, 397, 398f
  systemic, 395, 397, 398f
circumcision, 485
circumduction, 159f, 161t
circumflex artery, 370
cisterna chyli, 416, 416f
clavicle, 100f, 102, 105f, 124f, 127, 128f, 129, 134f
  acromial end of, 127
  sternal end of, 127
  surface anatomy of, 224f, 230f
clavicular notch, 123, 124f
clavodeltoid (clavobrachialis) muscle, in cat, 523, 523f, 524f, 525t, 526f, 527f
clavotrapezius muscle, in cat, 523, 523f, 524f, 525t, 526f
cleidomastoid muscle, in cat, 518f, 519, 519f, 519t
clitoris, 491, 493f, 494
  in cat, 572
clotting. See coagulation

clotting time, 354–355, 355f
coagulation, 351–352
  deficiencies of, 352
  process of, 352
  testing of, 354–355, 355f
coagulation time, 354–355, 355f
coarse adjustment knob, of microscope, 31f, 32
coccygeus muscle, 205, 208f
coccyx, 99, 101f, 119, 120f, 123, 123f, 137, 491
cochlea, 309, 310f, 311, 311f, 312, 312f, 315f, 316f
cochlear duct, 310f, 312, 312f, 316f
cochlear nerve, 309, 310f, 311, 311f
codominance, 505
cold receptors, 275, 277
colic flexure
  left, 458f, 459
  right, 458f, 459
collagen
  in bones, 93
  in connective tissue, 75
  in skeletal muscle, 167
collecting duct, 471, 474f, 475f
colliculi, in sheep, 270f, 271, 272
colon
  ascending, 458f, 459, 563f
  in cat, 542f, 559, 560f, 573f, 576f
  descending, 458f, 459, 462f
  sigmoid, 458f, 459, 462f
  transverse, 458f, 459, 563f
color blindness, 299, 301, 302f, 509, 510
color vision
  genetics of, 506t, 509, 510
  test of, 301, 302f
columnar epithelium
  pseudostratified, 69, 70f, 424, 488, 488f
  simple, 69, 70f, 494
common bile duct, 457
common carotid artery, 368f, 370f, 373f, 398, 399f, 408f
  in cat, 538f, 539f, 540, 540f, 541f
common hepatic duct, 457, 457f
common iliac artery, 398, 399, 400f, 550f
common iliac vein, 401, 402f, 550f
  in cat, 546, 547f
compact bone, 76t, 78f, 94, 94f, 95, 95f, 98f
complete dominance, 505, 509–510
compound light microscope, 29, 30f
computer-based data. See BIOPAC exercises
conclusions, 1
condenser, of microscope, 29, 30, 30f
conduction system, heart, 377, 379f
condylar canal, 109f
condylar joint, 156, 156f, 157t
condyles, 102
  of femur, 139f, 140, 144f, 158f
  occipital, 108, 109f, 117f
  of tibia, 140, 140f
cones, 289, 291

conjunctiva, 289, 291, 292, 295f
connective tissue, 69, 75–81
  adipose, 76t, 77f
  areolar, 76t, 77f
  blood as, 76t, 78f, 343
  bone as, 76t, 78f
  dense, 76t
  dense irregular, 76t, 77f, 89f
  dense regular, 76t, 77f
  elastic, 76t, 77f
  functions of, 75
  locations of, 75, 76t
  loose, 76t
  micrographs of, 77f–78f
  in respiratory system, 424, 424f
  reticular, 76t, 77f
  in skeletal muscle, 167, 168
  structure of, 75
  types of, 76t, 77f–78f
conus medullaris, 241
convergence reflex, 303
coracobrachialis muscle, 196, 196f, 198f
  in cat, 525f, 526f, 527, 528t
coracoid process, 127, 128f
cornea, 289, 290f, 291, 292, 294f, 295f, 299
corniculate cartilage, 423, 423f
coronal (frontal) plane, 9, 15f
coronal suture, 107, 108f, 109f, 110f, 116f
corona radiata, 495f, 498f
coronary arteries, 398
  in cat, 539, 539f, 541f
  left, 370
  openings of, 369f, 374f
  right, 368f, 370
  in sheep, 371, 372
coronary sinus, 368f, 370
  opening of, 370f, 374f
  in sheep, 372
coronary sulcus, 368f, 370
coronoid fossa, 129, 130f
coronoid process
  of ulna, 129, 130f
  of vertebrae, 108, 109f
corpora cavernosa, 485, 486f, 487, 488, 488f, 491
  in cat, 574
corpora quadrigemina, 254
  in sheep, 270f, 271, 272
corpus albicans, 491, 498f
corpus callosum, 253, 255f, 260f
  in sheep, 271, 272, 272f
corpus luteum, 491, 498f
corpus spongiosum, 485, 486f, 487, 488, 488f
corrective lenses, 299
Corti, organ of (spiral organ), 309, 311f, 312, 312f, 316f
cortical nephrons, 471, 472, 474f
cortical radiate artery, 474f
cortical radiate vein, 474f
corticomedullary junction, 474f
corticospinal tract
  anterior, 243f
  lateral, 243f
corticosteroids, 328

cortisol, 328
costal cartilage, 119, 123, 124f, 128f
Coumadin (warfarin), 354
coxa, 102, 137, 138f
coxal region, 17f
cranial bones. See skull
cranial cavity, 9, 12f, 110f, 118f
cranial nerves, 233, 253, 256–261
  in breathing control, 447–448
  mnemonic for, 257
  names and designations of, 253, 260f
  photograph of, 257f
  in sheep, 271–272, 271f
crenation, 50, 53–54
crest, of bone, 102
cribriform foramina, 110f
cribriform plate, 108, 118f, 282f
cricoid cartilage, 328f, 423, 423f, 556f
  in cat, 551, 553f
cricothyroid ligament, 423, 423f
  in cat, 554f
crista ampullaris, 319, 321, 321f, 322f
crista galli, 108, 110f, 118f
cross-matching, of blood, 361
cross section, 71, 71f
crown, of tooth, 454, 454f
crown-to-heel fetal height, 148, 149f, 150f
crown-to-rump fetal measurement, 148, 149t, 150f
cruciate ligament
  anterior, 158f
  posterior, 158f
crural region, 17f
CSF (cerebrospinal fluid), 233, 241
cubital fossa, 226f
cubital vein, median, 401, 401f
cuboidal epithelium, simple, 69, 70f
cuboid bone, 140, 141f, 145f
cuneate fasciculus, 243f
cuneiform
  intermediate (middle), 140, 141f, 145f
  lateral, 140, 141f, 145f
  medial, 140, 141f, 145f
cuneiform cartilages, 423
cupula, 319, 321f, 322f
curvatures
  stomach, 456, 456f
  vertebral, 119, 120f, 225f
cuspids, 454, 454f
cystic duct, 457, 457f
  in cat, 559
cytokinesis, 61–67, 61f, 62f, 499
cytoplasm, 39, 41f, 47f
cytoplasmic organelles, 39
cytoskeleton, 39
cytosol, 39

D

D antigen, 361–362, 364
data, analysis of, 1, 3
daughter cells, 61, 62f, 66
dead space, anatomic, 429, 434
deafness, tests for, 312–313, 313f

decibel (dB), 309
decidua basalis, 499, 501*f*
decussation, 241
deep artery of thigh, 399, 400*f*
deep brachial artery, 398, 399*f*
deep femoral artery, 399, 400*f*
  in cat, 543*f*, 544*f*, 545, 545*f*
deep femoral vein, in cat, 546, 547*f*
deep muscles of back, 205–206,
    206*f*, 209*f*
delta cells, 330, 330*f*, 338
delta waves, of EEG, 264, 265*f*
deltoid muscle, 196, 196*f*, 197*f*,
    203*f*, 204*f*, 222*f*,
    224*f*, 225*f*
deltoid tuberosity, 129, 130*f*
demilunes, 455, 455*f*
dendrites, 233, 234*f*, 235, 235*f*
dense (compact) bone, 76*t*, 78*f*, 94,
    94*f*, 95, 95*f*, 98*f*
dense connective tissue, 76*t*
  irregular, 76*t*, 77*f*, 89*f*
  regular, 76*t*, 77*f*
dens of axis, 121*f*, 123
denticulate ligaments, 241, 244*f*
dentin, 454, 454*f*
deoxyhemoglobin, 351–352
depression (movement), 160*f*, 161*t*
depressions, of bones, 99, 102
depth of field, of microscope, 33
dermal papilla, 88*f*, 89*f*
dermis, 87, 88*f*
  papillary region of, 89*f*
  reticular region of, 89*f*
descending colon, 458*f*, 459, 462*f*
  in cat, 542*f*, 559, 560*f*
descending (motor) tracts, 241, 243*f*
deuteranopia, 301, 302*f*
development, early or embryonic,
    499, 500*f*, 501–504, 501*f*
diabetes mellitus, 337–342
  insulin-dependent, 337
  insulin shock in, 338
  non-insulin-dependent, 337–338
  risk factors for, 338
  type 1, 337
  type 2, 337–338
diaphragm, 407, 421, 429, 447,
    449*f*, 462*f*, 556*f*
  in cat, 538*f*–540*f*, 542*f*–543*f*,
    546*f*–547*f*, 552, 553*f*,
    559*f*–560*f*, 560*f*
diaphragm of pelvis, 205, 208*f*
diaphysis, 94*f*, 95, 98*f*
diarthroses, 155, 157*t*
diastole, 378, 407
diastolic pressure, 407, 409, 409*f*
diencephalon, 254, 255*f*, 260*f*
  in sheep, 272
differential white blood cell count,
    346, 346*f*, 346*t*
differentiation, 61
diffusion, 49, 50
  extracellular concentrations and,
    49, 53–54, 53*f*, 59–60
  facilitated, 50
  rate of, 50
  simple, 50, 51*f*

digastric muscle, 188, 190*f*
  in cat, 518–519, 518*f*, 519*f*,
    519*t*, 558*f*
digestive system, 13
  cat, 557–563
  enzymes of, 453, 465–470
    pH and, 465–466
    temperature and, 465, 467
  organs of, 453–464, 462*f*, 563*f*
dimples/no dimples, 506, 506*t*,
    508*f*, 510
diploid number, of
    chromosomes, 505
directional terms, 9, 14*f*
dissect, definition of, 515
dissecting microscope, 35, 36*f*
dissection
  cat
    cardiovascular system,
      537–550
    digestive system, 557–563
    musculature, 515–536
    respiratory system, 551–556
    urinary system, 565–570
  eye, 292–294, 293*f*, 294*f*
  pig kidney, 472, 473*f*
  sheep
    brain, 269–274
    heart, 371–372, 371*f*, 372*f*
    kidney, 472
  techniques of, 515
distal, 9, 14*f*
distal convoluted tubule, 471, 474*f*
distal epiphysis, 94*f*, 95, 95*f*, 98*f*
distal interphalangeal joints,
    226*f*, 231*f*
distal phalanx
  foot, 140, 141*f*, 145*f*
  hand, 129, 131*f*, 135*f*, 136*f*
dizziness (vertigo), 319, 321–322
DNA, in cell nucleus, 39
dominant trait, 506–509, 506*t*,
    507*f*–508*f*
dorsal artery of foot, 399,
    400*f*, 408*f*
dorsal body cavity, 9, 10*f*
dorsal column, 243*f*
dorsal (posterior) horn, 241, 242*f*,
    244, 244*f*
dorsalis pedis artery, 399,
    400*f*, 408*f*
dorsal respiratory group,
    447–448, 449*f*
dorsal root, of spinal nerves,
    241, 242*f*, 243*f*, 244,
    244*f*, 246*f*
dorsal root ganglion, 236, 236*f*,
    241, 242*f*–244, 246*f*
dorsiflexion, 160*f*, 161*t*
dorsum region, 18*f*
ductus deferens, 485, 486*f*,
    487, 579*f*
  in cat, 574, 574*f*, 575*f*, 576*f*
duodenal papilla, 457*f*, 462*f*
  major, 457*f*
  minor, 457*f*
duodenum, 457*f*, 458*f*, 459
  in cat, 559, 560*f*

dupp sound, of heart, 378,
    378*f*, 380
dural venous sinus, 401, 401*f*
dura mater
  brain, 254*f*
    in sheep, 269
  spinal cord, 241, 243*f*, 244, 244*f*
dynagram, 176
dynamic equilibrium, 319, 321*f*
dynamometer, hand, 176

E

ear(s), 309–318
  bones of, 99, 107, 309, 310*f*,
    311, 311*f*
  external, 309, 310*f*, 311
  functions of, 309. *See also*
    equilibrium; hearing
  inner, 309, 310*f*, 311
  middle, 309, 310*f*, 311
  structure of, 309–312, 310*f*,
    311*f*, 312*f*, 315*f*, 316*f*
eardrum, 309, 310*f*, 311
earlobe, free *vs.* attached, 506,
    506*t*, 508*f*
early development, 499, 500*f*,
    501–504, 501*f*
early proliferative phase, of female
    reproductive cycle, 496
ear stones (otoliths), 319, 320*f*
eccrine sweat gland, 87, 88*f*, 89*f*
ECG. *See* electrocardiography
ectoderm, 499, 501
edema, 415
EEG. *See* electroencephalogram
efferent arteriole, 474*f*
efferent (motor) neurons, 167, 176,
    233–236, 234*f*, 235*f*, 247
Einthoven's triangle, 380
ejaculatory ducts, 485, 486*f*, 487
EKG. *See* electrocardiography
elastic cartilage, 76*t*, 78*f*
elastic connective tissue, 76*t*, 77*f*
elastic fibers, in connective
    tissue, 75
elastic recoil, in respiration, 429
elbow (olecranon process),
    129, 130*f*, 134*f*, 225*f*,
    226*f*, 231*f*
elbow joint, 156
electrocardiogram, 377
electrocardiography (ECG, EKG),
    377, 379–380
  after exercise, 389, 390, 390*f*
  baseline, 387
  BIOPAC exercise, 387–393
  calibration in, 388, 389*f*
  components of, 379*f*, 379*t*, 387,
    388*f*, 388*t*
  data analysis in, 390
  durations of, 379*t*
  electrode placement for, 379–380,
    379*f*, 387, 388, 389*f*
  findings in, significance of, 379*t*
  heart sounds and, 380–381, 381*f*
  lead values in, 387, 388*t*
  limb leads for, 380, 380*f*,
    388, 389*f*

lying down, 389, 389*f*, 390
  normal, 379*f*
  recording in, 389
  setup for, 388
  sitting up, 389, 389*f*, 390
electroencephalogram (EEG),
    263–268
  calibration in, 264, 264*f*
  date analysis in, 265–266
  recording in, 264–265, 265*f*
  setup for, 264
  waves or rhythms of, 263–264,
    265*f*, 266*f*
electromyogram, 175
electromyography (EMG),
    175–185
  calibration of, 178, 178*f*,
    180, 180*f*
  data analysis in, 178–179,
    181–182
  electrode placement for,
    177, 178*f*
  I-beam measurement in,
    179, 179*f*
  integrated, 176–177
  of motor unit recruitment,
    179–182, 181*f*
  of muscle fatigue, 179–182, 181*f*
  recording in, 178, 179*f*, 180–181
  setup for, 176–177, 180
  standard, 176–177
elevation, 160*f*, 161*t*
ellipsoid (condylar) joint, 156,
    156*f*, 157*t*
embryoblast (inner cell mass), 499,
    500*f*, 501
embryonic development, 499, 500*f*,
    501–504, 501*f*
EMG. *See* electromyography
emmetropia, 299
emphysema, 425
enamel, tooth, 454, 454*f*
endocardium, 369, 369*f*, 370*f*
endocrine glands, 325
  function of, 326–327, 330–331
  histology of, 326–330
  negative feedback control of, 326
endocrine system, 13, 325–336
endoderm, 499, 501
endolymph, 310*f*, 311*f*, 321, 321*f*
endolymphatic sac, 310*f*
endometrium, 491, 492, 493*f*, 494,
    496, 496*f*, 499, 500
endomysium, 167, 169*f*
endoneurium, 233, 236, 236*f*
endoplasmic reticulum, 39, 47*f*
  rough, 39, 40*f*
  smooth, 40*f*
endosteum, 94*f*, 95
enzymes, 22
  digestive, 453, 465–470
    pH and, 465–466
    temperature and, 465, 467
  lock-and-key model of,
    466, 466*f*
  pancreatic, 453, 465–466
eosinophils, 344, 345*f*, 346*t*
ependymal cells, 233

heart, *continued*
    sounds of, 377, 378, 378*f*,
        380–381, 381*f*
    structure of, 367–376, 368*f*,
        369*f*, 370, 373*f*, 374*f*
    valves of, 367, 369*f*, 370, 370*f*,
        374*f*, 377, 378, 378*f*
    wall of, 369*f*
heart murmur, 377
heart rate, 377, 410
heat receptors, 275, 277
height
    fetal, 148, 149*f*, 149*t*, 150*f*
    measurement of, 3–4, 4*f*
    and vital capacity, 431, 432*t*,
        433*t*, 441
hematocrit, 351–353, 353*f*
hematology analyzer, 343,
    355, 355*f*
hemispheres, cerebral, 253,
    254, 254*f*
    in sheep, 269, 270*f*, 271
hemoglobin, 343, 351–354
    pH and, 352, 355–356, 356*f*
    saturation of, 352
    structure of, 352
    testing for, 353–354, 354*f*
    in urine, 479, 481
hemoglobinometer, 353–354, 354*f*
hemolysis, 53–54
hemolytic disease of newborn, 362
hemostasis, 343, 352
Henle, loop of (nephron loop),
    471, 474*f*
heparin, 354
hepatic artery, in cat, 542, 544*f*
hepatic ducts, 457, 457*f*
    in cat, 559
hepatic (colic) flexure
    left, 458*f*, 459
    right, 458*f*, 459
hepatic portal vein, 401, 402*f*
    in cat, 546, 548*f*
hepatic vein, 401, 402*f*
hepatopancreatic ampulla, 457*f*
hepatopancreatic sphincter,
    457, 457*f*
hertz (Hz), 309
heterozygous, 505
hilum
    of kidney, 472
        in cat, 567, 568*f*
    of lung, 421, 422*f*, 423
hindlimb muscles, in cat, 528–533,
    529*f*, 530*f*, 531*f*,
    532*f*, 533*t*
hinge joint, 156, 156*f*, 157*t*
hip(s)
    bones of, 99, 100*f*, 102, 105*f*,
        137, 138*f*
    joint, 156, 159*f*
    muscles of, 211–219,
        212*f*–215*f*, 217*f*
        in cat, 528–533, 533*t*
homeostasis, endocrine regulation
    of, 325
homologous chromosomes, 505
homozygous, 505

horizontal acceleration, 319
horizontal (transverse) plane, 9, 15*f*
hormones, 325
    functions of, 325
    secretion of, 325
        adrenal, 328
        pancreatic, 330, 338
        pituitary, 326–328
        thyroid, 328
horns, spinal, 241, 242*f*, 244, 244*f*
humeroscapular joint, 156*f*
humeroulnar joint, 156*f*
humerus, 94, 101*f*, 102, 105*f*, 127,
    128*f*, 129, 130*f*, 134*f*
    anatomical neck of, 129, 130*f*
    distal features of, 129
    head of, 129, 130*f*, 134*f*
    proximal features of, 129
    shaft of, 129
    surface anatomy of, 225*f*
    surgical neck of, 129, 130*f*
hyaline cartilage, 76*t*, 77*f*
hydrochloric acid, in gastric juice,
    453, 456
hydrogen ions, in pH, 22–23, 22*f*
hydrometer, 480, 481*f*
hydrostatic pressure, 50
hydroxyapatite, 93
hymen, 491, 493*f*, 494
hyoglossus muscle, in cat, 519,
    519*f*, 519*t*
hyoid bone, 99, 100*f*, 107, 423*f*
    in cat, 554*f*
    muscles moving, 188, 190*f*
    surface anatomy of, 222*f*
hyperextension, 158*f*, 161*t*
hyperglycemia, 337
hyperopia, 299
hypertension, 407
hypertonic solutions, 49, 50, 52,
    53–54, 57*f*, 59–60
hyperventilation, effect of,
    449–450
hypochondriac regions, left and
    right, 16*f*
hypoglossal canal, 110*f*
hypoglossal (CN XII) nerve, 257,
    257*f*, 260*f*
    in sheep, 271*f*, 272
hypoglycemia, 337
hypophysis. *See* pituitary gland
hypotension, 407
hypothalamus, 254, 255*f*, 260*f*,
    325, 326, 327*f*
    in sheep, 272, 272*f*
hypothesis, 1, 3
hypotonic solutions, 49, 50, 52,
    53–54, 57*f*, 59–60
H zone, 168

I

I band, 168
I-beam measurement, 179, 179*f*
Ichikawa's test for colors, 301
ileocecal junction, 458*f*, 462*f*
ileocecal valve (sphincter), 453,
    458*f*, 459
    in cat, 559

ileum, 458*f*, 459
    in cat, 559, 560*f*
iliac artery
    in cat, 542, 543*f*, 544*f*, 545*f*, 566*f*
    common, 398, 399, 400*f*, 550*f*
    external, 399, 400*f*
    internal, 399, 400*f*
iliac crest, 137, 138*f*, 223*f*, 224*f*,
    230*f*, 231*f*
iliac fossa, 137, 138*f*
iliac spine
    anterior superior, 137, 138*f*,
        212*f*, 224*f*, 227*f*
    posterior superior, 137, 138*f*, 223*f*
iliacus muscle, 211, 212*f*
iliac vein
    in cat, 545*f*, 546, 547*f*
    common, 401, 402*f*, 550*f*
    external, 401, 402*f*
    internal, 401, 402*f*
iliocostalis muscle, 205, 206*f*, 209*f*
iliolumbar arteries, in cat, 542, 543*f*
iliolumbar vein, in cat, 546, 547*f*
iliopsoas muscles, 211, 213*f*
iliotibial tract, 212*f*, 214*f*,
    228*f*, 231*f*
ilium, 99, 137, 138*f*, 139, 144*f*
immunity, lymphatic system
    in, 415
implantation, 499, 500*f*
incisive foramen, 109*f*, 117*f*
incisors, 454, 454*f*
    in cat, 558
incomplete dominance, 505
incus, 107, 310*f*, 311, 311*f*
independent event, 509
inferior, 9, 14*f*
inferior articular processes, 123
inferior colliculus, in sheep,
    270*f*, 272
inferior mesenteric artery, 398,
    400*f*, 550*f*
inferior mesenteric vein, 401, 402*f*
inferior nasal concha, 108, 108*f*,
    115*f*, 118*f*, 290*f*
inferior nuchal line, 109*f*
inferior oblique muscle, 290*f*, 291
inferior rectus muscle, 290*f*, 291
inferior vena cava, 368*f*, 374*f*, 401,
    402*f*, 472*f*, 550*f*
    path of, 397, 398*f*
    in sheep, 371, 372*f*
inferior vertebral notch, 123
infrahyoid muscles, 188, 190*f*
infraorbital foramen, 108*f*, 115*f*
infraspinatus muscle, 196, 197*f*,
    198*f*, 201*f*, 204*f*
    in cat, 523, 523*f*, 524*f*,
        525*t*, 526*f*
infraspinous fossa, 127, 128*f*
infundibulum (pituitary stalk),
    326, 327*f*
    in sheep, 271*f*, 272
infundibulum of uterine tubes,
    492, 493*f*
    in cat, 571, 572*f*
inguinal canal, 485, 487
    in cat, 574, 575*f*

inguinal ligament, 206*f*, 400*f*, 402*f*
inguinal lymph nodes, 416, 416*f*
inguinal regions, left and right, 16*f*
inner cell mass, 499, 500*f*, 501
inner ear, 309, 310*f*, 311
innominate bone, 102, 137, 138*f*
insertion, of muscle, 157–161,
    168, 170*f*
inspiratory capacity (IC), 431*f*,
    434, 441
inspiratory reserve volume (IRV),
    431*f*, 434, 441
insula (insular lobe), 253, 255*f*
insulin
    in diabetes mellitus, 337
    function of, 337
    secretion of, 330, 337, 338
insulin-dependent diabetes mellitus
    (IDDM), 337
insulin resistance, 338
insulin shock, 338
insulin therapy, 337
integumentary system, 13, 87–92
    components of, 87, 88*f*
    functions of, 87
interarterial septum, 367
intercalated discs, in cardiac
    muscle, 84*t*
intercarpal joint, 156*f*
intercostal muscles, 447, 448*f*
    external, 209*f*, 429, 447,
        448*f*, 449*f*
    internal, 429, 449*f*
intercostal nerves, 449*f*
intermediate cuneiform, 140,
    141*f*, 145*f*
intermediate filaments, 39
internal acoustic meatus, 108, 110*f*
internal anal sphincter, 459
internal carotid artery, 398, 399*f*
internal iliac artery, 399, 400*f*
    in cat, 542, 543*f*, 544*f*, 566*f*
internal iliac vein, 401, 402*f*
    in cat, 546, 547*f*
internal intercostal muscle, 429,
    447, 448*f*, 449*f*
internal jugular vein, 401, 401*f*,
    416, 416*f*
    in cat, 545, 546*f*
internal oblique muscle, 205,
    206*f*, 207*f*
    in cat, 520*f*, 522, 522*f*, 522*t*
internal respiration, 421
interneurons, 233, 253
interphalangeal joints
    distal, 226*f*, 231*f*
    proximal, 226*f*
interphase, 61, 61*f*, 66
interstitial cells, 485, 487, 487*f*
intertubercular sulcus, 129, 130*f*
interventricular artery
    anterior, 368*f*, 370
    posterior, 368*f*, 370
interventricular septum, 367,
    370*f*, 374*f*
interventricular sulci, 368*f*, 370
    in sheep, 371, 371*f*, 372*f*
intervertebral discs, 119, 120*f*, 126*f*

intervertebral foramina, 119,
120*f*, 241
intervertebral notch, 122*f*
intestinal trunk, 416, 416*f*
inversion, 160*f*, 161*t*
involuntary muscle, 83, 84*t*
iodine test for starch, 23, 24, 26
iris, 289, 290*f*, 291, 292, 293,
294*f*, 295*f*
iris diaphragm, of microscope, 30*f*,
31, 32
irregular bones, 93, 94
ischial spine, 137, 138*f*, 139
ischial tuberosity, 137, 138*f*,
139, 491
ischiocavernous muscle, 205, 208*f*
ischium, 99, 137, 138*f*
Ishihara's test for colors, 301, 302*f*
islets of Langerhans, 330, 330*f*,
331*f*, 338, 339*f*, 340*f*
isotonic solutions, 49, 50, 52, 57*f*,
59–60
isthmus of thyroid gland, 328*f*

J

jejunum, 458*f*, 459
in cat, 559, 560*f*
joint(s), 155–166
ball-and-socket, 156, 156*f*, 157*t*
cartilaginous, 155, 157*t*
classification of, 155, 157*t*
condylar (ellipsoid), 156,
156*f*, 157*t*
fibrous, 155, 157*t*
function of, 155
hinge, 156, 156*f*, 157*t*
movement of, 155, 157–162,
158*f*–159*f*, 160*f*,
161*t*, 166*f*
pivot, 156, 156*f*, 157*t*
plane (gliding), 156, 156*f*, 157*t*
saddle, 156, 156*f*, 157*t*, 160*f*
structure of, 155, 156–157, 157*f*
synovial, 155–157, 155–162,
156*f*, 157*f*, 157*t*
jugular foramen, 108, 109*f*,
110*f*, 118*f*
jugular notch, 123, 124*f*, 224*f*
jugular trunk, 416, 416*f*
jugular vein
in cat, 538*f*, 540*f*, 541*f*, 545,
546*f*, 558*f*
external, 401, 401*f*
internal, 401, 401*f*, 416, 416*f*
juxtamedullary nephrons, 471,
472, 474*f*

K

karyotype, 64*f*
keratinized epithelium, 69
ketones, in urine, 479, 481
kidney(s), 472*f*, 570*f*
blood vessels of, 473, 474*f*
in cat, 542*f*, 544*f*, 560*f*, 565–567,
566*f*, 567*f*, 568*f*,
573*f*, 575*f*
functions of, 471
hilum of, 472

location of, 471
pig, 472, 473*f*
right, 471
secretions of, 471, 479
sheep, 472
structure of, 471–478, 477*f*
knee-jerk reflex, 247–248, 249*f*
knee joint, 156, 157, 158*f*
Korotkoff sounds, 409
kyphotic curve, 225*f*

L

labia majora, 491, 492*f*, 493*f*,
494, 578*f*
in cat, 572
labia minora, 491, 492*f*, 493*f*, 494
labyrinth(s)
bony, 311
membranous, 311
lacrimal apparatus, 289, 290*f*, 291
lacrimal bone, 108, 108*f*,
109*f*, 116*f*
lacrimal gland, 290*f*, 291
lacrimal sac, 290*f*, 291
lactation, 491, 494*f*
lactiferous ducts, 491, 494, 494*f*
lactiferous sinus, 494*f*
lacuna, of bone, 94, 95*f*
lambdoid suture, 107, 108, 109*f*,
110*f*, 116*f*
lamella, of bone, 94, 95*f*
lamellated corpuscle, 88*f*, 275,
276, 276*f*
lamina, 122*f*, 123
landmarks, external (surface
anatomy), 221–232
Langerhans, islets of, 330, 330*f*,
331*f*, 338, 339*f*, 340*f*
large intestine, 453, 458–459
in cat, 559, 574*f*
histology of, 459, 460*f*
structure of, 458–459, 458*f*
laryngeal muscles, 188, 190*f*
laryngeal prominence, 222*f*
laryngopharynx, 422*f*, 423, 455
larynx, 421, 422*f*, 423–424,
423*f*, 428*f*
in cat, 541*f*, 551–552, 552*f*, 554*f*
lateral, 9, 14*f*
lateral condyle
of femur, 139*f*, 140, 144*f*, 158*f*
of tibia, 140, 140*f*
lateral corticospinal tract, 243*f*
lateral cuneiform, 140, 141*f*, 145*f*
lateral epicondyle
of femur, 139*f*, 140, 144*f*
of humerus, 129, 130*f*
lateral excursion, 160*f*, 161*t*
lateral flexion, 159*f*, 161*t*
lateral horn, 241, 242*f*, 244, 244*f*
lateral malleolus, 140, 140*f*,
227*f*, 228*f*
lateral meniscus, 158*f*
lateral pterygoid muscle, 188, 189*f*
lateral rectus muscle, 290*f*, 291
lateral regions, left and right, 16*f*
lateral reticulospinal tract, 243*f*
lateral rotation, 159*f*, 161*t*

lateral spinothalamic tract, 243*f*
lateral sulcus, 253, 255*f*, 256*f*
lateral ventricles, of brain, 254
in sheep, 272, 272*f*
latissimus dorsi muscle, 196, 197*f*,
198*f*, 204*f*, 209*f*, 223*f*
in cat, 521*f*, 522*f*, 523, 523*f*,
524*f*, 525*f*, 525*t*,
526*f*, 527*f*
law(s)
of probability, 505, 509
scientific, 1
of segregation, Mendel's, 510
left, as directional term, 9, 14*f*
left bundle branch, 377, 379*f*
left hemisphere, 254, 254*f*
leg
anatomical definition of, 137
arteries of, 399, 400*f*
bones of, 137, 139–145,
140*f*, 144*f*
muscles of, 211–213, 215*f*,
216*f*, 219*f*
surface features of, 227*f*, 228*f*,
230*f*–231*f*
veins of, 401, 402*f*
length, measurement of, 2*t*
lens, of eye, 289, 290*f*, 295*f*
accommodation of, 300–301
aging and, 299
in dissected eye, 291, 292,
293, 294*f*
normal transparency of, 293*f*
opaque (cataract), 293, 293*f*
visual function of, 299
lens, of microscope, 29, 30*f*
cleaning of, 30
eyepiece (ocular), 29, 30*f*,
31, 31*f*
objective, 29, 30*f*, 31
lens paper, 29
lesser curvature, of stomach,
456, 456*f*
lesser omentum, 563*f*
in cat, 559, 560*f*
lesser sciatic notch, 137, 138*f*
lesser trochanter, of femur,
139, 139*f*
lesser tubercle, of humerus,
129, 130*f*
lesser wing of sphenoid bone, 108
leukocytes. *See* white blood cells
levator ani muscle, 205, 208*f*
levator palpebrae superioris
muscle, 290*f*, 291
levator scapulae muscle, 189*f*, 195,
197*f*, 201*f*
in cat, 520, 521*f*, 522*t*,
523*f*, 527*f*
Leydig (interstitial) cells, 485,
487, 487*f*
ligamentum arteriosum, 368*f*, 374*f*
in sheep, 371*f*
limb length, measurement of,
3–8, 3*f*
line (linea), of bone, 102
linea alba, 206*f*, 207*f*, 224*f*
linea aspera, 139, 139*f*

lingual frenulum, 454, 454*f*
in cat, 558
lingual tonsils, 415
lip(s), 454*f*
of cat, 558
lipids, tests for, 22, 23–24, 25
liver, 453, 457, 457*f*, 563*f*
in cat, 538*f*, 542*f*, 558–559, 559*f*
lobar bronchi, 556*f*
lobar bronchus, 421, 422*f*
lobe(s)
of brain, 253, 254*f*, 255*f*
in sheep, 269, 270*f*
of breast, 491, 494*f*
of liver, 457
in cat, 558–559, 559*f*, 560*f*
of lung, 421, 422*f*, 423
in cat, 538*f*, 552, 553*f*
of pituitary gland, 326, 327*f*
lobules
of breast, 494*f*
of lung, 423
of thymus, 417, 418*f*
locus, of gene, 505
long bones, 93–98, 94*f*, 95*f*, 98*f*
longissimus muscle, 205,
206*f*, 209*f*
longitudinal fissure, 253, 254*f*
in sheep, 269, 270*f*, 271, 271*f*
longitudinal section, 71, 71*f*
long thoracic vein, in cat, 545, 546*f*
loop of Henle (nephron loop),
471, 474*f*
loose connective tissue, 76*t*
lordotic curve, 225*f*
loudness, of sound, 309
lower esophageal sphincter,
455, 456*f*
lower limb
arteries of, 399, 400*f*
bones of, 102, 137, 139–145,
139*f*, 140*f*, 141*f*, 144*f*
development of, 148, 150*f*
muscles of, 211–219,
212*f*–217*f*, 219*f*
surface features of, 227*f*, 228*f*,
230*f*–231*f*
veins of, 401, 402*f*
lubb sound, of heart, 378, 378*f*, 380
lumbar artery, in cat, 544*f*
lumbar curvature, 119, 120*f*
lumbar nerves, 241
lumbar region, 16*f*, 18*f*
lumbar trunk, 416, 416*f*
lumbar vertebrae, 119, 120*f*,
122*f*, 223*f*
lumbodorsal fascia, in cat, 524*f*
lunate, 129, 131*f*, 136*f*
lung(s), 423–424
in cat, 538*f*, 539*f*, 541*f*, 542*f*,
552, 553*f*, 559*f*
circulation in, 398*f*, 424*f*
function model of, 430, 430*f*
hilum of, 421, 422*f*, 423
inflation of, 424
left, 422*f*, 556*f*
lobes of, 421, 422*f*, 423, 556*f*
lobules of, 423

lung(s), *continued*
　right, 422*f*, 556*f*
　smoking and, 425
　structure of, 421
　tissue of, 424–425, 425*f*
　volumes and capacities of,
　　430–435, 431*f*, 441–446
luteinizing hormone (LH), 326–328
lymph, as connective tissue, 76*t*
lymphatic capillaries, 415
lymphatic ducts, 415
lymphatic system, 13, 415–420
　immune function of, 415
　pathways of, 416, 416*f*
lymphatic trunks, 415, 416, 416*f*
lymphatic vessels, 415, 416,
　　416*f*, 417*f*
　afferent, 417*f*
　efferent, 417*f*
lymph nodes, 415, 416–417, 416*f*
　palpation of, 417
　regions of, 416–417, 416*f*
　structure of, 417, 417*f*
lymph nodules, 417, 417*f*
lymphocytes, 344, 345*f*, 346*t*, 415,
　　417, 418
lymph sinus, 417, 417*f*
lysis, 50, 53–54
lysosomes, 39, 40*f*
lysozyme, 289

M

macrophages, 415, 417, 418
maculae, 319, 320*f*
macula lutea, 289, 291, 292, 293*f*
magnification, total, 31, 31*t*
major calyx, 472, 473, 473*f*, 570*f*
　in cat, 568*f*
male reproductive system, 13,
　　485–490
　in cat, 572–574, 574*f*, 575*f*
　microscopic anatomy of,
　　487–488, 487*f*, 488*f*
　organs of, 485–487, 486*f*, 579*f*
malleolus
　lateral, 140, 140*f*, 227*f*, 228*f*
　medial, 140, 140*f*, 227*f*, 228*f*
malleus, 107, 310*f*, 311, 311*f*
mammary glands, 491
mammary region, 17*f*
mammillary bodies, 254, 257*f*
　in sheep, 271, 271*f*, 272, 272*f*
mandible, 108, 108*f*–110*f*,
　　115*f*–116*f*, 118*f*, 126*f*,
　　160*f*, 222*f*, 243*f*
　in cat, 518*f*, 551, 557
mandibular condyle, 108,
　　109*f*, 116*f*
mandibular foramen, 108
mandibular fossa, 108, 109*f*
manubrium, 99, 123, 124*f*, 127
marginal artery, 368*f*, 370
mass, measurement of, 2*t*
masseter muscle, 188, 188*f*, 191*f*
　in cat, 518–519, 518*f*, 519*f*,
　　519*t*, 558*f*
mastication, muscles of, 187, 188,
　　188*f*, 189*f*

mastoid fontanel, 147, 148*f*
mastoid foramen, 109*f*
mastoid process, 108, 109*f*, 116*f*,
　　222*f*, 231*f*
maxilla, 107, 108, 108*f*, 109*f*,
　　110*f*, 115*f*, 116*f*,
　　117*f*, 118*f*
maxillary sinus, 108, 422
maxillary sinuses, 107
measurements, 1–8, 2*t*
meatus, 102, 422, 422*f*
mechanoreceptors, 275
medial, 9, 14*f*
medial condyle
　of femur, 139*f*, 140, 158*f*
　of tibia, 140, 140*f*
medial cuneiform, 140, 141*f*, 145*f*
medial epicondyle
　of femur, 139*f*, 140
　of humerus, 129, 130*f*
medial excursion, 160*f*, 161*t*
medial longitudinal arch, 227*f*
medial malleolus, 140, 140*f*,
　　227*f*, 228*f*
medial meniscus, 158*f*
medial pterygoid muscle, 188, 189*f*
medial rectus muscle, 290*f*, 291
medial reticulospinal tract, 243*f*
medial rotation, 159*f*, 161*t*
median cubital vein, 401, 401*f*
median fissure, anterior, 242*f*,
　　244, 246*f*
median sacral artery, 398, 400*f*
median sacral crest, 123, 123*f*
median sulcus, posterior, 242*f*,
　　244, 246*f*
mediastinum, in cat, 552
medulla oblongata, 254, 255*f*, 260*f*
　in respiratory regulation,
　　447–448, 449*f*
　in sheep, 269, 270*f*, 271, 271*f*,
　　272, 272*f*
medullary cavity, 94*f*, 95, 98*f*
medullary rhythmicity area,
　　447, 449*f*
meiosis, 61, 491, 505
Meissner's corpuscle, 88*f*, 275,
　　276, 276*f*
melanin, 88–89, 89*f*, 506
melanocytes, 89–90
membranes, 9–20
membranous labyrinth, 311
Mendelian genetics, 505
Mendel's law of segregation, 510
meninges, spinal cord, 241, 242*f*,
　　243*f*, 244, 244*f*
meniscus
　lateral, 158*f*
　medial, 158*f*
menstrual phase, of female
　　reproductive cycle, 496
mental foramen, 108, 108*f*, 115*f*
mental region, 17*f*
mesenteric artery
　anterior, in cat, 542, 543*f*, 544*f*
　inferior, 398, 400*f*, 550*f*
　posterior, in cat, 542, 543*f*, 544*f*
　superior, 398, 400*f*, 550*f*

mesenteric vein
　anterior, in cat, 546, 548*f*
　inferior, 401, 402*f*
　posterior, in cat, 546, 548*f*
　superior, 401, 402*f*
mesentery, 458*f*, 459, 563*f*
　in cat, 559, 560*f*
mesoderm, 499, 501
mesovarium, in cat, 571, 572*f*
metabolic rate
　basal, of cell, 42–43, 43*f*, 48
　thyroid gland and, 330–331, 331*f*
metabolism, cellular, 40, 42–43, 48
metacarpal bones, 95*f*, 100*f*,
　　102, 127, 129, 131*f*,
　　135*f*, 136*f*
metacarpophalangeal joint,
　　156*f*, 226*f*
metaphase, of mitosis, 61, 61*f*,
　　62*f*, 66
metatarsal bones, 98*f*, 100*f*,
　　102, 137, 140, 141*f*,
　　145*f*, 227*f*
meterstick, 3, 3*f*
metric ruler, 3, 3*f*
metric system, 1–8, 2*t*
microfilaments, 39
microglia, 233
microhematocrit reader, 353, 353*f*
micrometer scale, in microscope,
　　33, 34*f*
microscope(s)
　basics of, 29–33
　care and use of, 29–38
　compound light, 29, 30*f*
　depth of field, 33
　dissecting, 35, 36*f*
　eyepiece power of, 31, 31*f*
　field of view, 32
　focusing of, 32–33, 32*f*
　lens-cleaning technique for, 30
　micrometer scale in, 33, 34*f*
　parfocal, 32
　parts of, 29, 30*f*
　slide preparation for, 33–34, 35*f*,
　　344, 344*f*
　total magnification of, 31, 31*t*
　working distance of, 32
microtubules, 39, 40*f*
microvilli, 40*f*, 281, 283*f*, 284, 284*f*
midbrain, 254, 255*f*, 260*f*
　in sheep, 271*f*, 272, 272*f*
middle cardiac vein, 368*f*, 370
middle cuneiform, 140, 141*f*, 145*f*
middle ear, 309, 310*f*, 311
middle ear bones, 99, 107, 309,
　　310*f*, 311, 311*f*
middle nasal concha, 108,
　　108*f*, 115*f*
middle phalanx
　foot, 140, 141*f*, 145*f*
　hand, 129, 131*f*, 136*f*
milk production, 491, 494*f*
minor calyx, 472, 473, 473*f*
　in cat, 568*f*
minute respiratory volume, 434
mirror, in microscope, 30, 30*f*
mitochondria, 39, 40*f*, 47*f*, 337

mitosis, 61–67, 499
　phases of, 61, 61*f*, 62*f*, 66, 67*f*
　in whitefish blastula, 63, 63*f*, 67*f*
mitotic phase, 61, 61*f*
mitral valve, 367, 369*f*, 370, 374*f*,
　　378, 378*f*
mixed nerves, 233
M line, 168
molars, 454, 454*f*
　in cat, 558
monocytes, 344, 345*f*, 346*t*
monosaccharides, tests for, 22, 23,
　　24, 25
mons pubis, 493*f*, 494
morula, 499, 500*f*
motor areas, of brain, 253, 256, 256*f*
motor association area, 256, 256*f*
motor cortex, primary, 256, 256*f*
motor neurons, 167, 176, 233–236,
　　234*f*, 235*f*, 247
motor spinal tract, 241
motor (descending) tracts,
　　241, 243*f*
motor unit, 167, 171*f*, 176
motor unit recruitment, 168, 176,
　　179–182, 181*f*
mouth, 453, 454–455, 454*f*
movement
　of eyes, 289, 319, 322
　of joints, 155, 157–162,
　　158*f*–159*f*, 160*f*, 161*t*, 166*f*
M (mitotic) phase, 61, 61*f*
MSG, taste of, 284, 287
multipolar neurons, 233
Multistix, 481
murmur, heart, 377
muscle(s). *See also specific types*
　abdominal wall, 205–206,
　　206*f*, 207*f*
　cardiac, 83, 84*f*, 84*t*
　cat, 515–536
　cells of, 83
　chest, 195–204
　deep back, 205–206, 206*f*, 209*f*
　eye, 289, 290*f*, 291, 292
　head and neck, 187–193
　hip, 211–219, 212*f*–215*f*, 217*f*
　lower limb, 211–219,
　　212*f*–217*f*, 219*f*
　pelvic outlet, 205–206, 208*f*
　respiratory, 429, 447–448,
　　448*f*, 449*f*
　shoulder, 195–204
　skeletal, 83, 84*f*, 84*t*, 167–174
　smooth, 83, 84*f*, 84*t*
　upper limb, 195–204
muscle fatigue, 176, 179–182, 181*f*
muscle fibers, 83, 167, 168, 168*f*,
　　169*f*, 176
muscle spindles, 247
muscle tissue, 69, 83–86, 84*t*
muscular system, 13
myelinated axon, 235*f*
myelin sheath, 233, 235*f*, 236,
　　236*f*, 237*f*
mylohyoid muscle, 188, 190*f*
　in cat, 518–519, 518*f*, 519*f*,
　　519*t*, 558*f*

609